W9-BNR-021

PRAISE FOR THE WORKS OF JONATHAN W. JORDAN

AMERICAN WARLORDS

"Mr. Jordan has done an admirable job of making these men come alive. He uses their own words to weave a good tale." —*The Wall Street Journal*

"An incredibly intimate account of the four men who led the nation to victory in the Second World War. Mr. Jordan's book should be required reading for all who seek greater insight into how wars are won. Superbly written, well researched, and highly interesting, *American Warlords* is in a class by itself."
—Jean Edward Smith, Pulitzer Prize finalist, winner of the Francis Parkman Prize, and *New York Times* bestselling author of *FDR* and *Eisenhower in War and Peace*

"Jordan delivers another page-turning chronicle of World War II. Small details and little-mentioned facts make this a highly informative look at four men in charge. . . . Readers will enjoy the intrigue, backstabbing, action, and diplomacy in this well-written book." —*Kirkus Reviews* (starred review)

BROTHERS RIVALS VICTORS

"Jonathan W. Jordan has pulled off a mission impossible: He has produced a groundbreaking study that transforms our understanding of all three men. . . . It is beautifully written and a classic work of popular biography and military history for the general reader. It is a pleasure to recommend it." —*The Washington Times*

"Humane, closely researched, and well-written . . . vividly conveys the mental and physical demands of high command." —*The Wall Street Journal*

"Anybody who believes that generals are just, rational men, imbued with a soldierly feeling of comradeship toward one another and an ingrained respect for their political superiors, will be shocked by this book."
—Michael Korda, *The New York Times Book Review*

ALSO BY JONATHAN W. JORDAN

Brothers Rivals Victors: Eisenhower, Patton, Bradley, and the Partnership
That Drove the Allied Conquest in Europe

Lone Star Navy: Texas, the Fight for the Gulf of Mexico,
and the Shaping of the American West

AS EDITOR
To the People of Texas: An Appeal in Vindication of His Conduct
of the Navy by Commodore Edwin Ward Moore

AMERICAN WARLORDS

HOW ROOSEVELT'S HIGH COMMAND
LED AMERICA TO VICTORY IN WORLD WAR II

JONATHAN W. JORDAN

NAL
CALIBER

NAL CALIBER
Published by New American Library,
an imprint of Penguin Random House LLC
375 Hudson Street, New York, New York 10014

This book is a publication of New American Library. Previously published in an NAL Caliber hardcover edition.

First NAL Caliber Trade Paperback Printing, May 2016

Copyright © Jonathan W. Jordan, 2015
Maps by Chris Erichsen
Photo credits appear on pages 590–91.
Penguin Random House supports copyright. Copyright fuels creativity, encourages diverse voices, promotes free speech, and creates a vibrant culture. Thank you for buying an authorized edition of this book and for complying with copyright laws by not reproducing, scanning, or distributing any part of it in any form without permission. You are supporting writers and allowing Penguin Random House to continue to publish books for every reader.

NAL Caliber and the NAL Caliber colophon are trademarks of Penguin Random House LLC.

For more information about Penguin Random House, visit penguin.com.

NAL CALIBER TRADE PAPERBACK ISBN: 978-0-451-41458-8

THE LIBRARY OF CONGRESS HAS CATALOGED THE HARDCOVER EDITION OF THIS TITLE AS FOLLOWS:
Jordan, Jonathan W., 1967–
American Warlords: How Roosevelt's High Command Led America to Victory in World War II/Jonathan W. Jordan.
p. cm.
Includes bibliographical references and index.
ISBN 978-0-451-41457-1
1. Roosevelt, Franklin D. (Franklin Delano), 1882–1945—Military leadership. 2. Presidents—United States—Biography. 3. Generals—United States—Biography. 4. Command of troops—History—20th century. 5. Eisenhower, Dwight D. (Dwight David), 1890–1969. 6. Stimson, Henry L. (Henry Lewis), 1867–1950. 7. Marshall, George C. (George Catlett), 1880–1959. 8. World War, 1939–1942—United States. 9. World War, 1939–1945—Campaigns. 10. United States—Politics and government—1933–1945. I. Title. II. Title: How Roosevelt's high command led America to victory in World War II.
E807.J645   2015
973.917092—dc23   2014036427

Printed in the United States of America
10 9 8 7 6 5 4 3 2 1

Designed by Elke Sigal

PUBLISHER'S NOTE
While the author has made every effort to provide accurate telephone numbers and Internet addresses at the time of publication, neither the publisher nor the author assumes any responsibility for errors, or for changes that occur after publication. Further, publisher does not have any control over and does not assume any responsibility for author or third-party Web sites or their content.

Penguin
Random
House

To Major Malcolm E. Jordan, U.S.A.F.

1939–2012

CONTENTS

$\backsim$

INTRODUCTION		XV
PROLOGUE	"THERE MUST BE SOME MISTAKE"	I

PART ONE | *Bringing the War Home*

ONE	"NEW POWERS OF DESTRUCTION"	II
TWO	THREE MINUTES	20
THREE	"THE HAND THAT HELD THE DAGGER"	29
FOUR	"FEWER AND BETTER ROOSEVELTS"	35
FIVE	THE NEW DEAL WAR	44
SIX	"ONE-FIFTY-EIGHT"	5I
SEVEN	THE PARABLE OF THE GARDEN HOSE	6I
EIGHT	INCHING INTO WAR	69
NINE	BEGGARS BANQUET	79
TEN	LAST STAND OF THE OLD GUARD	84
ELEVEN	YEAR OF THE SNAKE	92
TWELVE	*KIDO BUTAI*	100

PART TWO | *A New Doctor*

THIRTEEN	KICKING OVER ANTHILLS	III
FOURTEEN	"DO YOUR BEST TO SAVE THEM"	118
FIFTEEN	"O.K. F.D.R."	121

SIXTEEN   "THERE ARE TIMES WHEN MEN HAVE TO DIE"   133

SEVENTEEN   "INTER ARMA SILENT LEGES"   139

EIGHTEEN   ROLLING IN THE DEEP   148

NINETEEN   SHARKS AND LIONS   156

TWENTY   "LIGHTS OF PERVERTED SCIENCE"   164

TWENTY-ONE   MIDWAY'S GLOW   168

TWENTY-TWO   "THE BURNED CHILD DREADS FIRE"   175

TWENTY-THREE   CHAIRMAN OF THE BOARD   190

TWENTY-FOUR   ALONG THE WATCHTOWER   198

TWENTY-FIVE   GIRDLES, BEER, AND COFFEE   205

TWENTY-SIX   THE DEVIL'S BRIDGE   215

TWENTY-SEVEN   "HOLLYWOOD AND THE BIBLE"   221

TWENTY-EIGHT   "A WAR OF PERSONALITIES"   237

TWENTY-NINE   BLIND SPOTS   250

THIRTY   STICKPINS   256

THIRTY-ONE   THE FIRST CASUALTY   266

THIRTY-TWO   LANDINGS, LUZON, AND LADY LEX   271

THIRTY-THREE   "A VITAL DIFFERENCE OF FAITH"   276

THIRTY-FOUR   PLAINS OF ABRAHAM   280

THIRTY-FIVE   THE INDISPENSABLE MAN   286

THIRTY-SIX   "DIRTY BASEBALL"   291

THIRTY-SEVEN   VINEGAR JOE AND PEANUT'S WIFE   296

THIRTY-EIGHT   A RUSSIAN UNCLE   298

THIRTY-NINE   RENO AND GRANITE   316

FORTY   "CONSIDERABLE SOB STUFF"   325

FORTY-ONE   SORROWS OF WAR   334

FORTY-TWO   "DR. WIN-THE-WAR"   342

FORTY-THREE   HALCYON PLUS FIVE   356

FORTY-FOUR   HATFIELDS AND McCOYS   368

FORTY-FIVE   MR. CATCH   373

FORTY-SIX   TRAMPLING OUT THE VINTAGE   386

FORTY-SEVEN   OLD WOUNDS   397

FORTY-EIGHT   VOLTAIRE'S BATTALIONS   405

FORTY-NINE   COUNTING STARS   410

FIFTY   THE TSARINA'S BEDROOM   415

FIFTY-ONE   "O CAPTAIN"   429

PART THREE | *Swords, Plowshares, and Atoms*

FIFTY-TWO   TRUMAN   441

FIFTY-THREE   DOWNFALL   448

FIFTY-FOUR   "COME AND SEE"   454

FIFTY-FIVE   "THIS IS A PEACE WARNING"   465

EPILOGUE   469

SELECTED ALLIED CODE NAMES   475

ACKNOWLEDGMENTS   478

SELECTED BIBLIOGRAPHY   479

ENDNOTES   495

INDEX   592

MAPS

～

I.	THE WAR IN EUROPE	10
2.	THE ARSENAL OF DEMOCRACY	81
3.	THE PACIFIC THEATER	96
4.	THE EMPIRE OF JAPAN	110
5.	THE WAR FOR ASIA	171
6.	MACARTHUR AND THE NAVY	200
7.	NORTH AFRICA	216
8.	CHINA–BURMA–INDIA	252
9.	THE WAR FOR ITALY	274
IO.	OPERATION OVERLORD	359
II.	TWO ROADS TO TOKYO	375
I2.	DEATH OF THE GERMAN REICH	431
I3.	OPERATION DOWNFALL: THE INVASION OF JAPAN	450

War is not merely a political act, but also a real political instrument.

—CLAUSEWITZ

If war does come, we will make it a New Deal war.

—ROOSEVELT

I frankly was fearful of Mr. Roosevelt's introducing political methods . . . into a military thing.

—MARSHALL

To hell with Roosevelt and Marshall and the Army and the Germans and the Russians and the British! I want to get the hell out of this hole!

—U.S. INFANTRY PRIVATE

INTRODUCTION

᧐

T HERE WAS A TIME WHEN A LIBERAL DEMOCRAT, A CONSERVATIVE REPUB-
lican, a general who served both parties, and an admiral who served none
set aside profound differences and led America through history's greatest blood-
letting.

Through nostalgia's myopic lens, it is easy to see a united nation, its resolve
hardened by Pearl Harbor, swept inexorably to victory on the broad shoulders of
the GI, the wings of the B-29, and the buoyant spirit that brought the world
baseball, Duke Ellington, the Ford Model A and the Lone Ranger. A nation to
whom triumph came as naturally as manifest destiny. Yet these images tell only
a small part of the story.

The vast mural of World War II—waves of heavy bombers, marines raising
Old Glory, snaking lines of deuce-and-a-half trucks—has become part of the
American legacy. But that mural was not painted overnight. In 1939 Rosie was a
homemaker, not a riveter. Black sailors served as butlers, not gunners, and America
reposed its safety in a handful of green, ill-equipped divisions led by untested
middle-aged officers.

From May 1940 until the war's end, the American war machine lurched
forward, determined but not sure-footed, ensnared by material shortages and en-
meshed in bare-knuckle politics. To break the empires of Hitler, Hirohito and
Mussolini, liberals compromised with big business and Republicans compromised
with Democrats. The Army cut deals with the Navy, and both swallowed trade-
offs with unions, farmers, miners and factory owners. American generals and admirals horse-traded with their British cousins, and commanders of all branches
courted congressional chairmen, business leaders and, journalists.

In Washington's marble corridors, the United States entrusted four men with the prosecution of America's war. General George Catlett Marshall, the Army's top soldier, won the admiration of Churchill, Stalin and Truman. Admiral Ernest J. King, a Porthos of the sea, saw in the oceans the key to America's global power. Secretary of War Henry L. Stimson, an old-line Republican from old-moneyed Long Island, distrusted the rapidly changing world, yet he championed futuristic weapons to prevent future wars.

And over these men hovered Franklin Delano Roosevelt, a liberal Democrat thrown into a war where his friends became enemies, his enemies trusted allies. He had staked his legacy on domestic reform, yet found himself shaping the world alongside Josef Stalin and Winston Churchill. A devious, self-described "juggler," Roosevelt would shift his political base, draw his nation toward war, weld an alliance with a dictator and an imperialist, and found a global institution dedicated to peace.

Roosevelt, Marshall, Stimson, and King are now ghostly images of our past, men who speak to us through grainy black-and-white newsreels and scratchy archived recordings. We see them through a glass darkly: Roosevelt, a rakish cigarette holder clenched between broad white teeth, assures the nation the only thing it has to fear is fear itself. Marshall, a constellation of stars on each shoulder, stares inscrutably into the distance as he ponders global strategy. A mustachioed Stimson and a bald, scowling King, giants behind the curtain, stand in the background, barely remembered faces in a faded gray photograph.

But in 1941, these ghosts lived in a world bursting with fire and fear. A world unraveling along two seams, where America could peer over either shoulder and see bubbling lakes of red. A nation unaware that it was on the road to a golden age that would be purchased with rivers of blood, mountains of treasure, and years of suffering.

A road that would begin with a strange sound rippling over a tropical paradise.

"THERE MUST BE SOME MISTAKE"

⌇

I T BEGAN AS A LOW HUM, A SUNDAY MORNING RUMBLE FROM THE ISLAND'S north side. To the islanders, the sound announced another training exercise at Wheeler or Hickam. Or perhaps a flight of bombers winging in from distant California. "Must be those crazy Marines," one sailor muttered as he took in fresh air through an open porthole.

The wind brushed past the few clouds that had bothered to show up that morning. Oahu's golfers, sailors, housewives, and soldiers stirred themselves for a day much like the previous Sunday, or the Sunday before that, or any other Sunday they could recall. The Bears would be playing the Cards at Comiskey Park, the Black Cat on Hotel Street was open for the breakfast hangover crowd, and Waikiki theaters would be showing a Ty Powers–Betty Grable film that afternoon. Readers who caught the morning's *New York Times* couldn't miss the page one headline: "NAVY IS SUPERIOR TO ANY, SAYS KNOX."

But that hum, so commonplace to the islanders, was followed by an odd roll of distant thunder. Which, to the untrained ear, sounded much like practice artillery. Or bombs.[1]

· · ·

The general stepped onto his porch near Washington's Potomac River. He had finished his horseback ride on a sorrel named Prepared, and a lanky, thick-headed Dalmatian named Fleet trotted at his heels. He wiped his boots, entered the house, and headed for the shower.

As he was rinsing, his orderly announced an urgent call from the War

Department. Colonel Bratton wished to speak to him about a matter he could not discuss over the telephone.[2]

Toweling off and changing into his gray business suit, General George Marshall climbed into the back of his government-issue Plymouth and rode to the Munitions Building, a crowded office complex on Constitution Avenue near Washington's famed Reflecting Pool. He strode into his sterile second-floor office shortly after eleven a.m. On his desk sat a lengthy typewritten message intercepted from Tokyo.[3]

The 5,000-word cable addressed to Ambassador Kichisaburo Nomura sounded ominous, yet its meaning was unclear. Another intercepted cable, decoded that morning, directed Nomura to deliver the long message to the U.S. government at exactly one o'clock local time on the afternoon of December 7.

There was something about that one o'clock deadline making Bratton jumpy. It made Marshall jumpy, too.[4]

Marshall's blue eyes sifted the message. Frowning, he picked up his phone and called Admiral Harold Stark, chief of naval operations. They needed to warn the Pacific theater that trouble lay ahead.

"What do you think about sending the information concerning the time of presentation to the Pacific commanders?" he asked Stark.

"We've sent them so much already," Stark replied. "I hesitate to send any more. A new one will be merely confusing."

Marshall hung up. He thought for a moment, then pulled out a sheet of paper and scratched out a warning to his commanders in the Pacific. A few moments later, he called Stark back and read him the message.

"George," said Stark, "there might be some peculiar significance in the Japanese ambassador calling on Hull at one p.m. I'll go along with you in sending that information to the Pacific."[5]

·　·　·

Black plumes rose from Oahu's center as the attackers swarmed from the northwest, southwest, and east. Hundreds of them—Zeros, Vals, Kates—descended on their targets. They spit fire at scampering men, skimmed waves and dove on warships slow to realize that Pearl Harbor was under attack.

Explosions rocked the harbor as men in dungarees, khakis, and undershirts, some with helmets, some without, dashed for anything offering cover. As the air filled with inky smoke, the attackers broke into small formations and plunged onto their main victims: the moored giants lining Battleship Row. Antiaircraft guns barked, men screamed, and the tattoo of a hundred Brownings filled the air.

But the deep basso sound of torpedoes and bombs dominated the symphony of death.[6]

· · ·

To the clinking of fork and knife on White House china, Franklin Roosevelt chatted over one of Mrs. Nesbitt's bland lunches with his gaunt warhorse, Harry Hopkins. The two political veterans, like nearly everyone in Washington, had been watching the diplomatic picture unravel to the brink of war. Roosevelt's orders to hunt German U-boats in the Atlantic was a gauntlet thrown at Hitler, while in the Far East, conquest by Japan was followed by American economic sanctions. Sanctions spurred new conquests, which begat fresh sanctions. By Thanksgiving, autumn's circular dance had brought the two partners within a knife's edge of war.

At twenty minutes of two, an aide interrupted lunch to announce an urgent call from the secretary of the navy. Roosevelt took the black handset and listened as Frank Knox told him of a report the Navy had received from Honolulu. The Japanese were bombing Pearl Harbor.

Roosevelt listened, thanked him, and hung up. He turned to Hopkins.

"There must be some mistake," a wide-eyed Harry said when Roosevelt broke the news.

Roosevelt shook his head. It was just the kind of unexpected thing the Japanese would do, he said. His voice growing cold, he added, "If this report is true, it takes the matter entirely out of my hands."[7]

· · ·

The ancient *Utah* suffered the first mortal blow. As her crew raised the colors over her fantail, a formation of Kates screamed down onto tiny Ford Island. They skimmed the waves and long, cigarlike tubes fell from their bellies and plowed the water's surface.

The attackers climbed, and a massive explosion shook *Utah* to her keel. A jagged wound gaped from her hull, and the target ship swallowed salt water and listed hard to port. As the sea poured in, her thick starboard moorings fought to keep her deck above the insistent waves. The moorings lost.[8]

Succeeding bomber groups pointed their noses toward Battleship Row. A torpedo rocked *Oklahoma* from stem to stern and two more pierced her wounded side. She lurched to port, smoke billowing from her hatches as men leaped into the oily sea. Salt water flooded her iron viscera, and she listed until her starboard propeller rose over the water's roiling surface. As she slipped below the waterline, four hundred terrified men scrambled belowdecks, clambering through hatches and up ladders, racing the rising water, every man clawing for that priceless path to daylight.[9]

Oklahoma's sisters fought back, spitting AA shells into the sky as fast as gunners could shove them into smoking breeches. But the Japanese tigers pounced from every direction, strafing, dropping 800-kilogram bombs, skimming the water's edge as torpedo sights aligned angle and distance.

Above screams of men and machines, a violent blast shook *Tennessee*. Another jolted *West Virginia*, whose captain lay dying in her conning tower. With a convulsion that rattled the harbor, *Arizona* leaped out of the water, her magazine a fuming volcano. In nine minutes, she took eleven hundred men to the bottom.[10]

Smoke obscured vision as scorched shells cooked off, steam boilers exploded, and the harbor was swathed in thick, oily smoke. Men—ants scurrying over steel giants—swarmed in all directions, sprinting to action stations, diving for cover, swimming through blazing water, saving themselves. Sacrificing themselves.[11]

· · ·

Walking the halls of the Munitions Building, the old lawyer was feeling his age. It had been a week of conferences, memoranda, cabinet meetings, and telephone calls, and the tired statesman with the shock of white hair ached for a rest. If he could shake loose from Washington, get away to his home on Long Island, he thought, he could catch up on some sorely needed sleep.[12]

But he wasn't about to shake loose from Washington, or get home to Long Island, or catch up on his sleep. Things had gone from bad to worse—much worse—over the last twelve days. The president had rejected Japan's last offer, and intercepted cables from Tokyo implied the Emperor's diplomats were about to break off negotiations. The question on everyone's mind was not whether Japan would fight, but *when* and *where*.[13]

So Secretary of War Henry L. Stimson, Wall Street's old Republican stalwart, would await his leader's call.

On the morning of December 7, Stimson's thoughts turned to a draft message President Roosevelt would deliver to Congress on the crumbling picture in the Far East. The president also wished to discuss Tokyo's latest intercepted message, which seemed to herald a rupture in diplomatic relations.

Buckling his worn leather briefcase, Secretary Stimson made the six-block walk to the old State, War, and Navy Building next door to the White House. There he and Navy Secretary Frank Knox were ushered into the austere office of Secretary of State Cordell Hull, a colorful old-line liberal from middle Tennessee.

Stimson, Knox, and Hull were convinced that the Japanese were up to something. They mulled over what the president should tell Congress, but given the high stakes and ambiguity of Japan's position, they reached few solid conclusions. The "War Cabinet," as Stimson liked to call the group, broke up and went their separate ways. Stimson went home to lunch.

The clock's hands had swept past the lunch hour when Stimson peered over his reading glasses at an approaching aide. There was a phone call from the president, the aide said.

Stimson walked to the phone and picked up the receiver.

"Have you heard the news?" an excited voice asked.

"Well, I have heard the telegrams which have been coming in about the Japanese advances in the Gulf of Siam."

"Oh, no, I don't mean that," said Franklin Roosevelt, his voice rising. "They have attacked Hawaii! They are now bombing Hawaii!"[14]

●　　●　　●

The second wave flew in from the east and tightened up for the attack run. Battleship Row was in flames, but the pilots who swarmed over the burning ships meant to leave nothing alive. Diving, climbing, pitching, and banking, they sent bombs and bullets flying in every direction.

The battleships *Maryland*, *Tennessee*, and *Pennsylvania* belched clouds of black smoke, their hulls and structures a mangled mess. The wounded *Nevada*, her boilers churning, managed to slip her cables and limp past *Arizona*'s sinking corpse.

As she pulled forward, *Nevada* drew the attention of some Val dive-bombers, which dipped their noses and dropped like ravenous hawks. Six bombs found their marks. Explosions rocked the battleship's forestructure and bridge, and men were torn to shreds as flames drove back her fire crews. She steered hard to port and two tugs crashed into her side, heaving the shattered ship toward the relative safety of the beach.[15]

●　　●　　●

As the cruiser *Augusta* swayed on a gentle tide at Rhode Island's Narragansett Bay, her wiry fleet commander, Ernest J. King, settled down for his afternoon nap.

The admiral, known for his short temper, was recovering from another of his recent trips to Washington—visits to educate policy makers about strategy and U-boats and the limits of American naval power. "Hell, I've got to go down to Washington again to straighten out those dumb bastards once more," he would tell his staff, without a hint of irony.[16]

When he leaned his thin frame over *Augusta*'s bridge window, Ernie King looked every inch a fighting admiral. His nose was sharp as a corvette's bow. He had a bald forehead that resembled a destroyer's round turret, and his quick brown eyes squinted over a quicker tongue. Dressed in a blue jacket with gold braid, a cigarette resting between his fingers, the admiral was Jove, Mars, and Neptune to every man in his fleet.

King had worked at his desk all morning, then lunched with his chief of staff. He usually took a siesta in his finely appointed cabin, but on this day he had not slept long before a marine knocked on his hatch. When King answered, the marine handed him a note.

He read the message. Of course it could happen. He had done it himself three years earlier in a fleet exercise, when he launched a surprise carrier raid on Pearl. That time the umpires told him he had blown Wheeler and Hickam fields to hell. It looked like the umpires were right.

The next day King told his aide to pack his bags for another trip to Washington. Those dumb bastards were going to need help.[17]

. . .

The death rattles of *Utah*, *Arizona*, and *Oklahoma* mingled with howls from *West Virginia*, *Nevada*, and *California*. Near their scorched hulks lay the smaller ships: cruisers, destroyers, tugs in various states of disrepair, all smothered in wind-whipped smoke.

Zeros buzzed like mosquitoes over the wreckage of Hickam, Bellows, Kaneohe, Ford, and Wheeler fields, strafing targets of opportunity and pouncing on the few Hawks that managed to get airborne. The Emperor's pilots beamed as they saw the prostrate hulks of capital ships and neat lines of smashed American planes.

After two hours of carnage, the sword returned to its sheath. The Zeros banked north, covering the triumphant retreat of dive- and torpedo bombers, and the second wave disappeared into the western sky.

On ground and water, amid flaming seas, among the smoke, the oily stench, the groan of melting steel, lay the bodies of 2,402 dead and dying men. Another 1,247 lay scattered about the harbor, on stretchers, grass, or pavement, as nurses and medics fought to save them.[18]

. . .

Roosevelt's large hands closed around a typed memorandum, the first of many he would receive that day. In a few efficient sentences, the note described a scene of devastation: *Oklahoma* capsized, *Tennessee* burning, minelayer *Oglala* possibly lost. Airfields smashed. Planes gone.

He squinted at the paper, then wrote down the date and time it arrived on his desk.[19]

Franklin Roosevelt's handsome face took on a gossamer calm that afternoon as the world rushed into his study. While two desk telephones clanged like alarm bells, news flooded in on waves of messengers, each phone slip or memo more agonizing than the last. The surface fleet was burning on the water. The air fleet was burning on the ground. Casualty counts could not be verified for some time, but the death toll would be appalling. The battleship *Arizona*—into which a young Frank Roosevelt had hammered the ceremonial first bolt in 1914—lay beneath Pearl's choppy waves, her steel hull now a tomb for hundreds of silenced sailors.[20]

In a cyclone of conversations, phone calls, jostling couriers, advisers and cabinet ministers, Roosevelt sat at his cluttered desk, a cigarette between his fingers. He spoke in a voice that remained steady and controlled, and he gave clear, unhurried instructions to his lieutenants: Send for Marshall and Stark. Assemble the cabinet. Execute standing orders for war. Freeze Japanese assets. Place all munitions factories under guard.[21]

As his blue eyes watched the ghastly picture unfold, Roosevelt remained grave, shaken but businesslike. He made none of his usual small talk as he organized message slips into neat little piles on his desk. He called for his secretary, Grace Tully, and took a deep drag on a cigarette as she walked in with her stenographer's pad.

"Sit down, Grace," he said. "I'm going before Congress tomorrow."

He began dictating a rough draft of his message to Congress, to the American people, to the world.

"Yesterday, comma, December 7, 1941, comma, a date that will live in world history . . ."[22]

That evening, Roosevelt met with his cabinet in the second-floor study as the leadership of the House and Senate began gathering outside his door. His eyes had lost their sparkle, his humor was gone, and his jaw showed none of the confident, angular jut the world had known these past nine years. Through pursed lips he narrated the day's events. It was, he said, the "most serious meeting of the cabinet since the spring of 1861."

Describing the scene of destruction, Roosevelt's famous voice halted. The nation's great communicator could scarcely force himself to utter the words—words confessing a disaster that had fallen on his watch. Perhaps seven out of eight battleships had been lost, and many men had died that day.

Cabinet secretaries who hadn't heard the full report were dumbfounded. Those who knew listened in mortified silence. His emotions getting the better of him, Roosevelt twice stopped in mid-thought to bark at his navy secretary, "Find out, for God's sake, why the ships were tied up in rows!"

A red-faced Knox shot back, "That's the way they berth them!"

Roosevelt ended the meeting, then nodded to his study door, where the leadership of Congress awaited. Powerful men standing outside that door would demand to know what went wrong. And the American people, standing behind them, would demand to know how their president would avenge the dead of Pearl Harbor.[23]

PART ONE

Bringing the War Home

1940–1941

It is a terrible thing to look over your shoulder
when you are trying to lead—and find no one there.

—FRANKLIN ROOSEVELT

THE WAR IN EUROPE
DECEMBER 7, 1941

Axis Powers
Occupied by Germany
Axis Satellites
Neutral
Allied Territory

250 Miles

Atlantic
Ocean

U.S.S.R.

FINLAND

NORWAY

Moscow

SWEDEN

North
Sea

Baltic
Sea

German Eastern Front:
December 7, 1941

UNITED
KINGDOM

London

Berlin

Warsaw

GERMANY

POLAND

Belgium

BOHEMIAN
PROT.

Paris

SLOVAK REP.

FRANCE

AUSTRIA

HUNGARY

ROMANIA

Black Sea

SWITZ.

VICHY
FRANCE

YUGOSLAVIA

BULGARIA

ITALY

TURKEY

SPAIN

Corsica

Rome

ALBANIA

Sardinia

GREECE

Mediterranean Sea

Sicily

ALGERIA

TUNISIA

ONE

"NEW POWERS OF DESTRUCTION"

∽

May 1940

THE WARM, CLUTTERED STUDY ON THE SECOND FLOOR WAS THE CENTER OF Franklin Roosevelt's world. As White House staff flitted about, straightening up, polishing, and tending the first family's living quarters, the president dominated the oval-shaped room. Stamp books lay on small tables, sailing ship paintings vied with old maps for wall space, and knickknacks crowded the desks and mantelpiece of the "Oval Study." It was Roosevelt's office, parlor and sanctuary.

For a man confined to a wheelchair—for whom going up or down a flight of stairs was a three-man project—the Oval Study was, in a sense, the cerebral cortex of the U.S. government. The famous Oval Office in the adjacent West Wing served as the president's formal workplace, a setting for important meetings, press conferences, photo opportunities and diplomatic chats. But in the upstairs study, with its nautical theme and view of the Washington Monument, Franklin Roosevelt felt free to unwind over drinks, gossip with friends, and chart America's course.

On the evening of May 10, 1940, as Washington's spring warmth ebbed, Roosevelt was relaxing in his sanctum when an aide announced an urgent telephone call from Europe. Handset pressed to his ear, Roosevelt listened intently as Ambassador John Cudahy described German bombers swarming over the Dutch, Belgian, Luxembourg, and French skies. Before long German paratroopers would be blanketing airfields and fortresses. The Belgian king called up his reservists, the Grand Duke of Luxembourg fled to France, and the French army was bracing for blitzkrieg.[1]

Roosevelt had hoped the Nazi tide would break against the Anglo-French levee that stretched from the English Channel to the Swiss border. But the Germans sliced through the Allied lines with their armored spearheads, and the Luftwaffe rained death from above.

Hanging up the phone, Roosevelt leaned back in his seat and considered the possibilities: What if the Germans were thrown back? What if France collapsed? What if, as in the previous war, a long, bloody stalemate gripped Europe? Each scenario implied a dozen political calculations, only a few of which Roosevelt could see clearly from his upstairs study in safe, isolated America.[2]

At 2:40 a.m., his calculations exhausted, Roosevelt turned in for the night. His powerful arms shifted his frame into an armless wheelchair, and a handsome black valet named George Fields wheeled the president into his bedroom.[3]

FDR awoke five hours later to his usual morning routine: a bath and shave, then back to bed for breakfast. As he sat in his pajamas, a baggy sweater pulled over the top, he munched on toast and eggs—about the only thing Mrs. Nesbitt cooked properly—and devoured the *New York Times*, *Herald Tribune*, *Baltimore Sun*, *Washington Post*, and *Washington Herald*.

The sum of those sources confirmed the disaster Cudahy described the night before. Bombs smashed French fortresses. German panzers clanked toward Paris, and the battle for Western Europe was joined.[4]

At two p.m., a valet wheeled Roosevelt into the neoclassical Cabinet Room to face a tense collection of ministers. Seated around the long wooden table were Agriculture Secretary Henry Wallace, Treasury Secretary Henry Morgenthau Jr., Secretary of State Cordell Hull, Labor Secretary Frances Perkins, Interior Secretary Harold Ickes, Attorney General Robert Jackson, Commerce Secretary Harry Hopkins, and his war and navy secretaries, Harry Woodring and Charles Edison.

Little business was normally accomplished in Roosevelt's cabinet meetings. Part social hour, part pep rally, they were filled with presidential anecdotes, musings, and a smattering of departmental reports that none of the other secretaries cared about. Interior's Harold Ickes described to his diary one cabinet meeting in which the stern Secretary Perkins delivered a twenty-minute discourse on labor relations:

> As usual, only the President listened to her. Harry Hopkins wrote me a note. . . . Bob Jackson was nodding from time to time and at intervals he and Morgenthau were joking about something. Hull sat with the air of an early Christian martyr, with his hands folded, looking at the edge of the table without seeing it or anything else. . . . As usual, I studiously

avoided being caught by Perkins' basilisk eye. Henry Wallace was contemplating the ceiling.[5]

But this day was different. Talk centered on military questions, which should have been the province of Secretaries Woodring and Edison. But the afternoon's discussion was dominated by the enigmatic secretary of commerce.[6]

Harry Lloyd Hopkins was America's most unlikely-looking vice-regent. Balding, toothy, and wrapped around an emaciated frame that barely supported his weight, the native of Sioux City, Iowa, looked like some rumpled nebbish who had wandered off a White House tour group. He had cut his teeth heading local welfare agencies when he was drawn into the spinning orbit of Franklin Roosevelt, first as Governor Roosevelt's emergency relief director, then as the free-spending head of the Works Progress Administration during the New Deal's heyday, and finally, as secretary of commerce.

Harry Hopkins had no training as a diplomat, economist or military strategist, but he possessed a razor-sharp mind and uncanny judgment that friends admired and foes despised. FDR had even considered him as a potential successor to the presidency, a praetorian who could be trusted to safeguard the holy tenets of the New Deal.[7]

If Hopkins had a weakness—aside from horse races, late-night drinking, and bare-knuckle politics—it was his health. In 1937 a massive tumor had forced doctors to remove three-quarters of his stomach. His eviscerated body could barely digest fats and proteins, and doctors gave him four months to live. Harry lived, but his remarkable spark left him, often for months at a stretch, and one journalist likened him to "an ill-fed horse at the end of a hard day."[8]

In Roosevelt's kingdom, titles were deceptive. He gave Hopkins the Commerce Department because the strain of the WPA job overtook his physical stamina. Later, when Commerce proved too arduous, FDR made him "special assistant to the President," an undefined role advising his liege on matters ranging from bomber production and legislation to screenings of White House movies.

At FDR's request, Hopkins moved into the second floor of the White House. His proximity to the president, along with his loyalty and keen instincts, made him a formidable power broker at 1600 Pennsylvania. "The extraordinary fact," wrote speechwriter Bob Sherwood, "was that the second most important individual in the United States Government during the most critical period of the world's greatest war had no legitimate official position nor even any desk of his own except a card table in his bedroom. However, the bedroom was in the White House."[9]

. . .

That May afternoon, Hopkins sat in the Cabinet Room, glass-eyed and inert as Madame Perkins droned. But when Roosevelt invited the Commerce Department to have its say, the rumpled corpse in the gray suit sprang to life.

Rubber, Hopkins announced. Rubber and tin would be the keys to the next war. Rubber went into military hardware ranging from bomber tires to intravenous bags, gas masks to tank treads. It was a vital strategic asset, and 90 percent of it came from colonies of two nations Hitler invaded—French Indochina and the Dutch East Indies. The United States could smelt its own steel; it could buy bauxite from Africa and refine its own oil. But the humble rubber tree had become America's Achilles' heel.

Words and statistics tumbled out as Harry outlined problem and solution. A private company, funded by the Reconstruction Finance Corporation, would quietly buy up a year's supply of rubber, tin, and other strategic materials. Overt government purchases would augment the stockpile. America would buy what it needed to arm itself and the democracies—assuming it had time.[10]

While Hopkins rattled off the details of his materials stockpile campaign, a bulletin arrived from London: British Prime Minister Neville Chamberlain had just stepped down, and Winston Churchill, Britain's First Lord of the Admiralty, would probably be asked to form the new government.

Roosevelt had met Churchill two decades before, and had begun corresponding with him secretly in 1939. He nodded approvingly. Winston, he told his cabinet, was the best man England had, "even if he was drunk half the time."[11]

Six days later, senators and representatives packing the Capitol's House chamber stood as the sergeant-at-arms announced the arrival of the President of the United States. Roosevelt, dressed in a blue pin-striped suit, his tie crisply knotted, walked slowly down the aisle, his right hand on a cane, his left on the forearm of a burly Secret Service agent.

Beneath his dark trousers stood ten pounds of steel braces holding his polio-stricken legs erect.* In a sleight-of-foot illusion practiced over twenty years, FDR had trained himself to mimic walking by using his back and side muscles to heave

---

* Later researchers, pointing to Roosevelt's known clinical symptoms, have suggested that FDR actually suffered from Gullain-Barré syndrome. Others have noted test results consistent with poliomyelitis. Because polio was the accepted cause of Roosevelt's paralysis during his lifetime, that "diagnosis" is used here.

one leg up and forward, and then the other. Painful but passable, all it required was a steady arm to lean on and a strong cane.

Smiling through the discomfort, nodding genially to congressmen he passed, Roosevelt walked to the wooden dais, fully aware of his task: to shock a nation nestled in the comforting cocoon of isolation. To convince a hostile Congress that the time had come to arm America and its allies. To set production goals so high, every industrial leader would think on a scale never before envisioned. The job of the man encased in leg braces was to shake a nation out of a paralysis of the mind and get it running to safety.[12]

"These are ominous days," his measured cadence began. "Days whose swift and shocking developments force every neutral nation to look to its defenses in the light of new factors. The brutal force of modern offensive war has been loosed in all its horror. New powers of destruction, incredibly swift and deadly, have been developed; and those who wield them are ruthless and daring. No old defense is so strong that it requires no further strengthening and no attack is so unlikely or impossible that it may be ignored."

The greatest threat, he explained, was modern airpower. A flight from Greenland to New England took six hours; from West Africa to Brazil, seven; from Brazil to the Canal Zone, less than seven. And the preponderance of air-power lay in Adolf Hitler's hands. Germany had more warplanes aloft than all its enemies combined, and its assembly lines in the Ruhr, Silesia, and the Saar Palatinate could outproduce all competitors.[13]

To balance the scales, Roosevelt asked Congress for $1.8 billion in military appropriations. He called for a bigger army. He called for a two-ocean navy. He called for an industrial base capable of turning out fifty thousand warplanes each year.[14]

Fifty thousand planes? Every aerospace expert knew it was an impossible dream. The nation's entire aircraft industry couldn't turn out ten thousand planes in a year; normal production was about two thousand. Besides, the Army lacked fifty thousand pilots, training fields for fifty thousand cadets, and repair shops to keep fifty thousand planes flying. Roosevelt might just as well suggest that America would send a man to the moon. He had set an outlandish production goal that would have Hitler, Göring, and every other German laughing their *kopfs* off.[15]

But Roosevelt had immense faith in his people's ability to do whatever they were asked, *if* they understood why they were doing it. When he took office in 1933, he saw a nation ravaged by depression and ruin. Back then, stabilizing the banking, securities, real estate and labor systems seemed impossible. Restoring jobs while holding down ruinous inflation seemed impossible.

But Roosevelt also saw, beneath the dust, among broken cornstalks and idled coke furnaces, a resourceful people who could set things right if properly led.

"There are some who say that democracy cannot cope with the new technique of government developed in recent years by some countries—by a few countries which deny the freedoms that we maintain are essential to our democratic way of life," he told Congress. "This I reject."[16]

. . .

Franklin Roosevelt may have rejected the notion that democracy cannot cope with a totalitarian state, but the man charged with planning the next war saw a scale tipped heavily in Germany's favor. In 1920, Congress had demobilized much of the Army, and the Great Depression nearly finished off what Congress did not. By 1939, when Hitler's panzers rolled into Poland, America was a third-rate power. The Army consisted of fewer than 200,000 men, about a quarter of whom were fully trained. Counting men with modest National Guard training or doughboys of the First World War, America could scratch together perhaps 400,000 men who had marched in straight lines at one time or another.

Across the ocean, Germany boasted nearly seven million active and reserve soldiers, most of whom were veterans of campaigns in Poland, the Low Countries, and France. Goose-stepping alongside were another million Italians under dictator Benito Mussolini, veterans of *Il Duce*'s Ethiopian campaign.[17]

Even if the United States enlisted another million and a half men—putting the Army a scant five million behind Germany—those inductees would have nothing to fire, fly, or drive. For a generation, the business of America had been business. The America of 1940 excelled at making automobiles and ironing boards, Corn Flakes and Coca-Cola. Congressmen spent money on projects that would get them reelected, and there were few projects in the military budget that fit that sacred criterion. Electrical plants in Tennessee and dams in Nevada trumped range finders and dive-bombers on the great congressional shopping list.

Lacking reserves of ammunition and gasoline, the Army could scarcely drill its regulars, let alone equip the Guardsmen. The buck private of 1940 wore his father's helmet, laced his father's puttees, and shouldered his father's 1903 Springfield rifle. On maneuvers, trucks played the part of tanks, flour bags replaced grenades, and telephone poles filled in for artillery. "My God, we were carrying some wooden machine guns and all kinds of damn mortars that were nothing but logs," remembered Colonel Dwight Eisenhower, then a staffer with the Third Army. And Third Army was one of the lucky units.[18]

If the Army had one thing in its favor, it was a devoted core of well-educated officers. These middle-aged men formed a pyramid of regimental, divisional, corps, and army commanders. And on the apex of this pyramid stood one man whose job was to find the weapons, barracks, vehicles, and money his men needed to fight.

George Catlett Marshall was sworn in as the Army's highest-ranking officer on the day German troops crossed into Poland. Distinguished-looking at fifty-nine, his ruddy face had not yet broken irretrievably with the handsome features of his youth. His reddish hair, now a mellowing silver, hadn't retreated in disorder, and his sharp nose set off a prominent upper lip that gave way to a soft, kindly chin. In manner, he was respectful yet aloof, a straitlaced sphinx who let neither friendship nor enmity supplant his dispassionate judgment.

But when his ice-blue eyes found carelessness, sloth, or stupidity, his temper blazed like a howitzer barrage. He would usually bawl out the offending officer, or turn ice-cold, which, to the career officer, was worse. Marshall spent much of his adult life struggling to keep this lurking fire—a rage that physically rattled his heart—buried within his breast.[19]

Marshall's intensity welled from an upper-middle-class childhood in Pennsylvania's coal country. When he was a boy in the late nineteenth century, his family rode the ups and downs of Gilded Age capitalism from near-wealth to near-poverty. His sister remembered young George as an "ornery little boy" who dumped water buckets on her boyfriends and was apt to throw things like cake tins when he got mad. An indifferent student in high school, in the fall of 1897 Marshall decided to pursue a college education below the Mason-Dixon Line at Virginia Military Institute.[20]

Twenty years before Marshall's birth, VMI students had rallied to the colors at the Battle of New Market, and ten cadets had given their lives for the Confederate cause. Southerners who idolized Stonewall Jackson and Robert E. Lee took no liking to the lanky Pennsylvanian from a city called Uniontown. His Keystone State accent made him a natural target for hazing, and officially overlooked abuse took on a vicious character early in his first, "rat" year. During his second week at school, a gang of upperclassmen forced him to squat over a naked bayonet until ordered to stop. He obeyed until his strength gave out and he sank down onto the blade, leaving a bloody gash on his upper thigh and buttock.

Yet Cadet Marshall said nothing to the school's doctors who patched up his wound, and he kept his mouth closed to school investigators as well. His stoicism earned him a pass from hazing for the rest of the year, and during the next three years his military bearing evolved into a gentility worthy of a General Lee. Astute but not intellectual, he graduated fifteenth in a class of thirty-three. His self-discipline and strict adherence to regulations propelled him to first captain of his senior class, the institution's highest cadet rank, setting him on the road to a lieutenant's commission in the United States Army.

As a field soldier, Marshall marked time in common Old Army posts: the Philippines, China, the American West. But Marshall's superiors recognized his potential as a staff officer, and the Army began limiting his time as a field commander. In 1918, as a staffer with the First U.S. Army, he carved out a reputation as an organizational genius. After the Armistice, he spent several happy postwar years with his beloved wife, Lily, in Washington, where he served as aide to General John "Black Jack" Pershing, America's hero of the Great War.

Marshall's bucolic life was shattered in 1927 when Lily died while recovering from a thyroid operation. Heartbroken, Marshall buried his grief by throwing himself into his work. At the Infantry School in Fort Benning, Georgia, Marshall—now a colonel approaching fifty—led a drive to reform the school. He reorganized the infantry curriculum, improved living conditions for the school's soldiers, and widened his circle of civilian and military acquaintances. One of those new acquaintances was the charismatic founder of the new Center for Infantile Paralysis at Warm Springs, just a few miles east of Benning: New York Governor Franklin Delano Roosevelt. Another was Katherine Tupper Brown, a widow with three teenage children whom Marshall courted, then married in 1930.[21]

Three years later, the War Department placed Colonel Marshall in command of nineteen new Civilian Conservation Corps camps. The camps were part of a New Deal public works program to provide employment to young men by teaching them to plant trees, cut trails, and build erosion breaks in national forests. The job of training these young men was repugnant to most regular officers, but Marshall threw himself into the work. The CCC was an interesting, fresh assignment, and Marshall felt his work in training civilians to act in teams would help him diagnose and remedy the growing pains of a large citizen army that one day might be called to defend America's shores.

By the late 1930s, Marshall, now a brigadier general, found himself back in Washington at the War Department. As head of the Army's War Plans Department, then deputy chief of staff, Marshall completed his apprenticeship in that mysterious, fluctuating power known as Washington politics. Working the corridors of power in a double-breasted gray business suit, the one-star general meticulously courted legislators, journalists, and movers within the administration.*[22]

---

* Until Pearl Harbor, Army and Navy officers in Washington wore business suits rather than uniforms, to avoid attracting undue attention from budget-cutters and pacifist legislators.

. . .

When the time came for the president to select the Army's next chief of staff, few Washington insiders would have put money on the horse from Uniontown. For one thing, he was no politician. Marshall was terrible at remembering names, or pronouncing them correctly, and made a point of honor of refusing personal requests of senators and congressmen except when strictly merited.

Marshall also prided himself on straight talk, a practice repugnant to most politicians, including his commander-in-chief. At one White House meeting, FDR proposed a massive aircraft purchase program for the Army Air Corps. Far from being grateful, Marshall bluntly informed the president there was no point in buying planes when the Air Corps lacked pilots, mechanics, spare parts, training fields, and aerodromes to house and service them. Annoyed by the brigadier's impudence, Roosevelt gave Marshall the cold shoulder and abruptly ended the meeting.

Disagreeing with the president should have ended Marshall's career. But Marshall had won the backing of Harry Hopkins and the retired General Pershing, whose name commanded respect. The cadet who refused to rat out his upperclassmen had, over the years, built a reputation as honest, closemouthed, and loyal. So in April, FDR summoned him to the White House and told him he would become the Army's next chief of staff.[23]

The prospect thrilled Marshall, but he kept his exuberance in check. Sitting before the president, Marshall warned Roosevelt that as chief of staff he would have to tell him unpleasant things from time to time.

"Is that all right?" asked Marshall.

"Yes," said FDR pleasantly.

Marshall stared at him. "You said 'yes' pleasantly, but it may be unpleasant," he warned.

Unpleasant advice was a condition Roosevelt felt he could live with, for he believed he could trust Marshall to follow orders and keep his mouth shut once a decision was made. "When I disapprove his recommendations," he told Sam Rayburn, the powerful Speaker of the House, "I don't have to look over my shoulder to see whether he is going to the Capitol to lobby against me."

Four days later, President Roosevelt announced Marshall's appointment as U.S. Army chief of staff, effective September 1, 1939.[24]

TWO

THREE MINUTES

∽

THE DAGGERS OF WASHINGTON WERE EASILY MATCHED BY THE BAYONETS of Army politics.

Marshall was an old-style patriot, a D'Artagnan who served the Republic, not the party that happened to be in power at the moment. But he was not naive. To do his job effectively, the chief of staff had to be a lobbyist, accountant, lawyer, carnival barker, encyclopedia, and Wizard of Oz. He gave speeches to shore up the Army's public support. He provided data to Congress and the White House on appropriations and other laws affecting the Army, which were often hot political potatoes. He answered to the secretary of war—a political appointee—whose bailiwick extended to politicized aspects of Army administration, such as weapon procurement and training camp construction.

And some of the time, he was a soldier.[1]

If the sinews of war are money, as Cicero claimed, in 1939 America's sinews were thin and brittle. Congress had grown used to cutting Depression-era budgets, and in ten years Marshall's paycheck had been cut twice. The Army had 188,656 officers, nurses, and enlisted men the year Marshall took the helm, making it a third the size of the Belgian or Romanian armies. It was a drop in the bucket compared to the legions of Hitler and Mussolini spread over Europe.[2]

To turn Congress around, Marshall spent the spring of 1940 testifying before appropriations subcommittees and buttonholing committee chairmen. "After the World War," he told Congress, "practically everything was taken away from Germany in the way of materiel. So when Germany rearmed, it was necessary to produce a complete set of materiel for all the troops. As a result, Germany has an army equipped throughout with the most modern weapons that could be turned out, and that is a situation that has never occurred before in the history of the world." He warned, "If Europe blazes in the late spring or summer, we must put our house in order before the sparks reach the Western Hemisphere."

Now Europe was blazing.[3]

To put that house in order, Marshall first had to win over a president who was anxious to stop Hitler, but hesitated, in post-Depression times, to ask Congress to fork over big bales of cash to the military. Marshall, the outsider, would have to find the right way to approach Roosevelt to get things moving at the top.

The logical go-between, so far as the Constitution was concerned, was the secretary of war, Harry Woodring. But Woodring was one of Roosevelt's isolationist cronies, and Roosevelt had better cronies for Marshall's purposes. Bypassing Woodring, he made an appointment to see the secretary of the treasury.

Henry Morgenthau Jr., a Dutchess County neighbor of Franklin and Eleanor Roosevelt, was one of FDR's oldest friends. Lacking the financial cunning to take over his father's real estate business, Henry Jr. took up farming. His friendship with Roosevelt was the rock upon which his fortunes rested, and when FDR was elected president in 1932, Morgenthau hoped his friend would appoint him secretary of agriculture. FDR appointed Henry Wallace, a prominent Iowan, to that spot, but not long afterward, his treasury secretary fell ill and was forced to resign. Roosevelt asked Morgenthau to fill the vacancy.

It did not matter to Roosevelt that, as Gladys Guggenheim Straus joked, he had managed to find "the only Jew in the world who doesn't know a thing about money." Roosevelt was used to pulling the important levers himself, and he saw Morgenthau, like Secretary of State Cordell Hull, as a figurehead—a fungible man whose basic loyalty ensured that the New Deal's dogma would be zealously preached and observed.[4]

Morgenthau was also a man who could help Marshall.

On May 13, as Hitler's panzers were crashing through France, Roosevelt summoned Morgenthau, Marshall, Budget Director Harold Smith, Secretary of War Woodring and Assistant Secretary Louis Johnson to the Oval Office to discuss cuts in the Army's $657 million appropriations request. Since Woodring was an isolationist, and Marshall had little influence with Roosevelt, it was up to the treasury secretary to convince Roosevelt to resist any further budget reductions.

Morgenthau made the Army's pitch, but Roosevelt insisted on additional cuts. Morgenthau pressed his point, then argued with Roosevelt until the annoyed president cut him off. "I am not asking you, I am telling you," he declared.

"Well, I still think you're wrong," huffed Morgenthau.

"Well, you've filed your protest," said Roosevelt, signaling that the meeting was at an end.

"Mr. President," pressed Morgenthau, playing his last card, "will you hear General Marshall?"

"I know exactly what he would say. There is no necessity for me to hear him at all."

Disregarding the less-than-subtle hint, Marshall walked straight to Roosevelt's desk. Practically standing over the president, he asked, "Mr. President, may I have three minutes?"

"Of course, General Marshall."

In well over three minutes, Marshall let loose a machine-gun burst of facts, statistics, figures and logic. He described shortages that would dismember operations, logistical gaps that would sire defeats, and lack of funding that would undermine their shared duty to provide for the common defense.

He ended his fusillade, saying, "I don't quite know how to express myself about this to the President of the United States, but I will say this—that you have got to do something and you've got to do it today."

Set on his heels, Roosevelt said nothing. As he looked at the general bearing down on him with his blue eyes, Roosevelt finally grasped the enormity of the problem Marshall had been describing to Harry Hopkins, Henry Morgenthau, Congress, and anyone else who would listen.

He told Marshall to come back the next day with a list of what he needed. Before long, the sinews of war were thickening around Marshall's skeletal force.[5]

.   .   .

George Marshall's life settled into a rhythm pendulating between Quarters One, his redbrick residence at Fort Myer, and the old Munitions Building on Constitution Avenue. He awoke early and took a half-hour ride on Rosita, Prepared, or one of his other horses, sometimes alone, sometimes with Katherine or old friends like Colonel George Patton. Over breakfast, he would skim up to nine newspapers, and by 7:15 he would be sitting in the back of his government-issued Plymouth for the seven-minute drive to his office.[6]

Marshall started his workday behind his large mahogany desk, reviewing messages that had come in overnight from sources around the world. Eventually he established a morning briefing system that gave him the world's big picture. Using a series of maps and charts, Marshall's "G-2" (Intelligence) and "G-3" (Operations) staffs presented a panoramic view of the world at war, region by region. Marshall would often make decisions during the briefing, usually on his own authority, sometimes after consulting with the secretary, sometimes with a call to the White House. The briefings underscored the interconnected nature of the war's moving fronts, ensuring he would not become bogged down in local crises at the expense of global strategy.[7]

Napoleon once quipped, "An army crawls on its belly," but the U.S. Army crawled on paperwork. Mountains of it. It would have taken dozens of George Marshalls to respond to the requests pouring in from thousands of directions. When he assumed his post, sixty-one staff officers reported to the chief of staff, as well as thirty important and 350 unimportant unit commanders. After Marshall rewrote this unwieldy system, only five officers had unbridled access to the chief.[8]

Most of Marshall's daily contact was limited to his general staff secretary, three regular secretaries, and an aide. To preserve his energy for the most important matters, Marshall appointed a secretariat to act as his paper gatekeeper. Headed by Colonel Orlando Ward, Lieutenant Colonel Omar Bradley, and Major Walter Bedell "Beetle" Smith, the secretariat arrived a few hours ahead of the chief each day to sift through a mountain of fresh paperwork. These men worked through most of the Army's routine issues, and Marshall expected those issues to be handled without his day-to-day supervision. "Don't fight the problem; solve it," he told them.[9]

To those who worked for him, Marshall was all business. He seemed born without a sense of humor; if he had one, aides figured, he always left it in his car. Except for his chronic problem remembering names correctly, he was proper, direct, and unequivocal—efficient as a drill press, with the same human warmth. As his civilian secretary Mona Nason charitably put it, "He was a perfectionist himself, and he did the others the honor of expecting them to be perfect too."*

"Perfectionism" was one word for it. Descriptions among his staff usually ran something like, "He scared the hell out of the men." When Miss Nason offered to transfer with him in case he was moved out of Washington, Marshall, touched by her loyalty, remarked that no other staffer had offered to move with him.

"It's difficult for people to offer because you're so reserved, sir," she replied.

Marshall flatly replied, "I *have* to be, or they'd walk all over me."[10]

One of Marshall's most important lieutenants was a fellow Pennsylvanian named Major General Henry H. Arnold, head of the Army Air Corps and Marshall's assistant chief of staff for air matters. A bona fide aviation pioneer, Arnold

---

* It was some time before Nason mustered the courage to tell Marshall her last name was "Nason," not "Mason," and his aide's last name was "McCarthy," not "McCarty" or "Carty." Later, she informed Marshall that the name of his senior planner, Brigadier General Albert C. Wedemeyer, was pronounced "*Wedd*emeyer," not "*Weed*emeyer."

had been trained by the Wright brothers, and he had set three early altitude records. An avid—almost rabid—proponent of airpower, Arnold pushed for the development of a long-range strategic bomber and an expansion of the Air Corps into an autonomous branch of service. His zeal earned him the enmity of many Army staff officers, most line officers, all of the Navy, and even a few journalists like the AP's Steve Early, whose influence with Roosevelt as presidential press secretary nearly sank Arnold's career.*[11]

Known to friends as "Hap" for his boyish smile and genial personality, Arnold was also one of Marshall's few friends who could get away with office levity. He once sent a comic actor into Marshall's office posing as a pushy Polish émigré seeking a U.S. Army officer's commission. Annoying Marshall by casually referring to him as "colonel" and "captain" in broken English, the man eventually offered Marshall a fine Polish bride in exchange for a commission. As an enraged Marshall was throwing the man out of his office, Arnold burst through the door, howling with laughter.[12]

Although Marshall's formal job was to manage the Army, he kept a close eye on Washington's power brokers. He memorized the names of the key congressional committee chairmen, journalists, informal advisers, and White House staffers, and he assigned liaison officers to keep those players informed and pacified. Important ones he occasionally hosted for dinner at Quarters One, or took to lunch at Washington's Alibi Club. For Marshall, social engagements were all business; after one dinner party, he calculated that requests made to him that single evening required him to follow up with thirty-two letters and several radiograms.[13]

By 1940, General Marshall was becoming a power in his own right. Reflecting later, he believed his influence grew because "in the first place, they were certain I had no ulterior motives. In the next place, they had begun to trust my judgment. But most important of all, if Republicans could assure their constituency that they were doing it on my suggestion and not on Mr. Roosevelt's suggestion, they could go ahead and back the thing. [The president] had such enemies that otherwise the members of Congress didn't dare seem to line up with him. And that was true of certain Democrats who were getting pretty bitter."[14]

Washington power brokers, of course, included the press. The men and women armed with notepads and telephone lists might help or hurt the Army, but because they had space to fill for their readers, they were incapable of doing

---

* Marshall's legman, Beetle Smith, uncovered the backstory of Roosevelt's hostility to Arnold from Edwin "Pa" Watson, Roosevelt's military aide. Much of Roosevelt's ire, said Pa, could be traced to Steve Early. To protect Arnold, Marshall asked the secretary of war to pitch Arnold's nomination to permanent major general to the president personally.

nothing. Marshall gave these journalists his time and regularly brought them into his office for briefings, both on and off the record.

At these briefings, Marshall won over the Fourth Estate, not so much for what he said, but for the way he said it. At a later press conference that became part of the Marshall legend, he briefed some thirty veteran correspondents on the complexities of the war's larger picture. Rather than take questions one at a time, he asked the reporters to give him all their questions before he spoke. Some asked detailed, technical questions, while others wanted to know about broad, geostrategic issues.

After everyone had asked their questions, Marshall leaned back in his chair and began speaking. He spoke for forty minutes, without notes, on the war's various facets, going around the world in an integrated presentation that addressed every question from every reporter. As he worked into his narrative the answers to each question, he looked directly at the reporter who posed it. It was, said one journalist, "the most brilliant interview I have ever attended in my life."[15]

Marshall's twin obsessions were organization and personnel. He built a cadre of hardworking staffers who possessed the intelligence and self-confidence to make decisions without passing everything up to him. He spent hours weeding out the verbose, the inarticulate, and the indecisive. Taking a special dislike to self-promoters—who are as abundant in the Army as anyplace else—he looked for men who put the country's interest above their own.

One of his wartime planners, Brigadier General Dwight Eisenhower, vividly remembered the chief's views on promotions, a subject dear to every career officer's heart. "I want you to know that in this war the commanders are going to be promoted and not the staff officers," Marshall told the overworked staff officer. "You are a good case. General Joyce wanted you for a division commander . . . You're not going to get any promotion, you're going to stay here on this job and you will probably never move."

Eisenhower, his bald forehead turning crimson, sputtered, "General, I don't give a good God damn about your promotion. I was brought in here to do my duty, I am going to do that duty to the best of my ability and I don't care."

As Eisenhower got up and marched to Marshall's door, he looked back sheepishly at his boss. "By God there was a little quick of a smile," he recalled. He later reflected, "[Marshall's] obsession about disliking people that were self-seeking in the matter of promotion or anything else, it was really terrific."*[16]

---

* To Marshall, Eisenhower's answer was the correct one. A few weeks after the exchange, Eisenhower received his second star. Before long, Marshall would send him to England as the Army's European theater commander.

The men who survived Marshall's weeding-out process left their marks on the service a thousand different ways. When, for example, "Beetle" Smith learned of a small two-seater being peddled by the Bantam Car Company, he looked into it and made a three-minute pitch to Marshall. Marshall told Beetle to get fifteen samples and have them tested. The contraption, nicknamed the "jeep," was quickly snapped up by every branch of the Army as fast as Bantam, Willys, and Ford could roll them out.[17]

In the office, Marshall worked with an intensity that awed his associates. But the Army's business was a marathon, not a sprint. He was known for saying no one ever had an original idea after three o'clock, and to keep up his canter, he regularly went home for lunch, took an hour's nap, and left the office each day around four or five in the afternoon. Westerns, detective stories, and *The Saturday Evening Post* formed the core of his pleasure reading, and he enjoyed lighthearted plays and comedic musicals. His aides, including Lieutenant Colonel Frank McCarthy and his bodyguard, Sergeant James Powder, kept track of everything from notes of things he promised to do for congressmen to his travel kit, which they stocked with pocket novels, chocolate bars, and dime-store reading glasses the chief was forever misplacing.[18]

The few who knew Marshall outside the office saw a warmth and sense of humor that he withheld from his official family. Though childless, he had a soft spot in his heart for children, especially Katherine's teenage son, Allen Brown. His passion was gardening, he enjoyed Gilbert and Sullivan, he was a fine fishing and duck-hunting companion, and he chuckled at funny stories, even if he was inept at repeating them. He would insist that his staffers get away for occasional vacations, telling them, "I don't want tired men making decisions that affect human lives." When he came home from overseas trips, he was meticulous about writing wives of officers he knew to let them know how their husbands were doing.[19]

In the words of one journalist, George Marshall was "the most self-confident man who ever wore pants." But the self-confident general often fretted over wild schemes his commander-in-chief might dream up when left to his own devices. He later admitted, "I frankly was fearful of Mr. Roosevelt's introducing political methods of which he was a genius into a military thing. . . . You can't treat military factors in the way you do political factors."[20]

In 1939, FDR moved the Joint Army-Navy Board into the Executive Office of the President. This bureaucratic shell game, largely unnoticed by Congress or the public, bypassed the secretaries of war and navy and gave Roosevelt a direct line to the nation's military planners. FDR thus became the loom through which all strategic threads ran. *New York Times* military correspondent Hanson Baldwin, one of the few pressmen who noticed the shift, remarked, "The President is paying even closer attention than formerly to national defense problems and is assuming even more completely his prerogatives as commander-in-chief."[21]

Cutting out the secretary of war made Marshall uncomfortable. In Marshall's mind, even a machine as cumbersome as an army could function like a Swiss watch as long as duties were logically allocated, men were dedicated, and the organization was properly structured. It bothered him that the president was back-channeling information from the Army.

But strict organization was tiresome, even anathema to Roosevelt. As a victim of infantile paralysis at age thirty-nine, he realized life had a certain randomness, and he had learned to become comfortable with inconsistency, even enjoy it. He treated lines of responsibility with benign neglect, to be used when convenient and disregarded when they got in the way. He would dole out overlapping projects to sworn enemies, bypass chains of command, and refuse to bind himself to any precedent that might not suit him in the future. A master of the art of ignoring problems until they solved themselves, FDR encouraged dissent and talked out of both sides of his mouth. Lines would remain fuzzy, opportunities would be seized as they stumbled across his desk, and a film of unpredictability would shadow the American high command.[22]

Roosevelt's informality especially troubled Marshall. He winced at the president's "cigarette-holder gestures"—a wave of the hand and blithe assurance that things would work out all right. Worst of all, one never knew when the president was making small talk and when he was mining facts for a critical decision. "Informal conversation with the president could get you into trouble," Marshall remembered. "He would talk over something informally at the dinner table and you had trouble disagreeing without embarrassment. So I never went."[23]

Marshall also avoided visits to Springwood, Roosevelt's Hyde Park mansion. He kept his distance from FDR's inner circle, discouraged Roosevelt from calling him by his first name, and even tried to avoid laughing at the president's jokes. When Harry Hopkins suggested that Roosevelt would welcome Marshall occasionally dropping by for a martini in the presidential study, Marshall replied, "I'm at the president's disposal and he knows it, twenty-four hours of the day. But if I attempted to step out of character, then it would be artificial, and I just don't think that I can or should do it."[24]

Marshall knew that when someone approached the president with an unpleasant or difficult subject, Roosevelt could parry and feint like a French swordsman. "It was frequently said in those days by politicians who had seen Mr. Roosevelt that they never got a chance to state their case," he reminisced. "He was quite charming and quite voluble and the interview was over before they had a chance to say anything." The best Marshall could do was avoid prolonging FDR's homespun homilies by remaining silent until the president drew a breath, then jump into the problem.[25]

When Roosevelt did ask Marshall for his opinions, he gave them straight and to the point. He counseled subordinates to put written recommendations to the president in plain English, without flourish, preferably condensed to one page. "He is quickly bored by papers, lengthy discussions, and by anything short of a few pungent sentences of description," Marshall told them. "You have to intrigue his interest."[26]

Another problem that defied solution was FDR's penchant for talking to key decision-makers without having anyone spread the word about what he was saying. Marshall once complained to Hopkins, "The President at times sees Admiral Leahy, Admiral King, Arnold, or me and then the problem is, who summarizes what has occurred and provides a check to see that the necessary instructions are sent around." He warned Hopkins, "We may get into very serious difficulties in not knowing the nature of the President's revisions of the drafts of messages we submit to him. All of these things may easily lead to tragic consequences."[27]

THREE

"THE HAND THAT HELD THE DAGGER"

~ഗ~

A s Rommel and Guderian administered the French Army its last rites, Prime Minister Winston Churchill sent Roosevelt a "Most Secret" message vowing to carry on the struggle even if France fell. But to continue that struggle, he needed weapons to defend his island nation while his battered army, being driven off the Continent, licked its wounds. He needed forty or fifty destroyers, he said. He also needed torpedo boats. Several hundred fighter planes. Antiaircraft guns. Ammunition. Artillery sights. Raw steel.[1]

The request was hardly unexpected, but it arrived at a moment when the Army's pantry, if not bare, was thinly stocked. Marshall warned Henry Morgenthau, whom FDR had placed in charge of weapons sales, "The shortage is terrible, and we have no ammunition for antiaircraft and will not for six months. So if we give them the guns they could not do anything with them. . . . Antitank guns, the situation is similar, a shortage . . . 50-caliber, our situation is the same." The Army Air Corps was training pilots in mocked-up wooden boxes, and General Arnold pointed out that the delivery of even one hundred planes to Britain—about a three-day supply of RAF battle losses—would set the Air Corps training program back six months.[2]

Europe's democracies looked done for, and it was Marshall's job to build America's arsenal, not give weapons to allies on the brink of surrender.

Roosevelt mulled over Churchill's plea, fully aware that a wrong guess about British aid would cost the country dearly. He firmly believed the best way to fight Hitler was with American machines and British blood. But shipping off weapons that might be needed to defend the Western Hemisphere, should London fall, was one of the biggest gambles he had faced since the Hundred Days of 1933.

Agonizing over the possibilities, he admitted to Harold Ickes, "If I should guess wrong, the results might be serious." Beetle Smith put it more bluntly to a friend: "Everyone who was a party to the deal might hope to be found hanging from a lamp post."[3]

Eleanor Roosevelt once quipped, "The president never thinks. He decides." After three weeks of dithering, he decided. The guns would go to Britain. If industry stepped up production, and if he could keep America out of war a little longer, he could make good his losses in weapons. So he hoped.[4]

There were three hitches. First, naval warships posed special problems, because they constituted a floating steel wall against invasion. Isolationists vehemently opposed the sale of warships, and with Republicans and pacifist Democrats taking aim at him, the summer of 1940 was not the time to force the destroyer question. The warships would have to wait.[5]

Second, neutrality laws forbade the government from selling weapons to belligerents. Private contractors could sell them for cash, but arms merchants did not have warehouses full of 75-millimeter fieldpieces, tanks, or machine guns. Roosevelt couldn't muster the votes in Congress to repeal the neutrality statutes, so he asked his lieutenants to figure a way around them.[6]

Morgenthau, Marshall, Undersecretary of State Sumner Welles and Acting Attorney General Francis Biddle sent staffers scurrying through old laws to find a loophole. Biddle's lawyers, being lawyers, soon came up with a clever answer: If the weapons were declared "surplus" by the Army Chief of Staff, they could be sold to private corporations, which could in turn resell them to qualified buyers on a "cash-and-carry" basis. Morgenthau would ensure that the only qualified buyer would be Great Britain, and the U.S. Navy would ensure that no unfriendly country could show up to take delivery.

Biddle's theory was a beautiful piece of legalistic clockwork. It gave Roosevelt the power to lend Britain a hand, and it seemed legal—or at least legal enough. All that remained was to have General Marshall declare Churchill's aid "surplus," and have Secretary Woodring approve the shipment.

Unwilling to entrust this job to Woodring, an isolationist at heart, Roosevelt handed the whip to Morgenthau and told him to crack it without mercy. "Give it an extra push every morning and every night until it is on the ships," he commanded. Morgenthau obediently pressured Marshall to get the weapons into British hands.[7]

The Army was desperately short of everything, but Marshall did his best to fulfill the president's wish. After sending ordnance clerks to scour armories and warehouses, he tracked down large stocks of World War I rifles, old Lewis machine guns, French 75s, and other semi-obsolete equipment he probably could call "surplus" with a straight face. By giving the British the benefit of every doubt—and counting in the Army's inventory some weapons that were still on assembly lines—Marshall's staff calculated that the Army could part with half a million Enfield rifles, 35,000 machine guns, 500 fieldpieces, a stock of .30-caliber ammunition, and other odds and ends without endangering national security.

Marshall signed off, but with serious misgivings. To describe five hundred artillery pieces as "surplus" in 1940 was to stretch the definition like a worn-out inner tube. "It was the only time that I recall that I did something that there was a certain amount of duplicity in it," he later admitted.

Other men felt the same way. Assistant Secretary of War Louis Johnson told Roosevelt he wanted a promise that if he, as assistant secretary, were convicted of a felony in shipping the weapons to Britain, Roosevelt would pardon him. Roosevelt threw back his head and let out a hearty laugh. Johnson was not laughing with him.[8]

Once Marshall came across, the Army moved quickly. Top-priority cables went to arsenals around the country, and weapons were packed in gooey cosmolene and shipped to the Army's arsenal in Edison, New Jersey, for loading onto freighters bound for Liverpool. As the Germans closed in on Paris, Marshall's "surplus" sat in warehouses along Edison's Woodbridge Avenue, ready for loading onto His Anxious Majesty's ships. The last requirement to send them over the ship rails was authorization from the secretary of war.[9]

And there lay the final catch.

Harry Woodring, a former governor of Kansas, reflected America's isolationist heartland. A fixture at the War Department since 1936, Woodring was acutely concerned with the health of America's farms and factories. He wanted combines, not cannon, rolling off assembly lines, and he refused to authorize military aid to Britain unless the President of the United States ordered him to do it. Otherwise, he said, the weapons would stay in America, where they belonged.[10]

FDR again had to lean on his own functionaries. He ordered Woodring to sign the transfer orders, and under protest, Woodring grudgingly complied.

With the last Rube Goldberg link completed, British transports were loaded up and steamed for England, their holds bulging with weapons that would be pointed at Germans before long. America was wriggling out of neutrality's cocoon.[11]

The arms deal was cloaked in the darkest secrecy—or the darkest secrecy possible, given the number of soldiers, clerks, teamsters, stevedores, diplomats, and railroad workers involved. But a transfer of this size could not be kept under wraps for long, and Roosevelt knew he must get in front of the story. He decided to announce the sale at a June 10 commencement speech at the University of Virginia, where his son Franklin Jr. was graduating from law school.

Shortly before he and Eleanor boarded a train for Charlottesville, Roosevelt learned that Italy had just declared war on France and Britain. Mussolini wanted to share in the spoils of war, and Italian claims to French territory would wilt like overcooked linguini if he waited until France surrendered.

Sitting in the White House editing a draft of his commencement speech, Roosevelt borrowed a phrase from Bill Bullitt, his ambassador to France, that painted a picture of fascist Italy plunging a knife into the back of its next-door neighbor. State Department Undersecretary Sumner Welles was horrified at the draft's "stab in the back" language, and he argued vehemently that Roosevelt must delete the phrase from his speech. Reluctantly, FDR agreed.[12]

On the three-hour train ride from Washington, he mulled over the speech with Eleanor in his private railcar. He told her he wanted to speak candidly to the American people, without diplomatic niceties diluting unpleasant truths. His "stab in the back" phrase might rile State—and it would certainly infuriate the Nazis, who claimed they had been stabbed by Jews and Communists—but FDR felt the public should have the facts laid bare, at least in this instance. He asked what she thought.

Eleanor was her husband's social conscience, the angel of his better self. With her receding jaw, short-chopped hair, and tapered, oval eyes, she was not pretty, and their marriage, based on respect and necessity, had long lost the embers of passion. It was little more than a political partnership with the added wrinkle of shared offspring.

Over drinks with his buddies, FDR would make cutting remarks about Eleanor and the "she-men" and "squaws" who lurched to her progressive banner. But he appreciated Eleanor's vast mental energies, and he valued her as both a moral barometer and a political sounding board. However lifeless their marriage had become, Eleanor and Franklin Roosevelt were indispensable to each other.

Eleanor thought over her husband's question, though the answer to her was obvious. As the train clacked along the tracks, she replied, "If your conscience won't be satisfied unless you put it in, I would put it in."[13]

"On this, the tenth day of June 1940, the hand that held the dagger has struck it into the back of its neighbor," Roosevelt told the assembled throng of graduates, parents, and journalists.

Standing before a sea of black-gowned collegians, Roosevelt declared, "We will extend to the opponents of force the material resources of this nation. . . . All roads leading to the accomplishment of these objectives must be kept clear of obstructions. We will not slow down or detour. Signs and signals call for full speed ahead."[14]

It was a rousing speech, but isolationists were reading different signs and signals. FDR's foreign policy, bellowed North Dakota Senator Gerald Nye, was "nothing but dangerous adventurism." David Walsh, an isolationist Democrat from Massachusetts, thundered from the Senate floor, "I do not want our forces deprived of one gun, or one bomb or one ship which can aid that American boy whom you and I may someday have to draft." Charles Lindbergh, the hero who flew the *Spirit of St. Louis* across the Atlantic, accused Roosevelt of whipping up a "defense hysteria." He warned his countrymen that foreign invasion would become a threat only if "American people bring it on through their own quarreling and meddling with affairs abroad."[15]

Lindbergh and many smaller voices of isolation spoke for a following that had never bought into the German threat. Most Americans wanted to stay out of a European war, and a hit song of 1939, "Let Them Keep It Over There," summed up popular sentiment. With liberal pacifists, socialists, pseudo-fascists, and anti-Rooseveltians singing a Hallelujah Chorus of isolation, FDR would have to shift elements of his power base and forge a coalition outside the New Deal faithful.[16]

While much of his own party turned against him, a growing number of Republicans—hard-liners on foreign policy—were openly sympathetic. "When I read Lindbergh's speech, I felt it could not have been better put if it had been written by Goebbels himself," Herbert Hoover's former secretary of state, Henry Stimson, wrote to Roosevelt. Republican dailies like the *Boston Herald* and *San Francisco Monitor* threw their voices behind aid to Britain. Republican businessman Wendell Willkie, a contender for the presidency in 1940, told his party's keynote speaker that if the leadership "attempts to put the Republican Party on record as saying what is going on in Europe is none of our business, then we might as well fold up."[17]

Sensing a new alignment of the planets, FDR began courting leaders on the other side of the aisle. "The President's calling lists cut across party lines," wrote a *New York Times* journalist who shadowed the president for a day. "Republicans who opposed his policies in 1936 are almost as numerous as well-wishing Democrats in the parade across the White House lobby."[18]

•   •   •

As foreign conflict loomed large in Roosevelt's calculations, he preferred not to think about what to do with his obstinate secretary of war, Harry Woodring.

"Every time I try to fire him," Roosevelt joked with Supreme Court Justice Felix Frankfurter, "he says, 'My wife is expecting a baby and I want it to be able to say that I was born when my daddy was the secretary of war.'"[19]

Roosevelt's advisers felt Woodring had to go. But Roosevelt was, by nature, an optimist who believed every man was his friend until proven different. "His real weakness," Eleanor once told an interviewer, was that Roosevelt "had great sympathy for people and great understanding, and he couldn't bear to be disagreeable to someone he liked . . . and he just couldn't bring himself to really do the unkind thing that had to be done unless he got angry."[20]

But angry or not, there could be only one commander-in-chief. Roosevelt needed the War Department carrying out presidential policy, not creating policy of its own. He hated firing old friends, but events were forcing a change at the top.

He needed to find new friends.

FOUR

"FEWER AND BETTER ROOSEVELTS"

〰

T HE LAST REPUBLICAN WHOM FRANKLIN ROOSEVELT WHOLEHEARTEDLY SUP-
ported was Cousin Theodore. Since then, he had fought the Grand Old Party
in every election. With his victories over Herbert Hoover in 1932 and Alf Landon
four years later, FDR had become the opposition's Public Enemy Number One, its
John Dillinger, Ming the Merciless, tormentor and nemesis.

Along the way, however, Roosevelt had found a few Republicans who did not
see him with horns and a pitchfork, while there were more than a few Democrats
who did. His big blunders since 1936—a misfired attempt to pack the Supreme
Court, a purge of conservative Democrats—had cost him dearly. As New Deal
Democrats and liberal pacifists opposed military support for Britain, FDR began
reaching out to backbenchers whose credibility among their own kind was unim-
peachable.

At age seventy-two, Henry Lewis Stimson was an old-line patrician devoted to the
twin ideals of American exceptionalism and civic virtue. A graduate of Phillips
Academy, Yale University, and Harvard Law School, Stimson had served in some
public capacity under nearly every president since Teddy Roosevelt. He had been
a U.S. attorney under the elder Roosevelt, secretary of war under President Taft,
governor-general of the Philippines under Coolidge, and secretary of state under
Hoover.

Of average build and height, Stimson wore his white, bowl-cut hair parted in
the middle, the same way he had worn it as a young lawyer of the late 1800s. With
a salt-and-pepper mustache bristling over his upper lip, he resembled a throwback

to the robber baron days when his grandfather had run a Wall Street investment bank for tycoons like Jay Gould and Cornelius Vanderbilt.

A disciple of the first President Roosevelt, Stimson, like T.R., believed in a vigorous life spent in service to his country, God, and the Republican Party, more or less in that order. He hunted big game, rode big horses, played a bully game of tennis, and worked with an intensity that belied his age. At fifty, he had shipped off as an infantryman to fight the Kaiser and mustered out of the war with the rank of colonel. During his years out of government service, he testified before Congress on foreign policy matters and corresponded with government officials he had known over his long career.[1]

When he was secretary of state in the late 1920s, Stimson paid $800,000 to acquire Woodley, a Federal-style mansion near Rock Creek Park formerly owned by Martin Van Buren, Grover Cleveland, and other notables. After Hoover's defeat in 1932, Stimson and his wife, Mabel, moved back to their Huntington, Long Island, estate, Highhold, but kept Woodley to rent to well-bred friends such as Colonel George S. Patton Jr. and State Department Assistant Secretary Adolf Berle.[2]

Like Roosevelt, Stimson saw the new brand of imperialism as a clear and present danger. With an enemy like Hitler, he told Congress, it made no sense to wait until the foe has "killed off the last nation that stood between us and safety." On the other side of the globe, when Japanese troops overran the Chinese province of Manchuria in 1932, he became one of China's most ardent supporters in Washington. As outgoing secretary of state, he articulated the "Stimson Doctrine," which refused to recognize any treaty forced upon a nation through military conquest. As a private citizen, he worked with industrial elites to boycott trade with Japan.[3]

Though he shared Roosevelt's desire to halt the fascist tide, Stimson remained ambivalent toward the man. Stimson believed Roosevelt was prone to welsh on gentlemen's agreements. He wholeheartedly agreed with fellow Republicans who found most of the president's initiatives "half-baked and dangerous," and upon his departure in 1933, President Hoover warned Stimson that Roosevelt would misrepresent things told to him in private.[4]

But Roosevelt had decided to give the War and Navy Departments a political transfusion, and he would use opposition blood. He considered a short list of Republicans, including Alf Landon and New York Mayor Fiorello La Guardia.

Justice Felix Frankfurter, whom Stimson had hired out of law school decades earlier, suggested adding Stimson to the list. Roosevelt liked the idea, but worried that Stimson's age and health might rule him out. Another friend of Stimson's, ignoring medical ethics, went straight to Stimson's personal doctor and asked whether his patient had any significant health problems. Sidestepping those same ethics, the doctor replied that he would answer the question only because the answer was "no."[5]

Frankfurter relayed this private medical information to Roosevelt. FDR thought about it, but for the moment he did nothing. He had another man to consider.

William Franklin Knox had run the *Chicago Daily News* before tossing his fedora into the 1936 election as Republican vice-presidential nominee. He had risen through the American middle class, from gym teacher, grocery clerk, and street reporter to newspaper publisher and Republican Party leader. When America declared war on Spain in 1898, he flocked to Teddy Roosevelt's Rough Riders banner and took part in the famed charge near San Juan Hill. He again volunteered during the First World War, and came home wearing a major's oak leaves.[6]

While Knox idolized the Oyster Bay Roosevelt, he had no love for the Hyde Park variant. After FDR defeated Hoover in 1932, Knox turned his sharp pen against the New Deal. During the 1936 campaign, he attacked Roosevelt's domestic programs with a vengeance. As he told one Boston audience, "The country needs *fewer* and *better* Roosevelts."

After he recovered from the crushing defeat of 1936, Knox's interventionist beliefs drew him toward the "lesser Roosevelt's" orbit. When Japanese aircraft sank the gunboat USS *Panay* in the Yangtze River in 1937, he telegrammed Roosevelt to assure him of "my unequivocal support in any further measures you may find it necessary to take to maintain American self-respect and respect for America abroad in a world that has apparently gone drunk and mad." Three years later, he would have dropped the word "apparently."[7]

Like Stimson, Knox's Republican bona fides made him a perfect choice for a coalition cabinet. It didn't hurt that Knox's rival newspaper was Robert McCormick's *Chicago Tribune*, flagship of the McCormick-Patterson newspaper syndicate, which had bitterly opposed FDR since 1932. If Knox were willing to throw in his fortunes with a Democrat, he would make a solid addition to FDR's cabinet.

FDR had asked Knox to join him in the past. Each time Knox declined, explaining that unless war became imminent, he could not in good conscience join a Democratic administration.

Hitler's panzer lunge over the Meuse River changed everything. On June 19, as Congress was passing a massive bill funding a two-ocean navy, Roosevelt rang up Knox and offered him his choice of the War or Navy Department. Henry Stimson, he said, would get whichever department Knox didn't choose.

Knowing Stimson had been head of the War Department under Taft, Knox told Roosevelt he didn't know enough about the Army, so FDR gave him the job of navy secretary.[8]

It didn't matter to Roosevelt that Knox knew even less about the Navy than he did the Army, for Roosevelt intended to act as his own naval secretary. As commander-in-chief, Roosevelt would set naval policy himself; he wanted a functionary, not a policy maker. A superb coastal yachtsman, FDR often reminisced about his years as Woodrow Wilson's assistant secretary of the navy. He intended to direct naval affairs regardless of who his nominal secretary might be.[9]

That same day, White House telephone operator Louise Hachmeister tracked down Henry Stimson at his apartment in New York's Pierre Hotel. Moments later, FDR's voice came on the line. He wanted to know if Stimson would accept the job of secretary of war.

Stimson agreed, on condition that he would not be required to participate in partisan politics. Roosevelt couldn't have been happier to make this concession, and he sent the names of Knox and Stimson to the Senate for approval just as Republicans began gathering in Philadelphia for their 1940 national convention.[10]

The twin appointments stung GOP delegates and party faithful. The *New York Times* reported that Stimson and Knox were "virtually read out of the Republican party," and angry Republican senators prepared for a gritty confirmation battle. *Time* commented, "If there was an opportunity to debate calmly the merits of Republicans Stimson and Knox in a Democratic Cabinet, the opportunity disappeared in the feverish political atmosphere of Convention Week. Senatorial debate grew bitter [and] reached a new low in wild charges and venomous insinuations, punctuated with cries of warmongering from isolationists, and virtual accusations of treason."[11]

But isolationist cries were weakening in the heartland. Churchill's defiant speeches from London, FDR's "stab in the back" address, and the Nazi onslaught in Europe had swung a narrow margin of public support to Britain's side. In July, Elmo Roper's opinion pollsters found that more than two-thirds of the public favored some kind of aid to Britain. Mainstream newspaper editors came out sympathetically for aid to the democracies, and a whimsical cartoonist signing his name "Dr. Seuss" began lampooning Republican isolationists by drawing their party symbol as an amusing half-elephant/half-ostrich creature, burying its head in the sand. He called his animal a "GOPstrich."[12]

Aftershocks from this seismic shift among voters rattled Capitol Hill offices. Despite partisan rancor and the opposition of men like Democratic Senator Burton K. Wheeler of Montana, another isolationist firebrand, Stimson's appointment passed on July 9 by a vote of fifty-six to twenty-eight. Knox was confirmed the next day by an even greater margin.

Roosevelt's cabinet now included two new war chiefs carrying their second Roosevelt banner—Franklin's, not Teddy's.[13]

The new secretary of war found an indispensable ally in the Army's chief of staff. Stimson had first met Marshall in France during the Great War, where the two men rode horseback together and shared a mess. Even then, Marshall stood out in Stimson's estimation as a sharp mind and first-rate soldier, and when President Coolidge sent Stimson to the Philippines as governor-general, Stimson unsuccessfully tried to persuade Marshall to join him as his military aide.[14]

At the end of June, Stimson invited Marshall to visit him at Highhold, and on the twenty-seventh Marshall flew to Long Island's Mitchell Field and took the twenty-minute drive to Highhold. There he had dinner with Henry and Mabel Stimson, and the two men talked genially until midnight.

He breakfasted early and flew back to Washington, reassured that he and the new secretary would forge a good working relationship. Marshall enjoyed his brief stay. "They are both delightful people and their farm is charming," he told his wife, Katherine.[15]

The two men were well matched. Neither harbored ambitions to move into the White House. Stimson trusted Marshall to handle the Army's affairs with Congress, the president, the press, and the British, while Marshall deferred to Stimson on administrative matters. Whenever FDR cut Stimson out of the chain of communication, Marshall briefed him on what the president was doing. Their adjacent offices in the Munitions Building were connected by a door, and that door was, at least figuratively, always open.[16]

On workdays Stimson, like Marshall, was an "early to bed, early to rise" man. He would awaken at five or six in the morning, dictate his diary entries for the previous day, then leave for work by eight. He threw himself into the department's workload with the fierce intensity he gave his law firm clients, and like Marshall, he disliked tackling complex problems after four or five in the afternoon. He relied on a talented team of lieutenants that included Jack McCloy, another former Wall Street lawyer, and Robert Patterson, who wore the belt of a German soldier he had killed in the First World War.*

---

* Like most of Stimson's circle, McCloy was an anti–New Dealer. As a lawyer with New York's prestigious Cravath, Swaine firm, he successfully represented the Schechter Poultry Corporation in a landmark constitutional battle against the National Industrial Recovery Act, a fundamental pillar of Roosevelt's New Deal.

Stimson's nights were quiet. He avoided the Washington dinner party circuit and dined at home with Mabel and a few guests, mostly old upper-crust friends like the Pattons or the Frankfurters. He might play a round of lawn tennis with friends or colleagues, but usually he and Mabel spent their evenings reading together or listening to the radio. On weekends they escaped Washington's humidity by flying to Highhold, and among the faux rustic charms of his estate, he could unwind, visit old friends, play tennis and ride horses.[17]

. . .

For three centuries, the British Empire had stretched across the world's oceans. Unlike the great land empires of Alexander, Caesar, and Napoleon, Britannia's rule depended on mastery of a line of choke points strung like pearls on a necklace. In 1940, the most important of these pearls were the West Indies, Gibraltar, Cairo, India, Singapore, and Hong Kong. As long as each of these redoubts held, the United Kingdom could shift its resources to defend any threatened portion of her empire.

Yet even with these colonies, Britain could not fight a world war without the resources of North America. The empire depended on Canada and the United States for raw materials to feed its factories, and for fruits and grains to feed its workers. The Atlantic shipping lanes carried that food and material; should Hitler slice those arteries, the blood spilling onto the Atlantic floor would ensure England's death.

The Royal Navy's best weapon against the U-boat menace was the humble destroyer, the tough rat terriers built to corner and sink submarines. But when Stimson and Knox took the oath of office, half of Britain's prewar destroyer fleet had been sunk or damaged. By mid-June, the Royal Navy had only sixty-eight destroyers fit for combat.[18]

Finding little opposition in the mid-Atlantic, Hitler's gray wolves unleashed hell. U-boats and bombers based in occupied France, Norway, and the Low Countries sank 155 merchant ships between April and June 1940, faster than Britain could replace its losses.[19]

The United Kingdom could not survive at this pace. In his May letter to Roosevelt, Churchill had requested fifty destroyers. He urged his request again in June and July with growing urgency. He assured Roosevelt the Royal Navy could make do even with the old four-stack destroyers of World War I vintage. But he needed those warships, and he needed them now.

On naval matters, Roosevelt and Churchill spoke the same language. FDR had been assistant navy secretary at the time Churchill had been First Lord of the Admiralty, the equivalent of the U.S. naval secretary. Roosevelt wanted to throw a lifeline to his drowning ally, but because the sale of fighting ships would pose insurmountable political problems, he felt he had to decline. Warships guarded America's shores and produced a feeling of safety that howitzers and fighter planes did not. Senator Walsh, an isolationist, was chairman of the Senate Committee on Naval Affairs, and FDR could not afford to antagonize the isolationist wing at such a delicate time.

Yet time was a luxury that Britain could no longer afford. On July 31, an anxious Churchill pleaded anew for the destroyers, concluding, "Mr. President, with great respect I must tell you that in the long history of the world, this is a thing to do now."[20]

Roosevelt shut himself in the balmy White House, shirtsleeves rolled up to his elbows, and struggled to find a way to send Churchill those destroyers.* He told his cabinet, "The survival of the British Isles under German attack might very possibly depend on their getting these destroyers."[21]

But neither he nor his cabinet could figure out how to do it.

The answer, and obstacle, was Admiral Harold L. Stark, chief of naval operations and Marshall's opposite number in the Navy. Affable and scholarly, with round glasses anchored below a shock of thick white hair, Stark was the Navy's top strategic thinker and a man with the credentials to persuade Congress to go along.

He was also eager to please his commander-in-chief. But as he sat in his spacious office at Main Navy, the Navy's Washington command post, he shook his head. The entire fleet had only 230 destroyers from Manila to the Virgin Islands. How could the Navy Department sell a fifth of its destroyer fleet without crippling America's thinly spread defense?

Stark's objection was no abstract policy problem; to him, it was personal. Under a statute authored by Senator Walsh, Stark was required to certify in writing that any ships sold to a foreign government were "not essential to the defense of the United States." Five months earlier, when Stark asked Congress to put the old four-stacks back into commission, he had testified that those destroyers were essential to the nation's defense.

Stark felt he had been placed in a terrible position of having to break his word to Congress, or embarrass the president by refusing to go along. Like a martyr resigned to the stake, Stark suggested he should be relieved of command, rather than be forced to recant his congressional testimony.[22]

---

* Air-conditioning was available during Roosevelt's time, but Roosevelt insisted on turning off the air conditioners in his living quarters, certain that the artificially cooled air aggravated his chronic sinus congestion.

Roosevelt brought the question to his cabinet on August 2, and there Frank Knox unwrapped an idea that had been batted around some months earlier: Instead of selling the destroyers for cash, why not swap them for British naval bases in the Caribbean and Canada?

Cabinet members supported the concept, but most assumed congressional approval would be necessary. That meant pulling along members of the Republican minority in an election year—an impossible task. Through Stimson, Roosevelt quietly reached out to pro-British Republicans for support, but even Stimson could make no headway.[23]

Roosevelt decided the stakes were too high to allow the fate of Britain to rest in the hands of Congress and a conscience-stricken admiral. He asked Dean Acheson, a Washington lawyer and former treasury undersecretary, to look through the laws and opine whether he needed congressional approval for a ships-for-bases trade. Acheson prepared a memorandum concluding that the commander-in-chief could authorize the transfer if the new bases would, on balance, increase the nation's security. If the bases were more valuable than the destroyers, Admiral Stark could honestly certify that the ships were unnecessary to national defense.[24]

Two days later, Roosevelt convened a luncheon at the White House with Stimson, Knox, Morgenthau, and Undersecretary Welles. He told his lieutenants—two conservative Republicans and two liberal Democrats—that he would accept Acheson's advice and make the deal with Churchill. He would inform Congress after the bargain was struck. He would catch hell from both sides of the aisle, probably, but it had to be done.[25]

Roosevelt's patchwork of legalisms would hardly have convinced a neutral jurist. But FDR saw the issue in simpler terms: As commander-in-chief of the armed forces—and head of state—he could send warships wherever they would do the country the most good. In 1907, the elder Roosevelt had done the same thing when he sent the Great White Fleet around the world, and FDR felt he had a better case for action than Cousin Theodore had.[26]

To Roosevelt, legal technicalities were fine so long as they didn't conflict with either the common defense or the general welfare—the really important things that follow the Constitution's opener, "We the People." His attorney general, Robert Jackson, later remarked, "The President had a tendency to think in terms of right and wrong, instead of terms of legal and illegal. Because he thought that his motives were always good for the things that he wanted to do, he found difficulty in thinking that there could be legal limitations on them."[27]

Admiral Stark knew that serving Roosevelt sometimes required bending those legal limitations. Some years earlier, when FDR had ordered him to build a set of bases in South America under dubious authority, Stark remarked that he would do it, but "I'll be breaking all the laws."

"That's all right, Betty," Roosevelt joked, using Stark's Naval Academy nickname. "We'll go to jail together."[28]

Roosevelt offered Churchill fifty destroyers in return for ninety-nine-year leases on seven British possessions off the Canadian coast and in the Caribbean Sea. It was a stiff price to pay, but a desperate Churchill accepted. The details would be worked out between FDR and Prime Minister Mackenzie King of Canada.[29]

Roosevelt took Stimson with him to meet with the Canadian PM in upstate New York, near where the First U.S. Army was holding maneuvers. Because the secretary of the navy was not invited and the secretary of state was on vacation, Roosevelt negotiated the deal through his secretary of war. *"It is a funny situation,"* Stimson told his diary two days later. *"For the last few days I have been acting more as Secretary of State than Secretary of War."*[30]

FDR didn't care about procedural formalities. He announced the deal in a press conference two weeks later. The destroyers-for-bases agreement, he claimed, was "probably the most important thing that has come for American defense since the Louisiana Purchase." When asked if the Senate had to ratify his decision, a beaming president said the deal "is all over. It is done."[31]

With a wave of Roosevelt's cigarette holder, Churchill had his destroyers. But the decision, Stimson knew, courted risk. Where, he wondered, would that leave America?

FIVE

THE NEW DEAL WAR

FRANKLIN ROOSEVELT LOVED TREES. HE TOOK PRIDE IN SELECTING THE varieties his workers planted around Hyde Park. He talked trees with his neighbors, and in his travels up and down the eastern United States he studied the maples, pines, sassafras, poplars, magnolias, and oaks that filled the landscape from Campobello to Warm Springs.

He understood, better than any other American, how the growing war effort mimicked those arboreal sentinels. The roots of the nation's might lay buried in its farms, mines, factories, and homes. The thick trunk—Congress, the White House, the War and Navy Departments, and the hundred-odd civilian agencies that ran the mobilization effort—channeled resources drawn from those roots. From that unruly trunk flowed arms, ammunition, food, supplies, and men into the ground, air, and sea forces, logistical departments, civil affairs, and diplomatic and intelligence services.

Roosevelt, his budget director once observed, "was the only one who really understood the meaning of *total war*." He appreciated, better than most of his military advisers, how a fighter plane sent to China affected Japan's threat to Russia, Russia's war with Germany, and Germany's campaign against Britain. As Roosevelt told a group of reporters, "There is just one front, which includes at home as well as abroad. It is all part of the picture of trying to win the war."[1]

He liked to say, "If war does come, we will make it a New Deal war." Roosevelt had already organized masses of men to fight unemployment and inflation. If the Great Depression could be overcome by American determination and competent leadership, that same energy, he believed, could vanquish the tyrants of Europe.[2]

* * *

Yet America's power still lay dormant. In 1940, its arms production was a quarter of Germany's. Steel plants were producing a third of their capacity. Shipyards took years to produce warships, and raw materials like tin, bauxite, and rubber were in dangerously short supply. American industry held immense potential, but it would be years before speeches, laws, and government money could turn that potential into weapons fired by trained soldiers.[3]

FDR knew the New Deal was a four-letter word to most men running the war industries. In his 1936 campaign, he denounced business magnates as "economic royalists," a term that raised vague images of guillotines and Madame Defarge. The National Labor Relations Act alone—to say nothing of the minimum wage law, securities regulations, or taxes funding the National Recovery Act— guaranteed that the name "Roosevelt" would be cursed in the dark-paneled clubs where barons gathered to blow off steam and plot their defense against New Deal revolutionaries.[4]

Yet capitalism, like democracy, spreads power among many players, and some of those players agreed with Roosevelt's foreign policy. To bring these economic royalists into his fold, FDR established a seven-member board, christened the National Defense Advisory Commission (NDAC), to which he appointed William Knudsen of General Motors, Edward Stettinius of U.S. Steel, and Ralph Budd, chairman of the Chicago, Burlington & Quincy Railway.[5]

It was an inspired decision. Knudsen, Stettinius, and Budd, natural leaders of the anti-Roosevelt clique, had spent eight years as outcasts in a New Deal– dominated Washington. Finding themselves welcomed back as patriots, they began bringing over fellow titans of industry.

As with most of Roosevelt's decisions, a move in one direction was balanced by a move in the opposite. He ensured the left controlled the NDAC by giving progressives four of the board's seven votes. He appointed union leader Sidney Hillman, the Trotsky of organized labor, to give workers a voice on the committee. Leon Henderson, a New Deal zealot, was appointed to handle prices. University of North Carolina Dean Harriett Elliott would advocate for consumers, and Federal Reserve Board member Chester Davis would provide input on farm production. Most importantly, the board reported to the president, ensuring that if any serious threat to New Deal progressivism arose, the New Deal would win.

In the NDAC, Roosevelt established a board charged with forging the tools of war. But he was careful not to take too great a step at one time. He knew the public would be torn between arms production and consumer goods, and at his press conference announcing the NDAC's formation, he was quick to reassure voters that rearmament would not force Americans to sacrifice too much butter for guns.

"I think the people should realize that we are not going to upset, any more than we have to, a great many of the normal processes of life," he told a group of journalists. American women, he promised, "will not have to forgo cosmetics, lipsticks, ice cream sodas. . . . In other words, we do not want to upset the normal trend of things any more than we possibly can help."[6]

. . .

Women might not have to forgo lipsticks, but young, able-bodied men would give up a great deal more than that before America would be ready to fight. In late May, FDR asked Congress to approve a contingency plan for calling up the National Guard, and in early June War Department planners urged Marshall to ask Congress to increase the size of the Regular Army to 530,000 men. Those men, they suggested, would form a cadre to train Guardsmen and draftees when the time was right.

Into this lion's den Roosevelt trod with caution. As Camel butts piled in his desk ashtray, he plotted his course with the eye of an old sailor scanning the clouds for squalls. An early push for conscription, he concluded, would derail aid to Britain, because a draft in mid-1940 would exaggerate the Army's shortage of rifles, planes, and ammunition. Isolationists like Walsh, Wheeler, and Lindbergh would wail that British aid was leaving the new draftees without the tools they needed to do their jobs.

Roosevelt disagreed. Better, he thought, to send those weapons overseas to fight Hitler now, rather than have them worn out or broken during basic training.[7]

Weighing on Roosevelt's mind was a fog of fear settling over the public. A coalition of isolationists, liberals, academics, and pacifists had rapidly coalesced against the draft, believing a draft would necessarily lead to war. Women calling themselves "Mothers of the USA" donned black veils and marched before the Capitol to oppose compulsory service.

For now, FDR backed down. His problem, he admitted, was "to get the American people to think of conceivable consequences without thinking that they are going to be dragged into this war." He explained to one diplomat, "American mothers don't want their boys to be soldiers."* He quietly ordered his military advisers to plan for expansion, but cautioned them to say nothing that might alarm the public.[8]

---

* He added, "American mothers don't seem to mind their boys becoming sailors."

. . .

As Hitler tightened his grip on Western Europe, FDR's public silence about the draft threw Stimson and Stark into fits. It might take a year to turn a school-teacher into a tanker or boatswain—longer for some specialists—and America needed those men now. Believing the public would follow its leader, Admiral Stark candidly told Roosevelt, "You could do so much more if you would strike out and lead."[9]

But as Roosevelt told his speechwriter, Sam Rosenman, "It is a terrible thing to look over your shoulder when you are trying to lead—and find no one there."[10]

General Marshall, a veteran of the Civilian Conservation Corps mobilization, was in no hurry to strike out and lead. A giant conscripted army in the summer of 1940 would be unmanageable because the bewildered draftees would have no place to sleep, no chow to eat, and no one to salute. Marshall needed to train more sergeants and build more training camps before he could absorb a mass of con-scripts. He was pressing every officer, congressman, and contractor to lay this foundation, but even with his officers moving at full speed, he knew the Army would not be ready for some months.

Politically, Marshall saw conscription as a glass of milk that would sour if left on the table too long. No one could know when war would come; it might be in late 1940, perhaps next year—or perhaps never. But once the president called out the National Guard or conscripted civilians, the Army would have a treacherously short window in which to train and deploy those men. Before long, politically connected Guardsmen and their families would begin pressuring congressmen to send their sons and husbands home. If Congress gave way and released the draftees, the Army might become a gutted shell just as the crisis hit American shores.

Marshall also knew a political danger awaited an army drafting its citizens. Muckraker journalists waited in the wings to break stories of a palace coup at the first sign of military expansion, and isolationists in the heartland would make a receptive audience. So Marshall insisted that any move to boost the Army's size must originate with Congress, not the War Department. The Army, he insisted, would play the role of reluctant bride led to the altar.

"You might say," he mused later, "that the Army played politics in this period. That is a crude expression. Actually, we had a high regard for politics. We had regard for the fact that the president did not feel assured he would get the backing of the people generally and in the Middle West particularly and had to move with great caution."[11]

At the end of June, a Democratic senator and a Republican congressman introduced a bill instituting a peacetime draft. The public was now willing to talk about conscription. With Congress taking the lead, Marshall and Stark felt they could safely encourage passage of a selective service bill, coupled with a twelve-month federalization of the National Guard and appropriations for barracks, uniforms, and vehicles for the new draftees.[12]

The next step was to get a cautious president to strike out and lead.

After six weeks of prodding by Marshall, Stark, Stimson, and others, FDR publicly threw his weight behind the selective service bill. Sitting before a bank of reporters crowding his Oval Office on August 2, he took a question about the draft from the *Baltimore Sun*'s Fred Essary.

"Mr. President," said Essary, "there is a very definite feeling in congressional circles that you are not very hot about this conscription legislation and as a result, it is really languishing."

Roosevelt jumped on Essary's question with both feet. Declaring that selective service was essential to national defense, he said he hoped war would not come, but if it did come, the nation must be ready. "We figured out in 1917 that the selective training or selective draft was the fairest and in all ways the most efficient way of conducting a war if we had to go to war," he said. "I still think so, and I think a majority of the people in this country think so, when they understand it."[13]

He hoped they did. A peacetime draft was a supreme gamble in an election year. His traditional allies—youth organizations, New Dealers, organized labor—found military service repugnant. Conscription might provide the best hope for America's defense, but it also provided the best hope for Republicans looking to unseat their nemesis in the next election, should he run again.[14]

·   ·   ·

It had gone almost without saying that Roosevelt would decline to run for a third term, a venerated custom observed by George Washington and respected by every president since 1796. Conservatives, including Stimson, felt it would be a mistake for him to run again. At times FDR considered progressive stalwarts like Cordell Hull, Harry Hopkins, South Carolina Senator James Byrnes, and others as possible successors. But by the summer of 1940, Roosevelt had no disciple with the right combination of skills for a wartime presidency. The fall of France had brought war closer to American shores, and as FDR saw it, the country couldn't afford to change horses in the middle of a rapidly filling stream.[15]

There was also the lure of the office, though Roosevelt wouldn't admit it. He had been at the center of power for nearly eight years, and he had enjoyed almost every minute at 1600 Pennsylvania Avenue.

So in the wee morning hours of July 19, speaking from the White House broadcast room, Franklin Roosevelt explained his reasons for breaking Washington's sacred custom:

> Lying awake, as I have on many nights, I have asked myself whether I have the right, as commander-in-chief of the Army and Navy, to call on men and women to serve their country or to train themselves to serve and, at the same time, decline to serve my country in my own personal capacity if I am called upon to do so by the people. . . .
>
> Like most men of my age, I had made plans for myself, plans for a private life of my own choice and for my own satisfactions to begin in January 1941. These plans, like so many other plans, had been made in a world which now seems as distant as another planet. Today all private plans, all private lives, have been in a sense repealed by an overriding public danger. In the face of that public danger all those who can be of service to the Republic have no choice but to offer themselves for service in those capacities for which they may be fitted.[16]

There was at least one other man who felt he could be of service to the republic. While Roosevelt's subalterns were building the case for conscription, Republicans meeting in Philadelphia nominated Indiana businessman Wendell Willkie for president. The war in general, and the draft specifically, would be the election's great issue. The Republican candidate could take that issue off the table by endorsing conscription, or he could use it as a political blackjack.

To FDR's relief, Willkie took it off the table. In his acceptance speech he declared, "I cannot ask any American to put their faith in me without recording my conviction that some form of selective service is the only democratic way in which to assure the trained and competent manpower we need in our national defense." Willkie went even further, telling his audience he agreed with Roosevelt: The full material might of America must be brought to support the western democracies.[17]

Willkie was a political outsider with the horse sense to agree with a good idea, even if that idea came from his opponent. But horse sense and political sense are two different things, and Willkie's endorsement of the draft couldn't help his electoral prospects. He was failing to distinguish himself from his opponent, and he would be left playing "me, too" on two of the election's most important foreign policy issues.

Reading a transcript of his challenger's speech that night, Roosevelt beamed. "Willkie is lost."[18]

While the Republican nominee supported the draft, his fellow conservatives, Democratic isolationists, and pacifists fought a rearguard action to halt selective service. As the bill wound through both houses of Congress, they launched hit-and-run attacks with crippling amendments and procedural objections. The Senate came within one vote of prohibiting the use of the National Guard outside U.S. territorial limits, while Representative Hamilton Fish of New York, Roosevelt's implacable foe from Dutchess County, persuaded colleagues to defer the draft's implementation until after the election. Two congressmen became so bitter they exchanged curse words and blows in the House chamber.[19]

Marshall and Stimson spent long hours brokering compromises to defeat killer amendments, and they watched anxiously as the bill worked its way to the House and Senate floors in September. The deadlock was finally broken on the fourteenth, when the House and Senate passed a bill that Roosevelt, Stimson, Marshall, and Stark could support. Two days later, Roosevelt signed the bill in the Oval Office as Stimson, Marshall, and the chairmen of the House and Senate military affairs committees looked on, unsure whether to appear supremely pleased or stately and solemn.[20]

When Roosevelt lifted his fountain pen, sixteen and a half million men became eligible for military service. Marshall was now authorized to call up to 900,000 men annually to fill nine Regular Army infantry divisions, four armored and two cavalry divisions, and eighteen National Guard divisions. In the event the country needed a dramatic expansion—say, to five million or more—it was Marshall's hope to use these first draftees as the backbone of a larger force.[21]

That evening, General Marshall appeared on the CBS radio network. He announced, "For the first time in our history we are beginning in time of peace to train an army of citizen-soldiers which may save us from the tragedy of war."[22]

But to train that army of citizen-soldiers, Marshall would have to rein in a president intent on giving away his army's weapons. He was about to learn how hard that would be.

SIX

"ONE-FIFTY-EIGHT"

~

O N September 7, 1940, Reichsmarschall Hermann Göring launched his long-awaited bombing campaign to shatter British will to resist. For two months, swarms of Junkers and Heinkel bombers gathered nightly to drop ton after ton of incendiaries on London and other major cities. Civilians burned to death, or were crushed beneath the rubble of flats, cathedrals and storefronts. Air wardens directed terrified Londoners into Underground tunnels and fought blazes that lit the night sky. Ambulance drivers negotiated debris-choked streets in the dark, and flak gunners threw everything in their limbers at the bombers.

Though Roosevelt ordered Stimson and Marshall to send England every weapon they could spare, Stimson doubted there was much left to give. *"This is going to be a rather agonizing affair, because we have so little that we can give them, if anything,"* he told his diary.[1]

Three weeks into the Blitz, Roosevelt summoned his military advisers and Morgenthau to discuss weapons shipments to Britain. One item high on his list was Boeing's new B-17 "Flying Fortress" bomber. He wanted bombs dropping on Berlin, and he asked Marshall when he would ship those Fortresses to England.

An uncomfortable pause. Marshall replied that the Army Air Corps had forty-nine bombers fit for duty on the continental United States.

Forty-nine. *"The President's head went back as if someone had hit him in the chest,"* wrote Stimson.[2]

But FDR recovered quickly, and he would not be put off by mere numbers, even if those numbers came from generals. Military men sometimes had to be pushed, or they would wait and wait until they had everything just right. Since everything would never be just right, Roosevelt wanted someone he trusted, someone like Morgenthau or Hopkins, doing the pushing.

To Stimson and Marshall, the president was pushing the wrong way. An orderly plan for allocating weapons was needed, wrote Stimson, *"so that we will not make the decisions, these vital decisions, as to what we give or do not give to the British, too haphazardly and under the emotion of a single moment."* Frustrated with Roosevelt's extemporaneous donations, in early 1940 the Army's air chief, "Hap" Arnold, testified before Congress that Air Corps effectiveness was being adversely affected by arms shipments to Britain.

When a furious Henry Morgenthau learned of Arnold's testimony, he went straight to FDR, who blew his stack. At his next meeting with Marshall and his staff, Roosevelt warned Arnold that there were places to which officers who did not "play ball" might be sent, "such as Guam." For the next nine months, Arnold was persona non grata at the White House.[3]

Roosevelt would make his generals play ball. When he learned the Army had no existing planes to spare for Britain, he turned his attention to aircraft still in production. Before year's end, he announced that the United States would split production of new warplanes with the United Kingdom on an "even-Steven" basis.[4]

Marshall held the line against Roosevelt's largesse. In a follow-up meeting, he had his aides show the president a chart indicating that only a third of the planes scheduled for production that month had actually been produced. He dryly asked whether the British would get half the number the Americans intended to produce or half the number they actually produced.

Stung by the implication that he was selling every available plane to Britain, FDR looked sharply at Marshall.

"Don't let me see that chart again," he growled.

Unruffled, Marshall merely nodded. He had made his point.[5]

The chart disappeared, but Congress did not. Churchill wasn't giving away more bases, and the Justice Department concluded that bombers could not be sold to Britain without congressional authorization. That required Marshall to certify that B-17 Flying Fortress bombers were unnecessary to U.S. defense, a patently absurd proposition.

Running out of room to maneuver, Roosevelt suggested giving Britain a limited number of Forts to test under combat conditions. It would provide useful information for future design modifications, he claimed, which would enhance the nation's defense. In his diary that night, Stimson called it *"the only peg on which we could hang the proposition legally."*[6]

The peg gave way when Justice Department lawyers concluded that Roosevelt could not part with U.S. property without congressional permission, even for "testing" purposes. Stimson, a more experienced lawyer than anyone at Justice, then came up with another creative solution. Perhaps, he suggested, Marshall could certify that B-17 bombers were unnecessary if they were traded for B-24 Liberator bombers still on production lines but allocated to England under the "even-Steven" rule.

This peg was nearly as shaky as the last one, but a sheepish Marshall complied. Like Admiral Stark, he had personal misgivings about using contorted legalisms to send weapons to London. Marshall later confessed, "I was a little ashamed of this because I felt that I was straining at the subject in order to get around the resolution of Congress."[7]

Coming off the selective service victory, Henry Stimson trudged through the pains of an army in adolescent growth. Labor issues, housing shortages, production delays, and organizational flaws up and down the Army's structure consumed his working days and never seemed to be resolved.

What vexed Stimson more than anything was FDR's penchant for making far-reaching decisions with no apparent method beyond whim or instinct. As Stimson saw it, two-thirds of the government's problems came from *"the topsy-turvy, upside-down system of poor administration [by] which Mr. Roosevelt runs the government."* After a particularly tiresome discussion on war production, Stimson told his diary, *"Conferences with the President are difficult matters. His mind does not follow easily a consecutive chain of thought but he is full of stories and incidents and hops about in his discussion from suggestion to suggestion and it is very much like chasing a vagrant beam of sunshine around a vacant room."*[8]

Occasionally, however, the vagrant beam of sunshine surprised Stimson. Once he complained to FDR that the administration's price control board was retarding war production by holding prices artificially low. Roosevelt, an old hand at economic manipulation, knew that high wages and consumer-goods shortages were a recipe for crippling inflation. He told Stimson he did not want weapons production spurring a "boom of rising prices" before the war, because after the war he wanted no severe price drops disrupting the economy. The nation's long-term interest required prices to remain relatively stable, and that required government to regulate prices in the arms market.

Stimson, the grandson of a Wall Street investment banker, believed in the free market like he believed in gravity or an all-powerful God. But he admitted he had not considered the economic impact of the war over the long run. As secretary of war, he didn't have to. But the president had to think about it every day.

The more he saw of Roosevelt, the more Stimson admired the man. Some months later, after watching him digest a War Department study of tank production, Stimson mused, *"It is marvelous how he can give so much attention to a detail and to do it so well as he has done this. . . . He has spread himself out extremely thin but nevertheless he does carry a wonderful memory and a great amount of penetrative shrewdness into each of these activities."*[9]

.   .   .

There was one question neither Roosevelt nor Stimson could answer, because in 1940 there was no practical answer. The Regular Army, numbering nearly half a million men, included 4,700 black soldiers, two black line officers—one nearing retirement age—and four colored combat units led by white officers. Most black enlisted men were assigned as laborers in quartermaster, engineer, or infantry units.

Compared to its sister services, the Army was progressive. The Air Corps would not train black pilots. The Navy had no black Annapolis graduates, and no black marines; with the exception of six petty officers, black sailors onshore were assigned dock and warehouse duties. Shipboard assignments for a Negro sailor were limited to cook or valet—"seagoing bellhops," the black press called them.

Given the tremendous manpower expansion the war would require, it made sense to find ways to boost non-white enlistment. But the most Roosevelt would back in an election year was a Selective Service Act provision pledging to increase black military participation to the Negro proportion of the civilian population, about 10 percent.

The catch, buried within the statute's language, was a proviso that no man of any race would be inducted "until adequate provision shall have been made for shelter, sanitary facilities, water supplies, heating and lighting arrangements, medical care and hospital accommodations." Whites had these accommodations, but since segregated facilities would require construction of a new set of barracks, hospitals, and other infrastructure, the "separate but equal" philosophy prevailing since Plessy sued Judge Ferguson would ensure that the Army remained all but closed to new black enlistees.[10]

Civil rights leaders, led by the Urban League's T. Arnold Hill and union organizer A. Philip Randolph, called on Roosevelt to open military doors to greater Negro participation. Their demand put FDR in a vise between two groups he could not afford to alienate—moderately conservative whites and reliably Democratic blacks. Under the pressure of an election year, Roosevelt had no choice but to meet with Randolph and Hill.

On September 27, Roosevelt, Knox, and Stimson's assistant Robert Patterson met in the Oval Office with Randolph, Hill, and Walter White of the National Association for the Advancement of Colored People. As they talked, Roosevelt's secret office recorder documented the exchange.[11]

"Mr. President," Randolph began, "it would mean a great deal to the morale of the Negro people if you could make some announcement on the role the Negroes will play in the armed forces of the nation. . . . I might say that it is the irritating spot among the Negro people."

"Yeah, yeah," agreed Roosevelt.

"They feel that they're not wanted in the various armed forces of the country and they feel that they have earned the right to participate."

Roosevelt said he would add something about that in a speech he would give in the near future.

"It's a start," said Randolph.

"Hell, you and I know it's a step ahead," agreed Roosevelt.

FDR explained his next steps in his typically serpentine fashion. "Now, you take the divisional organization. What are your new divisions? About twelve thousand men."

"Fourteen. They vary—" began Patterson.

"Yes, and twelve, fourteen thousand men," said Roosevelt. "Now suppose you have in there one—what do they call the gun units? Artillery?"

"Batteries," said Patterson.

"What?"

"Batteries."

"One battery, with Negro troops and officers, in there in that battery, uh, like for instance from New York, and another regiment, or battalion—that's half a regiment—of Negro troops. They go into a division, a whole division of twelve thousand. And you may have a Negro regiment, you would, here, and right over here on the right in line would be a white regiment in the same division, maintain the divisional organization. Now what happens? After a while, in the case of war those people get shifted from one to the other. The thing we sort of back into."

As for the Navy, Roosevelt vaguely claimed that inroads were being made for black sailors outside the messmen's corps. Turning to Knox, he said, "Another thing I forgot to mention, I thought it about, oh, a month ago, and that is this: We are training a certain number of musicians on board ship, the ship's band. Now there's no reason why we shouldn't have a colored band on these ships, because they're darn good at it."

The dismay on the faces of the black leaders set Roosevelt on his heels.

"At worst, it will increase the opportunity," he said defensively. "That's what we're after."[12]

But Knox was not interested in opportunity. He had to work with the admirals whose officers would revolt, and he flatly declared that the Navy's special circumstances—men living and working in confined quarters—made integration unworkable.

"We have a factor in the Navy that is not so in the Army, and that is that these men live aboard ship," he said. "In our history we don't take Negroes into a ship's company." He added that it was a particular problem for sailors and officers from Southern states.

Trying to lighten the atmosphere, Roosevelt joked, "If you could have a Northern ship and a Southern ship it would be different."[13]

The civic leaders were not amused. But they left the Oval Office believing Roosevelt sympathized with them enough to act on their rather modest requests.

FDR left the meeting torn between his notion of equality and the demands of the moment. His conscience told him that if war came, the Negro soldier deserved the right to etch his place among the nation's war monuments. Colored troops had fought bravely at Battery Wagner, San Juan Hill, and the Meuse-Argonne, and he saw acceptance of black troops as a morally compelling step.

But the country was also facing a grave foreign threat, and he was facing an election year. The day Roosevelt met with Randolph, Hill, and White, General Marshall was writing Senator Henry Cabot Lodge Jr. that it was no time "for critical experiments which would have a highly destructive effect on morale," such as service integration. With Europe under Hitler's jackboot, America could not afford to weaken its anemic military power through a new social initiative.[14]

Stimson agreed. He was the son of an abolitionist who had fought for the Union, but Henry Stimson's job was to build an Army, not reform society or rectify inequities rooted in 300 years of slavery. He told his diary, *"This crime of our forefathers had produced a problem which was almost impossible of solution in this country and I myself could see no theoretical or logical solution for it at war times like these."*[15]

A stumble by the president's press secretary, Steve Early, triggered another lurch toward desegregation. When the September White House meeting produced no apparent results, black leaders turned to Eleanor Roosevelt. She began hounding the War Department to make concessions to the concept of equality, and aggressively followed up on complaints of discrimination by colored soldiers in training camps and surrounding towns.[16]

Pushed by Eleanor, the Army agreed to induct colored units into each major branch of service, including aviation training. It refused to integrate units below the brigade level, however. "The policy of the War Department is not to intermingle colored and white enlisted personnel in the same regimental organizations," it announced. "This policy has proven satisfactory over a long period and to make changes would produce situations destructive to morale and detrimental to the preparation for national defense."[17]

When the White House acknowledged the statement, Early gave journalists the misleading impression that the black leaders who met with Roosevelt agreed with the War Department's policy. The outraged trio called Early's statement a gross mischaracterization of their meeting with Roosevelt and denounced the Army's policy as "a blow at the patriotism of twelve million Negro citizens."

Black newspapers picked up the cry, which echoed in the mainstream press. Harlem's *Amsterdam News* lashed out at the "Jim Crow Army," while the *Kansas City Call* carried the story under headlines blaring, "Roosevelt Charged With Trickery in Announcing Jim Crow Army Policy."[18]

The attacks stung FDR, who would be counting on heavy black voter turnout in November. When Eleanor explained to her husband the predicament the civil rights leaders were in—they had to respond to Early's statement, or their constituents would accuse them of selling out—Roosevelt issued a clarifying statement of his own. There was no fixed policy regarding future units, he said, so the War Department's policy was not necessarily the way things would always be. "At this time and this time only, we dare not confuse the issue of prompt preparedness with a new social experiment, however important and desirable it may be."[19]

Roosevelt's clarifying statement actually made his position less clear. But as a sop to civil rights leaders, he announced the promotion of Colonel Benjamin Davis, 36th Coast Artillery Regiment, to brigadier general. He also had Stimson appoint Howard Law School's Dean William Hastie as civilian aide to the secretary of war on matters of race.[20]

Stimson was furious at the election-year concessions. He frumped to his diary, *"There is a tremendous drive going on by the negroes, taking advantage of the last weeks of the campaign in order to force the Army and the Navy into doing things for their race which would not otherwise be done and which are certainly not in the interest of sound national defense."* The root cause, he grumbled, was *"Mrs. Roosevelt's intrusive and impulsive folly."*[21]

Whether folly, political gamesmanship, or sound policy, the War Department made limited efforts to accommodate its commander-in-chief. Marshall formed a cavalry brigade from its two traditional colored regiments, the Ninth and Tenth, and he placed General Davis at the brigade's head. The Army Air Corps also began providing limited, indirect support for black pilots through the Civilian Pilot Training Program based near Tuskegee, Alabama.[22]

An annoyed Stimson doubted that the newly inducted men would develop the leadership skills required of combat officers, and he did little more than pay lip service to Dean Hastie's cause. During an unusually lighthearted cabinet meeting, he wrote, *"I had a good deal of fun with Knox over the necessity that he was now facing of appointing a colored Admiral and a battle fleet full of colored sailors. . . . I told him that when I called next time at the Navy Department with my colored Brigadier General I expected to be met with the colored Admiral."*[23]

. . . .

FDR had declared Wendell Willkie's prospects dead in August, but by late September the morbid campaign had a curiously strong heartbeat. After a soft start, Willkie began throwing hard rhetorical punches hitting on the evils of a third-term presidency, and blistered Roosevelt's warmongering foreign policy. Cheering supporters in Republican strongholds played up the theme with the fervor of a tent-revival choir. If Roosevelt won, Willkie told an audience in October, "You can count on our men being on transports for Europe six months from now."[24]

Willkie's wave was rising, and no one knew when it would crest. By October, Roosevelt's lead appeared vulnerable, and just as the Gallup organization showed Willkie gaining in strength, John Lewis of the Congress of Industrial Organizations spurned Roosevelt, his longtime ally, and endorsed Willkie as the lesser of two evils. When the defection became public, labor leader Sidney Hillman remarked after the meeting that he had never seen Roosevelt "so thoroughly scared."[25]

Against Willkie's rising fortunes, Roosevelt took a deep breath and plunged into the campaign's final weeks. It would be an uphill fight, because over his shoulders lay one of the heaviest political burdens a presidential candidate has ever borne: the imposition of a peacetime draft.

After passage of the Selective Service Act, American men between the ages of twenty-one and thirty-five assembled before their local draft boards. Each man was given a registration number, the magic sequence of digits that would determine when he would be drafted. On October 29, six days before voters headed to the polls, Franklin Roosevelt and Henry Stimson formally inaugurated the draft on the stage of Washington's War Department auditorium.[26]

The scene for the first lottery drawing emphasized the event's solemn nature. There were no military bands, no rousing speeches, no color guards shouldering guidons and rifles. Instead, President Roosevelt stood at a podium next to Secretary Stimson and made a few short remarks about the awful necessity that had forced the nation to defend itself. He did nothing to minimize the gravity of the occasion—though he studiously avoided the word "draft," substituting the word "muster," a term reminiscent of the patriot armies of George Washington and Old Hickory.

A large glass fishbowl containing hundreds of cobalt capsules was placed before Stimson. A blindfold, made of yellow linen cut from a chair used at the signing of the Declaration of Independence, was placed over Stimson's eyes. Stimson lowered his left hand into the bowl and withdrew the first capsule he touched. He handed the capsule to Roosevelt, who opened it and read the paper slip inside.

In a firm, low voice Roosevelt called out, "Drawn by the secretary of war, is serial number one-fifty-eight."[27]

The decision to hold the draft lottery one week before Election Day created a bitter division among the president's men. "Any old-time politician would have said [it] could never take place," said FDR's speechwriter Sam Rosenman. FDR's men had urged their boss to schedule the lottery for late November, and Roosevelt knew that photos of himself, standing before a fishbowl representing millions of young men, would be inserted into Republican flyers and imprinted onto the minds of voters heading to polling stations.[28]

But Roosevelt, like Willkie, felt he must be candid with the public. War might be coming, and men might die. The issues were too grave for political gamesmanship and too big to be swept under someone else's rug.

Democratic allies saw a wave of fear washing over the electorate, and party leaders begged him to assure mothers and fathers that their boys would be sent to training camps, not to war in France. Roosevelt refused to underwrite such a sweeping commitment. He promised he would not send them into "foreign wars" unless America was attacked, but that was as far as he would go.

On October 31, during a short campaign swing, Roosevelt included in a Boston Garden speech a line underscoring his peaceful intentions. Before a crowd of enthusiastic New Englanders, he told the mothers and fathers of America, "I have said this before, and I shall say this again and again: Your boys are not going to be sent into any foreign wars."[29]

In this speech, he did not include his customary phrase "except in case of attack." He felt the caveat went without saying. "It's clearly implied," he told an uneasy Rosenman. "If we're attacked, it's no longer a foreign war." Surely the American people would understand that by now.[30]

As Roosevelt and Eleanor took the train home to Hyde Park on the fourth of November, surveys showed Willkie pulling even. Yet Roosevelt smiled gamely as he drove to his polling station in the white-framed town hall. He chatted up the locals, most of whom he knew, and signed the register, listing "tree farmer" as his occupation. One reporter watching him wrote, "Nothing in the President's demeanor, at the polling place or in his home circle, indicated concern over the outcome of the election. In fact, he was extraordinarily jovial."[31]

He was less jovial as the evening's early returns trickled in. Willkie was no Alf Landon, and the prognosis looked grim. Sweating from anxiety, Roosevelt retreated to his dining room and ordered his Secret Service guard, Mike Reilly, to keep everyone out.[32]

His able assistant, Marguerite "Missy" LeHand, shuttled in fresh returns from chattering teletypes as FDR watched the final battle unfold. Willkie's early lead was disheartening, but as the evening sky deepened, the center of Roosevelt's battle line held. The big northern states, New York, Illinois, Ohio, and Pennsylvania, advanced across the electoral map, breaking for the incumbent. The labor vote, the black vote, the lower-class vote, and the foreign-born vote held steady on the flanks. When the smoke cleared, FDR's banner, singed and frayed, still flew over the battlefield.[33]

Around midnight, a visibly relieved Franklin Roosevelt rolled out to Springwood's porch to acknowledge a small crowd of local Democrats, reporters, and well-wishers. One supporter carried a homemade placard reading, "SAFE ON THIRD."[34]

He was. By morning, the results were history: FDR had won 54.7 percent of the popular vote, to 44.8 percent for Willkie.[35]

For a change, Roosevelt even carried his home district.

SEVEN

THE PARABLE OF THE GARDEN HOSE

᠁

M ARSHALL, STIMSON, AND THEIR RETAINERS SAT IN THE TREASURY BUILD-
ing's cavernous conference room, listening patiently as Henry Morgenthau
lectured before a large blackboard. With his bald pate, flat voice, and round
pince-nez glasses, Morgenthau resembled the Norman Rockwell image of a local
accountant auditing a small town's bookkeeping journals.[1]

Which was appropriate, except the ledgers did not belong to a small town.
They belonged to the British Empire.

Behind Morgenthau, written in long chalk columns, was a list of the United
Kingdom's liquid assets: dollars, bullion, foreign securities, mineral stockpiles—
everything but the crown jewels and Nelson's Column. The chalk lines showed
that the ripened fruits of two and a half centuries of conquest now totaled less
than five billion dollars. It was an embarrassing position for a cash-only arms deal,
since five billion dollars was about what Britain needed in military and food aid,
more or less immediately.[2]

The prospect of a British bankruptcy was deeply unsettling to the man who
lived in the house next door to Treasury, the juggler who had dreamed up one hy-
phenated scheme after the next—"cash-and-carry," "destroyers-for-bases," "even-
Steven"—to keep Britain afloat. Across the Channel, Hitler had snatched up half a
dozen big arms factories and many smaller ones, widening the production gap between the Axis and Britain. By December 1940, His Britannic Majesty was losing
the war one tank, one bomber, one bullet at a time.[3]

For once, even Roosevelt was stumped. Lending money and extending credit
were clearly things only Congress could authorize, and nobody believed Congress would loan Britain a penny. The Great Powers had been more than a little

delinquent in paying their debts after the last big war, and few congressmen had much appetite to tell constituents their tax dollars were being shipped off to Britain.[4]

Roosevelt needed some quiet, uninterrupted time to think about the problem, and in early December he scheduled a post-election vacation cruise. As always, the Secret Service kept the president's itinerary under wraps. To inquisitive journalists Roosevelt blithely announced that he would be shopping for Christmas cards on Christmas Island and hunting Easter eggs on Easter Island.

But his real destination was the Caribbean, where he could inspect his newly acquired bases from the fantail of the cruiser USS *Tuscaloosa*. And do a little thinking.[5]

The commander-in-chief was piped aboard ship on December 3, accompanied by his affable gatekeeper and military aide, Brigadier General Edwin "Pa" Watson; his personal physician, Rear Admiral Ross McIntire; Harry Hopkins; and the new First Dog, a small black Scottie named Fala.

Standing out under the boom of twenty-one-gun salutes, *Tuscaloosa* and her crew bore their passengers to Guantánamo Bay, Puerto Rico, and former British outposts in Jamaica, St. Lucia, and Antigua, which had been acquired in the destroyers-for-bases deal. As the graceful ship plowed south, FDR and his entourage spent the better part of each day fishing, chatting, playing small-stakes poker, and watching the placid islands drift by.

The fish didn't bite, even for the President of the United States—and despite emphatic advice on bait radioed to *Tuscaloosa* by author Ernest Hemingway. The biggest catch that week, a twenty-pound grouper, was hooked by Harry Hopkins, whose wasted body proved unequal to the reluctant fish. After a short struggle, Hopkins handed his rod to Dr. McIntire to reel in the catch. The ship's crew eventually landed the grouper, sparking a heated debate among McIntire, Hopkins, and General Watson over who held the honor of landing the cruise's biggest fish.[6]

Though the press speculated that Roosevelt simply wanted a break from the Battle of Britain, his thoughts drew nearer to the embattled isle. Navy seaplanes regularly approached *Tuscaloosa* with bags of dispatches from London describing bombing raids setting fire to cities, flattening entire blocks, and turning buildings into brickbat tombs for women, children, and old men.[7]

One of these dispatches, dated December 7, came from the "Former Naval Person," as Churchill liked to call himself. The message bore a bleak prognosis emphasizing the war at sea, not Blitz in the sky: "We can endure the shattering of our dwellings and the slaughter of our civilian population by indiscriminant air attacks and we hope to parry these increasingly," Churchill said. "The decision of 1941 lies upon the seas; unless we can establish our ability to feed this Island, import munitions of all kinds which we need . . . we may fall by the way, and the time needed by the United States to complete her defensive preparations may not be forthcoming."[8]

As Roosevelt sat on *Tuscaloosa*'s fantail, turning Churchill's words over in his mind, a dim thought took shape. The notion was only a nibble on his hook, but as he thought about it, he realized he had hooked a catch much fatter than Harry's thrashing grouper.

"I didn't know for quite a while what he was thinking about, if anything," said Hopkins afterward. "Then, one evening, he suddenly came out with it—the whole program. He didn't seem to have any clear idea how it could be done legally. But there wasn't a doubt in his mind that he'd find a way to do it."[9]

The "whole program" was simply to lend Britain everything it needed. Not with the idea of receiving immediate payment, but by getting everything back once the war was over—or being paid for items damaged or lost. Roosevelt would close America's "Cash Only" store and open a rental company.

Suffused with his new idea, FDR ordered the ship to set course for Washington. At his next press conference, he borrowed a metaphor from Harold Ickes to explain his solution in language the average American could understand.

"Suppose my neighbor's house catches on fire, and I have a length of garden hose four or five hundred feet away," he told a group of journalists.

> If he can take my garden hose and connect it up with his hydrant, it may help put out his fire. Now what do I do? I don't say to him before that operation, "Neighbor, my garden hose cost me $15; you have to pay me $15 for it." What is the transaction that goes on? I don't want $15—I want my garden hose back after the fire is over. All right. If it goes through the fire all right, intact, without any damage to it, he gives it back to me and thanks me very much for the use of it.

Of course, Roosevelt allowed, the hose might be damaged. In that case, the neighbor would have to buy him a new hose. Either way, the owner of the hose would be out nothing, and the fire would be put out, protecting his own home from the spreading flames.[10]

To the public, Roosevelt's homely metaphor smacked of common sense. To his advisers, it was brilliant in its audacity. Frances Perkins called it a "flash of almost clairvoyant knowledge and understanding." Speechwriter Bob Sherwood marveled at FDR's thought process. How he came up with the idea, wrote Sherwood, "Nobody that I know of has been able to give any convincing idea. . . . He did not seem to talk much about the subject in hand, or to consult the advice of others, or to 'read up' on it. . . . One can only say that FDR, a creative artist in politics, had put in his time on this cruise evolving the pattern of a masterpiece."[11]

To drive home his parable, Roosevelt took to the airwaves in an end-of-year fireside chat. The fireside chat, an informal radio address directly to the people, was a powerful tool that he used sparingly, for he knew the public would grow tired of his monologues if they heard them too often. While he had become famous for his fireside chats, he had given only about two per year since taking office in 1933. His reluctance to fill the air with speeches meant that when he did speak, his voice reached an audience of around fifty million listeners.

"My friends," his voice crackled from wireless speakers across the land, "this is not a fireside chat on war. It is a talk on national security; because the nub of the whole purpose of your president is to keep you now, and your children later, and your grandchildren much later, out of a last-ditch war for the preservation of American independence."

In measured tones he told families sitting around living rooms and kitchens that negotiated peace with Hitler's Germany was a forlorn hope. "No man can tame a tiger into a kitten by stroking it," he said. "There can be no appeasement with ruthlessness. There can be no reasoning with an incendiary bomb." The Nazis were overrunning Europe, and the Japanese were subjugating Asia. Should they rule the Eurasian landmass, he warned, "all of us in the Americas would be living at the point of a gun."

It was no time for business as usual. America's allies needed aid that only the United States could produce, and his new program, popularly called "Lend-Lease," would give them that aid. The nation, he declared, must become the "great arsenal of democracy."[12]

· · ·

Henry Stimson believed America could not remain a mere arsenal for long. In his finely appointed living room at Woodley, after a pleasant meal with Frank Knox, the two old colonels turned their thoughts to Europe. Sitting among Woodley's antique furnishings and Winslow Homer paintings, both men believed Churchill could hold out against Hitler, but Great Britain could never defeat Germany on her own, even with material support from the United States. The time had come to think about what the country would do when Britain could no longer be saved by "help short of war."[13]

Their focus on Germany was sharpened by a recent memorandum from Admiral

"Betty" Stark. In the late 1930s, American planners realized that the nation, alone or with allies, might be forced into war against coalitions of enemies. The Joint Army-Navy Board, America's highest military coordinating body, had grouped likely war scenarios into a color-coded set of plans code-named RAINBOW. The first variant, RAINBOW 1, assumed an invasion of the Western Hemisphere with no U.S. allies, while the last, RAINBOW 5, assumed the United States would fight Germany and Japan with the help of Britain and France.[14]

Noodling over the rainbows until two o'clock one morning, Stark felt it was time to rewrite the playbook. Beset by enemies on two sides of the globe, the United States could limit its defense to the Western Hemisphere (variant "A," or, "AFIRM," in the naval alphabet), split its forces between the Atlantic and the Pacific (variant "BAKER"), throw everything against Japan in the Pacific ("CAST"), or defeat Germany before turning against Japan ("DOG").

Stark thought that Germany would be the more formidable enemy. Because the Third Reich was economically self-sufficient and Japan was not, the defeat of Japan would have little impact on German fortunes. The defeat of Germany, by contrast, would inevitably lead to Japan's defeat, since the British and American navies would be free to concentrate their full might against Japan, blockade its Home Islands, and starve it into submission. If Germany and Japan declared war on the United States, Stark recommended Plan DOG.

Stark knew DOG would be an extraordinarily hungry puppy. Germany, Europe's major land power, could be subdued only by transporting an immense army to France under thick clouds of aircraft, and escorted by a huge, expensive combat fleet. It would require amphibious landings, the most difficult of all military operations. And it would require the United States to swallow the loss of the Philippines, which would not be able to withstand the Japanese surge while American forces were sent to Europe.

Nonetheless, the bookish admiral felt the "Germany first" strategy offered America her surest path to victory.

Talking over Stark's memo at Woodley that night, Knox and Stimson agreed. Knox forwarded the memo to President Roosevelt, who affirmed Stark's hound as America's strategy for the coming war. The decision made, the United States now had a blueprint for defeating Hitler, Hirohito, and Mussolini.[15]

But none of these men were at war with the United States. Not yet.

·   ·   ·

America rang in 1941 with fireworks, champagne, and kisses. But when the final strains of "Auld Lang Syne" faded, and the street sweepers began brushing away the empty bottles and paper streamers, Roosevelt's first order of business was the passage of H.R. 1776, a bill authorizing the president to transfer war materiel to "the government of any country whose defense the President deems vital to the defense of the United States."[16]

The battle over British aid was anything but certain, and the old, familiar menagerie sallied forth with old, familiar attacks. Isolationists said the bill would give FDR license to carry on a proxy war in Europe. Republicans said it would drag America into a direct war with Germany. Lend-Lease, Senator Wheeler claimed, was a New Deal farm subsidy that would "plow under every fourth American boy."[17]

As committee hearings in Congress heightened the political drama leading up to the floor debate, opponents of Lend-Lease lined up their own star witness: Roosevelt's recently resigned ambassador to the United Kingdom, Joseph Kennedy.

Joe Kennedy was a Boston Irish Catholic who had made his fortune in stocks, real estate, and liquor. After an interview with the *Boston Globe* in which he predicted the fall of democracy in Britain, Cordell Hull's State Department quietly asked him to resign.

Thoroughly embittered against the Roosevelt clique in general, and the State Department in particular, Kennedy gave isolationists the impression that he would come out foursquare against the Lend-Lease bill. The anti-Roosevelt coalition saw Kennedy as the big stick to kill British aid, and FDR's nemesis, New York Representative Hamilton Fish, scheduled the former ambassador as the opposition's star witness.[18]

When he caught wind of Kennedy's invitation to testify, FDR invited Kennedy to the White House for a morning chat in the presidential bedroom. Sitting in his wheelchair, shaving before a bathroom mirror, he motioned for Kennedy to sit down on the toilet seat next to him. Then Roosevelt did what he did best.

Smiling sympathetically, he fondly recalled the good work the two of them had accomplished together. He agreed that Kennedy had been ill-treated by Secretary Hull and others, and he promised Kennedy that once the Lend-Lease bill was passed, he would make sure the country took notice of the ambassador's laudable public service.

In seventy-five minutes, Roosevelt blunted Kennedy's wrath. When Kennedy appeared before Congress, he testified blandly that he opposed the bill's dilution of congressional authority, but he admitted that aid to Britain was America's best means of avoiding war. Isolationists listened in dismay as their biggest gun blew up in their faces, and Roosevelt's supporters could claim with a straight face that Kennedy essentially supported their bill.[19]

Roosevelt had his own big guns—Marshall and Stark, Morgenthau, Stimson. But after neutralizing Kennedy, he decided to counter with a true blockbuster: the former Republican nominee for president, citizen Wendell Willkie.

Stimson had once been Willkie's lawyer, and he remained on good terms with the Republican moderate. He assured FDR that Willkie supported the principles behind Lend-Lease, and FDR persuaded Willkie to fly to London to meet with Churchill and other British ministers. On his return, Willkie testified before a packed congressional committee room that without aid to Britain, London would fall and fascist dictators would be free to turn their guns against the Americas.

One red-faced isolationist on the congressional panel, Senator Bennett Clark of Missouri, reminded Willkie about his scathing attacks on FDR's foreign policy during his 1940 campaign. Looking the senator in the eye, Willkie replied, "I struggled as hard as I could to beat Franklin Roosevelt and I tried to keep from pulling any of my punches. He was elected president. He is my president now."

The room erupted into applause.[20]

With Willkie's cannonade, the opposition broke and the bill sailed through. On the afternoon of March 11, ten minutes after the bill arrived at the White House, Roosevelt placed his fountain pen on the official copy and signed it into law. Five minutes later Roosevelt directed Knox and Stark to send the Royal Navy naval guns and shells, three thousand gun charges, and two dozen PT boats. It was the tip of what would become a very large, very costly iceberg.

Lend-Lease was a smashing victory for the White House. It was also a victory for the hard-pressed United Kingdom. A grateful Winston Churchill later called it "the most unselfish and unsordid financial act of any country in all history."

But this unsordid act would take many months before it would influence the tides of war. And it would scatter seeds of bitter fruit that would ripen as America was called to fight.[21]

★ ′ ★ ●

George Marshall agreed with Admiral Stark that an Atlantic strategy—"Germany First"—made the most sense. He also believed America should avoid war with Japan at nearly any cost; if forced to fight in the Pacific, the country should restrict operations to the bare minimum needed to hold a defensible line.[22]

But to Marshall, grand strategy counted for little until he had an army to carry it out, and his mind was fixed on one immense task: to train and equip 1.4 million new draftees. That mission required construction of forty-six huge training camps in the East, in the South, and on the West Coast, located on the first of 44 million acres the Army would buy or lease. Everything Marshall needed was a rush job, since Congress had not thought to appropriate money for the new camps before passing the Selective Service Act. Civilian construction crews rotated night and day building roads, barracks, offices, and hospitals. Workers in sweat-stained overalls and faded hats graded fields, assembled radio towers and laid sewer and power lines, literally paving the way for a mass migration of citizens-turned-soldiers.[23]

The overstretched Quartermaster Corps, under Brigadier General Brehon Somervell, shouldered the burden of procurement, construction, and the embarrassing waste that accompanies every large government program. Congress, vigilant as ever for overspending in programs benefiting the poorly connected, found a plump target in the Army's construction budget. The Senate formed a special committee to investigate the nation's defense spending, and a former artillery captain, Senator Harry Truman of Missouri, took over as the committee's chairman.[24]

The Truman Committee, as it became known, set off a minor panic in the War Department. To Stimson and Marshall, the Truman Committee smacked of the Civil War–era Joint Committee on the Conduct of the War, an investigating body that radicals in Congress had used to hamstring Lincoln's generals. Stimson feared history would repeat itself, and it gave him no comfort that the new committee chair was a scion of Kansas City's famously corrupt Tom Pendergast political machine.[25]

Marshall had known Harry Truman since the First World War, and he held somewhat more faith in public servants like Truman. He felt the best approach to Truman's committee was to embrace it, not rebuff it. As he told one deputy, "A free and easy and whole-souled manner of cooperation with these committees is more likely to create an impression that everything is all right in the War Department than is a resentful attitude, and [it] must be assumed that members of Congress are just as patriotic as we."[26]

Marshall made a special effort to placate the temperamental senator. He met with Truman often, and he would leave meetings with his senior staff to take the senator's calls. A constructive atmosphere grew from the top down, and the Truman Committee became one of Marshall's closest allies in the Senate. By openly trusting the motives of Truman and his committeemen, Marshall had neutralized a monster before it could grow fangs.[27]

Which was fortunate, since, in the Atlantic Ocean, Hitler was baring fangs of his own.

EIGHT

INCHING INTO WAR

ᔆ

B RITAIN WAS LOSING THE BATTLE OF THE ATLANTIC, A STRUGGLE MEASURED not in conquered territory but in tonnage and rates of loss. From April to December 1940, German bombers and U-boats sank 878 merchant ships. In one convoy alone, 11 out of 41 supply ships went to the bottom, taking with them thousands of tons of badly needed food, ammunition, and weapons.

Worse was to come. As Admiral Dönitz's U-boats slaughtered merchantmen in the North Atlantic, the heavy cruisers *Scharnhorst* and *Scheer* ran down supply ships between Africa and South America. Merchantmen went down at three times England's capacity to replace them, and British imports dropped to two-thirds of their prewar levels.

That meant Britons were slowly starving. While basic foods could still be rationed, imports like citrus fruits and eggs were luxuries of the past. Planes, fuel, and antiaircraft guns were joining trucks, steel, and medicines on the Atlantic's dark floor. Churchill told his public in a radio address from London, "It is the Battle of the Atlantic which holds the first place in the thoughts of those upon whom rests responsibility for procuring the victory."[1]

The Atlantic also held first place in the thoughts of those on the sidelines. Admirals in Washington knew America could be pulled into war at any moment, and a reorganization of the peacetime fleet was overdue. In early February, with Roosevelt's consent, Admiral Stark formed the Atlantic Fleet and placed it under the command of a thin, bald sea dog who had been looking for a fight.[2]

Ernest J. King was sixty-two years old when Stark yanked him off the General Board, an old-folks home for senior admirals. When he assigned King command of the Patrol Force, a small fleet guarding America's Atlantic coastline, the news rustled gossip grapevines at officers' clubs from Newport to San Francisco. King, the rumor mill decreed, had been washed up. He was a good fighting admiral, but he had three strikes against him: he was combative with his fellow officers, he drank too much, and he was a carrier admiral.[3]

A descendant of lowland Scots, King hailed from a middle-class family in Lorain, Ohio. He attended the Naval Academy in Annapolis, and when the Spanish-American War broke out during his first cadet year, he finagled a sea assignment. He returned from his first cruise with a tattooed dagger on his right arm, an anchor on his left, and a bellicosity worthy of a bosun's mate. Gliding over deck and bridge in his tailored Brooks Brothers uniform, he argued with superiors, wielded naval regulations like a boarding cutlass, and accepted nothing short of perfection from his men.[4]

A voracious reader, King studied the campaigns of Napoleon, Jackson, and Grant as intently as he studied Nelson, Mahan, and Fisher. He penned thoughtful articles on shipboard organization, commanded a high-profile submarine salvage operation, and worked his way up the Navy's slippery ladder, through destroyers, cruisers, and submarines.

In the 1920s, when other officers looked at naval aviation as if the Wright brothers had suggested putting bicycles on ships, King embraced the air service. He commanded the carrier *Lexington* for two years and did a stint with the Bureau of Aeronautics at Main Navy, the naval headquarters on Constitution Avenue. In Washington, two blocks from an obscure Army colonel named George Marshall, King learned the inner mysteries of the Navy's bureaucratic machinery, Congress, and the unpredictable mustang called Washington politics.[5]

King once told a friend, "You ought to be suspicious of anyone who won't take a drink or doesn't like women." Ernie King was guilty of neither sin, and he earned a reputation as a man who played as hard as he worked. "He was the damnedest party man in the place," said one officer who saw him splice the main brace at club bars on many a weekend. "Ernie was the first guy there on Saturdays. . . . He joined the club because actually he was a great guy with the ladies and with liquor both."[6]

An agile dancer and conversationalist, he could be both solicitous and forward to the fairer sex. At dinner parties attractive women sat next to the "garter snatcher" at their own risk, for King's hands might spend as much time under the table as above.* His marriage to Mattie Egerton, a once-comely Baltimore so-

---

* One woman, whose knee evidently seemed a likely harbor for King's hand at a dinner party, allegedly scolded him, "I will have you know this is a tablecloth and not a bedsheet!"

cialite, had worn thin after producing six daughters and a son, and King's affairs in port and abroad were a matter of enthusiastic Navy scuttlebutt.[7]

King cared nothing for what people thought of him, and early in his career he decided he was not tough enough to be a great admiral. Determined to excise this career flaw, he drove his men with a fervor that would have done credit to Captain Bligh. He bullied colleagues and harangued subordinates. Aboard the carrier *Lexington*, rumor had it that the admiral's right hand was more sunburned than his left, the result of shaking his fist at pilots through the open bridge window. "There are two kinds of naval officers," he once told a friend, "good guys and S.O.B.s, and the quicker you learn to be a S.O.B. the better off you will be!"*[8]

When war broke out in Europe in 1939, King's career had been stranded in the horse latitudes. His last really interesting assignment was Fleet Problem XIX in 1938, a war game in which he launched a surprise carrier attack on Pearl Harbor, to what the umpires said was devastating effect.

Since then, he had been passed over for the Navy's top job, chief of naval operations, in favor of his old friend Admiral Stark. Some of King's friends blamed his stalled career on his drinking and petulance, while others saw the backroom hand of the "gun club," as the battleship admirals were known. Whatever the reason, the General Board, King's current assignment, was the Navy's traditional last stop before the glue factory.[9]

But his fortunes began looking up in the fall of 1940, when Admiral Stark called him into his office to discuss the Patrol Force. It was a small collection of ships, hardly worthy of the term "fleet." Most of the capital ships—carriers, battlewagons, and cruisers—were assigned to the Pacific and Asiatic squadrons. And the Patrol Force job would not entitle King to a promotion. He had worn the three stars of a vice admiral two years earlier, on a temporary basis, when he commanded the Aircraft Battle Force. He was back to two stars now, and as commander of the Patrol Force, he would remain a two-star rear admiral.

It didn't matter, for King was elated. The Patrol Force would get him out to sea, and out of the hell of being washed ashore. By the grace of God and Betty Stark, King was getting a chance to end his career aboard a warship's bridge. He snapped up the job and shipped out for what he knew would be his last big assignment.[10]

When his two-star pennant snapped over the ancient battleship *Texas* in December 1940, King lost no time showing his men who was boss. He told one

---

* After Pearl Harbor, King did not say, as popularly claimed, "When they get in trouble they send for the sons of bitches." But he told a friend he would have said it if he had thought of it.

subordinate, "I don't care how good they are, unless they get a kick in the ass every six weeks, they'll slack off." He whipped his squadron into shape, ordered wartime blackouts and drilled, drilled, drilled them until general quarters was running at a pace that would have made Lord Nelson smile.[11]

But looking over the hardworking Texans, he decided their appearance was not quite right. King harbored a fetish for naval uniforms, and in his mind he often tinkered with the Navy's look. Now, as squadron commander, he was in a position to remake his men's uniforms in his image. He issued a general order prescribing a new uniform for the Patrol Force: a thin white jacket atop heavy blue pants.

Of all the military services, the Navy is the most set in its ways. To the surprise of few besides King, the Texans loathed their new slops. "It's too hot in the legs, where you want to be cool, and too cool above, where you want to be warm," complained one intrepid watch officer.

"Well," said the bemused admiral, "it shows who can prescribe the uniform."[12]

Though Ernie King generally didn't give a damn what his men thought of him, after a few weeks of grumbling he reinstated the old look. But before long his sartorial obsession got the better of him. Concluding that Navy whites stood out against the ship's gray structures, he decided to camouflage his men by dyeing their uniforms light brown. Every man, King decreed, would have one set of his whites soaked in coffee, which he ordered brewed up in enormous cauldrons under the tearful eyes of the ship's galley cooks.

"At the first morning quarters after the completion of the task," recalled one officer, "the *Texas* crew had uniforms ranging in color from ecru to chocolate brown." Painfully aware that the Texans would be the laughingstock of the fleet once they reached the next port, King quietly dropped his experiment, and said not a word about it again.[13]

A few months into King's sea command, his code clerks received a signal from the Navy Department that caught the admiral's eye: The varied and dispersed naval forces of the United States would be consolidated into three separate groups.

The Navy works in mysterious ways, but it was not difficult for King to sniff out what was going on behind closed hatches. The incumbent commander-in-chief, U.S. Fleet, Admiral James Richardson, had butted heads with Roosevelt over the president's decision to move the bulk of the surface fleet from San Diego to Pearl Harbor. In no mood to hear the admiral's carping, FDR fired Richardson and reorganized the fleet into three components, each designed to deal with a different problem: the Atlantic Fleet, for defense of the sea-lanes to Britain; the

Pacific Fleet, to defend America's western frontiers; and the Asiatic Fleet, to contain the Japanese in China.[14]

While King mulled over who his Atlantic Fleet boss might be, he received a letter from Rear Admiral Chester Nimitz, the Navy's personnel head. King, Nimitz said, was being promoted. What's more, the president had decided to give him a fourth star as soon as another full admiral sailed into mandatory retirement, probably during the summer of 1941. Nimitz's letter left no question who would run the Atlantic Fleet; its new commander would be Vice Admiral Ernest King.[15]

King was thrilled. The job would put him on the front lines of an important war—a war still undeclared, but one where he could make a real difference.

A congratulatory letter from Secretary Knox followed Nimitz's cable. It was Knox's wish, he said, "to maintain the closest possible relationship. . . . I am still a great deal of a novice in this Navy business, and I am depending upon you men to help me along in my education."[16]

As King saw it, Knox's letter summed up exactly where the two men fit into the Navy's picture. And he would never let Knox forget it.

· · · ·

Roosevelt had reached a watery crossroads. He put the odds of war at about one in five, yet he had just won an election on the promise of no direct intervention in Europe. He then campaigned for Lend-Lease with the assurance that material aide to England was the best way to keep America out of war. How, he wondered, could he do anything more for Churchill after making such fundamental promises to his own people?[17]

The "how" emerged through another glacial shift in public opinion. On April 8, a Gallup poll announced that 41 percent of those surveyed favored convoy protection of British merchant ships. That was a good sign, for it suggested a strong and growing minority was following the president's lead.

But fully half the public opposed shooting at German U-boats, and Roosevelt vividly remembered public support for President Wilson drying up as the death toll mounted in the spring of 1918. The public might be with him now, but that consensus could easily fall apart like pound cake in a thunderstorm once casualty lists brought home war's human cost.[18]

In small, mostly unheralded steps, Roosevelt edged away from the crepe paper fiction of neutrality. On April 10, he huddled with Stimson, Hull, Knox, Stark, Morgenthau, and Hopkins to discuss how far into the eastern Atlantic he could send U.S. combat ships. Though he didn't feel that Congress would permit him to convoy ships to England, perhaps he could find a way around the neutrality laws by extending the Navy's territory, under the guise of defending the Western Hemisphere.

Calling for an atlas, FDR and his lieutenants pored over a map of the Atlantic. Looking between Africa's west coast and the pregnant bulge of Brazil, then moving up the map, Roosevelt concluded he could reasonably declare the twenty-sixth meridian to be the new American patrol zone. British ships west of that line would not fall under the Navy's full protection, but American warships would track any German submarines they found and report their locations to the British, who could steer their convoys clear of danger or send destroyers to sink the offending sea wolves.[19]

FDR knew the move would rile insolationists almost as much as it would inflame the Germans. Sniffing political fallout in the air, he told Knox and Stark there would be no public announcement about the new patrol zone. The Navy's actions, he said, would speak for themselves.*[20]

Not everyone on his team was comfortable with this characteristically Rooseveltian approach. As Henry Stimson saw it, the president was steering the country toward war without a fixed end in mind, weaving a giant tapestry with only the foggiest idea of what the image would look like. If Roosevelt wanted war with Germany—which Stimson fully supported—then as president, he should rally the public, announce his intentions openly, and ask Congress for a declaration. If not, he should stop provoking Hitler and risking full-scale war over an incident at sea. Stimson confided to his diary, *"I am worried because the President shows evidence of waiting for the accidental shot of some irresponsible captain on either side to be the occasion for his going to war."*[21]

Thinking as a former secretary of state, Stimson believed it was all the more important for Roosevelt to have a coherent foreign policy in the Atlantic because clouds were darkening over the Pacific. By 1941, Japan boasted the third-largest navy in the world, the world's largest carrier fleet, and fifty-one infantry and light armor divisions. In China, its rampaging army had driven back Generalissimo Chiang Kai-shek's Nationalist forces, and fresh divisions began sweeping down the China coast toward French Indochina. *"The Japs are getting up steam,"* a worried Stimson wrote. *"Every message that we get from every direction indicates that they are gaining momentum and determination to go ahead."*[22]

---

* FDR briefly considered sending U.S. submarines after the German battleship *Bismarck*. Robert Sherwood recalled him thinking aloud, "Suppose we order them to attack her and attempt to sink her. Do you think the people would demand to have me impeached?" The Royal Navy made the point moot when it sent the dreaded battleship to the bottom on May 27.

. . .

On April 24, Roosevelt summoned Stimson, Knox, Hull, and Hopkins to the Oval Office for a conference on the Pacific. Stimson, concerned about a shortage of ships in the Atlantic, begged him to transfer the big battleships from Pearl Harbor to the East Coast. The move would not affect the Pacific force, he said, because General Marshall had looked into Hawaii and "felt that Hawaii was impregnable whether there were any ships left there or not; that the land defense was simply sufficient, together with the air defense, to keep off the Japanese."

Knox agreed with Stimson. Hawaii was impervious to attack, so most of the Pacific Fleet could safely move over to the Atlantic.[23]

Roosevelt disagreed. The fleet at Pearl Harbor protected the entire southwestern Pacific, not just Hawaii, he said. The Navy might have to defend Singapore, Australia, New Zealand, or the Dutch East Indies. Besides, moving battleships to the Atlantic would not affect the U-boat menace, since the Navy was not presently fighting German subs. The Navy's job was peaceful. Its orders were simply to locate the aggressors and report back to Washington. As he had told the press, the Navy was on a fact-finding mission, more or less.[24]

As he listened, Stimson grew tired of Roosevelt's double-talk, of his pretense that the U.S. could support Britain without giving the fascists cause for war. The United States was acting as the eyes and ears of the Royal Navy, which was hunting U-boats, and that meant the U.S. Navy was hunting U-boats. America was in an undeclared war in the Atlantic.

To Stimson, Roosevelt's contorted logic seemed better suited to bargaining with precinct bosses than homicidal dictators. *"I wanted him to be honest with himself,"* he wrote after one meeting. *"To me it seems a clearly hostile act to the Germans, and I am prepared to take the responsibility of it. He seems to be trying to hide it into the character of a purely reconnaissance action which really it is not. However, as he fully realizes that he will probably get into a clash with the Germans by what he does, it doesn't make much difference what he calls it."*[25]

Sophistry never bothered FDR. Whatever Stimson, Marshall, Knox, and Stark might think, Roosevelt was the Boss, and the Boss was saying "no." The battleships would remain at Pearl Harbor, where they would be safe.[26]

. . .

In the Atlantic, Admiral King's ships had been thrown into the front lines of an undeclared war. But neither the president nor the Navy Department gave him any guidance. King continually pressed Betty Stark for something approaching coherent orders: When a German U-boat showed itself, should the Americans fire on it? What if a German commerce raider were in distress? To whom would his ships disclose the locations of U-boats? King was fond of saying, "We do the best we can with what we have," but his men couldn't do their jobs without being told what those jobs were.

Admiral Stark was sympathetic, for he had asked Roosevelt the same hard questions and had gotten nowhere. Stark complained to a friend, "To some of my very pointed questions, I get a smile or a 'Betty, please don't ask me that!' Policy seems something never fixed, always fluid and changing."[27]

But to King, the ultimate realist, the president's policy was damned well fixed—Roosevelt just didn't want to admit it. "FDR's basic idea," he concluded, "was to inch into war little by little."[28]

· · ·

In the spring of 1941 public opinion inched toward Britain, though the margin of support grew slower than spring corn. By mid-May, 52 percent of the public supported convoys protected by the U.S. Navy, though 79 percent wanted to stay out of war, and 70 percent felt FDR was either doing enough or too much for England.

Furthermore, the public shift was not mirrored in Congress. Fifty legislators, led by Republican Senator Robert Taft, pledged their "unalterable opposition" to convoys, while a nose count of the Senate showed forty Democrats and Republicans opposing convoy legislation, more than enough to sustain a filibuster. For now, Britain would have to fight the war alone.[29]

With popular support growing, cabinet hawks—Stimson, Knox, Ickes, and Morgenthau—urged Roosevelt to put the nation on a war footing. The public had gone as far as it would go on its own, they said. Any further movement would require presidential leadership.[30]

In late May Roosevelt decided to take a giant step toward war by declaring an unlimited national emergency. He would announce a basic plan to meet the emergency, and he hoped to convince the public that America's safety depended on stronger measures—even measures that courted war.

The declaration of an unlimited emergency activated enormous presidential powers. With the stroke of a pen, Franklin Roosevelt could expand the authorized size of the Army and Navy, compel factories to turn out weapons, and deploy troops overseas. He would assume the most dictatorial powers the republic would tolerate.

He vetted his speech to a large cast of contributors—Hopkins, Stimson and Knox, Hull and Welles—and he put the pens of Bob Sherwood and Sam Rosenman to work condensing information pulled from many different departments. But the speech was Roosevelt's, and as he revised draft after draft, he knew the public would cast—or hold—the decisive die.

"There's only a small number of rounds of ammunition left to use unless Congress is willing to give me more," he explained to Rosenman. "This declaration is one of those few rounds, and a very important one. Is this the right time to use it, or should we wait until things get worse—as they surely will?"[31]

As the clock's hands ticked toward 10:30 p.m. Washington time, listeners across the nation tuned in their radios to hear the president's words. FDR preferred a live audience when speaking, even to radio listeners, and he delivered this speech before an assembled group of cabinet members, high officials, and representatives of the Pan American Union who had gathered in the East Room. Sweating in the warm May heat, surrounded by flags of the western nations, Roosevelt gazed beyond the tuxedoed diplomats into the faraway faces of his real audience, men and women in homes, farms, and apartments across the land.[32]

He told the nation the struggle was not a European conflict, but "a war for world domination." He reminded listeners that he had promised never to send American boys to war except in case the nation were attacked. Yet the word "attack," he said, was less obvious than it had been in the days of George Washington.

"Some people seem to think that we are not attacked until bombs actually drop in the streets of New York or San Francisco or New Orleans or Chicago," he said. "But they are simply shutting their eyes to the lesson that we must learn from the fate of every nation that the Nazis have conquered. . . . We know enough by now to realize that it would be suicide to wait until they are in our front yard. When your enemy comes at you in a tank or a bombing plane, if you hold your fire until you see the whites of his eyes, you will never know what hit you. Our Bunker Hill of tomorrow may be several thousand miles from Boston."

In light of the German menace, he concluded, "I have tonight issued a proclamation that an unlimited national emergency exists and requires the strengthening of our defense to the extreme limit of our national power and authority. The nation will expect all individuals in all groups to play their full parts, without stint, and without selfishness, and without doubt that our democracy will triumphantly survive."[33]

After finishing his speech, a damp, elated Franklin Roosevelt was wheeled out to the cool air of the South Lawn's rose garden. There Eleanor and several friends, including songwriter Irving Berlin, had gathered for a wrap party for the big speech. Frolicking tunes of a Guatemalan marimba band lingered under the Jackson magnolia tree, and FDR unwound by swapping jokes and anecdotes with friends. After bantering with partygoers, he headed upstairs to the Monroe Room, where Berlin tickled the ivories to tunes like "Alexander's Ragtime Band."[34]

Roosevelt, Hopkins, and a few other intimates sang along with the great composer. None of them knew that in another month, the emergency Roosevelt had just declared would take a horrifying turn for the worse.

NINE

BEGGARS BANQUET

〰

A S DAWN BROKE ON THE TWENTY-SECOND OF JUNE, 1941, HITLER'S ARMIES lunged east. Spearheading a horde of Italian, Romanian, Hungarian, and Finnish allies, the Wehrmacht smashed through brittle Soviet resistance and drove into eastern Poland, Belorussia, and the Baltics. In two weeks the invaders reached the Dnieper River in Ukraine, and in less than a month they pushed within two hundred miles of Moscow. Towns were razed, villagers were massacred, and western Russia was drenched in blood.

In the invasion's first hours, Roosevelt ducked press inquiries by taking a motor visit to Maryland, and some observers surmised he was keeping his nose to the wind. Isolationists, Catholics, and conservatives felt the best course was to do nothing and let the two tyrants bleed each other white. The communist left, which had opposed aid to Britain while Stalin was at peace with Germany, without blushing reversed course and demanded support for any enemy of Germany. *"Perhaps [Roosevelt] was not able to make up his mind as to what our attitude should be,"* Harold Ickes told his diary. *"It would be just like him to wait for some expression of public opinion instead of giving direction to that public opinion."*[1]

Winston Churchill, by contrast, harbored no doubts. At London's Number 10 Downing Street, he explained to his private secretary, "I have only one purpose, the destruction of Hitler, and my life is much simplified thereby. If Hitler invaded hell I would at least make a favourable reference to the Devil in the House of Commons." That night Churchill proclaimed to Parliament, "Any man or state who fights on against Nazidom will have our aid. . . . It follows, therefore, that we shall give whatever help we can to Russia and the Russian people."[2]

. . .

"Whatever help" Churchill was prepared to give the Russian people included American tanks, American trucks, American bombers, and American steel. His generosity, in light of the vast scale of the German-Soviet War, placed an enormous new strain on America's high command.

At FDR's insistence, Harry Hopkins hounded the War Department to allocate massive quantities of equipment and raw materials to Russia, Britain, and China. Roosevelt ordered his lieutenants to give those shipments top priority, and he bridled at every sign of delay. "If I were a Russian I would feel that I had been given a run-around in the United States," he lectured one Lend-Lease official. "Please, with my full authority, use a heavy hand—act as a burr under the saddle and get things moving!"[3]

Just what, Marshall wondered, did the president think they had left? Lend-Lease was draining equipment sorely needed for training the men Roosevelt and Stimson had just drafted. The Air Force was short of every type of combat plane, and tanks, trucks, boots, and guns could not roll off assembly lines fast enough to equip the explosion of draftees, much less the Russians, Britons, and Chinese.

The production problem, as Marshall saw it, was due to Roosevelt's own impulsiveness—his inability to sit still long enough to think through the whole picture. Marshall complained to Morgenthau one afternoon, "First the President wants five hundred bombers a month and that dislocates the program. Then he says he wants so many tanks and that dislocates the program. The President will never sit down and talk about a complete program and have the whole thing move forward at the same time."[4]

Worst of all, Marshall knew this exodus of equipment would be for nothing. His G-2 (Intelligence) experts believed the Red Army would not last through the winter. Stalin's purged officer corps had made a very poor showing against the tiny Finnish army the year before; now that it was pitted against the varsity team, the Red Army was shattering.[5]

But the president put his foot down. Based on reports from Hopkins, who had met with Stalin, Roosevelt's instinct told him the communists would not only survive but would eventually swallow up Hitler's legions, as they did with Napoleon a century before. As he wrote to his ambassador to Vichy shortly after the invasion, "Now comes this Russian diversion. If it is more than just that it will mean the liberation of Europe from Nazi domination—and at the same time I do not think we need worry about any possibility of Russian domination."[6]

Henry Stimson agreed with Marshall, but he knew there was no point in

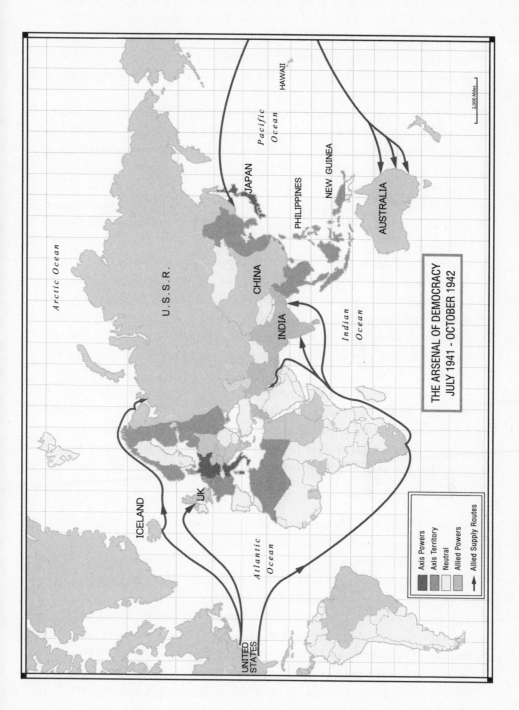

THE ARSENAL OF DEMOCRACY
JULY 1941 - OCTOBER 1942

Arctic Ocean

U.S.S.R.

JAPAN

Pacific Ocean

HAWAII

CHINA

PHILIPPINES

NEW GUINEA

INDIA

Indian Ocean

AUSTRALIA

ICELAND

UK

Atlantic Ocean

UNITED STATES

Axis Powers
Axis Territory
Neutral
Allied Powers
Allied Supply Routes

2,000 Miles

arguing with Roosevelt. On the last day of July he spoke with Marshall about Roosevelt's idea to ship scarce American P-40 Warhawk fighters to Vladivostok in the Far East, then let the Soviets fly them to European Russia. Marshall said a trans-Siberian flight was impossible for the single-engine pursuit planes; the engines would burn out over the long distance, and the planes would be lost or too worn-out to be of any use.

At the next day's cabinet meeting, a truculent Roosevelt gave Stimson hell for not moving enough equipment to the Red Army, especially those P-40 fighters. "I am sick and tired of hearing that they are going to get this and they are going to get that," he bellowed at Stimson. "The only answer I want to hear is that it is under way."

In his diary, Ickes called the exchange *"one of the most complete dressings down that I have witnessed,"* while Morgenthau, in his, thought Stimson looked *"thoroughly miserable."*[7]

With the restraint of a veteran diplomat, Stimson explained the Army's point of view. They had no fighter planes to spare, and any they sent Russia—if they did survive the journey—would arrive at the front too late to help the Red Army. *"But in his outburst today,"* Stimson told his diary, *"the President said we must get 'em, even if it was necessary to take them from troops and I felt badly about it. I didn't get half a chance with him. He was really in a hoity-toity humor and wouldn't listen to argument."*

His anger glowing, Stimson growled into his dictating machine, *"This Russian munitions business has shown the President at his worst. He has no system. He goes haphazard and he scatters responsibility among a lot of uncoordinated men and consequently things are never done. This time I got very angry over it for he had no business to talk that way in the Cabinet."*[8]

⋆　　⋆　　⋆

While the war for Russia gave Roosevelt and Churchill a respite in Europe, the threads of peace were unraveling on the other side of the globe. In response to Japan's war against China and moves into northern French Indochina, in the summer of 1940 FDR approved a limited embargo against Japan that halted shipments of aviation fuel and high-grade scrap metal.

Japan wasn't cowed, and talk from Tokyo grew belligerent. In October Roosevelt heard a rumor that the head of the Japanese Press Association said Japan would refrain from war only if the United States withdrew its military forces from Wake Island, Midway Island, and Pearl Harbor. "God!" he exploded to an aide. "That's the first time any damn Jap has told us to get out of Hawaii! And that has me more worried than anything in the world." He said he would not mention the remark in public, "because it'll only stir up bad feelings in this country, and this country is ready to pull the trigger if the Japs do anything."[9]

The biggest trigger America could pull would be oil. Japan imported 80 percent of its petroleum from the United States, and its national stockpiles would last only eighteen months in wartime. A petroleum embargo by the United States would cripple the imperial war machine.

Then again, it might force the Emperor to grab the oil-rich Dutch East Indies.

The question of Japan split FDR's cabinet. Stimson, Morgenthau, Hopkins, Ickes, and Assistant Secretary of State Dean Acheson demanded a hard line, while Hull and Welles saw an oil embargo leading to a war Americans didn't want.[10]

Divided councils had never bothered FDR, for they effectively gave him the deciding vote. For now, he cast his vote with Hull and Welles. He had not given up on a negotiated peace with Japan, and he felt America could not afford a distraction in Asia while Hitler remained the chief threat. "It's terribly important for the control of the Atlantic, for us to help keep peace in the Pacific," he told Harold Ickes. "I simply have not got enough Navy to go around—and every little episode in the Pacific means fewer ships in the Atlantic."

After a cabinet meeting that month, Ickes concluded, *"The President was still unwilling to draw the noose tight. He thought it might be better to slip the noose around Japan's neck and give it a jerk now and then."*[11]

The question neither Ickes nor anyone else could answer was: At what point would the noose get so tight that Japan would try to break free?

TEN

LAST STAND OF THE OLD GUARD

〰

A S SUMMER'S HEAT BAKED WASHINGTON'S BROWNSTONES, DRIVING RESIDENTS indoors, George Marshall saw dusk setting on the army he had labored so long to raise.

It had been a marvelous force, considering where he had started. Resistance to the draft had given way to acceptance; military service had become a fact of American life and had transformed the cultural landscape. More than 1.2 million inductees were wearing khaki, training in camps, and molding themselves, slowly but certainly, into a competent fighting force. The little absurdities of military life, which America was beginning to relearn from a generation before, became fodder for Abbott and Costello films, Bob Hope jokes, jukebox songs, and comic strips.

The Selective Service Act specified that draftees would serve for a presumptive period of twelve months. But there was a catch. As General Marshall pointed out during congressional hearings on the bill, the law allowed the government to keep draftees in service more than twelve months if the country faced an emergency.

Marshall's caveat went unheard in the din of an election year, and he sensibly did nothing to highlight it. The public was left with the impression that the administration would send the boys home after twelve months.

As the first draft class entered its eighth month of service, conscripts began asking their officers—and congressmen—if they would be returning to their jobs in the fall. Politicians demanded to know whether the Army would keep its "one-year promise," and the stage was set for another battle on Capitol Hill.[1]

Sending men home just as they had been fully trained would reverse the gains of two years and throw the Army back to 1939. In fact, the Army would be worse off than in 1939, since Marshall had cannibalized the few Regular Army divisions to spread seasoned officers among the new units. Two-thirds of all enlisted men and three-quarters of all officers would be released from service if the draft were not extended. If draftees went home, Marshall would have to reorganize the Army once again just to get it back to its paltry state of two years earlier. As he saw it, the country had no choice but to keep the new men in uniform a while longer.[2]

FDR was reluctant to join this fight. He had given Britain destroyers, rammed home Lend-Lease, and declared an unlimited national emergency; he had pushed his luck as far as luck would take him, at least for the moment. It was not until June 19, less than three months before the first draftees were to return home, that Marshall and Stimson finally wrung a lukewarm commitment from him to ask Congress to extend the conscription term. They agreed, however, that the push would be more successful if the Army took the lead with Congress, and let FDR quietly work the telephones. For this job, Roosevelt was happy to stay in the background.[3]

Because only a third of the Senate's members would come up for reelection in 1942, it was easier to persuade the upper chamber to go along with the draft extension. The House would be the real battleground, and in mid-July Speaker Sam Rayburn of Texas told Stimson an extension would not pass the House. They simply lacked the votes.

Some legislators despised FDR enough that it influenced their vote, though most simply despised the idea of losing reelection bids the next year. A Gallup poll showed that 45 percent of the country—more than 50 percent in the Rockies and the Midwest—opposed draft extension. Feeling was naturally strongest among Republicans, so General Marshall made a special effort to reach out to the GOP.[4]

He made little headway, in part because of bitterness built up over three electoral losses to a smug liberal Democrat. During one closed-door dinner meeting with Republican congressmen at the Alibi Club, a representative told Marshall, "You put the case very well, but I will be damned if I am going along with Mr. Roosevelt."

Incensed, Marshall spat back, "You are going to let plain hatred of the personality dictate to you to do something that you realize is very harmful to the interest of the country."

He pushed hard with the other invitees, but the dinner yielded a poor harvest. He later grumbled, "I had all the Republican congressmen in there and out of that I only got a few votes."[5]

In 1936, Sam Rayburn had warned his fellow Democrats, "When you get too big a majority, you're immediately in trouble." Now Roosevelt's party was in trouble. In early August, House Majority Leader John McCormack of Massachusetts counted forty-five Democrat votes against draft extension, with another thirty-five Democrats undecided. Republicans were nearly unanimous in opposition, so unless Stimson and Marshall could bring wavering Democrats into line, the Army would melt away come October.[6]

The bill was scheduled to hit the House floor in mid-August, so early in the month was the time to call the touchdown play. But the two running backs who should have been carrying the ball—Marshall, the respected front man, and FDR, twisting arms behind the scenes—abruptly left Washington without disclosing their destination.

To leave the capital the week before the biggest fight since Lend-Lease seemed unthinkable, yet Henry Stimson found himself alone in the lion's den. *"There is a wild rumor going around town tonight to the effect that the President is going to meet Churchill somewhere up near Canada,"* a puzzled Stimson told his diary on August 6. *"Perhaps it's true because for the first time Marshall failed to tell me where he went when he left last Sunday."*[7]

. . . .

To the clang of bells, the thump of feet, and the rattle of chains that punctuated the command, "Weigh anchor!" Admiral King's flagship, USS *Augusta*, slipped out of Newport, Rhode Island, and headed south to Long Island Sound. The stately cruiser steamed into the East River and obediently came to rest at the cluttered Brooklyn Navy Yard.

Not long after her arrival, a barge crammed with workmen pulled alongside. Carpenters began attaching odd-looking boarding ramps to *Augusta*'s hull while curious sailors looked on. Judging by its structure, the ramp was built to accommodate the least seaworthy of landlubbers. Someone who would have more trouble than most coming aboard. Someone important.[8]

The next day a destroyer hove to and transferred General Marshall, Admiral Stark, and a small galaxy of star-shouldered officers to King's flagship. *Augusta* weighed anchor and steamed east to Martha's Vineyard, anchoring in Menemsha Bight off Cape Cod. On the night of August 4, the President of the United States was wheeled up the new ramps with a small company of retainers.[9]

They were going to meet Great Britain's harbor pilot.

Steaming through Nova Scotia's choppy waters at twenty-two knots, *Augusta* and her escorts whisked their charges to Placentia Bay, near the village of Argentia on the island of Newfoundland. Roosevelt spent the next two days meeting with advisers and doing a bit of angling from *Augusta*'s fantail—catching, in the words of the official log, "an ugly fish which could not be identified by name, and which he directed be preserved and delivered to the Smithsonian Institute."[10]

As they awaited the arrival of Prime Minister Churchill and his staff, Roosevelt called his military chiefs into conference, where they discussed the protection of Atlantic shipping. In their discussions, Roosevelt provided little clear guidance on matters of cooperation, planning, or strategy, and a puzzled Marshall left the meeting bothered by the president's lack of answers to questions the British would ask. Churchill was sailing an awfully long way through submarine-infested waters, and he would want more than words of good cheer from the Americans.[11]

The ambiguity did not bother Roosevelt in the slightest. America was not at war yet, and FDR had lived his whole life offering glib half answers, encouraging words, and equivocal promises. He shunned rigid plans, and would not let his military chiefs—who had predicted the quick demise of Russia—discuss anything of significance with their British counterparts. He would set the agenda as he saw fit.

On the morning of August 9, *Augusta*'s lookouts spotted three British destroyers in the offing. They stood in for Placentia Bay, leading the way for the battleship HMS *Prince of Wales*, recently of the Birkenhead shipyard, and more recently of a death struggle with Hitler's *Bismarck*.

The pride of the Royal Navy made a grand entrance. Her battle scars accentuated her smooth skin, billowing pennants and proud turrets rising from her deck. As "The Star-Spangled Banner" wafted from His Majesty's Ship over the water, a veteran crew assembled in neat files along *Prince*'s rails. Nestled among the navy blue coats stood the prime minister, clad in the fraternal uniform of an Elder Brother of Trinity House.[12]

In a plain fedora and gray civilian suit, FDR stood to greet Churchill on *Augusta*'s covered deck, a cane in one hand, his son Elliott holding the other, his legs locked into steel braces concealed beneath his trousers. *Augusta*'s band struck up "God Save the King," and a murmur rippled through the crowd as Churchill made his way aft to meet the president.

Roosevelt's handsome face broke into a smile as Churchill drew near. "At last, we've gotten together," he said.

"Yes," said Churchill, an impish grin spreading over his pink jowls. "We have."[13]

Roosevelt took an instant liking to the English dynamo, and the brotherhood reached its high tide the first full day of the conference. It was a Sunday morning, and before a day of meals and meetings, divine services were held on the British battleship's afterdeck.

Roosevelt's inability to walk had never prevented him from working the levers of power. But to cross from an American cruiser to a British battleship in a tight anchorage tested the skill of both FDR and his naval pilots. In a series of tight moves, *Augusta*'s consort, the destroyer *McDougal*, pulled alongside the cruiser's bow, and the Augustines transferred Roosevelt to the destroyer. *McDougal* then hove around for a "Chinese landing" on the battleship's port quarter, inching her bow up to *Prince of Wales*' railing.

As *McDougal* touched home, her boatswain's mate, holding a heavy coiled rope, called out to figures gathered on *Prince*'s quarterdeck, "Hey! Will you take a line?"

A portly man in a dark suit darted forward and replied, "Certainly!"

The rope flew, and before his tars could run up to assist him, the prime minister of Great Britain caught and hauled in the line. The warships were secured, a bridge ramp was thrown up, and the officers of two navies held their breath as the President of the United States boarded Britain's towering flagship. On two feet.[14]

Walking was a form of torture that Roosevelt reserved for the most sacred occasions, such as convention speeches or annual addresses to Congress. His back and side muscles strained as he took each awkward step in his heavy steel braces. He walked over the bow ramp gripping Elliott's arm, his teeth clenched, his face plastered with a rigid, artificial smile.[15]

Churchill would have been delighted to step forward—to cut short Roosevelt's walk, perhaps take an elbow alongside Elliott. But he knew Roosevelt was a proud man determined to show that there was no obstacle so great it couldn't be overcome by unbending will. Nodding silently in support, Churchill stood on the ship's deck, surrounded by the trappings of empire, beaming in admiration as his partner walked the deck and took his place among the chairs set before a pulpit draped with American and British flags.

After the homilies, long-winded prayers, and remarks by U.S. and British naval chaplains, the service closed with hymns selected by the two leaders. The emotional crescendo reached its peak as the host rang out "Onward Christian Soldiers," "O God, Our Help in Ages Past," and an old FDR favorite, the naval hymn, "Eternal Father, Strong to Save."[16]

Since the days of the ancient Greeks, naval songs have moved the hearts of men buffeted by wind and wave. This service, held in a common tongue and offered to a common god, bound those gathered in common cause. "Every word seemed to stir the heart," Churchill wrote years later. "None who took part in it will forget the spectacle presented. . . . It was a great hour to live."[17]

Churchill, though not a religious man, delighted in the pomp and spectacle of ecclesiastical ceremony. He used to joke that he was not a pillar of the church, but rather a buttress—he supported it from the outside. Roosevelt, by contrast, was a devout Episcopalian, a true believer, and the incorporation of Christianity into Allied war aims moved him emotionally. Hours after the last notes died away and the altar was broken down, he told Elliott, "It was our keynote. If nothing else had happened while we were here, that would have cemented us. 'Onward Christian Soldiers.' We *are*, and we *will* go on, with God's help."[18]

The Americans might go on with God's help, but the British were looking for something a bit more tangible. Speaking with the lispy eloquence that had inspired millions, the prime minister in the next meeting starkly outlined the Empire's most critical needs: heavy bombers, medium bombers, seaplanes, medium tanks, destroyers (always), and antiaircraft guns. Ammunition. Food. Fuel. They also needed the Americans to take over convoy routes of the North Atlantic so the Royal Navy could fight the enemy in the Mediterranean and North Seas.[19]

The Americans were lavish with sympathy but tight with promises. Marshall, Stark, and General Arnold were short on nearly everything, and for months they had been splitting aircraft production down the middle. Now, with Russia elbowing aside the British and American armies, and China making fresh claims, the American chiefs could offer little more than what they had committed to provide before the voyage began.

"*[The British] did not appreciate that on top of this load we had to take care of the needs of China, Russia, British Colonies, Dutch East Indies,*" Arnold scribbled in his little pocket diary a few days later. "*Fortunately, we were able to get away without promising or giving away everything we had. As a matter of fact, we might have lost everything we owned, including our pants—but we didn't.*"[20]

Trolling for a fish that Churchill could take back to England, Roosevelt ordered Stark to provide naval protection for British convoys beginning September 1. King's Atlantic Fleet would move from reporting on U-boats to defending Commonwealth merchantmen.

Materially, it was not much. But to Churchill, this concession was more welcome than a fresh shipment of tanks or bombers. An undeclared war at sea, he knew, just might pull the United States into a declared war on land. And a declared war on land meant victory.[21]

War's tools and strategy interested Roosevelt, but only marginally. War is a milestone in the life of a nation, but that life goes on, in some form or other, after the guns fall silent. As head of state, Roosevelt had to make a difference to the ordinary man in ordinary times. Things like the National Recovery Act, or Social Security, or even his disastrous plan to pack the Supreme Court, drew Roosevelt's fertile mind. So with democracy's two leaders free of outside distractions,

Roosevelt pondered a sweeping declaration of Allied war aims to lay before mankind.

In centuries past, European alliances had been formed either to upset the old order or to defend it. Churchill was heir to an empire that traditionally defended it. Coalitions opposing Spanish monarchs, the Sun King, Napoleon, and the Kaiser had been forged to preserve balance on the Continent. In a way, restoring a world of constitutional monarchies—an order interrupted by communism and fascism—had been Churchill's great commission.

The United States saw the world differently, and would not shed its blood simply to preserve Europe's *ancien régime*. America needed a goal worthy of the sacrifice it would be asked to make, and Roosevelt, a son of Woodrow Wilson's idealism, channeled his beliefs into a statement of American principles. Anxious to bind America's fate to Britain's, Churchill agreed to a joint statement of Anglo-American war aims that became known as the "Atlantic Charter."[22]

It was breathtaking in tenor and scope. The Atlantic Charter pledged that the nations aligned against the Axis powers would seek no territory, encourage free trade, support democratic institutions, and work toward a permanent system of world security. The document was, in Churchill's words, a "profound and far-reaching" declaration of the ideals that would guide the Anglo-American alliance.[23]

On what was billed as another "fishing trip," each leader had come away with a big catch. Churchill won American protection of his merchantmen, binding America's fate to Britain's. With the Atlantic Charter, Roosevelt claimed the idealistic vision of his hero.

But by pledging his nation to a world of peace, FDR had irrevocably set his country on the path to war.

·   ·   ·

Henry Stimson missed the drama at Placentia Bay because he had a role to play in Washington. As Roosevelt, Stark, Marshall, and King shook hands with their British counterparts, Stimson was working phones and halls to convince balking congressmen that they should ignore the specter of voter retribution and extend the draft.

The old guard opposition—Republicans, conservative Democrats, isolationist groups, anti-Churchill Irishmen, and a hodgepodge of pacifists, Hooverites, intellectuals, and anti-Willkies—vilified Roosevelt as a three-term dictator and Stimson as his warmongering Iago. A few Republicans, like House Minority Leader Joe Martin of Massachusetts, sympathized with the bill, but most saw draft extension as a golden opportunity to give FDR a black eye that voters would remember going into the 1942 midterm elections.

With Roosevelt and Marshall AWOL from the capital, and Stimson hobbled by charges of extremism, it was up to Sam Rayburn, John McCormack, House Whip Pat Boland, and the renegade Republican Jim Wadsworth to scrape together enough votes to save the Army. Rayburn, McCormack, and Boland called on friends in the American Legion to push fickle Democrats, which brought in a few more votes, and Wadsworth eventually delivered twenty-one Republicans. But it would be a close call, and the gauntlet of a full House vote had yet to be run.[24]

The final vote, taken on the evening of August 12, fell to a sharply divided chamber after a long day of bitter debate. As summer rains lashed the Capitol's roof, the chief clerk called out names for a voice vote.

The abundance of "nay" votes indicated that Democrats, feeling the heat, were breaking ranks and deserting Roosevelt. But the total inched up in the administration's favor until the House tally clerk, Hans Jorgensen, sat before a list of 204 "aye" votes, 201 "nays," and 27 abstentions.[25]

When the count was announced, a flurry erupted around the Speaker's rostrum. Seeing the close margin for a bill he had thought would pass easily without him, Democrat Andrew Somers of New York leaped forward to change his "aye" to "nay." The ayes still had it, for the moment, by one vote.

But Sam Rayburn knew how to run his house. Unwilling to let one more vote slip away, the Texan announced that voting was over. To the howls of opponents, Jorgensen read the roll of 203 "aye" votes to 202 "nays," and Rayburn fought back complaints of procedural irregularities from the isolationists. After nearly twelve hours of legislative combat, the Speaker's gavel slammed down, and Rayburn's lieutenants rushed it to the Senate for approval.[26]

In a democracy, armies can die from political wounds more swiftly than from combat, disease, or desertion. America's army would not die that day. The battle won by Henry Stimson and Sam Rayburn on a rainy August evening ensured that America's soldiers and sailors would continue to defend the nation until Hitler was vanquished and Europe was safe.

But, as Wellington once said about Waterloo, it had been a damned near-run thing. And the men who so painstakingly planned for war in Europe were about to learn they had an enemy coming at them from a different direction.

ELEVEN

YEAR OF THE SNAKE

∽

I F AMERICA WERE DRAWN INTO WAR, SAID ROOSEVELT, IT WOULD BE BECAUSE Germany fired the first shot. Abraham Lincoln had refused to go to war until the Confederates opened fire—but he kept a Union garrison at Fort Sumter to ensure that sooner or later, the Confederates would draw first blood. Keeping a promise he made to Churchill at Argentia, FDR ordered Admiral King to begin escorting British merchant ships in the Atlantic, knowing the Germans would not tolerate the arrangement for long. Fort Sumter was open for business.[1]

Roosevelt took another long stride toward war on September 11, when in his second fireside chat of the year he announced a new policy toward U-boats in the Atlantic. The destroyer USS *Greer*, he announced, had been suddenly and deliberately attacked by a German submarine. The United States would therefore shoot German U-boats on sight. "When you see a rattlesnake poised to strike, you do not wait until he has struck before you crush him," he told his listeners. "These Nazi submarines and raiders are the rattlesnakes of the Atlantic. . . . From now on, if German or Italian vessels enter the waters, the protection of which is necessary for American defense, they do so at their own peril."[2]

The facts, as Roosevelt knew, were far more equivocal. *Greer* had been pinpointing the U-boat's location for a British depth charge bomber, and there was no hard evidence the German captain had even known *Greer*'s nationality.

But to FDR, freedom of the seas was a just cause for tightening the screws on Nazi commerce raiders. Details did not concern him.[3]

His fireside chat shored up public support for his anti-German stance. A Gallup poll in late September showed that 62 percent of the public favored taking a hoe to the rattlesnakes in the front yard. The same survey found 70 percent agreeing that the defeat of Germany was more important than keeping America out of war. After years of slow awakening, the public was ready to go to war with Germany.

The question was, when would Roosevelt ask them to draw the sword?[4]

. . .

Ernie King knew that if he took a hoe to the rattlesnakes, the snakes would bite back. A week after Roosevelt's speech, a U-boat sank the American freighter *Pink Star* off Greenland's coast. The attack sent to the bottom enough cheese to supply 3,500,000 Britons for a week, 420,000 quarts equivalent of powdered milk, and enough orange juice to provide 91,000 citizens with vitamin C for seven days. *Pink Star* was followed to her grave by USS *Kearny*, one of King's newest destroyers, and the destroyer *Reuben James*, torpedoed by the *U-552* off Iceland, taking down more than a hundred sailors. The United States was at war, whether Roosevelt admitted it or not.

The losses of *Reuben James* and *Kearny* were bitter pills for Ernie King to swallow. Having penned the "shoot on sight" orders, he spent some bleak hours in his cabin mulling over the circumstances that had led to the deaths of those men. *His* men.[5]

Yet the realist in King banished his sentimental side belowdecks. Divorcing personal feelings from professional judgment, he wrote a friend, "The *Kearny* incident is but the first of many that, in the nature of things, are bound to occur. It is likely that repetition will lead to open assumption of a war status. . . . I'm afraid the citizenry will have to learn the bitter truth that war is not waged with words or promises or vituperation but with the realities of peril, hardships, and killing— *vide* Winston Churchill's 'blood, sweat and tears.'"[6]

As the war at sea pitched and tossed, the public, not comprehending tonnage or sea-lanes or shipyard production, looked to its flesh-and-blood leaders to give a face to abstract policy. That November, readers saw King's stern face staring from a *Life* magazine cover under the eye-catching heading, "King of the Atlantic: America's Triple-Threat Admiral is the Stern, Daring Model of a War Commander."

But the realist in King also knew that with one year left until mandatory retirement at age sixty-four, America's triple-threat admiral would probably sail into his final port before war formally began. His naval career, launched during the Spanish-American War, was about to drop anchor without the great prize he had chased his entire professional life—Chief of Naval Operations.

With a heavy pen and a heavier heart, King began catching up with old acquaintances. He thanked them for their friendship over the years, and told them that once his tour with the Atlantic Fleet was over, some younger, luckier admiral would take the helm against Hitler.[7]

. . .

Or Hirohito. Driven by victories in China, insecurity over economic resources, and that intoxicating drug, national pride, Japan's reluctant prime minister, Prince Fumimaro Konoye, was pushed into an unyielding stance against its great Pacific rival, the United States.

In 1940, winds of nationalism swept into power Tokyo's hawks, led by a fire-brand war minister, General Hideki Tojo. The presence of Tojo and other militants in Prince Konoye's cabinet heightened anxiety among westerners that Japan was setting her sights on the resource-rich lands of Southeast Asia. Her conclusion of a pact with Germany and Italy in September 1940 fueled fears of Nippon hegemony over an "East Asia Co-Prosperity Sphere." As if speaking directly to Roosevelt, the tripartite treaty pledged each nation to aid any signatory "attacked by a power at present not involved in the European War or in the Sino-Japanese conflict."[8]

Swayed by its powerful military cliques, the Japan of 1941 coveted Europe's empires of 1900. Unable to see that the age of naked conquest had come and gone, Japan intended to remake Asia in her own image. Her troops had ravaged China's coast from Manchuria to the border of French Indochina. Her soldiers massacred, raped, tortured, and enslaved hundreds of thousands of Chinese and Koreans. Despite the backing of Stalin and Roosevelt, the two armies resisting the onslaught—the nationalist Kuomintang Army under Chiang Kai-shek, and Chinese communists under Mao Tse-tung—had been driven from the coast into China's vast interior.

To contain this new and violent empire, Roosevelt supplied Chiang with war loans and weapons. He shipped Lend-Lease materiel up the Burma Road to Chiang's base at Kunming, and arranged for American volunteers recruited from the Army Air Corps to fly Tomahawk fighters against the Japanese from a British airfield in northern Burma.*[9]

To the Chinese, the Year of the Snake carries connotations of suspicion, distrust, and cunning. The year 1941—a year of many snakes—brought Tokyo and Washington to the crisis point. While Japan seethed over U.S. embargoes and demands

---

* The American Volunteer Group, led by a sharp-chinned colonel named Claire Chennault, borrowed the RAF shark's teeth motif for their engine cowls and won lasting fame as the "Flying Tigers."

to give up hard-won conquests in China, Roosevelt revealed his quiet loathing of the Emperor's diplomats to journalist Quentin Reynolds. "They hate us," he told Reynolds. "They come to me and they hiss between their teeth and they say, 'Mr. President, we are your friends. Japan wants nothing but friendship with America,' and then they hiss between their teeth again, and I know they're lying. Oh, they hate us, and sooner or later they'll come after us."[10]

Roosevelt's scrap-metal embargo and the construction of a two-ocean Navy encouraged Japanese countermoves to seize oil, rubber, and minerals in the Dutch East Indies. When Roosevelt and his petroleum coordinator, Harold Ickes, found new ways of cutting exports to Japan without declaring a full-blown embargo—delaying export licenses, limiting port use, banning steel drum sales, and pulling tankers from the Pacific—planners at Imperial General Headquarters in Tokyo grew thoroughly alarmed.

A year earlier, FDR's advisers had predicted that a threat to Japan's economy could lead to war. Roosevelt's aim, as Churchill put it, was to "keep that Japanese dog quiet." But to keep the dog quiet, he would have to be firm, yet not so firm as to force Japan's hand. Roosevelt was walking a thin, wobbling tightrope, and he did not know how far he would get before his foot slipped.[11]

Like the United States, Japan was unsure of herself. From their planning desks in Tokyo, the Emperor's warlords saw two military options. First, Japan's army in Manchuria could strike north, into Siberia, while the Red Army was fighting for its life near Moscow. Stalin's forces were formidable, but his Far East army could not count on reinforcements from the overstretched Moscow front. A lightning blow might capture a buffer zone in Siberia, eliminate a traditional enemy, perhaps even force the collapse of Stalin's government.

Or Japan could strike south, into Indochina, the Dutch East Indies and Siam. This would leave the Emperor with an unsteady truce to the north, and probably war with the United States and Britain. But the payoff would be immense: Japan would have access to unlimited quantities of rubber, tin, oil, and rice.

Although U.S. cryptographers had not yet unlocked Japanese naval codes, they were able to read signals between Tokyo's foreign office and its embassies in Berlin and Washington. That intelligence gold mine gave Roosevelt important clues into the debate raging within the Emperor's council, which was split between her Pacific-minded navy and Asian-minded army. As Roosevelt told Harold Ickes, "The Japs are having a real drag-down and knock-out fight among themselves . . . trying to figure out which way they are to jump."[12]

They jumped—or at least hopped—on July 17, when Hirohito's government forced Vichy France to accept a joint protectorate over French Indochina. Japanese

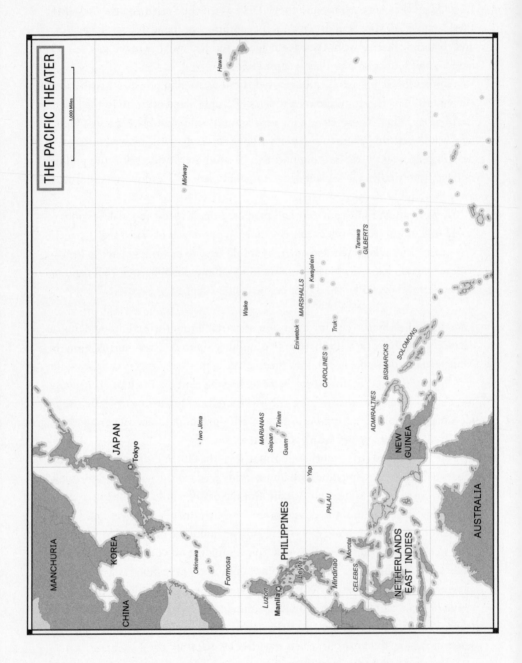

THE PACIFIC THEATER

1,000 Miles

MANCHURIA
CHINA
KOREA
JAPAN
Tokyo
Okinawa
Formosa
Iwo Jima
PHILIPPINES
Luzon
Manila
Leyte
Mindinao
Morotai
CELEBES
NETHERLANDS
EAST INDIES
AUSTRALIA
Yap
PALAU
MARIANAS
Saipan Tinian
Guam
CAROLINES
Truk
Eniwetok
Wake
MARSHALLS
Kwajalein
ADMIRALTIES
NEW GUINEA
BISMARCKS
SOLOMONS
Tarawa
GILBERTS
Midway
Hawaii

warships occupied naval bases at Hanoi, Cam Ranh Bay, and Saigon, putting the Empire in position to attack Western possessions anywhere along the East Asian rim. The Philippines, Malaya, and the Dutch East Indies watched anxiously as the katana blade was lifted, not knowing where it would fall next.[13]

Roosevelt was furious. Without waiting for an "I told you so" from cabinet hawks like Ickes, he issued an executive order freezing Japanese currency in the United States. To buy more oil, Japan would have to apply for cash from the Foreign Funds Control Committee, headed by Assistant Secretary of State Dean Acheson. Acheson, influenced by Ickes, believed the Japanese dog would growl least if oil was tightened to an uncertain trickle, and as Roosevelt slipped off to Argentia to meet Churchill, Acheson suspended action by the Foreign Funds Control Committee on Japan's requests until Roosevelt's return—effectively imposing a month-long oil embargo.[14]

· · ·

Looking ahead to a war that was growing more probable by the day, in the summer of 1941 the Joint Army-Navy Board prepared estimates of the manpower and equipment needed to defeat Germany and Japan. The "Victory Program," as the collage of memoranda, tables, and appendices was nicknamed, assumed the United States would hold to a "Germany-first" strategy, that the Soviet Union would be defeated in mid-1942, and that the United States would be ready to invade Europe no earlier than July 1, 1943.

The Victory Program forecast a ground force that would peak at 6.7 million men. The Air Forces would need 2 million additional men and 26,000 combat planes, while the Navy required 1.25 million men and 869 ships of various types. All told, Roosevelt's war chiefs predicted the country would need to put 10.8 million men in uniform before the war was over.[15]

Roosevelt knew that if the public peeled back the layers of this particular onion, it would be his eyes tearing up. His first thought, he told Stimson, was that the Joint Board estimates would ignite "a very bad reaction" if the public ever found out.

In December, the public did find out, when the estimates were leaked to the *Chicago Daily Tribune*, *Washington Times-Herald*, and other newspapers owned by anti-Roosevelt partisans Robert McCormick and Cissy Patterson. On December 4 the McCormick-Patterson syndicate published an in-depth article on the Victory Program under the blazing headline "F.D.R.'S WAR PLANS!" The papers painted the assumptions as a countdown for war, and claimed, "July 1, 1943 is fixed as the date for the beginning of the final supreme effort by American land forces to defeat the mighty German army in Europe."[16]

Stimson was apoplectic when he read the story. "What do you think of the patriotism of a man or newspaper which would take these confidential studies and make them public to the enemies of this country?" he asked journalists at a press conference. At the White House, he told Roosevelt the story was probably a violation of the Espionage Act, a crime carrying the death penalty. He hoped some guilty McCormick employee would be found and put on trial, to *"get rid of this infernal disloyalty we now have in America First and the McCormick family papers."*[17]

Roosevelt emphatically agreed, for Bertie McCormick and Cissy Patterson had been his most implacable critics in the press. He told Stimson they might be able to arrest those responsible for publication, but he wanted to make certain he was on solid legal ground first. He put J. Edgar Hoover's Federal Bureau of Investigation on the case, then turned back to the problems of the Far East.*[18]

.   .   .

The strain of dealing with a nation sliding into war ground at George Marshall's soul. His regimented life required him to spare a few minutes each day to refresh his mind, but the ceaseless demands of the Army, the War Department, the White House, the British, the Chinese, the Russians, and Congress never left his mind. Each day brought new fires to put out, leaving Marshall with precious little time to focus on big questions like global strategy. "The Army used to have all the time in the world and no money," he groaned. "Now we've got all the money and no time."[19]

One afternoon in late August, Marshall scrounged time for himself, when he and his wife, Katherine, took a rare retreat to their permanent home, a Leesburg, Virginia, brick house named Dodona Manor. It was the first time he had been home in three months, and his mission that day was to prune dead limbs from one of the apple trees scattered about his large yard.

When he was not preparing for war in Asia or Europe, Marshall's mind turned to gardening. He seeded his personal letters to family with references to lilacs, gardenias, roses, the condition of his oaks and elms, and his great pride, a compost heap that he turned religiously whenever he came home. On this inviting

* The source of the leak was never positively identified. Major Al Wedemeyer, who headed the staff work from the Army side, was suspected, though Marshall believed Wedemeyer to be innocent, and Wedemeyer ultimately went on to high command in the War Department and in China. Eventually Senator Burton Wheeler claimed an unnamed Air Corps captain brought him the report, which he had shown to the *Tribune*'s Chesley Manly. Because U.S. military secrets would be disclosed in any judicial proceeding, the McCormick employees responsible for the story were never brought to trial.

summer day, he had put his office in order long enough, he hoped, to give himself a few delicious moments to think about something other than war.

A pruning saw in hand, he managed to climb into one tree and shimmy over to a dead limb before an aide called him to the telephone. A German U-boat had been spotted prowling the Caribbean, threatening Dutch oil refineries on Aruba and Curaçao with its deck gun.

Marshall climbed down the tree and placed a call to the president at Hyde Park. When Roosevelt picked up, he asked the president to request permission from Holland's Queen Wilhelmina to install coastal guns and an air base on her islands. Roosevelt agreed. Problem solved.

He returned to his orchard, took up his pruning saw, and prepared to attack the dead branch again. He had not gone far up the tree before the aide returned: The president wanted more information about Aruba to give Queen Wilhelmina. Marshall climbed back down and dictated a cable that outlined the defense plan for Aruba.

He hung up, picked up his saw, and walked back to the waiting tree. He was halfway through the necrotic limb when, once more, the aide called him to the phone. Now the War Department was on the line, looking for specific guidelines about Aruba's coastal defense.

Marshall stared heavily for a moment, then sighed, "Call for my car. I am leaving for Washington."

He motored back to Washington, back to the Munitions Building, back to the endless rows of unpruned trees waiting in his office. Aruba and Curaçao got their coastal guns, the German sub was chased off, and Marshall's apple tree rode out the winter, a half-severed limb hanging lifelessly from its trunk.[20]

TWELVE

KIDO BUTAI

〰

AUTUMN 1941 FOUND AMERICA AND JAPAN WALTZING AROUND A BALLROOM laced with landmines. A secret request by Prince Konoye to meet with Roosevelt went unanswered—a victim of Hull, who wanted no meeting until all details had been worked out, and cabinet hawks, who wanted no discussions at all. Relations reached the breaking point in mid-October, when Japan's hawks forced Konoye out of power and persuaded Emperor Hirohito to replace him with the fiery General Tojo.[1]

FDR's war council saw extremism spreading through the imperial cabinet, making war, in Stimson's view, inevitable. *The Japanese Navy is beginning to talk almost as radically as the Japanese Army,"* he told his diary. *"We face the delicate question of the diplomatic fencing to be done so as to be sure that Japan was put into the wrong and made the first bad move—overt move."* The U.S. ambassador to Japan, Joseph Grew, warned Hull of a possible "all-out, do-or-die attempt, actually risking national hara-kiri, to make Japan impervious to economic embargoes abroad rather than yield to foreign pressure."[2]

While no one knew where this do-or-die attempt would be aimed, one place seemed likely: the Philippine Islands. Lying astride Japan's supply lines to the East Indies, the 1,500-island commonwealth begged for an attack by the Emperor's powerful air and sea forces. Japan's occupation of southern Indochina left the Philippines encircled on three sides, and in the event of war, U.S. naval plans assumed the Philippines would be lost. The plans called for the fleet to withdraw to Hawaii or British-held Singapore until the islands could be retaken, hopefully within six months. Until then, the Philippines would stand alone, and almost certainly fall alone.[3]

Lieutenant General Douglas MacArthur, the U.S. military adviser to Manila since 1935, had been haranguing Washington for years to fund the territory's defenses. Marshall, who had been stationed there as a young lieutenant, wanted to accommodate MacArthur, but by August 1941 the Commonwealth of the Philippines had just more than 22,000 men under arms, half of them enthusiastic but poorly equipped Filipino scouts. With Russia, Britain, China, and the Regular Army competing for a meager supply of weapons, the Philippines defenders had to do without.[4]

They had, at least, one of America's most distinguished generals to lead them. First in his class at West Point, Douglas MacArthur had built an unmatched reputation as commander of the Rainbow Division in World War I, superintendent of West Point, and the Army's youngest chief of staff during the Hoover years. Now, at age sixty-one, his face still retained the handsome features of his youth. Vain and brilliant, brave and selfish, MacArthur exuded a personal magnetism that few other men possessed, and he won the admiration and confidence of both American and Filipino soldiers.

His blind spot was civilian politics. As chief of staff under Hoover and FDR, he exceeded orders with the former and argued bitterly with the latter. He used brute force to eject protesting veterans during the Bonus March in 1932, earning the enmity of Washington's newspapers. After Roosevelt shipped him off to Manila to become the U.S. military adviser there, MacArthur, basing his analysis on a poll of *Reader's Digest* subscribers, confidently told the commonwealth's president that Roosevelt would lose the 1936 election to Alf Landon. He retired from the Army the next year and assumed the rank of field marshal of the Filipino forces.

Looking for some way to support the Philippines, in the summer of 1941 Marshall recommended MacArthur's recall to active duty as commanding general of the U.S. Army Forces in the Far East, and he scheduled U.S. reinforcements for the Philippines, which hopefully would arrive on Luzon in mid-December. He also scraped together a few heavy bombers to threaten Japan's sea-lanes, though without ample fighter and naval protection, long-range bombers would be a provocation more than a deterrent.[5]

Given American weakness in the Far East, Marshall hoped that Roosevelt could keep the peace with Japan. In a joint memo in early November, he and Stark reminded the president, "An unlimited offensive war should not be undertaken against Japan, since such a war would greatly weaken the combined effort in the Atlantic against Germany, the most dangerous enemy." In light of America's inability to wage a two-front war, they said, further Japanese thrusts into China, Thailand, or even the Soviet Union would not justify a U.S. declaration of war against Japan.[6]

Though loath to suggest foreign policy, Marshall privately felt that war with Japan must be avoided at almost all costs. On November 3 he privately told the Joint Board, "It appeared that the basis of the United States policy should be to make certain minor concessions which the Japanese could use in saving face." He suggested that these concessions might include "a relaxation on oil restrictions or similar trade restrictions."[7]

. . .

On November 4, Japan's foreign minister gave Ambassador Nomura three weeks to conclude an agreement that would guarantee Japan a stable oil and steel supply from the United States. Japan, in turn, would evacuate troops from French Indochina and most of China. Manchuria, the Mongolian border region, and the southern Chinese island of Hainan would remain under Japanese control. Tokyo's message warned, "This time we are showing the limit of our friendship; this time we are making our last possible bargain."*[8]

Nomura doubted Japan's concessions would be enough, for he knew China was Roosevelt's great sticking point. Nomura cabled Tokyo, "For the sake of peace in the Pacific, the United States would not favor us at the sacrifice of China. Therefore the China problem might become the stumbling block to the pacification of the Pacific and as a result the possibility of the United States and Japan ever making up might vanish."[9]

As Nomura wrote these words, he had no idea that the American high command was reading his dispatches. Cryptographers working on a top secret program named MAGIC intercepted and decoded traffic between Tokyo and its embassy as fast—sometimes faster—than Japanese embassy clerks. Two weeks later, American eavesdroppers were reading coded instructions to Nomura offering a six-month truce and reiterating Japan's final timetable: "If within the next

---

* Critical passages from Foreign Minister Shigenori Tōgō to Ambassador Nomura were sometimes poorly translated or paraphrased by overworked American cryptographers. For example, in the November 4 cable, Secretary Hull read the sentence, "This time we are showing the limit of our friendship; this time we are making our last possible bargain." According to Pulitzer Prize–winning historian John Toland, this sentence was more accurately translated as, "Now that we make the utmost concession in the spirit of complete friendliness for the sake of a peaceful solution, we hope earnestly that the United States will, on entering the final stage of the negotiations, reconsider the matter."

three or four days you can finish your conversations with the Americans; if the signing can be completed by the 29th (let me write it out for you—twenty-ninth) . . . if everything can be finished, we have decided to wait until that date. This time we mean it, that deadline absolutely cannot be changed. After that things are automatically going to happen."[10]

Things did happen. With Marshall's warnings ringing in his ears, Roosevelt told his counselors he would back off his demand for a complete withdrawal from China in favor of something less. But the distrustful Cordell Hull, seething over Japanese war preparations, opposed a softening of America's line on China. Hull merely assured the Japanese ambassadors that he was giving their proposals "sympathetic study" and would get back to them in due course.

Will Rogers had once quipped that diplomacy is the art of saying "good doggie" until you can find a rock. Hull saw the Emperor picking up a rock, and he had no intention of giving Japan six months to get ready to throw it. In closed-door meetings, his lispy Tennessee drawl would explode in a cloud of curses about the low skunks of the Japanese race. (At one point Roosevelt whispered to Labor Secretary Frances Perkins, "If Cordell Hull says 'Oh Cwist' again I'm going to scream with laughter! I can't stand profanity with a lisp.")[11]

But neither Roosevelt nor his lieutenants were laughing when, on November 25, the Office of Naval Intelligence reported that the Japanese Navy, led by Admiral Isoroku Yamamoto, was massing ships off southern China. Just as Nomura's instructions had forewarned, things were happening: the Japanese were about to lunge into Southeast Asia.[12]

Japan's mobilization changed everything. Roosevelt, Stimson wrote, *"fairly blew up—jumped in the air, so to speak, and said that he hadn't seen [the naval reports] and that that changed the whole situation because it was evidence of bad faith on the part of the Japanese that while they were negotiating for an entire truce— an entire withdrawal—they would be sending this expedition down there to Indo-China."* It was such a hostile act that the president, he wrote, wondered aloud *"how we should maneuver them into the position of firing the first shot without allowing too much danger to ourselves."*[13]

The next day Hull demanded a complete withdrawal of forces from Indo-china and China. Though his message was phrased as a reply, the Tojo cabinet, unwilling to erase the conquests of nearly a decade, considered Hull's proposal an ultimatum. Japan rejected it.[14]

Signs pointed to war, and that worried Marshall and Stark. On November 27 they sent Roosevelt a sober warning about Japan's next moves. "Japan may attack: the Burma Road, Thailand, Malaya, the Netherlands East Indies, the Philippines, [or] the Russian Maritime Provinces," they said. In light of U.S. weakness in the

Pacific, they reiterated, "The most essential thing now from the United States viewpoint, is to gain time."[15]

But time, like peace, was slipping away. A dispirited Hull saw nothing more that negotiations could do, and he consigned himself to the sidelines. He would play the diplomatic game out a few more days, but as Marshall and Stark were asking the president to buy them more time, Hull was telling Stimson, "I have washed my hands of it. It is now in the hands of you and Knox—the Army and the Navy."[16]

*   *   *

Those hands were too weak to restrain Japan, and naval intelligence soon reported Japan's main battle and transport fleets steaming around southern Indochina. Rounding the bend at Saigon meant they were heading to Thailand, Burma, or Malaya. Since the Philippines lay along their supply routes, Japan would also probably try to neutralize America's garrison there with a secondary attack. If war broke out, MacArthur's troops would bear the brunt of the storm.[17]

Stark's intelligence men had been tracking Japan's carrier fleet, a fast, mailed fist that the Japanese nicknamed *Kido Butai*, or "mobile force." The Americans were unable to decipher the Japanese Navy's operations code, designated JN-25, but Filipino listening stations indicated its carrier fleet was guarding Japan's home waters—evidently in reserve to parry a U.S. counterattack.[18]

During Thanksgiving week, General Marshall was away from Washington, in North Carolina, observing the final phases of the Army's 1941 maneuvers. In the general's absence, a worried Henry Stimson sent a warning to Army commanders in Manila, Hawaii, Panama, and California. His message read:

NEGOTIATIONS WITH JAPAN APPEAR TO BE TERMINATED TO ALL PRACTICAL PURPOSES WITH ONLY THE BAREST POSSIBILITIES THAT THE JAPANESE GOVERNMENT MIGHT COME BACK AND OFFER TO CONTINUE. JAPANESE FUTURE ACTION UNPREDICTABLE BUT HOSTILE ACTION POSSIBLE AT ANY MOMENT. IF HOSTILITIES CANNOT, REPEAT CANNOT, BE AVOIDED THE UNITED STATES DESIRES THAT JAPAN COMMIT THE FIRST OVERT ACT. THIS POLICY SHOULD NOT, REPEAT NOT, BE CONSTRUED AS RESTRICTING YOU TO A COURSE OF ACTION THAT MIGHT JEOPARDIZE YOUR DEFENSE.[19]

The next day, Stimson carried his G-2 staff intelligence summary to the White House. He told Roosevelt that Japan was in position to hit the Philippines, Thailand, the Dutch East Indies, or Singapore, the powerful British fortress on the Malay Peninsula. Roosevelt read the report carefully and told Stimson he saw only three courses of action: he could do nothing, he could present Japan with yet another proposal, or he could declare war. Stimson said he could see only two options, for doing nothing was no longer an option. Roosevelt agreed.[20]

At the White House later that day, Roosevelt, Stimson, Marshall, Hull, Knox, and Stark discussed Japan's next move. Roosevelt said he agreed with the G-2 assessment. Scanning his mental atlas of the world, Roosevelt suggested that Japan might have another target in mind: the Isthmus of Kra, the slim waist of the Malay Peninsula. Its capture, he pointed out, would threaten Singapore and cut the Burma Road, China's main supply route.

"This," thought Stimson, *"was a very good suggestion on his part and a very likely one."* He recorded the sobering discussion:

> *It was agreed that if the Japanese got into the Isthmus of Kra, the British would fight. It was also agreed that if the British fought, we would have to fight. And now it seems clear that if this expedition was allowed to round the southern point of Indo-China, this whole chain of disastrous events would be set on foot of going.*[21]

. . . .

FDR planned to attend Thanksgiving dinner at the one place, after Hyde Park, where his heart belonged: his infantile paralysis institute in Warm Springs, Georgia. He would not change those plans unless things took a turn for the worse. But to wring whatever time was left in a drying diplomatic sponge, he ordered Hull to tell Japan the United States would negotiate a new trade and raw materials agreement, and would unfreeze Japanese assets, in return for a complete withdrawal from China.

To Tokyo, withdrawal from China was as absurd as asking Washington to withdraw from the Philippines, and Roosevelt's last-minute overture was dead on arrival. On the morning of November 29, Hull telephoned Roosevelt to say that the fragile talks could be broken at any moment. Roosevelt, his voice low and serious, told Hull to call him after dinner that evening. If there was no improvement, he would leave Warm Springs early and return to Washington.[22]

Roosevelt spent the day out of sorts. He listened absently to the Army-Navy football game, took a half-hour swim, and presided over Thanksgiving dinner in the dining hall with the center's polio victims and staff. Over a finely trimmed gobbler and a table brimming with bowls of whipped potatoes, gravy, turnips, and pie, FDR somberly expressed his wish that all nations might have more to give thanks for at the end of 1942.

At nine o'clock that evening, Hull called back. He had just read an AP report of a speech that Tojo would deliver the next day. Tojo, the report said, would call upon the people of Asia to roll back U.S. and British "exploitation" of their homelands. The speech was the kind of fire-and-brimstone harangue that would make war inevitable, and Hull implored Roosevelt to return to Washington immediately.*[23]

Roosevelt, his face grave, told his aides to have the railway crew prepare the Presidential Special for an early return. Before leaving Warm Springs—twenty-six hours after he arrived—he bade a solemn good-bye to a small crowd of patients assembled in front of Georgia Hall. He closed his remarks by telling them, "This may be the last time I talk to you for a long time."[24]

The Presidential Special chugged back to a capital in the fading twilight of peace. He made a last-ditch attempt to buy Stark and Marshall more time, but cables intercepted from Tokyo told him it was probably hopeless. One ordered Nomura's staffers to destroy their coding equipment, while others referred to "final" deadlines to reach a settlement. Something was happening in the Imperial councils.[25]

American codebreakers deciphered Tokyo's diplomatic instructions rapidly, but MAGIC was such a precious asset it could be entrusted to only a few individuals. Those few—overworked, understaffed, overcompartmentalized, and short of translators—were unable to perceive much beyond a general bellicosity in the tone of individual messages.

Meanwhile, the *Kido Butai* went very, very quiet. Its operation codes abruptly changed, blinding U.S. naval intelligence, and no one was certain where it was. By Friday, December 5, the best Secretary Knox could tell the president was, "We expect within the next week to get some indication of where they are going."[26]

Unsure how much time was left in the hourglass, Roosevelt sent a personal appeal to Emperor Hirohito on the night of Saturday, December 6. In his letter,

---

* Tojo never gave the speech. The written message, attributed by news services to Tojo, was read by another person at a mass rally commemorating the first anniversary of the Sino-Japanese Basic Treaty. (Tojo had been expected to speak at the rally.) The wire service report, like some MAGIC decrypts, included poor translations that exaggerated the speaker's bellicosity.

he asked the Emperor to withdraw his army from Indochina. None of the nations threatened by Japan, he pleaded, "can sit either indefinitely or permanently on a keg of dynamite."[27]

The evening Roosevelt's message to Hirohito went over the airwaves, MAGIC men in Washington intercepted a lengthy fourteen-part message from Japan's Foreign Office to Nomura. The first thirteen parts were deciphered, translated, and circulated by special couriers to the president and the War, Navy, and State heads. FDR, Knox, and Stark received their copies of the thirteen parts that night—though for reasons no one could explain, General Marshall could not be found. His copy was locked in his office safe for the night.[28]

On the morning of Sunday, December 7, the last of the fourteen parts reached Washington. The final segment declared that attempts by Japan to resolve matters through negotiations were pointless. A follow-up message directed Ambassador Nomura to deliver all fourteen parts to the United States government at precisely one o'clock local time.

Marshall, called into the office after his Sunday morning horseback ride, talked over the message with his staff and prepared a warning to Army units in the Pacific. With Betty Stark's permission, he directed it to the Navy as well. He fired off the message a little before noon and directed Colonel Rufus Bratton to find out how long delivery would take. After checking with the Signals Corps staff, Bratton told Marshall the message would be delivered within thirty minutes.[29]

Marshall wanted the message delivered quickly, but junior officers in Signals had to be circumspect, for Japan could not know the United States had broken its "impregnable" diplomatic code. The telephone would be the fastest method of getting the warning off to Manila, the Panama Canal, Hawaii, and the West Coast. But it would also be the least secure. Security, as always, took priority, so they sent the message by telegraph—Western Union—to the West Coast.[30]

As Marshall's message buzzed over the telegraph wires, Japanese bombers, cruising through magnificent blue skies, came within sight of the lush slopes of Oahu.

PART TWO

A New Doctor

1942–1945

∽

If you or I begin fighting at the very start of the war,
what in the world will the public have to say about us?
They won't accept it for a minute. We can't afford to fight.

—GENERAL MARSHALL TO ADMIRAL KING, MARCH 1942

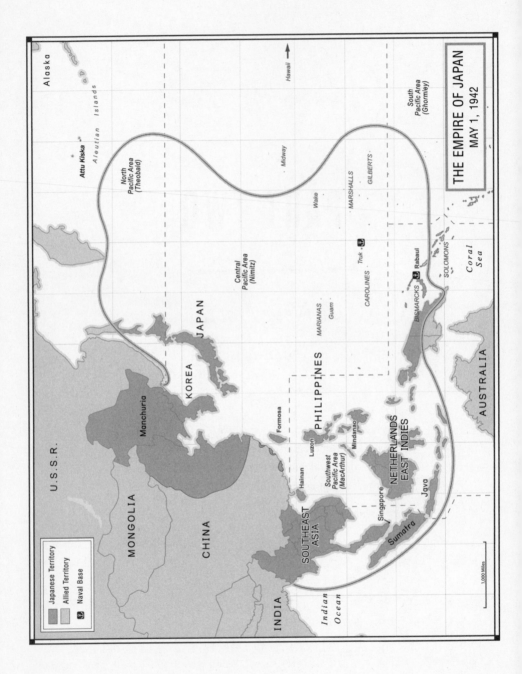

THE EMPIRE OF JAPAN
MAY 1, 1942

Japanese Territory
Allied Territory
Naval Base

U.S.S.R.

MONGOLIA

CHINA

INDIA

Indian
Ocean

Manchuria

KOREA

JAPAN

Hainan

Formosa

SOUTHEAST
ASIA

Singapore

Sumatra

Java

NETHERLANDS
EAST INDIES

Luzon

Mindanao

PHILIPPINES

Southwest
Pacific Area
(MacArthur)

AUSTRALIA

Alaska

Aleutian Islands

Attu Kiska

North
Pacific Area
(Theobald)

Central
Pacific Area
(Nimitz)

Midway

Wake

Hawaii

MARIANAS

Guam

CAROLINES

Truk

MARSHALLS

GILBERTS

South
Pacific Area
(Ghormley)

BISMARCKS

Rabaul

SOLOMONS

Coral
Sea

1,000 Miles

THIRTEEN

KICKING OVER ANTHILLS

P EARL HARBOR WAS A SMOLDERING WRECK AS FRANKLIN ROOSEVELT STOOD behind the House chamber rostrum.

"Yesterday," he began, "December 7, 1941—a date which will live in infamy—the United States of America was suddenly and deliberately attacked by naval and air forces of the Empire of Japan."

His low, cadenced voice reverberated through the chamber. Words like "unprovoked" and "dastardly" rang with controlled outrage, and in fewer than five hundred words he laid out the nation's case against Japan. He asked Congress for a declaration of war.[1]

The response was nearly unanimous. Republicans, Democrats, and independents of every shade and hue demanded vengeance for the men lying beneath Oahu's waves. In thirty-three minutes, the Senate and House passed resolutions declaring war against Japan, and the sole dissenting vote, by Montana pacifist Jeannette Rankin, was met with hisses from her colleagues.[2]

Leaders of opposition groups fell in line with their president. Herbert Hoover immediately declared his support for Roosevelt. John Lewis told the press, "When the nation is attacked, every American must rally to its support. . . . All other considerations are insignificant." Even Charles Lindbergh, America's most outspoken isolationist, called on his countrymen to line up behind their president.*[3]

---

* Roosevelt blacklisted Lindbergh from military service, but Hap Arnold felt the aviation pioneer could be useful as an aeronautics consultant. Lindbergh served honorably as a civilian adviser to the Army Air Forces, flying (unofficially) fifty combat missions in the Pacific and shooting down (unofficially) one Japanese aircraft.

Like the rest of the nation, Washington was transformed overnight. Men who had ridden buses in civilian suits turned up for work wearing Army and Navy uniforms. The president's Secret Service detail, swelling from eleven men to seventy, drove Roosevelt to Congress in an armored Cadillac once owned by Al Capone. Gas masks were issued to the president and his staff, black curtains hung over White House windows to foil Japanese bombers, and fireplaces were extinguished. "The house was chill and silent, as though it had died," recalled an old friend of Eleanor's. "Even Fala did not bark."[4]

With war a horrifying, bloody fact, FDR's job was to keep the nation's eye on the larger picture. While Germany remained, strategically speaking, the chief threat, to most of the public Japan was the aggressor. *Japan* had shed American blood at Pearl Harbor. *Japan* had sent *Oklahoma* and *Arizona* to the bottom. Sure, Germany had fired on American vessels at sea, but Roosevelt had never told them those incidents were grounds for war. It would be a hard sell to convince the public that Hitler was the greater villain.[5]

FDR reminded the American people that Germany was also their enemy. In a fireside chat two days after the Pearl Harbor attack, he spoke of America's duty to "reinforce the other armies and navies and air forces fighting the Nazis" as well as the "warlords of Japan."

That was the best he could do until, four thousand miles away, a cocky dictator in a double-breasted suit stepped to the Reichstag's dais and denounced Roosevelt as the criminal offspring of Jewish influence and democratic ineptitude. Referring to the leaked Victory Plan, Hitler claimed that the United States planned to attack Germany in 1943. "This man alone was responsible for the Second World War," *Der Führer* thundered. Germany, he declared, "considers herself to be at war with the United States, as from today."[6]

Deep inside, Roosevelt breathed a sigh of relief. Hitler had given him what he needed most: an undeniable reason to fight Nazi Germany.

On December 15, Admiral King found himself sitting before Navy Secretary Knox. The secretary told King that he and Roosevelt had worked out some changes in the Navy's command: Admiral Husband Kimmel, commander of the Pacific Fleet, would come home to face an investigation over the Pearl Harbor disaster. Admiral Chester Nimitz, the Navy's personnel chief, would take Kimmel's place. And the position of commander-in-chief, U.S. Fleet, or "CINCUS," would go to Admiral Ernest J. King.[7]

During his tenure as Atlantic Fleet commander, King had left an excellent impression on Knox and Roosevelt. Like Admiral Stark, King had known Roosevelt when "young Frank" was assistant navy secretary. King's fleet had done an admirable job in the Atlantic under the circumstances, and his arrangements during *Augusta*'s rendezvous with Churchill had been flawless.

Navy politics being what they were, King was cagey when Knox approached him about the CINCUS job. He made the obligatory denial of the crown—protesting halfheartedly that the decision should come from the chief of naval operations—and accepted the offer only after Knox told him that the matter had been settled by the president.[8]

Even for a jaded old salt of sixty-three, it was a heady revelation. Commander-in-chief, U.S. Fleet, put him at the pinnacle of the Navy's combat operations. He would be second only to the chief of naval operations in seniority. In some ways, CINCUS was even better than the CNO spot, since CNO was a planning position and CINCUS was a fighting man's job.

Getting down to brass tacks, King told Knox he would need some changes to make the arrangement work. First, the traditional headquarters of CINCUS—with one of the three main fleets—had to go. With a two-ocean war, he said, it made no sense to post himself far from the place where big decisions would be made. His headquarters should be at Main Navy in Washington, near the president.

Of course, said Knox. What else?

Next was his relationship to Admiral Stark. As CNO, Stark's job was to prepare war plans, provide strategic direction, and advise the president. CINCUS merely put the plans into action. But the line between strategic planning and war operations was a thin one, so Ernie King wanted something in writing that would divide the waters between himself and Betty. And that something needed the president's signature.

Knox assured King they would work something out. The last thing he and Roosevelt wanted was their top admirals tripping over each other.[9]

Third, King didn't like the acronym for his job. "CINCUS" sounded too much like an invitation to "Sink Us," which was no joking matter after Pearl Harbor. The new acronym for the fleet commander-in-chief would be "COMINCH."[10]

There was another duty King wouldn't take. He loathed the press and believed nothing good could come of talking to the papers. For years he had snubbed reporters and had turned down an offer for a press relations officer. Now that newsmen would be on his doorstep, sometimes literally, he wanted nothing to do with them.

Knox agreed to take most of this load off King's shoulders, though he said King would have to make important announcements to the press from time to time.[11]

King also wanted dispensation from testimony before Congress, except in the most crucial matters, like budgets. King understood his own limitations as a speaker, which were accentuated whenever he spoke alongside the Army's top general. "Marshall would sound off without any notes, and speak very well indeed," King later remarked. "I had trouble that way. . . . I had to use damned notes, and have things written up first, and read them. My education was defective. I can sound off all right and sometimes to the point, but it is a great help if you have been trained as a speaker."[12]

Knox agreed. King had testified a few times previously, and had been a poor witness. He was impatient with congressmen and too forthright for his own good. His voice was gravelly and halting, he stared down at prepared notes, and he looked uncomfortable when speaking. "I have come to the conclusion that good witnesses on the Hill are probably born," wrote one of King's congressional liaisons. "Admiral King was not born to be a good one."[13]

Finally, King requested authority over the Navy Department's technical bureaus—Navigation, Shipbuilding, Ordnance, and the like. The Navy needed one man who could coordinate strategy with production, he insisted. It was no good to plan a war when someone else controlled the supply of men and weapons.

Knox thought about it, and took that question to the Boss.

The Boss balked. The naval bureaus had been independent of sea operations for more than a hundred years. They were powers unto themselves when Roosevelt was assistant secretary a generation before, and it would take an act of Congress to change the Navy Department's basic architecture.[14]

Moreover, bureau chiefs were the men who petitioned Congress for money. FDR, who habitually diluted all authority except his own, didn't want any single admiral, including his new COMINCH, given that much power.

So Roosevelt refused King's request. But he promised King that any bureau chief who didn't cooperate with him would be fired. That, he said, should give King all the power he needed.[15]

King agreed, and on December 18 Roosevelt signed Executive Order 8984, which placed King in command of all American naval operations. With a typically Rooseveltian touch, it also expanded King's new powers by making COMINCH directly responsible to the president, throwing a wrench into King's relationship with Secretary Knox. The order then trimmed Admiral Stark's sails by providing that while the CNO would develop long-range war plans, short-range planning belonged to King.[16]

Once King saw the president's signature on Order 8984, he made one last move to complete his conquest. He told his aide to draft a polite note to his old friend Betty Stark. The note read, "I would appreciate your preparing a memorandum stating what functions and responsibilities the CNO should turn over to COMINCH."[17]

Main Navy, said one staffer, looked like an "anthill with the top kicked off." Admiral King spent the first four days of his new job kicking over anthills from a musty, unswept government office on the building's third deck.*[18]

If Marshall was the prototypical clean-desk man, Admiral King was the tornado-desk man. "His desk was something of a rat's nest," remembered his flag secretary. "He was a miserable housekeeper . . . the papers were six inches deep on his desk. His incoming basket was always overflowing."[19]

Like all senior commanders, King functioned through his staff, the men he depended on to make big recommendations, refine sweeping concepts, and draft orders that moved thousands of men across the globe. He often reminded his staffers, "If I make a mistake, it is likely to be a big one; don't let me make it if you can put me right."[20]

To avoid those big mistakes, King surrounded himself with a small core group of devoted, hardworking officers and drove them relentlessly, at the cost of a few heart attacks and one suicide. He replaced nine-to-five civilian typists with enlisted men, and rotated officers between staff and sea duty, so Main Navy wouldn't catch a debilitating case of the "Washington mentality."[21]

A special peeve of King's was paperwork, which he fought like it was Yamamoto's flagship. He ordered his staff to reduce anything addressed to him to a single typewritten page. "He didn't want any two-page memoranda," recalled one of his men. "If you sent him a memorandum that didn't have your signature at the bottom of the page, he'd throw it in the wastebasket, and you'd wonder where it was."[22]

King's replies were often sent on two-by-three-inch yellow paper slips, penned in his precise, heavy script and peppered with the quotation marks, underlines, and excessive punctuation that marked his writing style. He would return a memorandum with "OK K" if it was excellent, "No K" if not. If it were particularly bad, he might write, "?!?!?!—K." He would never provide an explanation. It was the job of King's bewildered officers to figure out why the boss didn't like it. The admiral wasn't a man to waste time or words on explanations to underlings.[23]

Apart from the few staffers whose careers he intended to advance, King remained aloof from his colleagues. He held no regular staff conferences, and allowed unlimited access only to his chief of staff, Rear Admiral Russell Willson;

---

* The Munitions Building, where the War Department was quartered, had "floors" and "doors." Main Navy had "decks" and "hatches."

his hard-boiled chief planner, Captain Charles "Savvy" Cooke Jr.; and his deputy, Rear Admiral Richard Edwards.[24]

He had a tongue like a bosun's lash. He made no small talk, confided in no one, and never mentioned his family or personal life. He had no social interests to speak of—at least, none he let on to those around him—and no interest in becoming friends with his brother officers. "He was the true lone wolf, and didn't give a damn whether anyone liked his attitude or not," recalled his intelligence chief. Another said, "He possessed little warmth or charisma, but captured one's loyalty and zeal with his own superior conduct and performance. He was just plain cold armor-plated steel."[25]

King's reputation permeated the chart room, the service's Holy of Holies. King's intelligence head recalled, "Within a few minutes of Admiral King's entrance into the room, everyone seemed to have evaporated into thin air; disappeared, flag officers as well as captains and the few commanders who were authorized. No one seemed to want to be where King was."[26]

Though his permanent office was on Main Navy's third deck, King knew he would be traveling to the West Coast, the Caribbean, and Hawaii, as well as around Washington. For that reason, he demanded three perquisites: a car, an airplane, and a flagship.[27]

The car and plane were easy. A vice president of the A&P Company loaned King his Cadillac for the war's duration, and for his personal aircraft, King yanked a beautifully appointed Lockheed Lodestar from the hapless chief of the Bureau of Aeronautics.[28]

The flagship was more difficult, since ships were expensive to maintain, and of questionable value to an admiral who rode a Cadillac to an office on Constitution Avenue. Initially, King appropriated the converted yacht *Vixen* from the commander of the Atlantic submarine fleet. But the yacht had been manufactured at the Krupp shipyard in Kiel, Germany—a pedigree the newspapers noticed—and King eventually decided he needed a more suitable, American vessel.[29]

A number of luxury yachts had been loaned by wealthy families to the Navy for war service. William Vanderbilt's beautiful yacht *Alva*, for instance, sailed into service as USS *Plymouth*, and Secretary Knox occasionally berthed aboard the former presidential yacht *Sequoia*.

From a stable of Thoroughbreds King chose *Delphine*, a 1,200-ton, 257-foot showpiece built for Detroit's Dodge family. King had *Delphine* repainted in wartime camouflage and fitted with an antiaircraft gun on her forecastle. After reviewing a list of imposing names drawn up by his staff, he had his yacht rechristened USS *Dauntless*.

Moored in the Anacostia River alongside the Naval Gun Factory, *Dauntless* served the nation valiantly as King's houseboat. The admiral would awaken every morning at 0700 for exercises and a shower, then take breakfast and read the newspaper—usually turning to the comic page first, his eyes zeroing in on his favorite strip, *Blondie*.

Breakfast concluded, he would proceed to the aft deck and stand at attention as the ship's colors were hoisted. A quick walk down the gangplank would take him to his waiting Cadillac, whose Marine Corps driver would let him off at the Mall near Main Navy. He would arrive at his office between 0831 and 0833 to begin his workday.

King's office routine rarely varied. Like Marshall, he would begin by reviewing the overnight dispatches before moving to antisubmarine, administrative, and logistics reports. At 0930, he would walk to Secretary Knox's office for his morning conference, and a half hour later he would be back at his desk, reviewing selections of the 35,000 pieces of monthly mail sifted through a funnel of staffers, secretaries, assistants, and deputies. Visitors were kept at bay until noon, and promptly at 1245 King would stop for lunch at a small mess room near his office, then return to work.

Despite his prewar reputation as a hard drinker, King's evenings were quiet. He would return to the pier in his car, salute the colors as they were retired for the night, then go below for dinner with the ship's duty officer. He swore off hard liquor—mostly—for the duration of the war, and he was rarely seen with anything stronger than the occasional glass of sherry or a beer. At night he relaxed by listening to Edgar Bergen–Charlie McCarthy routines on the radio, watching Spencer Tracy films and devouring Western novels by Zane Grey and Ernest Haycox.

King's weekends were usually spent working, though occasionally he would spend them at the homes of Navy friends near Washington—and sometimes at the homes of wives whose husbands were away on naval duty. One or more Sundays each month, he would ride to the Naval Observatory, his official residence, and spend the afternoon with his wife, Mattie, and any of their six daughters who happened to be in town.

Otherwise, he rarely saw his wife, his daughters, or his son, a midshipman studying at Annapolis. Once, when King's communications officer and wife paid a social call on Mattie King, Mattie politely inquired after her husband's health.[30]

FOURTEEN

"DO YOUR BEST TO SAVE THEM"

∽

NDER AGREED ALLIED STRATEGY, GERMANY WAS AMERICA'S MAIN ENEMY.
But Japan was hardly a footnote. Hours after the Pearl Harbor attack, a
Japanese air strike caught and destroyed more than half of MacArthur's heavy
bombers on the ground. Five days later, 3,000 soldiers of Japan's 16th Division
swarmed ashore on Luzon, the largest of the Philippine Islands, and just before
Christmas, General Masaharu Homma's Fourteenth Army landed 48,000 superbly trained soldiers at Lingayen Gulf, on the island's west side. MacArthur's
outnumbered force was corralled onto the tiny Bataan Peninsula and Manila
Bay's gateway island of Corregidor.[1]

With each passing week, the picture grew darker. Japanese infantry poured
down the Malay Peninsula toward Singapore, scooping up the remaining French
possessions and threatening the resource-rich Dutch East Indies. They captured
Guam, and laid siege to Wake Island and Hong Kong.

For Stimson and Marshall, the first order of business was reinforcing MacArthur.
Five days after Pearl Harbor, Marshall summoned MacArthur's former chief of
staff, Brigadier General Dwight Eisenhower, a man who knew the islands well.
Marshall moved Eisenhower into the second spot in the War Plans Division and
ordered him to find reinforcements for Luzon's garrison. "Do your best to save
them," he commanded.[2]

Japan's silk noose tightened around the Pacific, and Marshall knew that
without control of the seas, little could be done. Stateside soldiers couldn't swim
across the ocean, and the Navy lacked the surface ships to protect slow, vulnerable
troop transports from Japanese cruisers, subs, and bombers.

Desperate to succor the garrison, Marshall even sent cash to Australia to bribe any blockade-runners willing to make the Australia-to-Manila run with food and ammunition. But he found few takers at any price. As Marshall's agent in Australia admitted, "We were out-shipped, out-planned, out-manned, and out-gunned by the Japanese from the beginning."[3]

From Corregidor's Malinta Tunnel, an out-manned, out-gunned MacArthur sent a steady stream of radio messages pleading for troops, planes, and naval support. He played every card in his deep rhetorical deck: he reminded Washington of its moral obligation to protect its dependents, he badgered Marshall to show his written pleas for help to the president, and he railed against War Department plans to build a base in Australia at the expense of the Philippines.[4]

MacArthur's demands taxed the patience of War Department staffers, who lacked enough of anything to go around. *"Looks like MacArthur is losing his nerve,"* an overworked Eisenhower grumbled to his diary. *"I'm hoping that his yelps are just his way of spurring us on, but he is always an uncertain factor."* Frustrated at being stuck in Washington while war raged in the Pacific, Ike complained, *"In many ways MacArthur is as big a baby as ever. But we've just got to keep him fighting."*[5]

On the surface, Marshall and MacArthur kept a respectful relationship, though it was also distant and cold. Marshall had been a protégé of General Pershing, who had denied MacArthur's application for a Medal of Honor during the last war. The slighted MacArthur had long despised the "Chaumont crowd," as he called Pershing's clique, and when he was chief of staff in the early 1930s, MacArthur paid back "Pershing's man" by refusing to change orders assigning Colonel George C. Marshall to a teaching post with the Illinois National Guard, a career-killing backwater. Marshall did not earn his first star until after MacArthur left Washington for Manila.[6]

Marshall did not hold the past against MacArthur, though, and he did his best for the Philippines' defenders. When MacArthur complained that Admiral Thomas Hart, the Asiatic Fleet commander, refused to spare any submarines to ferry antiaircraft ammunition to Corregidor, Marshall sent Eisenhower to Admiral King's office with a message asking King to press Hart to relieve MacArthur's ammunition shortage. King curtly replied, "You may tell General Marshall that if any more drastic action is necessary than is represented in my usual method of issuing orders, such action will be taken."

The remark was typical King gruffness, and Eisenhower left Main Navy red-faced and fuming. But after Eisenhower left, King gave Hart a hard shove, and two days later he notified Marshall that two submarines laden with antiaircraft ammunition were departing for Luzon.[7]

The submarine shuttle was a tiny drop in a badly leaking bucket. Marshall gave MacArthur a consolation gift—a promotion to four-star general—but that was all he could do. Before that dismal December ended, MacArthur declared Manila an open city. His men, haggard and half-starved, settled into their last positions, on Bataan and Corregidor, and awaited whatever aid their country might send them.[8]

To those stranded men it would be a long wait, for an unexpected visitor to Washington was about to direct everyone's attention away from Douglas MacArthur and the Philippine Islands.

FIFTEEN

"O.K. F.D.R."

⌇

"I HAVE A TOAST TO OFFER," THE PRESIDENT SAID, AFTER WHITE HOUSE butlers had cleared the guests' dinnerware. Raising his glass over the long dinner table, he announced, "It has been in my head and on my heart for a long time—now it is on the tip of my tongue: 'To the Common Cause.'"

Roosevelt's guests, Secretary and Mrs. Hull, Lord Beaverbrook, Ambassador Halifax, and, most serenely, Prime Minister Winston Spencer Churchill, raised their goblets in reply and sipped.[1]

The prime minister and his entourage had crossed the Atlantic aboard the battleship *Duke of York* and arrived in Washington on December 22 for a series of top secret meetings code-named ARCADIA. As the city awaited winter's first snowfall, Churchill and Roosevelt began charting the war's course.

Pondering an immense global jigsaw puzzle in the family quarters that night, the two men sorted through the war's theaters like cards in a tarot reader's deck: British Isles defense, the Azores, the Burma Road, Russia, Brazil, bombers, the Pacific, U-boats, Spain, west Africa, northern France, Singapore. Each problem carried its own unique quirks, and like hungry puppies yelping for a tired mother, each theater cried out for an unsustainable share of food, money, weapons and fighting men.[2]

To Roosevelt fell the added duty of hosting his frenetic houseguest. The stocky prime minister wedged into the Rose Suite easily enough, but his servants and equipage kept the White House staff busy moving beds and dressers around the second floor. Custodians removed most of the furniture from the Monroe Room, which, temporarily, became the headquarters of the British Empire. Churchill had oversized maps of every theater mounted on the room's walls, and officious-looking gatekeepers moved symbols, pushpins, and lines around the maps as they received fresh dispatches from London. In his map room, Churchill could see at a glance the strategic picture in Asia, the Atlantic, the Mediterranean, and Europe.

Roosevelt, an avid stamp collector with a passion for geography, was mesmerized by Churchill's portable map room. He and the prime minister spent hours there discussing sea-lanes to Sumatra, Suez trade routes, the Brazil-Africa gap, and maritime choke points from Iceland to Murmansk.[3]

For FDR, it was an education in world war. To Eleanor, a pacifist at heart, it seemed like an overgrown, violent board game. "They looked like two little boys playing soldier," she remarked dolefully after stopping by one evening. "They seemed to be having a wonderful time, too wonderful in fact."[4]

Churchill kept Roosevelt awake into the wee hours of the morning discussing strategy, politics, and Britain's experience fighting the Germans. "The outstanding feature [of the conference]," Churchill later wrote, "was of course my contacts with the President. We saw each other for several hours every day, and lunched always together, with Harry Hopkins as a third. We talked of nothing but business, and reached a great measure of agreement on many points, both large and small."[5]

The conversations were accompanied by endless trays of cigars, liquor, and food. "[Churchill] ate, and thoroughly enjoyed, more food than any two men or three diplomats," remembered Roosevelt's Irish bodyguard Mike Reilly. "He consumed brandy and scotch with a grace and enthusiasm that left us all open-mouthed in awe." Roosevelt's secretary Bill Hassett, lacking Reilly's appreciation for Churchill's talents, told his diary, *"Churchill is a trying guest—drinks like a fish and smokes like a chimney, irregular routine, works nights, sleeps days, turns the clock upside down."*[6]

Churchill's upside-down hours exhausted Roosevelt and infuriated Eleanor, who would shuffle into the smoke-filled study several times a night with progressively less subtle hints about bed and the next day's work. The men ignored her. Around midnight, she would leave in a huff and they would talk until two or three in the morning.[7]

As First Lady, Eleanor was polite and proper to the Conservative Party leader. But as a die-hard progressive, her every fiber opposed nearly everything the British Empire stood for. Riding in the back of the president's touring car with Franklin and Eleanor one day, Churchill began thumping a pet theme of his: After the war, the two countries must form an Anglo-American alliance to meet the world's problems. To these ramblings Roosevelt merely nodded, repeating offhandedly, "Yes, yes, yes," as the car rumbled on.

Eleanor sat as Churchill prattled on about the Anglo-American postwar order. She held her tongue until she could hold it no more.

"You know, Winston," she interjected, "when Franklin says 'yes, yes, yes,' it doesn't mean he agrees with you. It means he's listening."

Churchill withdrew into sullen silence.[8]

But oftener Roosevelt did agree with Churchill, and Stimson and Marshall soon learned what could happen when the two political leaders sequestered themselves out of earshot of military advisers. On the afternoon of December 23, Churchill sent word that he was anxious to speak with Stimson on the subject of the Philippines. The next morning, Stimson and Eisenhower unrolled their maps before Churchill and showed the prime minister the vital points along MacArthur's battlefront. The PM, clad in pajamas and slippers, listened intently and dismissed the two men with his thanks.[9]

Later that day, a British secretary telephoned Marshall to request an addition to the day's military agenda. Britain proposed to divert a division of U.S. reinforcements slated for the Philippines to Singapore. Checking around, Marshall learned the idea had already been agreed to between Roosevelt and Churchill.

Marshall was aghast. He went straight to Stimson, and behind closed doors the two men blew their stacks. Churchill had a lot of nerve asking the president to interfere with a U.S. reinforcement operation. MacArthur was in desperate need of help, and Churchill was poaching what little help they could give him.

Brimming with indignation, Stimson rang up Harry Hopkins and told him what happened. He asked Harry to tell the president that if he persisted in overriding his war chiefs, "The president would have to take my resignation!"

Hopkins, used to his boss's erratic style, interrupted a meeting between Roosevelt and Churchill and came straight to the point. Taken aback, both men denied they had made any such deal. They said they had no intention of diverting troops from MacArthur's Philippines to Britain's Singapore. When Harry called Stimson back to tell him both men denied the rumor, Stimson read from a British memo reflecting just such a discussion. Hopkins admitted the paper belied Roosevelt's story, but it didn't matter, because the transfer was off.[10]

Later that day, Roosevelt made a cutting remark in Stimson's presence about inaccurate statements flying around about deals between himself and Churchill. Stimson held his tongue, but he would not trust Churchill so long as Singapore was imperiled. He told his diary that night, *This incident shows the danger of talking too freely on international matters of such keen importance without the President carefully having his military and naval advisors present. This paper, which was a record made by one of Churchill's assistants, would have raised any amount of trouble for the President if it had gotten into the hands of an unfriendly press. I think he had felt that he had pretty nearly burned his fingers."*[11]

It would not be the last time those fingers were nearly burned.

.   .   .

While questions of empire and politics belonged to Roosevelt and Churchill, the problem of crafting a military strategy to fit political goals fell to Marshall, Stark, King, and Arnold. Working alongside the Americans—and sometimes at cross-purposes—were the men who wielded the Empire's trident: Admiral Sir Dudley Pound, the good-natured First Sea Lord who, like Lewis Carroll's dormouse, tended to doze off in meetings; Field Marshal Sir John Dill, whom Marshall liked best among the Brits; and Air Chief Marshal Sir Charles Portal, an Oxford-educated airman whom Marshall and King agreed possessed the best mind of the lot.*[12]

One of the biggest stumbling blocks for the ARCADIA participants was American disorganization. Caught up in the calamities of Pearl Harbor, the Philippines, U-boats, and Germany's declaration of war—to say nothing of domestic problems with the draft, Lend-Lease, and weapons production—U.S. commanders had little time to ponder grand strategy. They were caught flat-footed by British questions about global logistics and were thrown into disarray over strategic nuances they hadn't time to study. "We were more or less babes in the woods on this planning and joint business with the British," remarked Brigadier General Thomas Handy, Eisenhower's planning deputy. "They'd been doing it for years. They were experts at it and we were just starting."[13]

Preparation for the conference was so haphazard that the first meeting in the Federal Reserve Building was held up when the embarrassed hosts realized the conference room Marshall's staff had reserved was too small to hold everyone. Marshall, mortified, scowled when he saw the tiny room and table, but before anyone could comment, the group's quick-thinking guide told them, "This room is for the overcoats and coats. The conference will be in the Board Room at the front." The chiefs obediently laid down their caps and coats on the table and were escorted to the magnificent boardroom, which was hastily opened for the gathering.[14]

The British sat together on the table's south side, quiet and tightly organized. They had rehearsed their arguments aboard the *Duke of York*, and they knew each service's strengths and problems intimately. The Americans couldn't find a crack

---

* Pound's lethargy was probably the result of a brain tumor undiagnosed until very late. Pound died in October 1943, and was replaced by Fleet Admiral Andrew B. Cunningham.

of daylight in the British shield wall when they sat down and began to talk strategy.[15]

Churchill intended to fight the Nazis the way Spaniards fought bulls. It was suicidal to attack a fresh, raging toro head-on, so Spanish matadors had developed a peripheral approach to wear the monster down before going in for the kill. Picadors stab and banderilleros jab, weakening the beast's neck muscles and tiring him out. Only when the bull is depleted, through blood loss and exhaustion, does the matador expose himself for the *tercio de muerte* and drive the sword home.

Through twelve days of formal meetings, the British chiefs trumpeted a similar strategy. They took as their guide a memorandum dictated by Churchill assigning primary importance to guarding America's war industries, protecting Atlantic convoys, and defending the British Isles. Churchill also advocated "closing the ring" around Germany through subversive activities, air bombardment, sea blockade, and propaganda.[16]

Early in the conference, the British chiefs presented a plan, code-named GYMNAST, that envisioned a landing in French Northwest Africa. To this they added a permutation called SUPER-GYMNAST, which included an American landing at Casablanca, on Africa's Atlantic coast.[17]

It was this "closing the ring" bit that Marshall didn't like. A military strategy premised on German political collapse seemed dangerously optimistic. Britain had not succumbed to hammering by the Luftwaffe, and Stalin's communists, battered to the gates of Moscow, were still fighting savagely. Why did Churchill think Germany would collapse when Hitler had the entire Continent under his thumb? To Marshall, a frontal assault through northern France was the only path to victory.

For similar reasons, Marshall saw North Africa as a sideshow. An invasion there would do little to relieve the hard-pressed Russians, and it would draw off troops, bombers, and ships needed for an invasion of northern France. Counting noses on the American side, Marshall found he had strong backers in Stimson, Stark, and Arnold. So he took his case for northern France to the one man whose vote counted most.[18]

FDR agreed with Marshall on the importance of coming to grips with Germany directly, but he expressed the concept a shade differently. According to Marshall's notes of their meeting, Roosevelt "considered it very important to morale, to give the people of this country a feeling that they are in the war, to give the Germans the reverse effect, to have American troops somewhere in active fighting across the Atlantic." That somewhere, Marshall felt, should be northwestern France, though FDR would not be more specific than "somewhere across the Atlantic."[19]

The divergence between the president and his army commander was a subtle one. But to an eye as keen as Churchill's, it was a crack of daylight between the generals and their commander-in-chief. He would return to that subject again, when the time was right.

. . .

On Christmas Day, General and Mrs. Marshall hosted Admiral and Mattie King and the British chiefs for a Christmas dinner. When the Marshalls learned the morning before that Christmas was also Sir John Dill's birthday, Katherine Marshall sent the general's orderly, Sergeant James Powder, to scour Washington pastry shops for an appropriate cake. The cake and candles secured, Powder managed to find a five-and-dime store open that morning, where he bought some cheap American and British flags to decorate the cake.

The dinner was a delightful gathering. On closer inspection, the guests noticed tiny markings on the flags that gave them a good chuckle: "Made in Japan." Fruit and eggs were plentiful—compared to England, at least—and Sir John remarked, to Marshall's astonishment, that it was the first birthday cake he had been given since he was a small boy.

The only mishap, unnoticed by Marshall's companions, was the absence of a Western Union singing telegram ordered by Dill's wife, Nancy. The singer was turned away from the Marshall home by Secret Service men, whose job evidently included foiling assassins dressed as Western Union singers.

But the dinner was otherwise a success, and the Allied war chiefs returned to work thankful for the short respite from a world plunged into blood.[20]

. . .

Though questions of grand strategy flared up during the conference, the first job of the Arcadians, taken up after Sir John's birthday celebration, was to bring order to an organizational mess. The cousins discussed actual fighting in only a few places—the North Atlantic, the Middle East, the Southwest Pacific, and the Philippines—and large questions turned on the command structure at the top.

"I am convinced," Marshall told the chiefs on Christmas Day, "that there must be one man in command of the entire theater—air, ground, and ships. We cannot manage by cooperation. Human frailties are such that there would be emphatic unwillingness to place portions of troops under another service. If we make a plan for unified command now, it will solve nine-tenths of our troubles."[21]

The other chiefs weren't so sure. Air Marshal Portal countered that in London, forces were allocated at the direction of the "highest authority"—meaning

Churchill—while Sir Dudley Pound expressed skepticism about Marshall's "supreme commander" concept. The British ran things by committee and consultation, and it was as unthinkable for a Royal Navy admiral to place his ships under a landlubber as it was for a field marshal to let an admiral lead an infantry assault.[22]

Perhaps, thought Marshall, if the British saw something on paper, they'd have fewer doubts. He went back to his office and told Eisenhower to draft a letter of instruction for a supreme commander for the American, British, Dutch, and Australian spheres in Asia and the Pacific. Military staffers, always on the lookout for new acronyms, promptly dubbed the new theater "ABDA."[23]

Admiral King was lukewarm on Marshall's proposal. He favored a unified command under a naval officer for coastal and island areas, but as a general principle he did not want a theater commander standing between himself and his fleet admirals—especially if that commander wore an army uniform. He told one colleague, "I have found it necessary to find time to point out to some 'amateur strategists' in high places that unity of command is not a panacea for all military difficulties!"[24]

To win over the U.S. admirals, Marshall made a round of lobbying calls on Main Navy. In Admiral Stark's office he met with Stark, King, Knox, and a bevy of lesser admirals who knew King's views and were obediently suspicious of the Army. But King was changing his vote. Initially ambivalent, he had seen Roosevelt leaning toward the "supreme commander" concept, at least for the Far East, and he was reluctant to buck the president.

At length King said he agreed that the Australasian region required a single commander for all forces. "When King said this," one staffer recalled, "all the other Navy people smiled and concurred." Stark, whose power had been eclipsed by King's, gave Marshall his blessing, and the Americans at last had a united front.[25]

But getting past the U.S. Navy was only the first phase. The Royal Navy, Stimson wrote, "kicked like bay steers" when they read the proposal. During a plenary meeting where Churchill argued against Marshall's supreme commander structure, Lord Beaverbrook, Churchill's Lend-Lease emissary, slipped Harry Hopkins a scribbled note that read, "You should work on Churchill. He is open-minded and needs discussion."[26]

Seeing an opening, Hopkins arranged a morning meeting to allow Marshall to plead his case to the prime minister.

Marshall walked through the door of the Rose Suite to find the head of the British Empire in his usual morning state, sitting in bed in his dressing gown, his pink face jovial and alert despite a bout of chest pain the night before.* Churchill greeted the general with his usual enthusiasm, and waving his cigar like a maestro's baton, he invited Marshall to make his sales pitch.[27]

As the recumbent warlord listened, Marshall paced up and down along Churchill's bedside, launching into the basic reasons a supreme commander was vital to future success. With air, ground, and sea forces each pushing its own viewpoint, someone would have to keep all three services marching in the same direction, at the same cadence.

Churchill listened carefully, then interrupted him sharply. What could an Army officer possibly know about commanding ships in battle? he asked.

"What the devil does a naval officer know about handling a tank?" Marshall retorted. Marshall's point—driven home fast and forcefully as he paced—was not to ask admirals to command tank battalions, but to find a man with good all-around judgment to balance the needs of competing forces without favoritism.

Churchill was loath to concede the point. He lapsed into historical analogies, reciting instances where "supreme commanders" had failed, and coalitions ruled by committees had been victorious.

Marshall cut him off. "I told him I was not interested in Drake and Frobisher, but I was interested in having a united front against Japan, an enemy which was fighting furiously," he said later. The Army needed sea transport, the Air Force land bases, the Navy anchorages. All would be selfish, and all needed an arbitrator on the spot. If the Allies didn't have a unified command, he warned, they would be finished.[28]

Churchill's round face soured. He tossed the covers aside, swung his stubby legs off the bed, and padded off to the bathroom, where a warm tub awaited him. Marshall waited patiently as the PM splashed around in deep thought for a few minutes.

Even with America in the war, Churchill knew he had to proceed gingerly, for the American public was bloodthirsty against Japan, and not yet fully invested in the war against Germany. *"Marshall remains the key to the situation,"* Churchill's personal doctor, Lord Moran, told his diary. *"The PM has a feeling that in his quiet, unprovocative way he means business, and that if we are too obstinate he might take a strong line. And neither the PM nor the President can contemplate going ahead without Marshall."*[29]

Stepping out of the tub and toweling off, Churchill announced his verdict.

---

* Unbeknownst to Churchill, his chest pains were symptoms of a heart attack.

The Americans, he said, would have their way. He gestured to the doorman to show in his chiefs of staff. Marshall had won this round.[30]

When Churchill forwarded Marshall's idea for a single theater commander to the British War Cabinet and Lord Privy Seal in London, they wired back to ask who would be giving the theater commander his instructions. With a shrug, Churchill referred the question to the military chiefs.[31]

That question bothered Roosevelt, who did not like the idea of a gaggle of Wellingtons running their own wars across the globe. The war would be run from Washington, and Roosevelt wanted to make certain that every "supreme commander" knew who the *supreme* commander was. On December 29 he asked Admiral King to lunch with him at the White House, and the two men discussed who would command the supreme commander.

King suggested a war council for the ABDA theater, on which the chiefs of the Australian and Dutch forces would be invited to deliberate with U.S. and British. FDR told King to consult with Marshall and the British, then come back with a recommendation.[32]

King gathered the chiefs, and in a hurried session the British proposed using "existing machinery"—meaning the British and American chiefs of staff—to command the ABDA theater. The Anglo-American chiefs would issue broad directives to the supreme commander and allocate troops and supplies to his theater. The British chiefs in London would consult with the Australians and Dutch where their interests were affected, but only the two great powers would sit on the committee. The committee would be known as the "Combined Chiefs of Staff," while the U.S. and British service heads, when conducting business separately, would be known as the "Joint Chiefs of Staff."[33]

Thinking it over, Marshall said he was prepared to accept the British proposal. King fell into line, and Stark and Arnold went along. The Combined Chiefs forwarded their proposal to Churchill and Roosevelt, who had one important change. FDR wanted the "chiefs of staff" council based in Washington—not split between London and Washington.[34]

Churchill balked. He wanted Washington to run the Pacific war and London to run the European war, and as a compromise, he suggested having committees in both capitals. But Roosevelt insisted on having the council in Washington; he wanted the levers of power at his elbow, not divided between himself and Churchill.

But once again, Churchill was prepared to give up control over the Combined Chiefs to keep America focused on the war against Germany, and he would not risk a rift at this delicate stage in the alliance. After arguing his point with FDR and getting nowhere, Churchill capitulated. The war would be run from Washington.[35]

. . .

The U.S. "Joint Chiefs of Staff" was not a body created by law or executive order. It was an association presumed to include, at a minimum, Marshall, King, and Stark—and possibly Lieutenant General Arnold, though he was merely Marshall's assistant for air matters. At Marshall's request, Arnold showed up for the group's regular Tuesday luncheons in the Public Health Building on Nineteenth Street. Since the Navy had two votes—King and Stark—not even King questioned Arnold's right to be there.[36]

Hap Arnold's status was important to Marshall, who envisioned the emancipation of the Air Forces from Army control some day in the future. To ensure that he was accepted as a full-fledged member of the new Joint Chiefs, Marshall asked Roosevelt's secretary, Marvin McIntyre, to include Hap's name in a press release announcing the president's military advisers. "I tried to give Arnold all the power I could," Marshall said later. "I tried to make him as nearly as I could Chief of Staff of the Air."[37]

Arnold's rise coincided with Marshall's grand reorganization of the Army, and it fit with FDR's determination to keep the important military men reporting directly to him. In March 1942, President Roosevelt signed an executive order placing Marshall in more or less absolute command of the Army Ground Forces, Army Air Forces, and Services of Supply. Brushing Stimson aside, the executive order also provided that the duties of the secretary of war "are to be performed subject always to the exercise by the President directly through the Chief of Staff of his functions as Commander-in-Chief in relation to strategy, tactics, and operations." In other words, FDR was sidelining Henry Stimson on questions of strategy.[38]

. . .

Admiral King had a logical mind, a mind fascinated by laws and regulations. Much of his daily thought—from working complex crossword puzzles in his cabin to crafting clear fleet orders at his desk—was devoted to the power of the written word. That unsentimental, logic-driven brain refused to accept ambiguous organizational structure, especially at the top of the command chain. So shortly after the ARCADIA conference ended, he began pressing Roosevelt for an official written charter to make the Joint Chiefs of Staff a legal, authoritative body.[39]

FDR scotched the idea. As a rule, he disliked rules. He intended to run the war out of the White House as he saw fit, and the last thing he wanted was some stuffed-shirt lawyer telling him, or his military chiefs, that they had exceeded their authority or were limited in what the president could order them to do. The Joint Chiefs of Staff would remain exactly what Roosevelt wanted it to be—a useful, pliable implement of the commander-in-chief's will.

For that, he did not need a piece of paper. When the Joint Chiefs sent him a draft executive order establishing their authority, he blithely declared that a legal charter might "impair flexibility in operations," and ignored the matter. The group's chairman later wrote, "I have heard that in some file there is a chit or memorandum from Roosevelt setting up the Joint Chiefs, but I never saw it."[40]

Marshall had once been appalled at Roosevelt's informal style of war management, and it was now the British turn to adjust to the shock of FDR's methods. The Combined Chiefs, like the Joint Chiefs, had no precedent, formal authority, or, grumbled Sir John Dill, much formality. The British Army's representative in Washington complained to London, "There is the great difficulty of getting the stuff over to the President. He just sees the Chiefs of Staff at odd times, and again no record. . . . The whole organisation belongs to the days of George Washington."[41]

Ernie King was as uncomfortable as Dill was. During the first meeting of the Combined Chiefs, he voiced his concern that the group had no legal authority to act. Their only "charter"—a reference in a memorandum they wrote earlier that year—had never been formally approved by the heads of state, and even that piece of paper did not explicitly extend beyond the ABDA theater.

Siding with King, the American and British chiefs drafted a comprehensive charter for Churchill and Roosevelt to sign. They sent it to the White House, where it was marooned in a sea of paper, probably as Roosevelt intended.

But the chiefs persisted, and in successive meetings, the Combined Chiefs and Stimson separately asked Roosevelt's military aide, Pa Watson, to ask the president to approve their charter.[42]

Roosevelt ignored them. The current structure appeared to be working, and it was not in Roosevelt's nature to tie his own hands with a document spelling out what his lieutenants could and could not do.*

King pleaded that to issue orders to theater commanders, he needed the imprimatur of the heads of the United States and Great Britain. So two months after

---

* Churchill evidently saw it the same way, for he never put his signature to any organic document either.

receiving the draft, an exasperated Roosevelt pulled out his fountain pen and jotted in the letter's margin the words "O.K. F.D.R."[43]

The Allied command structure finally had the official backing of the United States.

On January 14 Britain's war leaders, less Sir John, who would stay in Washington, packed their bags and bade good-bye to their compatriots. During their short time on Mount Olympus, the headstrong leaders had agreed upon the major issues: Germany first, a committee to run the war, and a supreme commander in charge of the Southwest Pacific and Asia. The Americans and the British had tested each other's mettle, and had forged personal relationships that promised to make future decisions easier.

But while they agreed at the highest levels, they would soon find devils scampering through many complex details—details growing into sharp focus as the war grew more violent, difficult, and uncertain.

SIXTEEN

"THERE ARE TIMES WHEN MEN HAVE TO DIE"

E RNIE KING WASN'T A MAN TO MAKE SNAP DECISIONS. HE CONSIDERED PRO-
blems from every angle, and gave his men plenty of time to think until they
found the right answer.

But in the opening months of the war, time was scarce and answers were few.
In the Atlantic, U-boats were wreaking hell on shipping to England and Russia,
and King lacked the destroyers, cutters and corvettes to protect them. In the Pacific, he had three fleet carriers to Japan's ten, and a handful of damaged battleships to engage a dozen of Yamamoto's.

The big fighting ships were only the most conspicuous part of the U.S. Fleet,
a complex organism swimming in many seas and oceans. In early 1942, King's
most desperate need was not new warships, but combat aircraft to protect the
ships he had. The planes, in turn, needed antiaircraft guns to protect them while
they were on the ground. The guns, in turn, needed shells. Shells could be moved
only by cargo ships, and cargo ships needed combat ships to protect them. King's
navy was a great wheel that could not roll forward until the whole circle was
complete.[1]

Pressure to roll that wheel forward clanged like a ship's bell in King's ears.
In his heart he was a pugilist, not a desk admiral, and he wanted his fleet
wreaking destruction on the enemy, not merely defending what he had. As he
summed up his philosophy of war, "No fighter ever won his fight by covering
up—by merely fending off the other fellow's blows. The winner hits and keeps on
hitting even though he has to take some stiff blows in order to be able to keep on
hitting."[2]

．　　．　　．

During the ARCADIA meetings over Christmas, the Combined Chiefs had agreed that the places where Japan was throwing stiff blows—from the Philippines to Malaya—should be lumped into a single theater, "ABDA." Britain's Indian theater lay to the west of ABDA; everything north of ABDA was Chinese territory, and everything to the east was the Pacific theater.

At the time, the Allied chiefs assumed that Britain could hold Malaya from its great fortress of Singapore. The Dutch, they hoped, could hold the long islands of Sumatra, Java, and New Guinea, forming a rough barrier dividing the Pacific and Indian oceans. The chiefs also recognized that if Japan captured the "Malay Barrier," supply lines between American-supplied Australia and British-supplied India would be severed. So as a contingency plan, they made a "gentlemen's agreement" to split the theater if the enemy reached Sumatra.[3]

The enemy did. On February 15 Singapore fell, and Japan's armies invaded Sumatra and Java, driving back the Dutch colonial defenders. The Allied leaders knew the collapse of the Malay Barrier would only be a matter of days, and when it fell, the ABDA theater would be split into two. General Wavell, whose headquarters was in India, would be cut off from Australia, the Pacific, and his supply line to America.

Reviewing the situation over lunch, Roosevelt and Hopkins concluded that the latest Japanese onslaught required the Allies to split ABDA into two theaters: a "China-Burma-India" theater, for everything of west of Malaya, and a Pacific theater for everything east.

King agreed, for he felt it was a logical next step. With an eye toward shutting the British out of the Pacific, he suggested treating the Australia–New Zealand area, along with the other Pacific islands, as a U.S. sphere of activity. Churchill, Roosevelt, and the Combined Chiefs, seeing this as a logical division of labor, agreed. Britain would take the lead in India, its historical dominion, and the United States would run the Pacific war.[4]

．　　．　　．

As MacArthur's trapped remnants dug in on the Bataan Peninsula, General Homma's Fourteenth Army closed in on Manila, and the capital city's residents fled to the mountains. Japan's Third Fleet cut off all supplies from the sea, and MacArthur's famished men huddled in their bunkers day after miserable day as waves of Betties and Vals pounded them from the sky.[5]

Stimson and Marshall had both served in those islands, Marshall as a soldier and Stimson as the territory's governor-general. The two men sweated bullets to aid the beleaguered defenders, but without a Navy to provide transport, a million men in America would do MacArthur no good.[6]

Every soldier above the rank of lieutenant, Stimson said, knew the Philippines could not withstand a determined assault. Early naval war plans called for a withdrawal from the Philippines to a defensible line in the Marshall and Caroline islands. Stark's Plan Dog memorandum assumed the Philippines would be lost, then recaptured, hopefully in about six months. These plans, drafted atop government-issue desks in Washington, were a logical response to an untenable situation, though logic gave little comfort to the soldiers, marines, and nurses marooned on the Alamo of the Pacific.[7]

It was also cold comfort to Philippine President Manuel Quezon. Ill with tuberculosis, trapped in the claustrophobic tunnels of Corregidor, Quezon grew despondent as he watched the slow death of his country. He ranted to MacArthur, accusing the United States of abandoning his people, and in a message forwarded by MacArthur, he asked the American government to grant the Philippines immediate independence. The new nation would declare itself neutral, disband its army, and prohibit all foreign forces, Japanese and American, from remaining on the island.*[8]

In his cover message to Marshall, General MacArthur vaguely endorsed the idea of neutrality. He suggested that no military advantage would be lost if the Philippines were neutralized, and he emphasized the hopelessness of the political and military situation. "The temper of the Filipinos is one of almost violent resentment against the United States," he wrote. "Every one of them expected help and when it has not been forthcoming they believe they have been betrayed in favor of others. . . . In spite of my great prestige with them, I have had the utmost difficulty during the last few days in keeping them in line. If help does not arrive shortly, nothing, in my opinion, can prevent their utter collapse and their complete absorption by the enemy."[9]

Henry Stimson had known war and peace from the front lines of both. As war secretary, he had a duty to protect the men on Corregidor and Bataan. Yet he knew the other nations of Asia—the Dutch colonies, the Burmese, the Australians—were looking to the United States for deliverance. America was in no position to save the Philippines, much less the rest of Asia, but neutralizing the islands would be a body blow to the Allied cause—a virtual surrender of American territory.[10]

Stimson walked into Marshall's office, where he found Marshall and Eisen-

---

* Under the Tydings-McDuffie Act of 1934, the Philippines would be granted independence in 1946.

hower leaning over a table, drafting a proposed response to MacArthur. Over Marshall's mahogany table, the three men wrestled with an awful decision: to accept defeat, or to order thousands of Americans and Filipinos to fight to their deaths.

It was a sobering moment for the two very different men—one, a hot-blooded patrician, the other a cold military calculator of blood, treasure, and machines. But different as they were, the two leaders arrived at the same place. Weighing carefully the lives of twenty thousand soldiers against the needs of the nation, Henry Stimson reached a conclusion often accepted by politicians but uttered by few: "There are times when men have to die."[11]

Roosevelt read Quezon's message, then looked at Stimson and Marshall. To refuse Quezon's plea would condemn thousands to their deaths, a fact driven home every time he received an updated casualty report from General Marshall. But to neutralize the Philippines would be a tacit admission that the United States would abandon its duty before it would stomach heavy losses. It would signal Tokyo and the rest of Asia that the United States was unwilling to bear the cost of victory.[12]

"We can't do this at all," said Roosevelt, his voice firm.

Marshall knew it was a call that only a man with his eye on the largest of pictures could make, a decision by a man who had asked his people to accept a draft and place their sons in harm's way. Now he would have to ask some of those sons to die. To Marshall, the liberal Democrat from upstate New York, the dilettante yachtsman whose career had centered on unemployment, social services, and economic reform, had become a decisive war leader. "I immediately discarded everything in my mind I had held to his discredit," Marshall said years later. "Roosevelt said we won't neutralize. I decided he was a great man."[13]

So the garrison would go down fighting; MacArthur would be allowed to arrange for the surrender of Filipino forces, but not Americans. In his reply, Roosevelt explained his decision to MacArthur. "As the most powerful member of this coalition we cannot display weakness in fact or in spirit anywhere," he said. "It is mandatory that there be established once and for all in the minds of all peoples complete evidence that the American determination and indomitable will to win carries on down to the last unit."

To Quezon, Roosevelt vowed, "As long as the flag of the United States flies on Filipino soil as a pledge of our duty to your people, it will be defended by our own men to the death."[14]

A few, he hoped, would not die under that flag. In a follow-up message Roosevelt offered to evacuate MacArthur's wife and son, along with the Commonwealth's top officials. MacArthur cabled Marshall, "I am deeply appreciative of the inclusion of my own family in this list, but they and I have decided that they will share the fate of the garrison." He declared he would fight on the Bataan Peninsula until driven off, then wage a final stand on the island of Corregidor.[15]

No one who knew Douglas MacArthur doubted his personal bravery. If he said he would die on Corregidor, he intended to die there. After checking with several of MacArthur's associates, Marshall concluded that only a peremptory order from the president would compel MacArthur to relinquish his command and save himself.[16]

The specter of an abandoned MacArthur led into captivity, or his broken body—and those of his wife and son—placed on display like hunting trophies, haunted Marshall, Stimson, and Roosevelt. The general was a national hero, a striking media personality, and a decorated battlefield commander.

He was also a rallying figure for Republicans like Wendell Willkie, who called for MacArthur's return to command all American forces in the Pacific. Willkie's cry was taken up by a public starving for a hero. In an article titled "Bring Home MacArthur!" *Time* gushed: "With the air of a man who has been groping for a word and hears it unexpectedly from a passerby, the U.S. echoed Willkie in the press, in Congress, on the street."[17]

Roosevelt knew MacArthur was an egomaniac and a political problem. But with the Japanese pounding on the gates of Bataan, he had to decide whether to make a martyr of the conservative icon or remove him to Australia for future use.

On Sunday evening, February 22, Roosevelt called King, Hopkins, and Marshall into his office, where the four men debated the fate of the general for half an hour. After thrashing the question out from a number of angles, Roosevelt made his decision: MacArthur would leave the Philippines and establish a new headquarters in Australia. Marshall would wire orders to Corregidor in the name of the president, and assure MacArthur that Roosevelt would urge Britain and Australia to accept him as commander of all Allied forces in the Southwestern Pacific Area.[18]

• • •

On a moonless night two weeks later, a tiny squadron of torpedo boats revved their engines and slipped past a Japanese naval squadron blockading Manila Bay. Aboard they carried a seasick MacArthur, his wife, Jean, his young son, Arthur, and a handful of staffers who would form the core of his new command. When the boats reached the southern island of Mindanao, MacArthur's party transferred to waiting B-17 bombers for a rough flight south to Australia.[19]

Upon his arrival at Terowie, South Australia, the general's train was set upon by a pack of reporters and townspeople, who jostled for a glimpse of the famous general.

Flashing a Hollywood smile to the crowd, he declared, "I came through and I shall return."[20]

SEVENTEEN

"INTER ARMA SILENT LEGES"

～

AS COMMANDER-IN-CHIEF, ROOSEVELT HAD TO SEND AMERICAN TROOPS wherever they were needed to win the war. As president, he had to keep the citizens at home feeding the voracious war machine. That meant production.

In his annual message to Congress in January, he declared, "We must raise our sights all along the production line," and for 1942 he raised these sights impossibly high: 60,000 planes, 45,000 new tanks, six million tons of merchant ships, and 20,000 antiaircraft guns. "These figures," he told Congress, "will give the Japanese and the Nazis a little idea of just what they accomplished at Pearl Harbor."[1]

Having started his draft speech with production targets supplied by his advisers, Roosevelt blithely penciled through the numbers and boosted them as he thought proper, asking no input from either military or industrial experts. His numbers were not based on what had been possible in the past; they were based on what America needed to defeat three enemy empires that had swallowed nearly one-third of the world's strategic resources and population.

The experts were appalled by Roosevelt's naïveté. Incredulous War Department officers lamely joked that their president had "gone into the numbers racket." When Harry Hopkins asked Roosevelt whether he should announce targets so unrealistically high, the president dismissed Harry's concern with another cigarette-holder gesture: "Oh—the production people can do it, if they really try."[2]

To ensure that they would really try, in January he shook up his Office of Production Management, the agency responsible for materials rationing and coordination of production. He moved GM's William Knudsen out and replaced him with Sears, Roebuck's Donald Nelson. Nelson's revamped agency, called the War Production Board, would make final decisions on raw materials allocation and set production priorities—subject, as everything else was, to Roosevelt's orders.[3]

Nelson and his fellow board members would be sorely strained in the coming months as America's factories were pushed to industrial feats never before imagined. And it would not only be the giants—the Fords and the Grummans and the U.S. Steels—pitching in. Jukebox maker Rock-Ola was given a contract to make M-1 carbines. Typewriter assembler Smith-Corona milled out bolt-action rifles. A barge company owned by Andrew Higgins of Louisiana began making plywood landing craft with steel ramp bows. Any shop that could turn out textiles was a supply of uniforms, tents, socks, and coats, and any electronics company was a source of generators, headsets, and radio tubes.

The demand for food, weapons, medicine, housing, and raw materials did more than shake the war industry. It transformed the genetic makeup of America. Sons of farmers moved to the big cities to work in factories and shipyards. Sharecroppers became carpenters and white-owned industries hired record numbers of Rosie the Riveters and black and Hispanic workers. A mass migration pushed American families from the rural midland to the industrialized coasts and Great Lakes regions, changing forever the nation's political and social maps.[4]

The war's emerging bottleneck was the shortage of shipping. The Army was training draftees at an unprecedented rate, and by the end of the year Marshall would have 1.8 million men in khaki. But it took 144,000 tons of shipping space to move a single infantry division, nearly twice that to move an armored division. America would need to produce an extra eighteen cargo ships every month, dedicated solely to the military, to send a force of 750,000 overseas by the end of 1942. "The war effort of the United States," Marshall predicted, "will be measured by what can be transported overseas."[5]

. . . .

During Churchill's visit, Roosevelt had kept a quietly covetous eye on the British map room, and after the PM and his entourage left, Roosevelt asked his naval aide, Captain John McCrea, to create a similar room where he could review his most secret dispatches and see at a glance where the battle lines stood.[6]

With the help of a young naval lieutenant, an actor named Robert Montgomery, McCrea delivered. He set up a small, efficient map room on the White House ground floor across the hall from the doctor's office. A sign outside the door read "NO ADMITTANCE," and meant it. Few officers besides Map Room staff were regularly admitted, and the only civilians allowed inside without special permission were Harry Hopkins and FDR's secretary Grace Tully.[7]

Within this sanctum, an aide recalled, "The walls were covered with fiber-board, on which we pinned large-scale charts of the Atlantic and the Pacific. Up-dated two or three times a day, the charts displayed the constantly changing location of enemy and Allied forces. Different shape pins were used for different types of ships, a round headed pin for destroyers, a square head for heavy cruisers. For the army we had a plastic cover with a grease pencil to change the battle lines as new dispatches came in."[8]

With a touch of the dramatic, an aide bought special pins for the locations of the Big Three leaders: FDR's pin was shaped like a cigarette in a holder, Stalin's like a briar pipe, Churchill's like a cigar. The floor was left uncarpeted so Roosevelt's wheelchair could roll freely, and maps were arranged so FDR could see them easily from a sitting position.[9]

In keeping with Roosevelt's obsession with secrecy, the Map Room also served as his clandestine communications center. Messages of the highest importance—cables from Churchill or Stalin, or the most important MAGIC intercepts—were routed there. To keep what he said hidden, even from his own military, FDR sent most outgoing wires through Navy channels and received incoming messages through the Army, ensuring that the White House maintained the only complete set of his correspondence with key leaders.*[10]

Roosevelt had spent his life looking at maps of far-off places—places he could see in his mind's eye, even if, at his age, it was unlikely he would ever visit them. To a man unable to walk, the stylized features depicted on stamps and maps—peninsulas, islands, straits, and bays—sparked his imagination. They were old friends he greeted with the delight he lavished on longtime Hyde Park neighbors. He understood, as few others did, that these obscure geographic features drove strategy, and strategy dictated how America would parcel out its blood, men and machines.

But the average American knew nothing of the railways of Burma or the atolls of the Pacific. For decades the nation had looked inward, from Jersey's Boardwalk to the Rockies, from Fisherman's Wharf to the Mighty Mississip. To give the public an idea of why American sons would be sent to fight for places like Java, Persia, or Midway Island, he needed to acquaint the public with the scope of this vast new war.

---

* Roosevelt and Churchill had a transatlantic telephone line that enabled them to talk directly with each other. The phone line was, however, vulnerable to enemy interception, even with the primitive scrambler devices available in 1942. Most critical messages were sent in coded cables or radiograms.

He would do something no president had done before. He scheduled a fireside chat and ordered his press people to ask listeners to have a map of the world open when he spoke to them. "I'm going to speak about strange places many of them have never heard of," he told his speechwriters. "I want to explain to the people something about geography—what our problem is and what the overall strategy of the war has to be. I want to tell them in simple terms of A-B-C so that they will understand what is going on and how each battle fits into the picture. . . . If they understand the problem and what we are driving at, I am sure that they can take any kind of bad news right on the chin."[11]

Intrigued by the president's request, the public responded enthusiastically. Citizens descended in droves on booksellers, libraries, and general stores to buy or borrow world maps. People who had never heard of Malaya or Leningrad or the Azores bought atlases for their families. With maps pinned to walls and spread over laps, sixty-one million adults tuned in on the night of February 23 to hear the sonorous voice of their president.[12]

"My fellow Americans," that voice began. "This war is a new kind of war. It is different from all other wars of the past, not only in its methods and weapons but also in its geography. It is warfare in terms of every continent, every island, every sea, every air lane in the world."

FDR walked his audience across the globe, starting with China and moving to the South Pacific, the Mediterranean, and the North Atlantic. He explained, in basic terms, how industries at home and maritime lanes at sea sustained the Allied war effort.

"Until our flow of supplies gives us clear superiority we must keep on striking our enemies wherever and whenever we can meet them, even if, for a while, we have to yield ground," he said. "The object of the Nazis and the Japanese is to separate the United States, Britain, China, and Russia, and to isolate them one from another, so that each will be surrounded and cut off from sources of supplies and reinforcements. It is the old familiar Axis policy of 'divide and conquer.'"

In Roosevelt's populist vision of total war, the woman in the factory or the child collecting scrap aluminum was as much a warrior as the marine at Midway or the soldier on Bataan. At every level, he said, the nation must match and exceed the effort of the enemy. "From Berlin, Rome and Tokyo we have been described as a nation of weaklings—'playboys'—who would hire British soldiers, or Russian soldiers, or Chinese soldiers to do our fighting for us," he said. "Let them repeat that now! Let them tell that to General MacArthur and his men. . . . Let them tell that to the boys in the Flying Fortresses. Let them tell that to the Marines!"[13]

•   •   •

As everyone knew, there was one group of American men who would not be flying bombers or fighting with the Marines. After Pearl Harbor, newspaper reports told of an unidentified Negro mess attendant at Pearl Harbor on the battleship *West Virginia* who, under strafing fire from Japanese Zeros, seized an idle machine gun and fired back until the weapon ran out of ammunition.

The unnamed messman was obviously a war hero at a time when heroes were badly needed, but the Bureau of Navigation, the Navy's personnel department, had an ironclad policy against employing colored sailors outside the messmen's corps—cooks or stewards. Keeping with naval tradition, the man would not be publicly recognized for wandering into the field of combat, the exclusive province of white officers and sailors.[14]

As the largest of the military services, the Army was the most visible focus of the nation's growing pains, and Stimson felt the War Department was getting more than its share of criticism over exclusion of black citizens from the military. In mid-January he complained to his diary, *"The Navy has successfully evaded taking any Negroes into the naval forces or marines. Consequently this raises the pressure on us because it makes us obliged to take more than the proportional ratio between white and colored people in the general population."*[15]

Stimson was, however, resigned to placating civil rights leaders, Eleanor Roosevelt, and the band of "do-gooders" that sometimes seemed to run the White House. But he was determined to make Knox's Navy play by the same rules. *"I am insisting that we shall create colored divisions and use them,"* he wrote in his diary.

> *I am also insisting that we shall give special attention to the training of colored officers. I am very skeptical about the possible efficiency of such officers but, as it has been determined that we shall have them, I propose that we shall educate them to the highest possible standards and make the best we can of them. . . . I am planning to take up with the President the misbehavior of his own pet arm—the Navy—which has been acting like a spoiled child in the matter.*[16]

Pressured by Stimson and others, FDR wrote to Knox, "I think that with all the Navy activities the Bureau of Navigation might invent something that colored enlistees could do in addition to the rating of messman." Two weeks later, Knox sent Roosevelt a report by the General Board underscoring the need to segregate colored and white sailors, except where colored sailors served as personal attendants to white officers. "Men on board ship live in particularly close association," the board solemnly declared. "How many white men would choose that their closest associates in sleeping quarters, in mess, be of another race?"

Incorporating colored sailors would depress morale and erode discipline, the board concluded. If that was discrimination, "it is but part and parcel of similar discrimination throughout the United States, not only against the negro, but in the Pacific States and in Hawaii against citizens of Asiatic descent." Efforts to overcome this discrimination in the Navy, said the board, were "political" problems instigated by both Republicans and Democrats "to gain the support of the negro vote."

The board therefore recommended limiting colored enlistees to the messmen's branch. If the president refused, they recommended throwing open the doors to all jobs the Navy offered to men of any race—something they were confident no sane president would do.[17]

The old refrain of morale, discipline, and the good of the service had always worked before. Except around the edges, Roosevelt had never kicked when the admirals dropped their kedges and refused to budge.

But this time it was different. The nation was at war, and the board hadn't shifted its sails to match the changing winds. The president was prying open factory doors to people of all races, and the Army was training black aviators and platoon leaders. Black families were moving from the rural South to the industrial North, and they would become a political force in the years ahead.

The board also made a serious mistake in calling Roosevelt's preferences "political." To Roosevelt, the admirals were using that word in a loaded, partisan sense, and he bristled at the implied accusation of playing politics with the nation's defense.[18]

FDR was tired of hearing the Navy buck the tide of progress. As he once told banker Marriner Eccles, "To change anything in the Na-a-vy is like punching a feather bed. You punch it with your right and you punch it with your left until you are finally exhausted, and then you find the damn bed just as it was before you started punching."[19]

This time, the feather bed would give. Roosevelt shot back a reply to Knox declaring the board's report unacceptable. He castigated the admirals for calling his position "political" and lectured the secretary, "Officers of the U.S. Navy are not officers only but are American citizens. . . . It is incumbent on all officers to recognize the fact that about 1/10th of the population of the United States is composed of members of the Negro race who are American citizens." He ordered the admirals to go back to their offices and prepare a better report.[20]

A month after Roosevelt's letter to Knox, the Navy Department's press office released the name of Doris Miller, the "unidentified Negro messman" whose valor at Pearl Harbor had earned him the Navy Cross. The story of this John Henry and his .50-caliber hammer energized civil rights leaders and pushed Roosevelt to shove the Navy a little more into line.[21]

But a little was as far as the incrementalist in FDR would push, and he would not force the Navy to open every hatch to black sailors. In its second report, the General Board recommended offering Negro enlistees general service duties—clerks, radiomen, ammunition handlers, and segregated Coast Guard cutter crews—so long as they stayed off oceangoing ships except as messmen. Knox forwarded the Navy's recommendation to Roosevelt with the suggestion that he "announce boldly that nothing will be done to impair morale by introducing a racial problem in the Navy while the war is in progress."[22]

It was far less than black leaders had hoped for, and a good deal more than the admirals wanted. But with a war to win, it was enough for Roosevelt.

•   •   •

While Knox's admirals fought over the Negro problem in Washington, Stimson faced a Japanese problem on the West Coast. According to the 1940 U.S. census, 112,353 persons of Japanese ancestry lived in the three Pacific coast states. Of these, 71,484 were American-born sons and daughters of immigrants, or *Nisei*.[23]

Since Pearl Harbor, Major General John DeWitt of the Western Defense Command, headquartered in San Francisco, had been worried about sabotage. A single-minded man, DeWitt believed a number of Japanese individuals, some of them citizens, some not, had been acting as spies, fifth-column agents and saboteurs. In late January, Canada began evacuating Japanese from sensitive areas within British Columbia, and DeWitt asked the War Department to do the same for a long list of industrial and population centers, including San Diego, Los Angeles, San Francisco, and Puget Sound.[24]

DeWitt's request reached sympathetic ears. Two years earlier, Roosevelt had told Stimson that German spies had stolen U.S. plans for the Norden bombsight, and in 1941 Marshall informed Stimson of a plan by Italian saboteurs to wreck ships in American harbors.

To Stimson, the question of loyalty was a natural one. He had traveled to Japan and negotiated a naval treaty with the Emperor's ministers years before. He felt he knew the Japanese about as well as any Occidental, and he believed Japan would not hesitate to use fifth-column agents on the western seaboard. There had been no acts of sabotage—yet—and many young men of Japanese extraction were training for the war in Europe. But with the West Coast populated by so many Japanese, attacks against military and industrial targets were only a matter of time.[25]

"All along the West Coast, the presence of enemy aliens became a suddenly, sinisterly glaring fact," warned *Time* magazine. News services reported FBI raids on houses with signal flags and radio transmitters. Columnist Walter Lippmann wrote in the *Los Angeles Times*, "It is a fact that the Japanese navy has been reconnoitering the Pacific Coast. . . . It is a fact that communication takes place between the enemy at sea and enemy agents on land. These are facts which we shall ignore or minimize at our peril."[26]

Whatever the facts, flames were fanned by political opportunism. California Attorney General Earl Warren and Governor Culbert Olson asked General DeWitt to expel the Japanese from California, while congressmen from coastal states called for the evacuation of persons of Japanese lineage.[27]

The blend of war fever, racial prejudice, and uncertainty produced an anti-Japanese flash pot on the West Coast. A reporter for *Life* wrote, "Officials found themselves torn between a violent popular outcry for tough treatment of Japs, and perverse apathy toward the even more numerous and equally dangerous Germans and Italians." One *Nisei*, booted from his job as an architectural draftsman, lamented, "What really hurts is the constant reference to us as 'Japs.' 'Japs' are the guys we are fighting. We're on this side and we want to help. Why won't America let us?"[28]

Stimson would have felt better had there been, say, a riot or a violent strike by Japanese Americans or their white antagonists. That would justify federal intervention. But relocating 100,000 peaceful residents because of their race seemed repugnant to a lawyer who venerated the Constitution. He cautioned DeWitt not to commit the War Department to any kind of mass relocation without his personal approval.[29]

Then he began having second thoughts. To his diary he reminded himself, *"The people of the United States have made an enormous mistake in underestimating the Japanese."* With an elder's disdain for the younger generation, Stimson concluded that the greater threat came, ironically, from the American-born *Nisei* more than from their immigrant parents—who, he figured, were probably of better character than their spoiled children. On February 10 he revealed his deeply conflicted mind to his diary:

> *The second generation Japanese can only be evacuated either as part of a total evacuation, giving access to the areas only by permit, or by frankly trying to put them out on the ground that their racial characteristics are such that we cannot understand or trust even the citizen Japanese. The latter is the fact but I am afraid it will make a tremendous hole in our constitutional system to apply it.*[30]

Better, he concluded, to rip that "tremendous hole" in the Constitution than risk another catastrophic attack. He regretted driving innocent people from their homes, but in true emergencies—and there were few emergencies greater than the present one—individual rights must be subordinated to the common good. It was the same deplorable principle that compelled old men to pull fresh-faced young men from their families, put them in uniform, and send them to die.

With that, Stimson recommended the forced relocation of men, women, and children of Japanese extraction. On February 11 he telephoned the White House and put the question directly to Roosevelt. FDR told Stimson to take whatever action he thought best. Citizens or aliens, Japanese, German, or Italian—the War Department had carte blanche to remove whomever it wanted, as military necessity demanded. Roosevelt's only limitation was, "Be as reasonable as you can."[31]

Neither Roosevelt nor Stimson was quite comfortable with their decision. But, as Stimson reminded himself, the ancient Romans used to say, *"Inter arma silent leges."* Accepting that the law was at least quiet—if not actually silent—in times of war, Roosevelt tore open the constitutional hole. On February 19, he signed Executive Order 9066, authorizing the secretary of war to designate any area as a military zone and remove such persons as he may designate from those areas.[32]

The next day, Stimson gave DeWitt his orders: All of coastal California, half of Washington and Oregon, and southern Arizona were designated as military zones. Any Japanese found there were to be removed and sent to "relocation camps" in Rocky Mountain states.[33]

Roughly 110,000 Japanese would be subject to the War Department's orders, and the Army began building internment camps ringed with towers, fences and barbed wire. Japanese families began moving in early May, and how long they would remain in those rough tar-paper shacks depended on how fast Roosevelt and his warlords could win the war.*[34]

But to win the war, Marshall and King would first have to settle a fight between the Army and the Navy.

---

* Since removing all 140,000 Japanese residents from the Hawaiian Islands was impractical, given the shipping shortage, most Japanese Americans there were permitted to remain at liberty near Pearl Harbor and other Hawaiian military installations, while West Coast Japanese Americans, more than 2,000 miles from the war zone, were interned.

EIGHTEEN

ROLLING IN THE DEEP

A DMIRAL KING HAD BEEN KEPT WAITING, AND HE HATED BEING KEPT WAITING. He sat in Marshall's antechamber, watching one cigarette after another burn to the butt. He sat an intolerably long time—the sort of wait a commander forces on a subordinate to show him who is boss—and Ernie King sure as hell wasn't Marshall's subordinate. With a sour look, he got up and walked back to his office at the other end of the Main Navy–Munitions Building complex, leaving a terrified reception secretary in his wake.

Marshall didn't know Admiral King was waiting for him, for at that moment he was placating a bad-tempered Australian foreign minister named Herbert Evatt, who had been mouthing off in the press about MacArthur and Australia getting the short end of the war effort. Marshall wore down Evatt's fury through patience, diplomacy, and firmness, but not before King's own patience expired.[1]

King had first met Marshall at a football game back in 1938. Three years later, he gave some of Marshall's lackeys a good chewing-out, and since then the two men had butted heads. "I had trouble with King because he was always sore at everybody," Marshall later told an interviewer. "He was perpetually mean."

Touchy as King was about rank, sparks occasionally flew over accidental breaches of naval protocol. At Placentia Bay in 1941, King bitched out General Arnold for coming aboard his flagship without first asking permission. In early 1942, when Arnold's stenographer sent King a message incorrectly addressed to "Rear Admiral King," the letter came back unopened with a heavy arrow pointing to the word "Rear."[2]

King even bit back at Army compliments when he considered them presumptuous. Once, Major General Mark Clark sent King a telegram from England praising the work of a Navy captain on his planning staff and recommending the man's promotion to rear admiral. Marshall later told Clark, "Let me tell you what happened. King and I were having a terrible argument on something. We couldn't agree at all. I was trying to get him in a good humor to get him softened up for this thing and I thought I had, and then the door opened and he came stomping into my office, put the telegram on my desk and said, 'Read that.' I read it, and King said, 'Who the hell is Clark??'"[3]

Marshall's staffers loathed the admiral. In his diary, Lieutenant General Joseph Stilwell called him a *"high-powered 'rejector'"* who *"won't co-operate or listen or help."* Eisenhower groused to his diary *"One thing that might help win this war is to get someone to shoot King. He's the antithesis of cooperation, a deliberately rude person, which means he's a mental bully. . . . Of course Stark was just a nice old lady, but this fellow is going to cause a blow-up sooner or later, I'll bet a cookie."*[4]

Service rivalries died hard, putting a further strain on relations between soldiers and sailors. As Henry Stimson later observed, chauvinism began in the academies—West Point and Annapolis—and reached a fever pitch every year with the Army-Navy football game. Because their officers rarely interacted in peacetime, most generals and admirals lacked the familiarity that breaks down distrust. The Navy, Stimson remarked, "seemed to retire from the realm of logic into a dim religious world in which Neptune was God, Mahan his prophet, and the United States Navy the one true church."[5]

To Ernie King, George Marshall could be "a very agreeable man when he wanted to be," but he didn't always want to be. "Sometimes I think he's stupid and other times, very good," King later mused. "His basic trouble is that like all Army officers, he knows nothing about sea power and very little about air power." King considered Marshall the second-smartest man among the Joint Chiefs, after himself, but that did not mean the two headstrong men would see eye to eye. "We had many fights," he recalled sourly.[6]

And Marshall, in King's estimation, was about the best the Army had. Arnold, King claimed, "didn't know what he was talking about." King told his biographer that to form a sound strategy, "You have to use imagination and horse sense. It always seemed to me that the Navy was better equipped with strategical insight than the Army. The Air Corps didn't know a damn thing about it."[7]

After King marched out of Marshall's antechamber that March day, Marshall figured he and King needed to clear the air between them. Picturing King building up a four-pipe head of steam, he scurried the two blocks from the Munitions Building to Main Navy's entrance. He presented himself to King's gatekeeper, and the flag secretary showed him in with pointed formality.

Taking a seat opposite the scowling admiral, Marshall said he hadn't any intention of being discourteous. He explained that he was tangling with a difficult Australian, and couldn't show the man out when King had arrived.

Frowning, King said nothing.

"If you or I begin fighting at the very start of the war, what in the world will the public have to say about us?" Marshall asked. "They won't accept it for a minute. We can't afford to fight. So we ought to find a way to get along together."

King stared at him, saying nothing. It seemed a minute or two passed before he opened his mouth.

"You have been very magnanimous in coming over here the way you have," he said with heavy, clumsy formality. "We will see if we can get along, and I think we can."[8]

.   .   .

Getting along would not be easy. The Allies had agreed that offensive action would be directed against Hitler, not Tojo, until the Third Reich was slain. But as American, British, and Dutch territories fell to the samurai sword, King grew anxious to stabilize the crumbling Pacific line. That would require more than just sitting back and staving off Japan's blows. As he told Secretary Knox, "[You have to] hold what you've got and hit them where you can, the hitting to be done not only in seizing opportunities but in making them."[9]

Because every island that fell would carry a heavy cost in ships and blood to recapture, in mid-February the admiral floated the idea of occupying a few minor islands near Australia with Army troops. To Marshall, King's proposal sounded like an offensive in disguise. The Army was supposed to be moving east, not west, and he wrote King on February 24, "In general it would appear that our effort in the Southwest Pacific must for several reasons be limited to the strategic defensive for air and ground troops." He politely asked King to flesh out his ideas in detail, but reminded him of the strategy the Allies had set for the war: offense against Hitler, defense against Japan.[10]

On March 2, King asked Marshall to consider a Pacific strategy of hopping along a series of islands that sheltered Australia's northern coast: the New Hebrides, the Solomons, and the Bismarck Archipelago. King intended to use marines and naval air to assault the beaches and wrest the islands from their defenders.

Once the islands were secure, the Army would follow up for occupation duty. The capture of these islands, he said, would safeguard the great Allied base in Australia and pave the way for future operations against Japan.[11]

Nothing doing, thought Marshall. King was proposing a full-blown offensive in the Pacific while Hitler was master of Europe. Moreover, the idea of the Army following in the footsteps of the Marines, guarding ammo dumps and ports, was as repugnant to Marshall as it surely would be to MacArthur.[12]

To King, Marshall was dangerously short-sighted. The Allies could make a few vital inroads to cover Australia while still fighting the Germans to the east— and for the moment, they were not fighting the Germans anywhere much except in the Atlantic Ocean, King's other domain.

Three days after sending his plan to Marshall, King went directly to the president. In a memo to Roosevelt, he suggested that a limited offensive in the Pacific would be desirable. Australia and New Zealand, he reiterated, were the keys to the Pacific War—"'white men's countries' which it is essential that we shall not allow to be overrun by Japanese because of the repercussions among the non-white races of the world." To safeguard Australia and expand the Allied perimeter, King proposed opening a "very few" lines of advance (repeating the underlined words "very few" four times in his memo, in case they might somehow be overlooked). King's very few lines of advance would include a drive to the Bismarck and a push toward the South China Sea.[13]

Roosevelt brightened as he read King's proposal. He wanted to show the public tangible progress in the Pacific, and aggressive action would give Churchill's people hope for the Far East—a feeling as scarce these days as fresh eggs in London.[14]

A pleased Admiral King went back to his planners and told them to begin drafting an advance through the Central Pacific. He was careful not to give the Army any idea of what he was cooking up.

When the Japanese sliced the ABDA region in two, the Allies had agreed to put the Americans in charge of the war east of Singapore. But the vast Pacific theater, which ran from the China coast to California, could not possibly be managed by a single commander.

On March 7, Marshall proposed a plan drafted by General Eisenhower that divided the Pacific theater into two regions. Under Eisenhower's proposal, the Army would handle everything in the Southwest Pacific, from Australia to the Philippines, while the Navy would get everything else. The Army, not the Navy, would take responsibility for the drive up the Hebrides, Solomons, and Bismarck islands that King had staked out in his memorandum to Marshall four days earlier. MacArthur, Eisenhower suggested, would command whatever naval, marine, and air forces were required for this first island-hopping campaign.[15]

King blew up when he read Eisenhower's plan. The small islands near Australia and New Guinea, he argued, must be attacked and defended by the Navy, which possessed the only trained amphibious forces. He recognized that Australia, the larger Dutch islands, and New Guinea were big enough for the kind of land war the Army was used to fighting, so he suggested giving MacArthur Australia and the big islands. Nimitz must have the smaller islands, regardless of where they lay.[16]

King's strategic conceptions, while generally sound, often became lost in his strident delivery. But Marshall looked beyond King's buzz-saw personality and crafted a compromise. The Navy would have the New Hebrides and most smaller Pacific islands. MacArthur would take the Australia–New Guinea area. Marshall only asked that King slot the Philippines within MacArthur's jurisdiction, "for psychological reasons."[17]

King agreed, and on March 16 the Joint Chiefs settled the matter. Admiral Chester Nimitz would command the "Pacific Ocean Areas" as an Allied theater commander. The Pacific Ocean Areas would be divided into three regions: the Central Pacific, under Nimitz; the South Pacific, under Vice Admiral Robert Ghormley; and the North Pacific, under Rear Admiral Robert Theobald. MacArthur would run the Southwest Pacific, which began at 160 degrees longitude, near the Solomon Islands. He would command ground and air forces within his area, as well as any Navy units *temporarily* placed under his command by Admiral King.[18]

And everyone would take orders from the Joint Chiefs.

So they hoped.

*   *   *

"The only thing that ever really frightened me during the war was the U-boat peril," Winston Churchill later wrote.

Churchill had good reason to be frightened during the spring and summer of 1942, for the Battle of the Atlantic was going badly for the Allies. During the first three months of 1942, Hitler's roaming U-boats—only a dozen or so at any given time—sent 237 ships to the ocean floor. Their happiest hunting grounds shifted from the eastern Atlantic, near Britain and Iceland, to the U.S. coastline. From Maine to Miami, city lights threw out an incandescent invitation to U-boat captains, who were grateful to find clear silhouettes of unescorted oil tankers and merchantmen passing by.* The gray wolves slaughtered the lambs faster than the Allies could replace them, and the president and public demanded that something be done.[19]

What that something might be remained unclear, for there were simply not enough escort destroyers to go around. When convoys were organized in one area, wolf packs simply moved to other waters, such as the Gulf of Mexico, where hunting was safe and prey abundant. The first German submarine was not sunk until March 1942. By May, six months into the war, the Allies had lost more than 360 merchant ships; the U.S. Navy, Royal Navy, Army Air Forces, and Coast Guard managed to sink a combined total of eight subs in the western Atlantic and Caribbean. Unfortunately for King, Hitler's shipyards were turning out eight new subs about every ten days.[20]

In his office, his chart room, and his cabin at night, King's thoughts turned to the challenges of protecting ships carrying the provisions Britain and Russia needed to survive. The first problem, he well knew, was a simple shortage of escorts. Fixated on big capital ships, the prewar Navy had paid scant attention to small craft and subchasers. When war broke out it had no corvettes, few planes, and few destroyers—and those few ships and planes were stretched across the Iceland-to-Murmansk run, the Atlantic, three U.S. coasts, and the Pacific from Alaska to Australia. The only solution was to wait for more of those small but precious fighters to slide off the docks, and a system of escorted convoys would not be fully operational until mid-May.[21]

A second problem was a shortage of bombers. Heavy and medium bombers, with their long range, big payloads, and radar units, made excellent sub hunters. But bombers were owned by the Army, and Army pilots were not trained to patrol water and protect shipping. Army bombardiers might hit a large, stationary factory or bridge, but they couldn't hit a crash-diving submarine with a depth charge.

King intended to rectify that. He wrote to General Arnold and asked for 1,300 bombers, B-24 and B-25 type, to conduct coastal patrols in the United States and Australia. Arnold refused, telling King he didn't have the planes to spare. Besides, he said, the B-24 was a long-range, strategic bomber, and the Air

---

* When the Navy proposed dousing city lights, a cry went up from Atlantic City to Miami about ruining the tourist season. Eisenhower's War Plans Division also opposed blackouts, contending that few sinkings actually occurred in lighted stretches of coastline. Lights were not blacked out until April 18, 1942, after the material equivalent of three large war factories had been sunk.

Force was the proper branch to carry out long-range missions. Congress placed coastal air patrols in the hands of the Army, not the Navy, and that was where they belonged. If King wanted an antisub air group, the Air Force would supply the bombers, but they would remain under Air Force command.[22]

Stimson heartily agreed. The Army had spent years developing bases, training programs, and logistical infrastructure to support its bomber fleet. One could not simply hand bombers over to sailors and assume they could do the job competently, any more than airmen could pilot a warship. The most Stimson would approve was a temporary proposal to give "unity of command" authority to admirals of the Eastern Sea Frontier over the First Bomber Command. But direct command of the bombers would remain with the Army.[23]

The more Ernie King saw of Henry Stimson, the less he liked him. King conceded that Stimson was an able statesman, but when he got an idea into his head, he simply couldn't let it go.

"Stimson was more than firm," King groused years later. "He was stubborn. Marshall had a hell of a time with him, just as Lee had with Jefferson Davis. In fact, Lee managed Davis, but Marshall had great trouble with Stimson." King added, "I've been thinking many times about Secretary Stimson, and why he did not like the Navy. Probably he went to sea once and got in a storm. That's the way Hoover was. He went in a battleship to South America, got seasick, and never liked the Navy."[24]

Neither had King forgiven Stimson for protecting Marshall over Pearl Harbor, while Betty Stark took the fall. He felt Marshall was just as culpable as Stark, yet Stark was being mothballed while Marshall was running the land war.

"The Army did not really tell the exact truth about what happened there," King muttered in his later years. The unshakable feeling that Marshall had escaped punishment, partly through Stimson's intervention, galled King to no end.[25]

Turning his broadsides back to Arnold, King sent the air chief another lecture in what, for him, was a king's ransom of words:

All of us—no matter what uniform we wear—must go to work to win the war. I stand on the ground that whenever the use of land planes will enable Naval air units more effectively to perform their tasks, they should have land planes. Whenever the Army air forces can make use of torpedoes, dive-bombers, etc.—as developed and used by the Navy—they should be free to make use of them. . . . I think that it is high time the trend toward a separate air force should be given up—and that we face the realities of the situation with which we are confronted."[26]

* * *

The stalemate continued through the summer with no resolution. Trying to prod King into accepting an Army-run coastal air command, Marshall replied with a testy letter of his own:

> The losses by submarines off our Atlantic seaboard and the Caribbean now threaten our entire war effort. . . . Of the 74 ships allocated to the Army by the War Shipping Administration, 17 of them have already been sunk. Twenty-two percent of the Bauxite fleet has already been destroyed. Twenty percent of the Puerto Rican fleet has been lost. Tanker sinkings have been 3.5 percent per month of tonnage in use.
>
> We are aware of the very limited number of escort craft available, but has every conceivable improvised means been brought to bear on this situation? I am fearful that another month or two of this will so cripple our means of transport that we will be unable to bring sufficient men and planes to bear against the enemy in critical theaters to exercise a determining influence on the war.[27]

King countered with an elaborate defense of the Navy's efforts to stop the U-boat menace. In the end he admitted, however, "Our east coast system is far from invulnerable and we may expect the Germans to return to this area whenever they feel inclined to accept a not-too-heavy risk."[28]

As ships went down in the Atlantic, ownership of the bombers remained stuck in a tug-of-war between the Navy's top admiral and the secretary of war. In the Pacific, Admiral King was quietly planning his first spectacular blow of the war.

NINETEEN

SHARKS AND LIONS

∽

THE SEAS PITCHED AND TOSSED AS USS *HORNET* TURNED INTO FORTY-KNOT headwinds. The carrier's long, flat deck rolled, and officers doubted that the plump bombers lining the runway would even make it into the air. Nothing that size had ever been launched from a carrier, and sailors pictured themselves fishing half-drowned bomber crews from billowing waves.

The flight deck officers, their yellow jackets adding a festive look to the runway, took the measure of the ship's pitch and raised their arms to signal "go." Glancing out the bomber's side window, Lieutenant Colonel James Doolittle, the group's leader, pushed his throttle levers, and the two big engines roared. He released the brake, and plane, crew, and bombs lurched forward.

As it rolled to the deck's edge, Doolittle's B-25 waddled so sluggishly it seemed impossible that it could get airborne by the time the deck ended and the ocean began. Doolittle heaved back on the stick, and the twin-tailed plane yanked up, nearly stalled, then went nose down toward the water.

Then, inexplicably, the ten-ton monster clawed its way into the air. Like an ungainly bumblebee, it slowly gained altitude, and within an hour Doolittle was circling overhead with fifteen other bombers. The bumblebees, airborne now, flew west. Destination: Japan.[1]

The Doolittle raid was the child of Admiral King's bloodlust and the creative impulses of Captains Francis "Frog" Low and Donald Duncan, two of King's planners. The raid required the approval of Marshall and Hap, of course—the bombers belonged to the Army—and President Roosevelt had to authorize the mission. Knox and King cautioned Roosevelt not to ask too many questions, however, and he dutifully backed off. It would not be until the mission was well under way that more than seven people knew the complete details of what was going on.[2]

The bombers reached Japan's shores on the afternoon of April 18 and emptied their bays on Tokyo and five other cities. Oil refineries, an aircraft plant, and a Japanese ship went up in flames. Smoke billowed into the sky, and the bombers jogged through antiaircraft fire for the safety of China and Vladivostok.*

Back home, news of the "Doolittle Raid" electrified the public. After taking it on the chin for four months, America had landed a blow of its own. Headlines screamed, "AMERICAN PLANES BOMB TOKYO," and editorial cartoonists set their pens to drawing beefy American arms shoving oversized bombs down the throat of an apelike, bespectacled Tojo. Even Henry Stimson, a skeptic of the plan, admitted that the raid produced *"a very good psychological effect in the country both here and abroad."*[3]

The Doolittle Raid was the kind of dramatic statement Roosevelt loved to make, at least when there was nothing else for him to do. When asked by reporters where the bombers had been launched from, he gave them an inscrutable grin. Referring to the mythical realm in Hilton's *Lost Horizon*, he replied, "They came from our new secret base at Shangri-La!"[4]

. . . .

But Shangri-La, wherever it might be, was too far from Berlin to launch a Doolittle raid against Hitler. Before the ARCADIA meetings with the British, Marshall and Stimson had been so busy fighting fires in the Pacific, sending aid to America's allies, and mobilizing for war that they had little time to focus on *where* to fight Hitler. When the conference ended, Marshall ordered his planners to take a hard look at America's options. As they looked, British talk of Africa and the Middle East began to sour in their stomachs.

Marshall and Eisenhower believed the fastest way to win the war was to assemble overwhelming forces in the British Isles in 1942, then launch an invasion of northwestern France the next year. To men seeking the "decisive battle," that holy grail of military strategists, the logic was inescapable. *"We've got to go to Europe and fight,"* Brigadier General Eisenhower told his diary. *"We've got to quit wasting resources all over the world—and still worse—wasting time."*[5]

To carry the war to Germany, Eisenhower planned three operations, code-

---

* All sixteen planes crash-landed or the crews bailed out. The crews of two planes, ten men total, were drowned or captured. Three of the eight captured men were executed, one died in captivity, and four were freed at war's end.

named BOLERO, SLEDGEHAMMER, and ROUNDUP. BOLERO, like Ravel's symphonic score, was a steady buildup of land, air, and naval forces in the United Kingdom. SLEDGEHAMMER was an emergency invasion of the French coast in 1942, to draw German forces away from Stalin if it appeared the Red Army was going down for the count. ROUNDUP, to be launched in the spring of 1943, was the main invasion of France. After careful study, Marshall approved Eisenhower's proposals, and he presented them to Roosevelt at a White House luncheon on March 25, with Stimson, Knox, King, Hopkins, and Arnold listening supportively.[6]

As Roosevelt listened, he instinctively began tossing out possible operations in and around the Mediterranean. Marshall and Stimson pulled hard on the wheel to steer him back to France. *"He looked like a man going off on the wildest dispersion debauch,"* wrote Stimson that night. *"But, after he had toyed a while with the Middle East and the Mediterranean basin, which last he seemed to be charmed with, Marshall and I edged the discussion over to the Atlantic and held him there."*[7]

Stimson could see only one strategy. From his time as an artilleryman in France, and from his wide reading over the decades, Henry Stimson believed with the zeal of a religious convert that concentration of force (BOLERO), and violent execution (ROUNDUP), were the "proper and orthodox" means of fighting the Germans. Paper cuts in the Mediterranean would do nothing except siphon off men and ships needed for the smashing blow in France.[8]

Blunt force was the nature of Henry Stimson's personality. He lived life on a big scale. He reveled in vigorous exercise, traveled the world, and hunted big game. In arguments before court, he preferred the rhetorical sledgehammer to the rapier. When presenting a point to Roosevelt, he eschewed understatement and unleashed Verdun-like barrages of evidence and argument from lengthy, bombastic memoranda.[9]

To set ROUNDUP in motion, there were two rapids Stimson and Marshall would have to cross. The first was America's ally. The British had their fingers in a hundred colonial pies, and their eyes were drawn to India, the East Indies, the Middle East, the Mediterranean, and the connecting sea-lanes. Those regions protected His Britannic Majesty's interests, but they would not win the war. Stimson feared that Churchill and his generals, if not kept on a short leash, might fritter away Allied strength on interesting but indecisive theaters.[10]

The second problem was Roosevelt. FDR supported a decisive invasion of Northwest Europe, and in early 1941 he had told Marshall and Stark that "we must be ready to act with what we [have] available." But that was before Pearl Harbor. With casualties mounting in the Pacific and vital equipment scarce, Roosevelt dreaded a premature invasion that invited disaster. Like Churchill, he began looking for alternatives to a *battle royale* in 1942.[11]

⋆　　⋆　　⋆

On Stimson's advice, FDR ordered Hopkins and Marshall to fly to England, their mission being to win over the British to a European invasion. The two men, accompanied by a few of Marshall's staffers, arrived in London on April 8 and got down to business with the British high command.[12]

The general who would be negotiating strategy with Marshall was General Sir Alan Brooke, Chief of the Imperial General Staff. Four years younger than Marshall, Brooke had commanded an artillery unit in the First World War and a corps during the Second. He stood slightly shorter than Marshall, and Brooke's narrow face was dominated by large owl-like eyes, a heavy, patrician nose, and a frown of the sort that might be found on some rare and exotic bird, such as a gray hornbill or a self-confident bald ibis. Nicknamed "Colonel Shrapnel" behind his back by intimidated staffers, Brooke was abrupt and somewhat rude by nature. His personality had been hardened by a long military lineage, the death of his wife in a car accident, and the recent horrors of Dunkirk. When pressed in conference, he would fire back with the rapid cadence of a Vickers gun and overwhelm his opponent with facts and figures, not all of which were necessarily germane to the dispute.[13]

Colonel Shrapnel held little regard for the Johnny-come-lately Yanks who knew nothing of the German fighting man, and his first impression of Marshall was of a pleasant man lacking military sense. As Brooke later remarked in his diary, Marshall *"is, I should think, a good general at raising armies and providing the necessary links between the military and political worlds. But his strategical ability does not impress me at all!!! In fact, in many respects he is a very dangerous man whilst being a very charming one!"*[14]

When Marshall outlined American plans to Britain's Defence Committee on April 9, he ran into fierce opposition to SLEDGEHAMMER, a risky operation for which the British would foot most of the bill. Having buried their dead in the Balkans, the Middle East, France, Belgium, and the Orient, British ministers were in no mood to pay the ferryman again.

The British view, as Marshall might have expected, was both global and flexible. To the Britons, Northwest Europe was a heavily defended corner of a much larger battlefield. In peripheral theaters like the Near East, a region anchored by Tobruk, in Libya, and Alexandria, in Egypt, the Allies could drain manpower from Hitler—much as the Spanish and Portuguese, aided by British gold, had drained the blood of Napoleon's Grande Armée 130 years before.

Complicating perceptions on both sides was the British experience in the First World War. Harking back to the slaughter of 1915, Lord Moran told Marshall the Americans were fighting the ghosts of the Somme. It became an article of faith in American circles that the last war was dictating Britain's strategy for the present one.[15]

But the British were not just plowing Flanders fields anew. In their darkest hours of 1940, when Hitler had threatened England with invasion, they had considered every facet of a Channel crossing from the receiving end. As they planned air, sea, and land defenses against Hitler's landing barges, His Majesty's subalterns developed a keen appreciation of the naval and air obstacles to a cross-Channel assault. Those obstacles, which had protected their island nation, worked in both directions.

Vice Admiral Lord Louis Mountbatten, Britain's combined operations chief, told Marshall the Allies lacked enough landing craft to deliver sufficient troops. Air Chief Marshal Portal cautioned that the RAF could not provide adequate air cover. SLEDGEHAMMER was, in the British view, an impossibility for 1942.[16]

They appeared much more sympathetic toward ROUNDUP, which would not take place until the spring of 1943. With Churchill taking the lead, the British war chiefs said they were in agreement with the Americans about ROUNDUP as the centerpiece for the next year, and they asked the Americans to proceed energetically with the BOLERO buildup in England. At a meeting with the War Cabinet on April 14, Churchill spoke glowingly of a cross-Channel effort and said nothing about North Africa. The nations, he said, would "march ahead in a noble brotherhood of arms." Churchill's burly military adviser, General Hastings "Pug" Ismay, recalled, "We were all rather carried away by the idea of millions of Americans falling into England and charging into the Channel and I thought that Winston was carried away emotionally with this great brotherhood in arms."[17]

Delighted at the emphatic British support, Marshall cabled Stimson that the prime minister had declared himself in complete agreement with the plan. The year 1943 would see the great invasion of Hitler's continental fortress.[18]

Yet lying behind the agreeable British facade was fear that a more candid face would drive Marshall and King toward the Pacific. The brothers in arms therefore gave Marshall what he wanted and sent him home to ship men, weapons, and aircraft to England. "We knew pretty well there wasn't a hope in hell of ROUNDUP, but we didn't say so," Ismay admitted later. "I think Marshall and Hopkins went back feeling that we were all sold on it."

This impression, he added with British understatement, would have "unfortunate consequences."[19]

·   ·   ·

While Marshall argued strategy in the Atlantic, King's intelligence analysts deduced that a Japanese invasion force was preparing to hit Port Moresby, a major

Allied port on New Guinea's southern coast near Australia. Moresby, they concluded, was a logical next step for Japan, for its capture would allow Admiral Yamamoto to threaten northern Australia by sea and air.

In mid-April, King ordered Admiral Nimitz to meet him in San Francisco at Twelfth Naval District Headquarters for a strategy conference. On April 25 the two admirals talked there for a few hours, then continued their discussion over lunch at San Francisco's Bohemian Club. Figuring Yamamoto would launch his attack around May 3, Nimitz proposed hitting him somewhere in the Coral Sea, off Australia's northeast coast. Vice Admiral William Halsey's carriers *Enterprise* and *Hornet*, returning from the Doolittle Raid, would steam southwest to join forces with Rear Admiral Frank Jack Fletcher, commanding the carriers *Yorktown* and *Lexington*. Once the enemy was spotted, Fletcher and Halsey would give battle.[20]

King was uneasy about the plan. He liked Nimitz but didn't entirely trust him. The Texan was a solid, marlinspike officer, but as a former Bureau of Navigation man, he was also the sort of officer King called a "fixer" or "trimmer"—a man willing to compromise, one who trimmed his sails and avoided the hard way that King instinctively sought. "Damn, if I could only keep him tight on what he's supposed to do. Somebody gets a hold of him," he told a lady friend.

King didn't trust Nimitz to stay the strategic course, and he didn't want Nimitz courting serious risks with the Navy's few precious aircraft carriers. Before leaving on trips to San Francisco, he would occasionally grouse, "Have to fly out and straighten Nimitz out again."[21]

In late April it was unclear to King whether Halsey would arrive at the Coral Sea in time to tip the odds in the Navy's favor. King suggested that Nimitz add some heavy battleships to his task force, just in case, but Nimitz brushed off the suggestion. Battleships, he reasoned, would be too slow, too vulnerable to carrier attacks, and too difficult to keep fueled and provisioned. He would engage the enemy with carriers alone.

King flew back to Washington. With some reservations, on April 27 he approved Nimitz's plan.[22]

In a confused carrier engagement in the Coral Sea on May 7 and 8, Fletcher's task force struck Admiral Shigeyoshi Inoue's invasion fleet as it approached Port Moresby. Torpedo and dive-bombers from *Yorktown* and *Lexington* tore down on Inoue's carrier spearhead. They sank one light carrier, *Shōho*, and badly damaged another, the heavy carrier *Shōkaku*. American fighters shot up the air complement of a third carrier, *Zuikaku*, so badly that she would be out of action for two months.

It was a muddled result at best, but the Battle of the Coral Sea gave the Americans something to celebrate: a victory. The invasion fleet had been turned back, and for the first time since the war began, the U.S. Navy had sunk a Japanese ship larger than a destroyer.[23]

The Americans did not come away unscathed, however. King's beloved carrier, USS *Lexington*, had been set upon by Japanese torpedo bombers and took two tin fish in her side. She survived the battle, briefly, but the Americans were forced to divert her planes to *Yorktown* and abandon ship. With a heavy heart, Admiral Fletcher sent the broken thoroughbred to the ocean floor with five torpedoes. *Yorktown,* her flight deck pierced by heavy bombs, limped home to Pearl Harbor to repair the damage.[24]

Smarting over *Lexington*'s loss, and anxious to keep the news from Yamamoto, King clamped down on word of the carrier's fate. He threw a tight news blackout over the sinking, and even concealed the loss from the British; he told Admiral Pound only that *Lexington* had been "damaged" during the battle.[25]

By mid-May, King was growing nervous about the Pacific situation in general and his fleet carriers in particular. Japan had nearly twice as many as he did, and the next big American flattop under construction would not have a champagne bottle broken across her bow until December. On May 14, he sent Nimitz a cable again urging caution: "Loss of Lexington represents one fifth our carrier strength in Pacific. . . . At present stage of our carrier building program we cannot afford to swap losses with this ratio."[26]

Because Yamamoto still held the upper hand, King's intelligence staff believed he would again strike toward the South Pacific—if not at Port Moresby, then at New Caledonia or the Fijis. King wired Nimitz to suggest that he put *Lexington*'s surviving planes ashore in Australia and Hawaii and pull his ships under the protective umbrella of land-based fighters. On May 17 he advised Nimitz to employ "strong attritional tactics and not repeat not allow our forces to accept such decisive action as would be likely to incur heavy losses in our carriers and cruisers."[27]

But Nimitz, like Yamamoto, was a poker player willing to gamble when he held better than two pair. His codebreakers at Pearl Harbor, laboriously decrypting the Japanese Navy's JN-25 code, convinced Nimitz that Yamamoto's target was Midway. To the Texan, that foreknowledge was better than three jacks and a king.

Nimitz acknowledged that the Aleutians, Moresby, or Oahu would be left naked if his carriers were concentrated around Midway. But he hadn't enough ships to defend every place at once, so hard choices had to be made. He placed great faith in his intelligence men, and as the Japanese fleet left its home waters in late May, Nimitz leaned over the table and put all his chips on Midway.[28]

So did Yamamoto. As a small island nation, Japan could not afford "attritional tactics" against a nation with oil, steel and prodigious industrial capacity. As King urged caution on Nimitz, Yamamoto's air fleet, centered around four large carriers, covered a massive invasion force steaming east. His plan was a three-pronged gamble: A decoy task force would stab at Alaska's Aleutian Islands; his battleship and transport armada would capture the American base at Midway; and a carrier force would destroy the American fleet from the north.[29]

Shortly after leaving their bases, the Japanese fleets went silent, just as before Pearl Harbor. The men of Main Navy had no idea where they were, and Nimitz had only three carriers, including the wounded *Yorktown*. But he had land-based air support from Midway, and he was confident he had read Yamamoto's cards correctly. "We are actively preparing to greet our expected visitors with the reception they deserve," he radioed King on May 29.[30]

As the Americans steamed west to meet the world's mightiest fleet, King could only hope Nimitz was right.

TWENTY

"LIGHTS OF PERVERTED SCIENCE"

ᔟ

HENRY STIMSON WAS A THROWBACK TO THE AMERICA OF 1900. THE SON OF A robber baron banker, he distrusted much of the newfangled world of the 1940s. As he once consoled an old cavalry officer who wrote to complain of the mechanization of the horse service: "[Some of my] oldest and choicest recollections are pervaded with the smell of horse sweat and saddlery. . . . Nearly all of us have to see some of the things we love remorselessly replaced in the modern mechanized world—a world that is just as repellent to me as I think it is to you."[1]

But wealth and social connections brought Stimson a wider perspective than most men, and at Highhold and Woodley he often reflected on scientific advances of the "modern mechanized world." His cousin Alfred Loomis, a patron of electronic and nuclear research, regaled him with advances in navigation technology, sound waves, and physics over after-dinner conversations at Highhold, or at the Loomis laboratory at Tuxedo Park. Through their talks, Stimson took an immediate interest in developments like radar, which were changing the way wars would be fought. And one project caught his eye like no other.

Shortly after war broke out in Europe, FDR appointed an "Advisory Committee on Uranium" to follow up on some theories suggested by physicists Albert Einstein and Leó Szilárd. The committee's purpose was the development of a superbomb fueled by fissile uranium, an explosive potentially far more powerful than combustibles like TNT or black powder. Such a weapon, scientists theorized, could destroy a harbor, perhaps even level an entire city. Nuclear science wouldn't exactly throw open the gates of hell, but it might pry them far enough apart to defeat Hitler.

There were a lot of "ifs" to the concept, and to many in uniform it seemed like a Buck Rogers idea. But Roosevelt dreaded a world in which the Nazis developed an atomic superweapon before the Allies did. Four years earlier, German scientists had split the atom, and Germany boasted some of the world's most respected minds in the field of theoretical physics. In a speech to the House of Commons, Winston Churchill had warned of a Nazi "Dark Age," made all the more sinister by the "lights of perverted science." Perverted or not, Roosevelt intended to have that light switch in Allied hands before Hitler could find it.

Nine days after Pearl Harbor, Roosevelt established a "Top Policy Committee" to organize the development of a nuclear weapon, referred to in hushed whispers as the "S-1" section. Three months later, the committee handed S-1 to the Army Corps of Engineers, which was well organized for the massive infrastructure and manpower needs the project would require.

Marshall oversaw the creation of a new headquarters with the innocuous name of "Manhattan District." He placed the "District" under command of Colonel Leslie J. Groves, an abrasive, tenacious engineer who had been the driving force behind the Army's massive pentagonal office building in Arlington, still under construction.[2]

S-1's cost mushroomed like a splitting atom as entire towns were built to house, feed, and provide laboratories for scientists, construction workers, and their families. Moving the project under Army jurisdiction put a tremendous strain on Marshall's funding, and since S-1 was one of the war's two most important secrets—the other being Allied decryption of enemy messages—Marshall could not simply tell Congress why he needed massive new funds. He would have to look under every War Department sofa cushion to find money he could divert for the Manhattan project.[3]

As Manhattan's costs ballooned, Marshall found he could push creative accounting only so far. "I obtained the first money," he said, "by taking twenty percent of the appropriations concerning such matters as the development of bombs, artillery and kindred matters—which was legal." As the project grew in size and complexity, however, the voracious baby required hundreds of millions of dollars, not merely millions. Marshall would have to find new ways to fund a project no one was supposed to know about.[4]

Stimson's credibility with Congress proved a godsend. He telephoned Senator Harry Truman, the committee chairman investigating Army-Navy waste, and asked him to back off an investigation into two vaguely described war plants. "That's a matter which I know all about personally, and I am one of the group of two or three men in the whole world who know about it," he told Truman.

"I see," said the senator.

"It's part of a very important secret development."

"Well, all right then—"

"And I—"

"I herewith see the situation, Mr. Secretary," Truman interrupted. "You won't have to say another word to me. Whenever you say that to me, that's all I want to hear."

"All right," said Stimson.[5]

He hung up on cordial terms with the senator, and Marshall and Stimson breathed another sigh of relief. Truman, they felt, would meddle no further. Thanks to Stimson's straight-shooting diplomacy, Senator Truman would have no reason to know about the atomic bomb's existence until the blast appeared in the papers.

⋆ ⋆ ⋆

As Groves and his scientists worked on one unconventional weapon, George Marshall set men to work on a very different one. The Army was swelling with draftees, and few of those citizen soldiers would have the natural toughness, the moral indifference—the hatred—to slaughter their way to Berlin or Tokyo.

Enlisted morale was usually left to unit commanders, but those officers were not usually inspiring speakers. Patton was colorful, but so crude that his message often got lost in translation. MacArthur was too aloof to fraternize with his men, and the lower levels—company-grades on whom the assignment of motivating men got dumped—were usually dull, inarticulate, or so bombastic that the men didn't take them seriously. Since becoming assistant chief of staff, George Marshall had kept a careful watch on complaints from the rank and file, and pointless speeches was one gripe he had been hearing since the first months of the draft.[6]

The nation had plenty of Hollywood artists who could stir emotions on a mass scale, and Marshall found ways to integrate directors, producers, actors, and studio heads into the war effort. One of these men, a director named Frank Capra, had signed up for the Army shortly after Pearl Harbor. The ticket-buying public loved his films; *Mr. Smith Goes to Washington*, *Lost Horizon*, and the Oscar-winning *It Happened One Night* were huge box office successes. So not long after Mr. Capra of the Screen Directors Guild became Major Capra of the Army Signal Corps, Marshall called him into his office for an hour-long talk.

"Within a short time," Marshall said, his blue eyes locked on the director, "we will have a huge citizens' army in which civilians will outnumber professional soldiers by some fifty to one." The unknown, he told Capra, was whether those citizens would fight with the same ferocity as Hitler's supermen and Tojo's samurai.

To Marshall the answer was obvious. "Young Americans, and young men of all free countries, are used to doing and thinking for themselves," he said. "They will prove not only equal, but superior to totalitarian soldiers *if*—and this is a large *if*, indeed—they are given answers as to *why* they are in uniform, and *if* the answers they get are worth fighting and dying for."

Marshall asked Capra to create a documentary series called *Why We Are in the War*. What he wanted, he said, was a film every soldier entering basic training could watch. He wanted it interesting, accurate, and informative.

The short, handsome director tightened up as he listened to Marshall stress the importance of the project. He was a feature filmmaker, a storyteller, not a propagandist or a historian, and he was being asked to shape the attitudes of millions of trainees. "General Marshall," he said at last, "I think it's only fair to tell you that I have never before made a single documentary film."

"Capra," said Marshall, "I have never been chief of staff before. Thousands of young Americans have never had their legs shot off before. Boys are commanding ships today, who a year ago had never seen the ocean before."

In Marshall's world, precedent meant nothing. Every man and every woman in the war effort were doing things they had never done before, for the sole reason that their country needed them to do it. Capra understood. Infused with the zeal of a new project, he promised to make Marshall "the best damned documentary films ever made."[7]

Capra proved as good as his word. His film *Prelude to War*, the first installment in a seven-episode series titled *Why We Fight*, skillfully wove in clips of Nazi and Japanese propaganda films. Featuring the narrative talents of screen icon Walter Huston, *Prelude to War* showed Marshall's soldiers in bold terms what they were fighting against. The movie included short cameos by FDR and Stimson and concluded with a stirring statement by General Marshall about America's war aims.

The movie was an instant hit with the troops, and Stimson delighted in giving advance screenings of Capra's films to important War Department visitors as each new installment was finished. Released to the general public in 1943, *Prelude to War* won an Academy Award for the year's best documentary.[8]

With the *Why We Fight* series, Marshall enlisted the popular media as an ally in the cause of victory. But as Yamamoto's fleet sailed toward Midway, Admiral King was about to learn that the popular media could threaten his war in the Pacific.

TWENTY-ONE

MIDWAY'S GLOW

∾

"U.S. KNEW ALL ABOUT JAP FLEET, GUESSED THERE WOULD BE A FEINT AT ONE BASE, REAL ATTACK AT ANOTHER."

So SCREAMED THE *WASHINGTON TIMES-HERALD*'S HEADLINE FOR SUNDAY, June 7, 1942. In a two-day battle, the newspaper reported, Nimitz's planes sank two or three Japanese carriers. The article claimed the Navy's admirals had known the strength of Yamamoto's battle fleet days before his ships converged on Midway Island. That invasion fleet was turned back, and the Americans had won a "momentous victory" in the Pacific.[1]

The *Times-Herald* article was generally accurate, which was why King exploded like a powder magazine when he picked up his Sunday paper. "U.S. KNEW ALL ABOUT JAP FLEET" pointed everyone—American and Japanese—to but one conclusion: the Navy's cryptographers had broken JN-25.

In all his years in uniform, King had never seen such treasonable publication of American military secrets—in this case, the most vital secret of the Pacific War. That very day he held a rare press conference and misled journalists about the source of the Navy's prognostication. He told them that since the Doolittle Raid and the Battle of the Coral Sea, the Navy had been expecting a Japanese counterattack, to "save face." Midway seemed the most likely target, which was why Nimitz's carriers had been ready for them.[2]

He then spoke off the record in a voice that told every reporter there would be hell to pay. The *Times-Herald*'s report, he said, "compromises a vital and secret source of information which will henceforth be closed to us. The military consequences are so obvious that I do not need to dwell on them, nor to request you be on your guard against, even inadvertently, being a party to any disclosure which will give aid and comfort to the enemy."[3]

Someone sure as hell gave aid and comfort to the enemy, and King figured that someone was Robert McCormick, owner of the *Chicago Daily Tribune*, a sister publication of the *Times-Herald*. The *Tribune* ran the same Midway story under the even more treasonable headline "NAVY HAD WORD OF JAP PLAN TO STRIKE AT SEA." McCormick, fumed one of King's staffers, was a "goddamn traitor. We're going to hang this guy higher than Haman."[4]

All King needed was a rope, a foreyard, and a drumhead court. With the blessing of Secretary Knox—whose newspaper, the *Chicago Daily News*, was the *Tribune*'s archrival—King launched an investigation that quickly uncovered the source of the leak. The late *Lexington*'s executive officer, Commander Morton Seligman, inadvertently gave access to the Japanese order of battle to a Mc-Cormick reporter, who used the information in a short article on the Japanese fleet. A *Tribune* editor who knew nothing of the Navy's codebreaking work decided to run the story under an eye-grabbing headline suggesting—correctly, as it turned out—that the Navy was decoding Japan's naval signals.[5]

FDR's attorney general, the affable-looking Francis Biddle, prepared an indictment under the Espionage Act of 1917. But when the decision was put to King, the admiral reluctantly concluded that the Navy could not afford to provide prosecutors with evidence, for fear of leaking additional secrets to the enemy through a public trial. McCormick and his employees would escape the wrath of Roosevelt, Knox, and King.*[6]

· · · ·

Victory at Midway gave the Americans some desperately needed breathing room. In a battle decided by luck as much as skill, Yamamoto lost the carriers *Akagi*, *Kaga*, *Soryu*, and *Hiryu*, cutting his mobile airpower in half. The United States lost the old *Yorktown*, but with the new *Essex*-class carriers in production on East Coast shipyards, the U.S. Navy was far better equipped to make good its losses.

---

* FDR considered other ways of shutting down the *Tribune*, such as cutting off its paper supply from Canada. Later, when he learned that the *Tribune* applied for a license to circulate a newspaper for American servicemen in England, Roosevelt cabled his ambassador to London, "I have wired Former Naval Person [Churchill] expressing earnest hope that application of Chicago Tribune will not be granted. I have told him that this so-called newspaper prints lies and misrepresentations in lieu of news." Churchill, who had suppressed publication of the *Communist Daily Worker* the year before, happily denied the *Tribune*'s application.

As Midway's echoes died and the battle's outcome became clear, Marshall and King knew that Nimitz's masterstroke had opened a road that would end in Tokyo with Japan's surrender.

But they needed to agree on where that road began.

The islands of the Pacific dot the ocean like constellations, some in tight clusters and others far removed from their neighbors. With Tokyo as the ultimate goal of the Pacific war, two island strands beckoned to American planners.

The first strand runs from Australia through the Southwestern Pacific. It begins with New Guinea and the Solomon, Bismarck, and Admiralty islands, which lie along New Guinea's northeastern coast. From the Admiralties, the path banks north to the Philippines, then to the Chinese island of Formosa (called "Taiwan" by the locals), then finally to the Japanese Home Islands. With a growing ground and air force under MacArthur in Australia, the southwestern route seemed the surest way to move a large land force into striking position against Japan. It would also be the more costly route, because the Americans would first have to neutralize the imposing Japanese base at Rabaul, on the Bismarck island chain.

The second strand runs through the Central Pacific across small islands suitable only as air and naval bases. It stretches from Hawaii to the Gilbert and Marshall Islands, the Caroline Islands, and the Mariana Islands of Guam, Saipan, and Tinian. Using the Marianas as a base, Nimitz could project America's sea and air power against the Philippines, Formosa, or Japan itself. Like the southwestern route, an imposing enemy naval base stood in the way—the island of Truk, between the Marshall and Caroline chains.

This second approach appealed to Admiral King, as smaller islands of the Central Pacific could be taken by marines until they reached the big islands of the Philippines, Formosa, or Japan; they would need the Army's help for those. As King saw it, the Navy could take the smaller islands far more cheaply than the Army could slug its way through New Guinea. Many enemy strongpoints could be bypassed, isolated from their supplies, and starved into submission.[7]

On June 8, General MacArthur unveiled his plan to capture Rabaul, the Japanese citadel bristling with six airfields, an immense naval anchorage, hundreds of fighters, and 100,000 defenders. Rabaul's capture would eliminate the single greatest Japanese threat in MacArthur's theater and clear the way for his advance north to the Philippines. All he needed, he told Marshall, was an amphibious marine division to take the beaches, and the loan of two carriers for air support.[8]

For Marshall, Pacific strategy was a dicey balancing act. His heart was with MacArthur, but the Navy would supply the transport and escort ships. He also

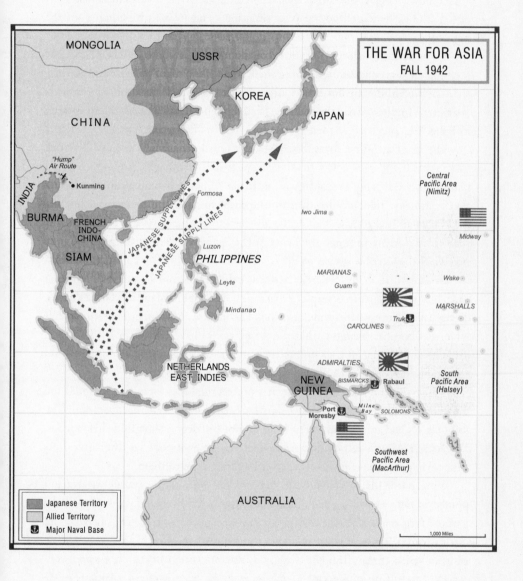

THE WAR FOR ASIA
FALL 1942

MONGOLIA

USSR

CHINA

KOREA

JAPAN

Central
Pacific Area
(Nimitz)

"Hump"
Air Route

INDIA

● Kunming

BURMA

FRENCH
INDO-
CHINA

Formosa

Iwo Jima

Midway

SIAM

JAPANESE SUPPLY LINES

Luzon

PHILIPPINES

MARIANAS

Guam

Wake

JAPANESE SUPPLY LINES

Leyte

MARSHALLS

Mindanao

CAROLINES

Truk ⚓

NETHERLANDS
EAST INDIES

ADMIRALTIES

NEW
GUINEA

BISMARCKS ⚓ Rabaul

South
Pacific Area
(Halsey)

*Milne
Bay*

SOLOMONS

Port
Moresby ⚓

Southwest
Pacific Area
(MacArthur)

AUSTRALIA

Japanese Territory
Allied Territory
⚓ Major Naval Base

1,000 Miles

knew he would get nowhere with the Navy unless MacArthur kept his mouth shut while Marshall worked out a deal with King. On June 10 he cautioned MacArthur, "Until I have had the opportunity to break ground with the Navy and British [here] please consider all this personally confidential, not discussing it at present with Navy, British, or Australian officials." MacArthur's reply acknowledged "the extreme delicacy of your situation," and he held his tongue for the moment.[9]

Those delicate negotiations with King hit a snag the next day, when King told Marshall he was considering landings he knew MacArthur would object to. King characterized his forthcoming operations as "primarily of a naval and amphibious character," implying that the first big Pacific offensive would have an admiral at its helm.[10]

The next day Marshall endorsed MacArthur's Rabaul plan.[11]

Considering the Navy's record thus far, Ernie King felt he had a right to dictate Pacific strategy. The Navy had studied the problem of Japan for twenty years, and the Navy had backed Marshall's calls for the ROUNDUP invasion of France. It was time for the Army to repay the favor. If the Army didn't want to play ball, the Navy had the forces to go it alone.[12]

In late June, King ordered Nimitz to begin planning operations against Tulagi, a small Japanese seaplane base near the larger island of Guadalcanal. Tulagi lay slightly within the Southwest Pacific area, MacArthur's territory, but King told Nimitz to assume that the Army would have no role in the landings; Tulagi would be a Navy show. He then informed Marshall that Nimitz would command the conquest of the Eastern Solomons.[13]

Marshall, who had assured MacArthur that any attack on Tulagi would fall under Army command, insisted that MacArthur must lead the operation. The Joint Chiefs had agreed that everything west of 160 degrees longitude belonged to MacArthur's Southwest Pacific Area; because the Solomons lay almost entirely west of that imaginary line, MacArthur had first claim to them.[14]

King pushed back violently. In his logic-driven mind, a lawyer's agreement pinned to degrees of longitude meant nothing; what mattered, he said, was the nature of the operation. The Tulagi landing was amphibious, so the Navy should run the campaign, regardless of where the islands happened to lie. To this King added a veiled threat. The Navy, he said, was prepared to take the Eastern Solomons, "even if no support of Army forces in the Southwestern Pacific is made available." He wanted those islands, and he would have them.[15]

While King argued with Marshall from Main Navy, MacArthur pushed him from distant Brisbane. He warned Marshall that any assault on Tulagi posed a grave risk if it were not combined with a sister operation to neutralize Rabaul, which would naturally send air and sea forces to Tulagi's defense. Not content to let military logic decide command questions, he also peppered Marshall with warnings of a sinister naval conspiracy against the Army. "By using Army troops to garrison the islands of the Pacific under Navy command," he wrote, "the Navy retains Marine forces always available, giving them inherently an army of their own." He claimed that during the Hoover administration, when he was chief of staff, he had learned of a Navy plan to win "the complete absorption of the national defense function" by relegating the Army to basic training, garrison forces, and supply. King's power grab for operations in the Pacific, he implied, was another brick in that intolerable wall.[16]

A weary Marshall took stock of both sides' positions. The Army had the ground and air forces, but King had the ships. This pointed to an eventual compromise. Statesmen when they had to be, the two chiefs agreed to split the offensive and give both Nimitz and MacArthur a chance to run the show. Admiral Ghormley, Nimitz's South Pacific subordinate, would direct "Task One" of the offensive, against Tulagi, the nearby island of Santa Cruz, and any surrounding islands necessary for Tulagi's capture. MacArthur would play a supporting role by providing air cover and garrison troops, just as King had asked.

Once Tulagi had been taken, Ghormley would hand over the reins to MacArthur for "Task Two," the western Solomons and northeast New Guinea. Once those stepping-stones were crossed, MacArthur would get his shot at Rabaul, designated "Task Three." All orders, King suggested, should come from the Joint Chiefs, "whose authority cannot be questioned by either principal—General MacArthur on the one hand or Admiral Nimitz on the other."[17]

The compromise was the best Marshall could do. To work around the geographic boundary problem, King and Marshall agreed to trim MacArthur's theater by one degree—sixty nautical miles—to put Tulagi within Ghormley's territory.[18]

Marshall believed it was time for the Army to reward the Navy's willingness to compromise by supporting Ghormley wholeheartedly in the fight for Tulagi. He told MacArthur he expected him to "make every conceivable effort to promote a complete accord throughout this affair." He acknowledged there would be irritations and minor problems with Army-Navy cooperation, but in view of the importance of the mission, he said, MacArthur must make every effort to suppress his team's disgruntlement and cooperate cheerfully with the Navy.[19]

The peace brokered by Marshall and King began unraveling almost as soon as it was settled, but strangely enough, it was not for reasons of Army-Navy jealousies.

Ghormley and MacArthur, taking a closer look at Japanese defenses, gulped hard when they considered what they were up against. After calculating the response from Rabaul if they landed on Tulagi—which MacArthur had warned about—they both expressed "the gravest of doubts" over prospects of success. Ghormley and MacArthur recommended postponing the operation.[20]

King was furious with Ghormley. The Navy had just beaten the enemy to a pulp at Midway, and King had used a great deal of clout with the Army to put Ghormley in command of the Tulagi invasion. Now Ghormley was giving the Army reason to doubt the Navy's capabilities.

He also roundly cursed MacArthur, whom he suspected of giving up on Tulagi because he couldn't run the whole show. "[MacArthur] could not understand that he was not to manage everything," King later frumped to an interviewer. "He couldn't believe that. Of course he was absolutely against going into Guadalcanal, and he said so."

With a sneer King reminded Marshall that three weeks earlier MacArthur had boasted that he could run all the way to Rabaul with a couple of carriers. Now, having looked at a map, MacArthur seemed unsure whether he, Ghormley, and Nimitz could take a few small, badly supplied islands thousands of miles from Tokyo.[21]

In the first week of July, as orders for Tulagi were being distributed, the picture shifted dramatically. Word reached Washington that the Japanese had landed on Guadalcanal, Tulagi's next-door neighbor, and had begun constructing an airfield that would play hell with MacArthur's men. It might even sever the supply lines connecting the United States with Australia. As coolies staked out and graded Guadalcanal's rough runway, Admiral King realized that Guadalcanal would be as important as Tulagi, perhaps more so.

It was a race against time, and King grew frantic with Ghormley for not pushing his men faster. An airfield on Guadalcanal would be a disaster, and Ghormley was dragging his heels. *Now*, King bellowed, was the time to press the advantage of Midway.[22]

TWENTY-TWO

"THE BURNED CHILD DREADS FIRE"

⌇

WHEN MARSHALL AND HOPKINS FLEW TO LONDON SIX WEEKS EARLIER, Churchill and his War Cabinet embraced the BOLERO buildup and the ROUNDUP invasion for 1943. Or so Marshall thought when he flew home.

But he couldn't shake a nagging feeling that Churchill's assurances were not what they had seemed. Before long, FDR received a cable from the prime minister referring to "complications" with ROUNDUP that Admiral Mountbatten would explain to him in person. In his message, Churchill returned to his North African theme, insisting the Allies "must never let GYMNAST pass from our minds." To Marshall, that kind of strategic thinking didn't sound good.[1]

Roosevelt had been doing some strategic thinking of his own. Since December 1941, when the Red Army turned back the Wehrmacht from the gates of Moscow, the Russian front had become the war's great slaughterhouse, a vast, corpse-strewn wasteland where Germans and their satellite soldiers were being butchered in stomach-churning numbers.

They were butchering Russians in stomach-churning numbers, too, and keeping Stalin from being forced into a separate peace with Hitler was, in FDR's view, the most crucial job the Allies faced. In a sermon he would preach time and again to the Joint Chiefs, he announced, "At the present time, our principal objective is to help Russia. It must be constantly reiterated that Russian armies are killing more Germans and destroying more Axis materiel than all the 25 united nations put together."[2]

On May 6 Roosevelt sent Marshall and King a memorandum summarizing his thoughts on global strategy. The Pacific would remain a "holding theater," though he wanted bomber bases pushed within range of Japan's Home Islands.

China and Burma would remain quiet for the moment. For Germany, he wanted action in 1942, not later. "I have been disturbed by American and British naval objections to operations in the European Theatre prior to 1943," he said. "I regard it as essential that active operations be conducted in 1942."[3]

As Allied strategy was batted around among the Munitions Building, the White House, Number 10 Downing, and Whitehall that May, a bald, stern-faced man known as "Mr. Brown" was ushered into the White House. He would be spending a few nights there, and valets unpacking Mr. Brown's bags were startled to find a loaf of black bread, a log of sausage, and a pistol among his clothes and shaving kit.

Mr. Brown, known to the rest of the world as Soviet Foreign Minister Vyacheslav Molotov, had been sent by Stalin to lobby for an invasion of Western Europe. Nicknamed "Iron-Ass" by Lenin for his ability to outsit opponents at the conference table, Molotov played his role with the meticulous diligence of a man whose life depended on it.[4]

It was an extraordinary mission for an old revolutionary who, eighteen months earlier, had sat across the table from Adolf Hitler. In a stark example of the "enemy of my enemy" principle, the Bolshevik who had signed a treaty with fascists was now asking capitalists to help him defeat his former ally.

FDR believed he could keep Stalin fighting Hitler. He had been the first U.S. president to recognize the Soviet government, he had been much more temperate in his anticommunism than Churchill, and he was the source of much of Stalin's war materiel. "I know you will not mind my being brutally frank when I tell you that I think I can personally handle Stalin better than either your Foreign Office or my State Department," Roosevelt wrote Churchill in March. "Stalin hates the guts of all your top people. He likes me better, and I hope he will continue to do so."[5]

On the morning of May 30, Roosevelt gathered Hopkins, Marshall, King, and Molotov to the White House. Speaking through an interpreter, Molotov asked his western allies to launch an invasion that would draw off forty or more German divisions from the Russian front. The next several months, he emphasized, would be crucial for the Soviet Union's survival. Unblinking through his round spectacles and showing no hint of human emotion, Molotov asked whether the United States and Britain intended to open a second front.

When the question was translated, Roosevelt turned to Marshall and asked if he could assure Stalin that the western allies were preparing a second front.

"Yes," came Marshall's reply.

Roosevelt told Molotov he could inform Stalin "we expect the formation of a second front this year."[6]

The addition of the phrase "this year" alarmed Marshall, who immediately began to backpedal. A second front, he cautioned, depended on available transport ships. He told Molotov it would not be possible to give the Soviets a second front in 1942 and simultaneously fulfill U.S. Lend-Lease obligations. America had men and the munitions at home, but it lacked the ships to send divisions to Calais while sending supplies to Murmansk and Arkhangelsk.

When Molotov demanded his full measure of supplies *and* a second front, Marshall pushed back. "What do you want," he asked testily, "a second front or Murmansk? It isn't possible to provide both."

King seconded Marshall's objection, and the meeting broke up as the two sides went back to study their cards.[7]

The next day Roosevelt summoned King, Marshall, and Hopkins to his office. Worried that his response to Molotov had been dangerously vague, Roosevelt said he wanted to assure Stalin that a second front would be open for business by summer's end.[8]

Marshall warned Roosevelt that promising a second front in 1942 would create titanic problems. Transportation and air cover over France were not up to the task in 1942. The British would also object vehemently, since they would have to supply most of the divisions that year.

Hopkins agreed with Marshall, but Roosevelt overrode them both. He cabled Churchill that he was "more anxious than ever that BOLERO proceed to definite action beginning in 1942," and the next morning he assured Molotov that the Americans did intend to open a second front that year. A satisfied Molotov packed his bags and flew home with pistol, sausage, and a commitment for a second front.*[9]

Shortly after Molotov departed, Steve Early, the president's press secretary, called Hopkins to discuss the wording of a press release that referred to a second front in 1942. Hopkins checked with Marshall, who again objected to any reference to "1942." ROUNDUP, he reiterated, would only be feasible in 1943.

Hopkins passed along Marshall's objections, but Roosevelt again demurred. He wanted a second front opened in 1942, and like "Iron-Ass" Molotov, Roosevelt

---

* Swallowing every impulse of training and instinct, a reluctant Secret Service never confiscated Molotov's pistol. Eleanor remarked, "Mr. Molotov evidently thought he might have to defend himself, and also he might be hungry."

refused to budge. The press release, issued June 11, read: "In the course of the conversations full understanding was reached with regard to the urgent tasks of creating a Second Front in Europe in 1942."[10]

. . .

Admiral Lord Louis Mountbatten, once Prince Louis of Battenberg and now Churchill's combined operations specialist, arrived from London the week after Molotov's visit. Young, Hollywood handsome, rich, and gregarious, he was well liked by the Americans, and Churchill occasionally used "Dickie" Mountbatten, for tricky negotiations with his most important ally.

Over a private dinner with Roosevelt, Mountbatten told FDR that the Allies lacked the specialized landing craft needed to mount a meaningful invasion of France that year. An invasion in 1942 would not draw off an appreciable number of German divisions from the Russian front, and it would probably end with American and British bodies floating in the English Channel.

Roosevelt understood the danger. It was magnified by the prospect of a Soviet defeat, which would make Germany invulnerable on the Continent. "My nightmare," he told Mountbatten, "would be if I was to have a million American soldiers sitting in England, Russia collapses and there'd be no way of getting them ashore."[11]

Stimson and Marshall worked furiously to counter whatever snake oil Mountbatten was selling. "We were largely trying to get the president to stand pat on what he had previously agreed to," Marshall later told an interviewer. "The president was all ready to do any side show and Churchill was always prodding him. My job was to hold the President down to what we were doing."[12]

To drive Mountbatten's message home, Churchill scheduled another visit to Washington, with a side trip to Hyde Park, to confer with Roosevelt. Before Churchill's arrival, FDR held a White House conference with Stimson, Knox, and the Joint Chiefs, where he dusted off GYMNAST. Marshall pushed hard for BOLERO, and Stimson, always supporting Marshall, argued vehemently against Churchill's African detour.

King uncharacteristically refused to stick his neck out. He saw ROUNDUP as an Army fight, and he kept his thin lips snapped tight, except when he agreed with Roosevelt. *"King wobbled around in a way that made me rather sick with him,"* Stimson grumbled to his diary. *"He is firm and brave outside the White House but as soon as he gets in the presence of the President he crumbles up."*[13]

Stimson and Marshall shuddered to think what Roosevelt and Churchill would do when, far from the guiding hands of their experts, the two men secluded themselves at FDR's Hyde Park mansion later that week. They were both scions of old and noble families—though not especially wealthy families, anymore—and both were accustomed to wielding immense power in their own domains. *"I can't help feeling a little bit uneasy about the influence of the Prime Minister on him,"* Stimson wrote. *"The trouble is Churchill and Roosevelt are too much alike in their strong points and in their weak points. They are both brilliant. They are both penetrating in their thoughts but they lack the steadiness of balance that has got to go along with warfare."*[14]

· · ·

Winston Churchill sat in FDR's study at Springwood, the Roosevelt family mansion in Hyde Park. The two men had spent part of the day riding around steep grades overlooking the Hudson River in FDR's lever-operated Ford convertible, an exhilarated FDR behind the wheel. The Ford darted between trees and around the winding roads of Hyde Park's grounds, and as it rolled close to the river's steep embankments, an uneasy Churchill sat quietly in the passenger's seat, mighty grateful that the hand-controlled brakes functioned as advertised.*[15]

Inside the house, they held a long and leisurely discussion among the portraits and mementos of Roosevelt's youth. As thoughts became words, and words became new thoughts, SLEDGEHAMMER began to crack. Churchill presented insoluble problems for an invasion that year: lack of landing craft, lack of troops, lack of air cover. Lack of British enthusiasm. His Majesty's government, he warned, would not support any landing on the French coast that courted a serious risk of another defeat. Unless the Allies had the means to land in France, and stay there, they mustn't make the attempt.[16]

By disregarding Marshall's advice about Molotov, Roosevelt found himself wedged between two strategic boulders. On one hand, Churchill was right; an invasion of France posed immense risks for 1942. On the other, he had just promised Stalin that the Allies would open a second front in 1942, and they could not renege on that commitment.

He also had to consider public reaction at home, for while the spear may be a military weapon, where that spear will be pointed is political. Roosevelt could preach the gospel of Germany First, but the big battles—Wake, Bataan, Coral Sea, Midway—had been fought against Japan. If the nation's fighting men were

* The Ford's hand-control system, like Roosevelt's wheelchair, was an ingenious contraption of FDR's own design. A left-hand lever pushed halfway engaged the clutch; pushed all the way, it engaged the clutch and the brake. A ratchet and lever attached to the steering column enabled Roosevelt to lock the accelerator at preset speeds with his right hand while holding the steering wheel.

only fighting in the Pacific until the middle of 1943, Americans might want to finish the war against Japan before opening a new can of worms in Europe.

Before leaving Hyde Park, Roosevelt concluded that he must put soldiers into battle against Germany before year's end. But he had no idea where to fight. On Saturday, June 20, he sent a message to Marshall and King asking for a list of places where U.S. ground troops could fight Germany before mid-September. He wanted their answer presented at the White House the next day.[17]

On a muggy Sunday morning, as the thermometer began its long climb into the nineties, Roosevelt sat perspiring in his White House study with Churchill and "Pug" Ismay, Churchill's military adviser. Churchill, playing on FDR's desire for action in 1942, favored North Africa, and he mounted a vigorous attack against SLEDGEHAMMER, the only American alternative for that year.[18]

Roosevelt and Churchill hadn't spoken for long when Captain McCrea, FDR's naval aide, unobtrusively entered with a pink slip of paper. He handed it to Roosevelt, who quickly read it and asked McCrea to hand it to Churchill.

Tobruk, the message said, had fallen to the Nazis. The last bastion of British power in Libya, the Singapore of North Africa, was now in German hands. General Claude Auchinleck's troops were streaming east in retreat toward Egypt, and the Royal Navy was preparing to evacuate Alexandria.

Churchill's jowly face turned pale. Perhaps thirty thousand Commonwealth troops had been led into captivity, and Egypt and the Suez Canal were exposed to the snapping fangs of Rommel's panzers. With Japanese troops driving through Burma toward the Indian border, an Axis linkup in the Middle East was a very real, very dangerous possibility. This second epic defeat for the year might even prompt a no-confidence vote in Parliament, bringing down Churchill's government.[19]

"For the first time in my life, I saw the prime minister wince," said Ismay.

"I can't understand it," Churchill sputtered. Tobruk had withstood a six-week siege a year earlier. Now it had fallen in thirty-six hours.

Had the British Army lost the will to fight? The men in the room hoped not. "Defeat is one thing," Churchill wrote. "Disgrace is another."[20]

Roosevelt appreciated the implications of the disaster, both for the war and for Churchill, the steady helmsman of Britain. Tobruk had capped off an *annus horribilis* not endured by Britons since 1776, and it was unclear whether his government would survive. Animated by the same instinct that impelled him to find jobs for the jobless, or to help children stricken with polio, he leaned forward in his chair and asked the grieving minister, "What can we do to help?"

The Englishman was touched by that simple question, a gesture of personal support from a man who sensed what he was going through. "Nothing could exceed the sympathy and chivalry" of his friend, he later said. "There were no reproaches, not an unkind word was spoken."

Even the flinty General Brooke was moved. For the rest of his life, he remained "impressed by the tact and real heartfelt sympathy that lay behind these words. There was not one word too much or too little."[21]

A grateful Churchill asked Roosevelt for three hundred of America's new Sherman medium tanks. Roosevelt sent for Marshall. He passed along Churchill's request.

Weighing the needs of his army against Britain's war in the Near East, Marshall replied, "Mr. President, the tanks have just been issued to the 1st Armored Division. It is a terrible thing to take weapons out of a soldier's hand, but if the British need is so great, they must have them."[22]

Marshall produced the tanks, along with one hundred 105mm howitzers. To this he added a few groups of aircraft, and even looked into the loan of General Patton's Second Armored Division.[23]

Grateful for FDR's assistance, Churchill returned to GYMNAST. In that humid oval room, Marshall pleaded the case for BOLERO, SLEDGEHAMMER, and ROUNDUP, while Churchill provided the strategic rebuttal.

For once, Roosevelt was unsure of himself. For the moment, he was inclined to hold Churchill to his ROUNDUP bargain, and he showed Churchill a letter from Stimson forcefully opposing GYMNAST. Whatever its merits as a diversionary tactic, North Africa would divert resources from the big cross-Channel effort. "When one is engaged in a tug of war," Stimson wrote, "it is highly risky to spit on one's hands even for the purpose of getting a better grip."[24]

But he still reached no firm conclusion. With unusual diffidence, Roosevelt agreed with Churchill to invade France in 1942—if an invasion were feasible. If not, the two men agreed to find an alternative. The Allies would continue to study SLEDGEHAMMER until September 1, after which time the weather would rule out an invasion for the year, and BOLERO would forge ahead. In the meantime, other alternatives, including GYMNAST, would be investigated.[25]

The agreement in Roosevelt's study, like so many of his decisions, was a polygamous marriage of convenience, politics, and strategy that gave everyone the impression they got what they wanted. Stimson and Marshall came away convinced that a second front in France was the presumptive mission, at least for 1943; because GYMNAST would deplete troops, air, and ships needed for ROUNDUP, they believed there would be no landing in Africa. It was a simple matter of military logic.

Churchill, on the other hand, understood the wheels within wheels of FDR's political mind. Like a musician anticipating the next note, Churchill sensed that what Roosevelt wanted most was an invasion in 1942. It was the time, not the place, that counted most. The president had agreed to support an invasion of France in 1942 if it were feasible, and Churchill knew damned well it would not be feasible. If Africa could offer Roosevelt what France could not, he could be brought around.[26]

Stimson and Marshall soon learned there was no such thing as a safe bet with politicians as nimble as Churchill and Roosevelt. On July 8, Churchill cabled FDR, "No responsible British general, admiral, or air marshal is prepared to recommend SLEDGEHAMMER as a practicable operation for 1942." Placing his stubby finger on the solution, he added, "I am sure myself that GYMNAST is by far the best chance for effective relief to the Russian front in '42. This has all along been in harmony with your idea. In fact it is your commanding idea. Here is the true second front in '42. Here is the safest and most fruitful stroke that can be delivered this autumn."[27]

To Marshall and Stimson, Churchill was repudiating a bargain they had just struck. Stimson complained to his diary, *"The British war cabinet are weakening and going back on Bolero and are seeking to revive Gymnast—in other words, they are seeking now to reverse the decision so laboriously accomplished when Mr. Churchill was here a short time ago."*[28]

Marshall was sick of going back to the garden and yanking out the weeds of North Africa. Every time he thought he had pulled them up and buried them at the bottom of his compost heap, he would look back, and there they were, pushing through the soil.

He realized, too late, that he had been outflanked. Roosevelt had been telling the Joint Chiefs since May that for political reasons, he wanted a battle across the Atlantic in 1942 to keep the public focused on Germany as the greater threat. But the men in uniform hadn't found him one. They "failed to see that the leader in a democracy has to keep the people entertained," Marshall said years later. "That may sound like the wrong word, but it conveys the thought. The people demand action. We couldn't wait to be completely ready."[29]

In desperation, Marshall turned to Ernie King.

Three months earlier, Brigadier General Eisenhower had suggested that if the British would not go along with SLEDGEHAMMER-BOLERO, the Americans should turn their full might against Japan. "Unless this plan is adopted as the eventual aim for all our efforts," he advised Marshall, "we must turn our backs upon the Eastern Atlantic and go full out, as quickly as possible, against Japan!"[30]

Now Marshall seemed ready to take Ike's advice. At a meeting of the Joint Chiefs on July 10, he made an impassioned plea to shift American efforts to the Pacific if Churchill did not back down on BOLERO, SLEDGEHAMMER, and ROUNDUP. He pointed out that a Pacific strategy "would tend to concentrate rather than scatter U.S. forces; that it would be highly popular throughout the U.S., particularly on the West Coast; that the Pacific War Council, the Chinese, and the personnel of the Pacific Fleet would all be in hearty accord; and that second only to BOLERO, it would be the operation which would have the greatest effect toward relieving the Russians."[31]

Admiral King was happy to sign on to Marshall's gambit. Though he accepted the "Germany First" policy intellectually, he knew GYMNAST would drain tremendous resources from the Pacific war. Besides, he doubted the British had ever been enthusiastic about a landing on the Continent.[32]

With Stimson's backing, a helpful nod from King, and the secret concurrence of Sir John Dill, Marshall sent Roosevelt a memorandum signed by himself and King. It read:

> Our view is that the execution of GYMNAST, even if found practicable, means definitely no BOLERO-SLEDGEHAMMER in 1942 and that it will definitely curtail if not make impossible the execution of BOLERO-ROUNDUP in 1943. We are strongly of the opinion that GYMNAST would be both indecisive and a heavy drain on our resources, and that if we undertake it, we would nowhere be acting decisively against the enemy and would definitely jeopardize our naval position in the Pacific. . . . If the United States is to engage in any other operation than forceful, unswerving adherence to BOLERO plans, we are definitely of the opinion that we should turn to the Pacific, and strike decisively against Japan; in other words, assume a defensive attitude against Germany, except for air operations; and use all available means in the Pacific.[33]

Marshall's willingness to forge ahead in the Pacific was not so solid as his letter implied. As he told Roosevelt in a follow-up memorandum, his real goal was to "force the British into acceptance of a concentrated effort against Germany." Only if this proved impossible, he said, would he recommend the United States, "turn immediately to the Pacific with strong forces and drive for a decision against Japan."[34]

.   .   .

As he sat in his Hyde Park study on a warm Sunday morning, Roosevelt turned over Marshall's memo in his large hands. He knew a radical shift to the Pacific was the last thing Stimson and Marshall really wanted. Even King understood the big picture, and the Navy could not have been entirely comfortable about backing off Hitler, even temporarily.

Well, the personal views of his service chiefs did not concern him. The commander-in-chief, not his military staff, would decide grand strategy. Roosevelt saw no wisdom in giving Germany a free hand for another year or two while Nimitz and MacArthur chased Yamamoto around the Pacific, and he would not let it be said that he had turned his back on an ally with whom he had so passionately and publicly bound America's fortunes.

Like the head of any other government agency, Roosevelt knew admirals and generals would kick from time to time. A president sometimes had to stick pins in them to get them moving in the right direction, and as commander-in-chief, Roosevelt had become a pretty good pin-sticker. He picked up the phone and called the War Department.[35]

If the Joint Chiefs wanted a major Pacific offensive, he said, then their commander wanted to know exactly what they had in mind. He ordered them to send him a statement setting forth precisely what they planned to do in the Pacific and why they wanted to do it. He wanted to know which islands they planned to assault, estimated landing dates, estimated numbers of ships, planes, and men they needed, and where they planned on finding those ships, planes, and men. He wanted to know how the "Japan First" effort would help the Russian and Middle East fronts and how the global picture would be improved by throwing everything against Japan rather than Germany. And he wanted it in Hyde Park by the end of the day.[36]

Marshall rushed back to Washington from his home in Leesburg. At the Munitions Building, he, Stimson, Arnold, King, and a buzzing hive of staffers worked furiously for two and a half hours preparing a draft plan. But the best they could come up with was a rough outline of the major strongpoints to be overcome between Midway and Tokyo. They could not promise that a shift to Japan would improve the picture much globally; all they could say was that a drive against Japan was better than a drive into the African desert.[37]

Roosevelt's anger flashed as he read the reply signed by Marshall, Arnold, and King. It was July 1942, seven months into the war, and if the American chiefs sincerely wanted to go to the Pacific, they would have given him better answers on short notice.

No, he decided, their proposal was just a ploy to squelch GYMNAST. With evident disgust, he set the memorandum aside, picked up his fountain pen, and scratched out a curt note to his military chiefs:

I have carefully read your estimate of Sunday. My first impression is that it is exactly what Germany hoped the United States would do following Pearl Harbor. Secondly it does not in fact provide use of American Troops in fighting except in a lot of islands whose occupation will not affect the world situation this year or next. Third: it does not help Russia or the Near East.
Therefore it is disapproved as of the present.
    —Roosevelt C in C

To drive home his point, he called his aide, Captain McCrea, and handed him a five-page telephone message to read to Marshall and King. The message concluded, "I am unwilling to continue with Bolero on the full basis unless we are going to do Sledgehammer in 1942. If we cannot, then we must attack at another point. Gymnast might not be decisive but it would hurt Germany, save the Middle East and make Italy vulnerable to our air power. The war will be lost this year in Europe and Africa not in the Pacific. I think we are doing well in the Pacific."[38]

Back at the White House, "Roosevelt C in C" told Stimson he was committed to BOLERO, but he objected to the Pacific business. It was, he said, "a little like taking up your dishes and going away." Stimson admitted the Pacific option was mostly bluff, but said it was a necessary bluff, to force the British to remain true to BOLERO and ROUNDUP.[39]

As Stimson left Roosevelt's office, Marshall came in and, in Stimson's words, had a "thumping argument" with his commander-in-chief. The general argued vehemently that an African expedition would deplete the force needed for northern France in 1943 while dangerously shortchanging the Pacific. A Pacific strategy, by contrast, would prevent Japan and Germany from joining hands in the Middle East. *"Between us the President must have had a rough day on those subjects,"* Stimson remarked.[40]

Unfortunately for Marshall and Stimson, the code names became confused when leaders used BOLERO, SLEDGEHAMMER, and ROUNDUP interchangeably. They were muddled further when the warlords spoke of semi-official variants like SUPER-GYMNAST, Maximum ROUNDUP, and ROUNDHAMMER. Stimson claimed the British were backing off BOLERO when it was really SLEDGEHAMMER they objected to. Marshall thought Brooke and Churchill opposed ROUNDUP, because in his mind GYMNAST made ROUNDUP impossible—though it was only

SLEDGEHAMMER that the British opposed. The meaning of the operations thus became lost at times in a fog of code words used imprecisely by the politicians. Or, in the case of Churchill and FDR, disingenuously.[41]

To cut through this fog, Roosevelt ordered Harry Hopkins to take Marshall and King to London. Together, they were to thrash out an agreement with the British. FDR was firm only about a clash with Germany in 1942, and he told Hopkins over dinner, "I do not believe we can wait until 1943 to strike at Germany. If we cannot strike at SLEDGEHAMMER, then we must take the second best—and that is not in the Pacific."[42]

Marshall and King drafted instructions for themselves and submitted them to Roosevelt for his signature. Roosevelt, still miffed over their Pacific bluff, scratched "Not Approved" across the top with his pen, then rewrote their orders. "If SLEDGEHAMMER is finally and definitely out of the picture, I want you to consider the world situation as it exists at that time and determine upon another place for U.S. troops to fight in 1942," he stressed. Four times in his three-page instruction letter he demanded a major operation in the year 1942, and as with his previous letter, he signed the letter, "Commander-in-Chief."[43]

· · ·

Marshall and King arrived in London on July 19, and for three days they fired every shell in their limbers to convince their hosts to conserve men and ships for ROUNDUP. Marshall even tried to convert SLEDGEHAMMER from a sacrifice play into a full-fledged invasion for the fall—a foothold that would be pushed in full the following spring.[44]

The British found Marshall's arguments absurd. The Empire had spent two years just trying to survive the war. Now they were trying not to lose it. They had lost too many long bets at Singapore, Crete, Tobruk, and Norway, and had no appetite to push their luck again. It was not yet time, in British eyes, to win the war, especially on a shoot-the-moon wager.

"The burned child dreads fire," King put it.[45]

A four-sided standoff emerged. Marshall and King insisted on an invasion of northwestern France in 1942. The British chiefs agreed with the Americans on the *where*, but violently disagreed over the *when*. Churchill agreed with the Americans on *when*, but disagreed with the military men over *where*; he wanted North Africa, or perhaps even Norway, either of which could be invaded in 1942. Roosevelt had his own *when*, but he didn't care *where* the Allies landed, so long as U.S. soldiers were shooting at German soldiers before year's end.

Cables buzzed back and forth between Hopkins and Roosevelt as the stalemate hardened. The two old friends concocted a private code for the negotiators, taken from the names of FDR's acquaintances at Hyde Park: Marshall was "Plog," the name of a Roosevelt family superintendent; King was "Barrett," a local farm manager; Churchill was "Moses Smith," a man who rented a farm on Roosevelt lands; and Brooke was "Mr. Bee," the caretaker of FDR's hillside cottage.[46]

Moses and Mr. Bee refused to budge, and after a rearguard battle, General Plog and Admiral Barrett were cornered by a reality as bleak as it was inescapable: An invasion of France was simply impossible before late September or October. The landing would be too little, too late, to be of any real use to Stalin, and if they attempted it, their men might be trapped on the French coast and driven into the sea.

"Moses Smith" ended the debate on July 22 by laying the question before the British War Cabinet, which unanimously voted down a cross-Channel invasion that year. FDR gave Marshall and King a peremptory order: the Americans were to agree to an operation somewhere to be commenced no later than October 30—not coincidentally, five days before the congressional midterm elections.

They had fought like Horatius at the bridge, but it was no use. Out of ammunition, surrounded, and staring defeat in the face, Marshall and King surrendered.[47]

A downcast George Marshall returned to his hotel suite at Claridge's, sat down at his desk, and took up a pencil. *What*, he asked himself, *would be the least dangerous diversion? The Middle East? Norway? West Africa?*

Scratching out several options, he settled on French North Africa, the British preference. A limited foray there would meet Roosevelt's requirements. ROUNDUP would be dead for 1943, because GYMNAST would eat up the needed shipping, but he felt the buildup in England might continue on a smaller scale, in case some unexpected opportunity sprang up on the Continent.[48]

As Marshall assembled his thoughts, Admiral King walked in. Marshall outlined his view of the lesser evil, and King, to Marshall's surprise, supported him without qualification. "It is remarkable now," Marshall told an interviewer, "but King accepted without a quibble. Usually he argued over all our plans."[49]

The next day Marshall presented his proposal to the British chiefs and asked for their endorsement. The Combined Chiefs of Staff approved Marshall's official summary in a bureaucratic-sounding memorandum designated "CCS/94."

But before they could forward CCS/94 to the two leaders, a cable arrived from Washington. Roosevelt *"evidently had a sharp attack of strategy the previous evening,"* one British staffer told his diary. He proposed *"combining GYMNAST with Dakar and God only knows where else"* for a fall invasion. Hopkins fired off a cable to Washington asking Roosevelt to hold off until the Combined Chiefs finished their plan.[50]

The War Cabinet also threw a last-minute wrench into the delicate deal by disputing a line in Marshall's proposed summary, which noted that GYMNAST in 1942 would make ROUNDUP impossible for 1943. "They didn't want it used against them politically if they prevented ROUNDUP in 1943, thus delaying the freeing of Europe," Marshall later explained. "I blew hell out of that and said unless the cabinet agreed I wouldn't go along."[51]

A compromise, softening Marshall's sullen indictment of GYMNAST, was eventually worked out. GYMNAST was rechristened TORCH, a merciful end to a code name that had caused such bitter feelings among allies.[52]

Before the Americans left London on July 25, the British threw them a bone. In return for scrapping SLEDGEHAMMER, they consented to an American commander in the North African theater. Marshall returned to Claridge's and had an aide summon General Eisenhower.[53]

The chief of staff had a bath drawn. He had undressed and immersed himself for a few minutes when he heard a knock on the lavatory door. It was Eisenhower. Not one for formalities—Marshall enjoyed dropping big news on subordinates in an offhanded way—Marshall brought Ike up to speed through the bathroom door, and told him an American would command the invasion of North Africa. That American, if Marshall and King had their way, would be General Eisenhower. The two men chatted for a moment, and Eisenhower left Marshall's bathroom doorway a happy man.[54]

In eight months following Pearl Harbor, Eisenhower had earned Marshall's confidence. The Kansan understood logistics and transportation, and he grasped the political realities that Marshall and Stimson faced daily in Washington. And unlike MacArthur, Eisenhower was not the type to scream for more, more, more.

As a measure of his trust, Marshall encouraged Eisenhower to take the Army's top-level talent with him to Africa. Before long, Generals George Patton, Lloyd Fredendall, Lucian Truscott, and Terry Allen, along with Colonel Walter Bedell Smith, the Army's ablest staff officer, would select trails that thousands of fighting men would follow. Whatever his personal feelings, TORCH was to be the Allied centerpiece for the year, and Marshall was determined to give it every opportunity to succeed.[55]

· · ·

A dejected Stimson said nothing publicly about his disagreement with Roosevelt, but it did not take long for hints of dissent to begin creeping into the newspapers. An angry FDR felt pressure to deny the implication that he was ignoring the advice of his military experts, and at a cabinet meeting on August 7, he told his ministers he had never overridden the advice of his military men, unless the Army and Navy disagreed and required a tie-breaker.[56]

Knox and Stimson, who knew better, held their tongues. But Stimson left the cabinet meeting with the feeling that Franklin Roosevelt, for all his positive qualities, was either self-delusional or a casual liar. Against the unequivocal advice of Admiral King, Generals Marshall and Arnold, and Secretaries Stimson and Knox, the president had vetoed the Pacific proposal and let the British kill SLEDGEHAMMER-ROUNDUP. Roosevelt, Stimson concluded, had a *"happy faculty of fooling himself and this was one of the most extreme cases of it that I have ever seen."*[57]

On the following Monday he drafted a letter to the president. In correct yet vaguely defiant tones, he reminded Roosevelt that he had overruled the advice of his military chiefs concerning North Africa. TORCH, he reiterated, would delay the liberation of France until at least 1944.[58]

Stimson took his draft to Marshall, who urged him not to send it. The letter would only stir up a hornet's nest, to no good result. Besides, he said, as the president's military adviser, he felt that he himself should be the one to lodge a protest, if a protest were needed. If Marshall let Stimson go to bat for him on a military argument, he said, it would appear as if he were "not being manly enough to do it himself."[59]

Stimson believed the president had been guilty of self-deception. He was, in Stimson's mind, an amateur making a rash decision about military matters where lives would be spared, ruined, and forever altered. Marshall said that Roosevelt knew exactly what he was doing. The upcoming midterm elections made him hungry for a military victory over Germany, and Roosevelt wanted tangible progress to show the public before November 3, 1942.[60]

But intentional or not, Roosevelt would do a lot more meddling. King and Marshall were about to see a new face in the president's inner circle.

TWENTY-THREE

CHAIRMAN OF THE BOARD

∽

Like many things in Roosevelt's war, change at the top began with the Navy. After Pearl Harbor, he appointed Ernie King as Commander-in-Chief, U.S. Fleet, the Navy's top fighting admiral, while Betty Stark remained Chief of Naval Operations—the man running long-term planning, shore logistics, budgets, and everything else that didn't go "boom."[1]

King found the arrangement unworkable. He believed the fuzzy, fluid line between strategic planning and battle planning required a single man to direct weapons and manpower to the fleet captains who used them. He told Secretary Knox he would step down as COMINCH in favor of Stark, if that was what Roosevelt wanted, but either Stark or King had to go. Knox took King to see Roosevelt, and in his low, slow cadence, King explained that the dual monarchy was breaking down.[2]

It was another decision that could only be made by the commander-in-chief, and like many of Roosevelt's decisions, it came down to his feeling for personalities. Stark was a good, intellectual administrator, but King was a fighter, and in war the Navy needed fighters. Besides, in the wake of Pearl Harbor, with its investigations and calls for scapegoats, Roosevelt wasn't about to place Betty Stark in charge of the war at sea. Roosevelt gave Betty a Distinguished Service Medal and packed him off to Europe, out of easy reach of Congress, and out of King's few remaining hairs. Admiral Ernest J. King, he announced, would assume the duties of both COMINCH and CNO.[3]

King told his staff to draft an executive order for Roosevelt's signature defining his new duties, and Roosevelt signed King's order into law as Executive

Order 9096 on March 12. With a smiling Frank Knox looking on, Admiral King gingerly placed his fingertips on the corner of a Bible and took the oath of office of chief of naval operations. When he released the Bible, he became the most powerful admiral on earth.[4]

Ernie King had little beyond mild amusement—with a liberal sprinkling of contempt—for the civilians who ran the Navy Department. Some he could safely ignore. Assistant Secretary Ralph Bard, for instance, the department's head of civilian personnel, invited King to lunch time and again, while King excused himself, always pleading "press of business." Bard finally gave up. "I didn't have time to educate those people," King muttered.[5]

But even the Navy's undisputed heavyweight champ couldn't ignore everyone. Secretary Knox was one of those "too-important-to-ignore" functionaries— which, to King, was most unfortunate. The secretary was, in his eyes, a civilian interloper of the worst kind, a landlubber-politician who had no idea how to run a navy at sea or ashore. To King, Knox and his undersecretary, a Dutchess County politico named James Forrestal, were incompetents to be tolerated but not encouraged. In 1942, a mutual friend passed on to King one of Secretary Knox's complaints that King wouldn't tell him about the Navy's war plans. King snorted, "Why should I? The first thing he does is to tell the reporters everything he knows."[6]

Each morning, King would arrive at Knox's chart room on Main Navy's second deck for an 8:30 a.m. conference of the bureau chiefs, the Marine Corps commandant, and the COMINCH department heads. Before one meeting, when the only men present were Knox, King, and Captain McCrea, the president's naval aide, King and Knox began quietly arguing over a matter at Knox's desk. As their voices grew terse and snappish, McCrea, standing on the far side of the room, heard Knox bellow, "Admiral King, that matter has been settled! I don't want it raised again. I trust you understand that the final word has been said!"

McCrea glanced up to see Knox glaring at King. King, his face flushed, glared back at Knox, furious but unable to tell Knox what he really thought of him. Soon other conferees arrived, and the argument was overtaken by the meeting.

The next morning, King summoned McCrea to his office overlooking Constitution Avenue. Standing stiffly, his back to the large picture windows, King asked, "You were at the secretary's conference yesterday, were you not?"

"Yes, sir," said McCrea.

"You heard what the secretary said to me?"

"I did, sir."

"I want to inform you that I am going to make an issue of his remarks," said King. "I have never been so spoken to since I can remember. I just want to make sure that you heard what he said to me and how he said it."

King obviously planned to take the matter up with the president, and with unaccustomed temerity, McCrea told the admiral that Knox had been within his rights. More importantly, McCrea said, the president had enough problems with strategy, Allied relations, domestic politics, the economy, and war production. He didn't need to be dragged into a personal spat between his navy secretary and senior admiral.

The lava in King's core began to bubble up from the mantle. As it rose through the crust, King's face turned the shade of crimson that his staffers had learned to dread. But he said nothing and spun away from McCrea and faced his window for a long, measured minute.

Then he turned back slowly, eyeing McCrea like he was measuring him for a coffin.

"Good day," he said, biting off his words.[7]

Thinking further, he let the matter drop.

As he balanced the twin duties of CNO and COMINCH, Admiral King learned how difficult Roosevelt could make life. While their relationship was cordial and constructive, FDR's experience as a small-craft yachtsman and former Navy assistant secretary gave him license, he felt, to meddle in details of naval assignments, promotion, warship construction, and strategy. King's jobs.

Roosevelt's affinity for the Navy was reflected in his choice of aides. His naval aide, Captain McCrea, was an experienced surface ship commander in line for one of the big new *Iowa*-class battleships in production. McCrea regularly attended staff meetings with King and Knox, and he was an efficient conduit of information between the White House and Main Navy.

FDR's military aide, Major General Edwin "Pa" Watson, was his poker-playing buddy, fishing-trip companion, gatekeeper, and all-around Man Friday. A merry knight of the round table with a talent for letting people down easy, Pa Watson was long on personality, jokes, and cologne, but thin on strategic insight or military information.

FDR didn't care. He trusted Marshall on Army matters, so he didn't need a true military aide. He probably didn't need a naval aide, either. "You know, Betty," he once told Admiral Stark, "I don't know much about the Army. I sort of have to take what General Marshall tells me, but I do know a lot about the Navy."[8]

Often Roosevelt's meddling in the Navy's business was harmless. He defended his prerogative to name commissioned vessels, and after one ceremony he sent Knox a note of complaint:

> *Will you tell the Navy Band that I don't like the way they play the Star*
> *Spangled banner—it should not have a lot of frills in it?*
> *F.D.R.*[9]

Other instances were more serious. In May 1942, King submitted to Knox a one-page list of ships he wanted for that year and the next. The list, totaling nearly 1.7 million tons of displacement, summarized the tools King believed he needed to win the war at sea.

King refused to clear his official requests with the White House in advance. But, in an unusual display of caution, he told Knox, "It is considered advisable that Presidential authority be obtained for initiating legislation for this combatant ship building program." Knox agreed.[10]

"I wish you and Admiral King would talk with me about this proposed new building program," a mildly irritated Roosevelt wrote Knox when he saw the list. "I should like especially to talk over the desirability of building 45,000 ton aircraft carriers, and the possibility of cutting the size of 27,000 ton aircraft carriers by four or five thousand tons and putting the saved tonnage into aircraft carriers of approximately twelve to fourteen thousand tons. . . . Also, I should like to discuss the relative advantages of 13,600 ton heavy cruisers vs. the 11,000 ton heavy cruisers."[11]

Roosevelt's fixation on heavy cruisers, light carriers, and small, sometimes useless specialty craft would dislocate King's sea strategy, just as his snap decisions on bombers and tanks disrupted Marshall's land strategy. He told himself he would not meddle in naval tactics, but he made friendly suggestions that came awfully close. Like an angler who knows a good fishing spot, he advised Admiral King to seek out U-boats around Sierra Leone, and in February he told King that a recent raid by the Italian Navy was "damned good," adding, "I wish you would turn loose your most imaginative people in War Plans to tell me how you think the Italian Navy can be effectively immobilized by some tactics similar to or as daring as those used by the Italians. I can't believe that we must always use the classical offensive against an enemy who seems never to have heard of it."[12]

As meddler-in-chief, FDR indulged a penchant for approving projects that walked in through back channels. Shortly after King signed his memorandum asking for specific ships, Henry J. Kaiser, an industrial mogul with no experience in shipbuilding, approached the Navy's Bureau of Ships with an offer to construct escort aircraft carriers. When the guardians at the bureau politely showed Kaiser the door, he went to Roosevelt.

Soon King, under direct order from the president, grudgingly accepted Kaiser's proposal. "Kaiser had the craziest ideas about machinery," King recalled. "I didn't agree with the damned things, but FDR decided."

In many cases, King or Knox detailed senior officers to figure out ways to kill Roosevelt's more harebrained schemes, and those men became quite good at smothering his larks with paper. But every now and then, Roosevelt's restless mind would come back to threaten King's barony of the sea.[13]

.   .   .

As disasters poured in from the Far East in the winter and spring of 1942, Congress raised a storm over the American command structure. Defeats in the Philippines, Wake, Guam, the Java Sea, and elsewhere seemed to point to a failure at the top, and the mushy Rooseveltian structure became the focal point of blame. By March 1942, seven bills to restructure the Army, Navy, and Air Force command were floating about congressional calendars, and every committee chairman, it seemed, had his own ideas about how to run the war. Wendell Willkie delivered a speech in Boston calling for an overlord of all armed forces, a sort of "secretary of defense." The name he put forward was Douglas MacArthur.[14]

With Admiral Stark's exit from Washington, General Marshall moved up as the top man of the Joint Chiefs. He held the same four-star rank as Admiral King, but because he was made Army chief of staff in September 1939, he was technically senior to King. "He used that, all right," King said later, "but I had no kick about that."[15]

What King *did* have a kick about was the Navy's isolation among the chiefs. When Stark was in the club, it made sense to let Marshall bring in General Arnold, so each service would have two votes. With Betty gone, the Navy could be outvoted by the Army and Army Air Forces at Marshall's direction.

To level the playing field, King suggested bringing in a Marine general. Marshall, like MacArthur before him, saw the Marines as a threat to the Army's preeminent role as the nation's land force, and supposedly remarked once, "I'm going to see that the Marines never win another war." King's suggestion went nowhere.[16]

Marshall, an admirer of clean lines of command, at first believed a single head of the Army and Navy should lead a "Joint Staff." King violently opposed the notion, as he knew who wanted to become that single head. "I feel quite sure that Marshall, right from the beginning, thought he was going to manage everything," King said later. "If he had had his way there would have been no Joint Chiefs of Staff."[17]

Marshall, like King, was no MacArthur. He harbored no inner ambition to become the republic's man on horseback. Constitutionally, the president filled that role, even though no president had actually directed military operations since George Washington rode out to put down the Whiskey Rebellion. Marshall felt the president would best be served by retaining himself and King as military advisers, instead of creating a "super-general" to rule over all chiefs.

The problem, as Marshall saw it, was not too much ambition by the military, but too little guidance from the civilians, meaning Roosevelt. The Joint Chiefs needed to know what the president was thinking, and for that they could not depend on Roosevelt. Notoriously secretive, Roosevelt would not let Marshall's senior staffers take notes during their meetings. Pa Watson, the Army's official window to the White House, was chronically unfocused, and depending on who happened to be hanging around the White House, on a given day presidential orders to the Joint Chiefs might go out through Stimson or Hopkins or McCrea, or some friendly senator or random messenger. Or they might not go out at all.

Roosevelt's nonchalance gave the tightly regimented Marshall fits. Used to following meticulously drafted orders, Marshall was bothered that the man he served sent his instructions as they popped into his head. He complained to former senator James Byrnes: "After Cabinet meetings, Mr. Stimson invariably makes some pencil notes and dictates a memorandum which is circulated over here, with relation to any matters that may concern the War Department. Possibly Mr. Knox does the same thing in the Navy Department. However, we have had cases where their impressions varied as to just what the President desired."[18]

To Marshall, the service heads needed a "chief of the joint general staff," to facilitate the flow of information between the White House and the military professionals, and looking across the Atlantic to Vichy France, Marshall found the perfect candidate. In February, he asked Roosevelt to tap Admiral William D. Leahy, U.S. ambassador to Vichy, as the senior commander among the American service chiefs.

When Marshall first mentioned the idea, Roosevelt balked, telling Marshall, "I am my own chief of staff." Marshall replied that Roosevelt couldn't possibly devote enough time to the Joint Chiefs with all his other commitments. He bluntly told the president, "You are not Superman." The Joint Chiefs needed someone to represent the White House, someone who could be trusted not to favor one service over another. The Army, Marshall declared, trusted Admiral Leahy.[19]

Affable, of medium height, with gray hair and a nose tilting "four degrees left rudder," Leahy had been chief of naval operations before Stark took over. He was an experienced, likable diplomat and a friend of the president. As a Navy man, he would also be acceptable to Admiral King.[20]

Except that he wasn't. While King liked Leahy, and owed some of his position to him, he didn't want an intermediary dropping anchor between himself and the president. King didn't think much of Leahy professionally, either, and he saw the droop-jowled ambassador as a yes-man, an old Bureau of Navigation "fixer" who indulged in habits that annoyed King—like working his way to the center of every group publicity photo. To King, Leahy was a man with little bark and less bite, and with the country at war, America needed as much bite as it could get.[21]

Bill Leahy returned from France a heartbroken man. His wife of thirty-eight years had died in Paris that spring from surgical complications after a hysterectomy, leaving Leahy, like Marshall years before, despondent. FDR ordered him back to Washington, and invited him to lunch on July 7. Having reignited his personal connection with Leahy, FDR's reservations about creating the new job subsided, and two weeks later he appointed the admiral "Chief of Staff to the Commander-in-Chief."[22]

Roosevelt was doing something unprecedented, something that gave his own title of "commander-in-chief" more than just a symbolic sound. But Leahy's position was symbolic, at least until Roosevelt, Leahy, or someone else came up with an intelligible job description. When newsmen asked whether Leahy would run the general staffs of the Army, Navy, and Air Forces, Roosevelt confessed, "I haven't got the foggiest idea." When they pressed him for something more than the foggiest idea, he explained that Leahy's job would be to do the president's legwork. Leahy would read, summarize, and index memoranda and such, to save the commander-in-chief time he could devote to other matters.

When journalists asked Roosevelt if the Senate would confirm Leahy's appointment, FDR shook his head. There would be no Senate confirmation, he said. Leahy was an active-duty naval officer and Roosevelt was the commander-in-chief of the Navy. It was, he said with a shrug, "an order to duty. Just a naval order, that's all."[23]

By the time Admiral Leahy settled into his new wheelhouse, regular Tuesday meetings of the Joint Chiefs were conducted as formal affairs. They met after a group luncheon, and during the meetings they would refer to each other as "General" or "Admiral," just as they did around their subordinates.[24]

Leahy ran the weekly Joint Chiefs meetings like the smooth diplomat he was, providing the oil that doused more than a few sparks in the secret meetings. As the president's man, he needed less eloquence than Marshall and less stubbornness than King, and he could afford to make his points indirectly, rather than with sixteen-inch broadsides. He sometimes played dumb, reminding everyone that he had never attended the Naval War College.

"Well, George," he might say after one of Marshall's lengthy speeches, "I'm just a simple sailor. Would you please back up and start from the beginning and make it simple, just tell me step one, two, and three."

A naval staffer recalled, "Well, Marshall or Arnold, whoever it was, kept falling for this thing, and they would back up and explain to this 'simple old sailor.' And as they did it—which is what Leahy knew damned well would happen—and went through these various steps, they themselves would find out the weakness or misconception or that there was something wrong with it. So he didn't have to start out by saying: 'This is a stupid idea and it won't work.'"[25]

Marshall would, on occasion, return to the idea of formalizing the position of chairman of the Joint Chiefs of Staff. But Roosevelt was uninterested, for he had all the authority he needed to get whatever he wanted from his chiefs. When Marshall sent him a draft bill establishing the position, seniority, and pay of Joint Chiefs chairman, Roosevelt let the idea suffocate in the overstuffed files of the White House. He and his chiefs had more important things to think about.[26]

Like Guadalcanal.

TWENTY-FOUR

ALONG THE WATCHTOWER

꿍

K ING GREW UNSETTLED AS HE LOOKED AHEAD TO OPERATION WATCHTOWER, the invasion of Guadalcanal. Admiral Ghormley probably had enough marines to take Tulagi and Guadalcanal as scheduled on August 7, but once those marines landed, he knew Japanese bombers and warships on Rabaul would counterattack, bombarding and strafing without mercy. Then the real battle would begin.[1]

King had blustered to Marshall about how the Navy would go it alone if the Army didn't want to help. But as D-day in the Solomons drew near, he realized the Navy lacked the air strength to fend off the swarms of bombers that would soon be streaming out of Rabaul. Admiral Jack Fletcher, his carrier commander, had three large flattops converging on the islands—*Saratoga*, *Enterprise*, and *Wasp*—but carrier planes were not the kind of force Fletcher could scramble day after day to fight Zeros, Vals and Kates.[2]

He needed the Army after all.

Hat in hand but head held high, King approached Marshall for more air and antiaircraft support from MacArthur. Putting his case to the general forcefully and without blushing, he reminded Marshall that in London, Marshall had told the Combined Chiefs that TORCH, if approved, would require fewer planes than BOLERO would use over the next several months. This would free up as many as seven hundred bombers, which could be put to good use in the Pacific. It was time, King said, for the Army to lend the Navy a hand.[3]

Marshall mulled over King's request. He had no personal problem with supporting Ghormley's Tulagi operation, but the TORCH estimates he had relied upon had been pretty rosy, and it didn't look like all those bombers would be freed up.

Moreover, when he'd told the British that cancellation of ROUNDUP would allow the Americans to divert three bomber groups to the Pacific, he hadn't figured anyone would actually ask him for those planes.[4]

He hadn't figured on Admiral King.

Marshall checked with Hap Arnold. Arnold, keeping one eye fixed on his strategic bomber fleet in Europe, protested to Marshall that TORCH would gobble up every American plane coming off the assembly line. The Army, he warned, could not afford to give the Navy any additional bomber groups.

So the answer was "no." Marshall would need the Navy's support for TORCH, ROUNDUP, and MacArthur's campaign against Rabaul, however, so for the moment, Marshall was unwilling to start a war with the Navy. He decided to wait a while before sending King his reply.[5]

· · ·

On the morning of August 7, the ramp doors dropped and the U.S. Marines stormed the beaches. Japanese defenders on RINGBOLT—the Allied code name for Tulagi—fought back fiercely once two leatherneck battalions waded ashore.

CACTUS, known to locals as Guadalcanal, seemed an easier catch. Hitting the defenders with complete surprise, eleven thousand devil dogs of the First Marine Division forced their way ashore and secured their principal objective, Lunga Point airfield, within two days.

The Japanese fleet was quick to respond. A squadron from Rabaul slipped through the Solomons "slot" in the early hours of August 9 and caught the Allies by surprise off tiny Savo Island, northwest of Guadalcanal. In a furious night attack, the Imperial Navy sent four Allied cruisers to the bottom, killing more than a thousand sailors in the worst ship-to-ship defeat in U.S. naval history.

The blow fell just as Admiral Fletcher was withdrawing his carriers to open water, which left Admiral Richmond Kelly Turner's invasion fleet without air cover. Fearful of another surprise attack, Turner withdrew his transport and supply ships, leaving the marooned leathernecks to deal with Japanese bombardment, counterattacks, and empty bellies.[6]

· · ·

An officer never knocked on King's cabin door—not if he knew what was good for him. Aboard *Dauntless* an unwritten order forbade aides from disturbing the admiral after he turned in for the night. Since there was nothing King could do from his flagship, good news, bad news—anything but an immediate summons from the president—would have to wait until morning.[7]

The hand that knocked on the door in the wee hours of August 12, a hand

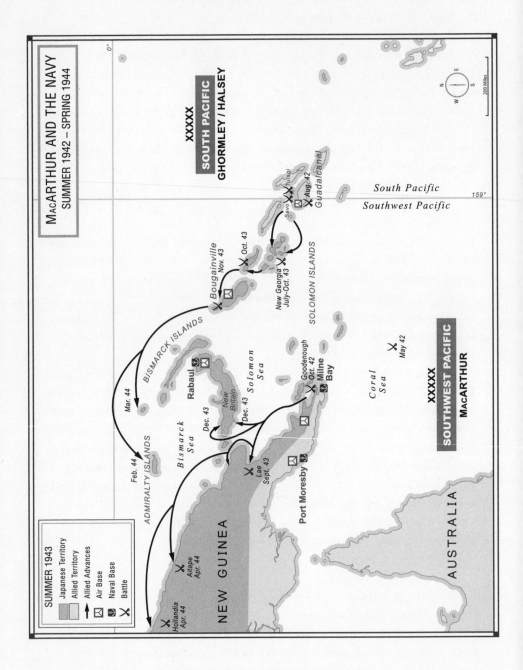

MacARTHUR AND THE NAVY
SUMMER 1942 – SPRING 1944

SUMMER 1943
- Japanese Territory
- Allied Territory
- Allied Advances
- Air Base
- Naval Base
- Battle

XXXXX
SOUTH PACIFIC
GHORMLEY / HALSEY

XXXXX
SOUTHWEST PACIFIC
MacARTHUR

South Pacific
Southwest Pacific
159°

Guadalcanal
Aug. '42
Tulagi
Savo
New Georgia
July–Oct. 43
SOLOMON ISLANDS
Bougainville
Nov. 43
Oct. 43

BISMARCK ISLANDS
ADMIRALTY ISLANDS
Feb. 44
Mar. 44
Rabaul
New Britain
Bismarck Sea
Dec. 43
Solomon Sea
Dec. 43
Goodenough
Oct. 42
Milne Bay
May 42
Coral Sea

Hollandia
Apr. 44
Aitape
Apr. 44
NEW GUINEA
Lae
Sept. 43
Port Moresby

AUSTRALIA

N
200 Miles

belonging to King's flag secretary, was shaky. But knock it did, and the disheveled admiral murmured as the officer turned on the light.[8]

"Admiral, you've got to see this," the flag secretary said. "It isn't good."[9]

King read the long-delayed message from Admiral Turner. Allied losses at Savo Island were four cruisers sunk, plus one cruiser and two destroyers damaged. "Heavy casualties," the report read. Supply and escort ships were pulling out of Guadalcanal. No enemy warships reported sunk.[10]

"I can't thank you for bringing me this one," King said bitterly. He stared at the message in disbelief.

"They must have decoded the dispatch wrong," he said at last. "Tell them to decode it again."

The secretary scurried off, and King slumped back on his bed.[11]

Ernie King prided himself on seeing the picture through the enemy's eyes, realistically and without the emotion that clouded other men's vision. But to King, Japan's strategic mentality was as dark as the Java Sea, and he found himself caught off guard by the ferocity of Japan's response. He had known a counter-attack was likely, a raid almost certain, but the Japanese were throwing far more at Guadalcanal than King thought rational. "While we expected the Japanese to react, we did not expect as violent a reaction as actually came," said one of King's confidants later. "The Japs dropped everything they were doing everywhere else, and piled into the Solomons."[12]

For Ernest J. King, the Savo Island battle was the blackest night of the war. Hundreds, possibly thousands of men had been lost. His great Solomons offensive was stalling, and marines who had landed in the first waves were stranded, their reinforcements pulling back, their only air cover—Army air cover—weak.[13]

Hauling himself out of bed the next morning, King rode to Main Navy and sent a terse message to Ghormley demanding an explanation. He dispatched a team of investigators to determine who, if anyone, was to blame, and ordered his staff to throw a news blackout on the disaster.[14]

As King was sorting through the red waters of the Solomons, deciding what to tell the president, a note arrived from the White House. It read:

Dear Ernie:—

You will remember the "sweet young thing" whom I told about Douglas MacArthur rowing his family from Corregidor to Australia—and later told about Shangri-La as the take-off place for the Tokyo bombers.

Well, she came to dinner last night and this time she told me something. She said, "We are going to win this war. The Navy is tough. And the

toughest man in the Navy—Admiral King—proves it. He shaves every
morning with a blowtorch."
   Glad to know you!
      As ever yours,
      F.D.R.

   P.S. I am trying to verify another rumor—that you cut your toenails
*with a torpedo net cutter.**[15]

In the wake of Savo Island, King's focus—besides replacing lost ships and men—was to keep Japanese intelligence officers from updating their order of battle charts to write off four Allied cruisers. That meant battening down some tight hatches on the news.[16]

As he had done after Coral Sea, King went to extraordinary lengths to ensure that no one knew of American losses in the Pacific. Several weeks after Savo Island, the Army-Navy Joint Planning Board invited the respected *New York Times* correspondent Hanson Baldwin to Washington to share his impressions after a recent tour of the South Pacific. While briefing some thirty senior Army and Navy officers on his observations, Baldwin happened to mention the temporary loss of the battleship *South Dakota*, which left only one U.S. battleship in the South Pacific.

To his astonishment, one of King's planners, Captain Charles Brown, leaped to his feet, roaring, "I object to that! I object to that!"

Everyone stared at Brown.

"This is top secret information!" Brown continued, not noticing the puzzled looks. "Admiral King has given the strictest orders that no one is to know about this!"

Baldwin was dumbfounded, and everyone, even Navy men, looked at Brown in bewilderment. Baldwin said he hadn't told anyone about the losses—except, of course, the Army-Navy planning board gathered together in the conference room that day.

When Brown continued to protest, it dawned on Baldwin what the captain was saying. When King ordered that no one was to know about *South Dakota*, he meant *No One* was to know—not even the board that formulated war plans for

---

* The blowtorch joke dated back to 1940, when an electric company executive told King, "They tell me you're so tough you shave with a blowtorch." Not long afterward, the executive sent King a miniature blowtorch crafted by Tiffany's. A later acquaintance sent him a brass crowbar engraved "A Toothpick for a Blowtorch."

the Joint Chiefs of Staff. The Army and Air Forces "had no idea of what had happened at Guadalcanal or Savo," Baldwin later wrote. "And these same officers were supposed to be planning! That, I think, is a hell of a way to run a war."[17]

It was a hell of a way to run a war. But as Japan piled reinforcements onto Guadalcanal's sandy west side, Admiral King faced a more immediate problem: he had caught the tiger by the tail, and neither he nor his planners could figure out how to turn it loose.

The Navy was becoming stuck in the Solomons, and it could not afford to remain stuck there for long, as both supply lines to Australia and America's military reputation were wrapped up in the success of WATCHTOWER. So five days after Savo Island, King followed up his request for bombers that Marshall had been ignoring for two weeks. Forced to respond, Marshall offered King one of the fifteen groups he had told the British he could transfer to the Pacific if ROUNDUP were canceled. But he would go no further.[18]

To Marshall, King was the kudzu wandering through his global strategy. Giving to Ghormley meant taking from MacArthur—and taking from Eisenhower, whose TORCH operation would go forward in just over two months. In Europe, Marshall was trying to convince a reluctant British ally to invade France in 1943, and Arnold was trying to amass a strategic bomber force in England. A departure of more bombers, fighters, and transports to the Pacific might give the British an excuse to postpone ROUNDUP indefinitely.[19]

Though it meant risking defeat in the Solomon Islands, Marshall, Leahy, and Arnold agreed that TORCH must remain the top priority. Marshall would help Ghormley by diverting some bombers he had planned to send MacArthur, and he would order MacArthur to give Ghormley as much air support as he could spare. Otherwise, he would stick to the "Germany First" plan.[20]

Though Guadalcanal was a Navy show, MacArthur knew Marshall might tighten the same ligature on him when it came time to hit New Guinea and Rabaul. In language calculated to explode on the Joint Chiefs like a time-delay bomb, MacArthur warned that America would face a repetition of the disasters of the war's early months unless the president and chiefs matched the Japanese on air, ground, and sea.[21]

Caught between shrill warnings from MacArthur and the demands of TORCH, Marshall scraped together an infantry division to send to the hard-pressed Pacific. He ordered Arnold to look for more planes, and passed MacArthur's plea for more support to Roosevelt.

He also asked Arnold to schedule a trip to the Pacific, to see what things looked like on the ground. Knowing the airman would be unpopular with the beleaguered marines and sailors, he advised Arnold before leaving: "Listen, don't get mad, and let the other fellow tell his story first."[22]

That was all Marshall could do for the Pacific War.

★ ★ ★

"Arnold is naturally discouraged by the raids that have disrupted all his efforts," Stimson told his diary on September 2.

Arnold had not been referring to German air raids over England, or Japanese raids on Guadalcanal. His prized bomber fleet—the strategic force that would wreck Hitler's industrial base—had become the target of King's shrill demands for aircraft. Before he left for the Pacific, Arnold complained to Stimson, "King never lets up. He has not receded one inch from his demands upon us and I prophesy that he will eventually get them all." The air warrior confided to his crumpled pocket diary, *"King for some time has tried to get more planes for the Pacific; tries subterfuge and cunning. Navy is trying to run a land war, relying on Army Air and Marines to put it across, but does not want to have its battleships or cruisers meet Jap Navy or get sunk."*[23]

At a Joint Chiefs meeting on September 15, King and Arnold plunged into a bitter argument over King's calls for more aircraft. Arnold argued that by April 1943, the Allies would have more planes in the Pacific than the Japanese would. But the marines being strafed on Guadalcanal didn't give a damn about April 1943, for without air support, they might not last that long. King needed those planes now.

The two commanders got nowhere, and after they had spun their wheels, Leahy finally suggested that the chiefs postpone their discussion until Arnold had a chance to see the Pacific for himself. Leahy managed to calm tempers, but he couldn't change the basic math: neither Arnold nor King had enough bombers to go around.[24]

King spent the next few weeks hectoring Marshall for more air support from Army assets in the Southwest Pacific, meaning MacArthur. Marshall passed along King's request, which drew a typical MacArthurian reply—pessimistic and rife with scorn for the Navy: "If we are defeated in the Solomons, as we must be unless the Navy accepts successfully the challenge of the enemy surface fleet, the entire Southwest Pacific will be in gravest danger." It was imperative, he said, "that the entire resources of the United States be diverted temporarily to meet the critical situation; that shipping be made available from any source; that one corps be dispatched immediately; that all available heavy bombers be ferried here at once . . ."

MacArthur's list went on. And on.[25]

TWENTY-FIVE

GIRDLES, BEER, AND COFFEE

⟡

N LATE 1942, MIDTERM ELECTIONS AND FIGHTING IN THE SOLOMONS LEFT AN invisible scar upon each other.

"In the two weeks before election the loss or retention of a jungled island 5,000 miles away may thus materially affect the character of the next Congress and the government of many a State," *Time* remarked on October 26. "These are the weeks when political candidates are at the mercy of chance." One Democratic strategist predicted, "The loss of the Solomons, if we do finally lose them, is going to set this country afire. Hell's fire, the people will be mad."[1]

War was intruding on politics, just as politics intrudes on war. As the elections drew near, antiadministration forces began shelling the White House with the usual salvos of late-campaign hyperbole. They latched onto the bloody stalemate in the Solomons, where soldiers, marines, and sailors were battling a fierce Japanese counterattack, and they accused Roosevelt of failing to put real muscle into the Pacific war. They claimed MacArthur had been denied supreme command in the Pacific theater to squelch his chances as a presidential candidate, and demanded to know why MacArthur wasn't getting more men and supplies to fight Japan in New Guinea.[2]

FDR had no intention of allowing a festering sore like Guadalcanal to infect the congressional elections. In late October, he ordered the Joint Chiefs to ensure "that every possible weapon gets into that area to hold Guadalcanal." Marshall and Arnold obediently scraped up a squadron of B-24 bombers with partly trained crews and sent them to New Caledonia. When Marshall told Roosevelt that shipping was the impediment to sending more weapons, Roosevelt ordered the War Shipping Administration to release twenty more cargo ships. What was good for the First Marine Division was good for the administration, and Roosevelt would make sure both had a fighting chance come November.[3]

A few days before the election, FDR was called to referee a fight between Elmer Davis, head of the Office of War Information, and Admiral King. Davis, charged with keeping the public informed of the war's progress, wanted to release news of the loss of the carrier *Hornet*, which had been sunk the day *Time* wrote of politicians being at the mercy of chance. King, naturally, opposed telling the public anything that would allow Japan's intelligence men to write *Hornet* off their order of battle charts, and refused to allow Davis to release the news.

Balancing military security against the upcoming elections, FDR again struck a compromise. "If the announcement of the loss were withheld, the Republicans would say it was on account of the election," he told his secretary, Bill Hassett. But taking King's objection into account, he agreed to announce only the loss of a carrier, not the class of carrier or its name.

Davis pressed his case. The families of sailors on other aircraft carriers would worry unnecessarily about their fathers, sons, and husbands if the carrier's name were kept from the public. FDR held his ground. The lives of the living, he reasoned, could be jeopardized if the Japanese knew the U.S. Fleet was down one large, fast carrier. Davis would get letters and calls, but *Hornet*'s fate would remain, to an extent, under wraps.[4]

The landings in Africa, if successful, would give the Democrats a big political boost, and FDR wanted that boost before November 3, not after. During an early meeting with Marshall on TORCH, Roosevelt had folded his hands and pleaded in mock supplication, "Please make it before Election Day."

Yet planners on both sides of the Atlantic recognized that an October date was unrealistic, given the transportation schedules required to move 60,000 men, water, food, weapons, and equipment from North America to North Africa—to say nothing of two other task forces embarking from England. When Eisenhower informed Washington that the landings would take place on the morning of November 8, Roosevelt was disappointed, but he went along without a word of complaint. The needs of the operation took first priority, and D-day for TORCH would remain in the hands of soldiers, not politicians.[5]

After D-day, however, a great many questions would remain in the hands of politicians, and some of those politicians would be French. There were an estimated 120,000 French troops in Morocco, Algeria, and Tunisia, and if those soldiers turned their guns on the invaders, the Allies might be driven into the sea. Even if the Allies managed to shoot their way through the French Army, they would take horrendous losses more profitably spent against Rommel than against the French.

If, on the other hand, the French joined the Allies, they could shut down Rommel's supply ports in Tunisia and force the Desert Fox to surrender—or starve.

The burden of persuading those Frenchmen to change sides would fall on Eisenhower's shoulders, and at Roosevelt's direction, Robert Murphy, America's consul general in Algiers, flew to London disguised as a lieutenant colonel to brief Eisenhower on his options.* The Allies, Murphy said, could align themselves with one of three faction leaders: Charles André Joseph Marie de Gaulle, leader of the French resistance in Dakar; Admiral Jean Louis Xavier François Darlan, commander of the French Navy and Vichy's second-in-command; or Henri Honoré Giraud, a former French Army commander who, Murphy said, had the stature to rally the French officer corps.[6]

After talking it over with Murphy, Eisenhower decided Giraud was the horse to back. De Gaulle's anti-Vichy broadcasts on BBC radio made him *un traître infâme* to most French officers, and Darlan, though powerful, was a dyed-in-the-wool anti-Semitic fascist. Giraud, said Murphy, was the compromise choice most likely to convince his countrymen to lay down their weapons—or better still, fire them at Germans and Italians.[7]

・　・　・

The problem of governing conquered lands had plagued the U.S. Army since the days of the Mexican War. Military men are not, as a general rule, adept politicians, and basic training doesn't cover subjects like municipal administration or international law. So in the summer of 1942, Stimson and Marshall asked reservist Brigadier General Cornelius Wickersham, a partner at an old Manhattan law firm, to head a new school of military government in the picturesque hills of Charlottesville, Virginia.[8]

As Marshall and Stimson envisioned the school as a way to avoid local mistakes that had ignited the Filipino revolt after the Spanish-American War. Like the infantry school at Fort Benning, the School of Military Government would be an advanced institution for officers tasked with rebuilding shattered governments on a democratic foundation.[9]

Liberals within the administration saw a right-wing demon in the making. Secretary Ickes, Undersecretary Acheson, and Harry Hopkins began hearing rumors of

---

* Marshall selected Murphy's disguise, reasoning, "No one ever pays attention to a lieutenant colonel."

a college of little MacArthurs—men, the stories said, sympathetic to a coup d'état against Roosevelt, in which Wickersham would be anointed president and Marshall his vice president. Bureaucrats in State, Treasury, and Interior began raising a ruckus.[10]

The accusations stung Marshall, who knelt at the altar of civilian rule. War in a democracy involves more than flanking movements and aerial bombardment, and George Marshall understood this better than any man in any nation's uniform. He explained his philosophy to John Hilldring upon Hilldring's appointment to head the Army's civil affairs branch:

> We have a great asset and that is that our people, our countrymen, do not distrust us and do not fear us. Our countrymen, our fellow citizens, are not afraid of us. They don't harbor any ideas that we intend to alter the government of the country or the nature of this government in any way. This is a sacred trust that I turn over to you today. We are completely devoted, we are a member of a priesthood really, the sole purpose of which is to defend the republic.[11]

Stimson dispatched his troubleshooter, Assistant Secretary Jack McCloy, to assuage the State Department. Before long, Acheson was satisfied that the Army was neither trying to supplant the Department of State, nor building a shadow government on American soil.

But alarm bells rang anew as Election Day approached. Stimson told his diary after one cabinet meeting, *"The cherubs around the White House still scented danger, and when they found that the school had enrolled a man by the name of Julius Klein (the same name as an old intimate friend and adviser of Herbert Hoover's), they thought the worst had happened and rushed to the President with the bad news. Apparently he swallowed the story and I found a rather sharp memo from him wanting to know what the Army was doing and by what authority it was doing it without going to him."*[12]

Three days later, Roosevelt, Stimson, Ickes, and other cabinet members argued for an entire afternoon over Julius Klein's role in the School of Military Government. Stimson wrote, *"I had all the typical difficulties of a discussion in the Roosevelt Cabinet. The President was constantly interrupting me with discursive stories which popped into his mind while we were talking, and it was very hard to keep a steady thread."*[13]

Hearing Stimson's report of the Julius Klein discussion, Marshall was astounded. "Mr. Stimson came back from a cabinet meeting that lasted almost all afternoon and I discovered that they had been discussing entirely this particular officer whose name was the same as that of an intimate of a onetime Republican leader," he recalled. "Mr. Roosevelt was very bitter about this matter and Mr. Stimson was very much stirred up over it."

Marshall looked into the matter and found that the Julius Klein at the School of Military Government was an entirely different fellow from the one Herbert Hoover had known. Evidently the man who consumed an entire afternoon of the wartime cabinet's attention had never met Hoover in his life. "I was shocked when I found out how long the cabinet spent—all afternoon—warring over this thing," Marshall said.[14]

Victory is paid for in blood and materiel. By October 1942, the Marines and the Navy had supplied most of the first, and the public supplied the second. Financed by defense bonds and taxes, war production for the year soared to $58.7 billion, elbowing its way into nearly a third of the nation's industrial capacity.

In May, FDR modified his list of production goals to make them less extemporaneous and bring some of them into line with recommendations of the Joint Chiefs. He reduced a few items (400,000 machine guns instead of an even half million), and added a few things to his "Must-Have" list, such as landing craft.[15]

Virtually everything on Roosevelt's shopping list had to be taken from consumer production, from cars to cans to clothing. For example, the War Production Board, the agency charged with finding enough cotton, steel, and everything else, issued "General Limitation Order L-85," which regulated the size and look of civilian clothing. To comply with Order L-85, men's apparel designers cut suits with narrow lapels and cuffless trousers, while women's designers introduced "mini" skirts, low-neckline dresses, and two-piece bathing suits. Other consumer goods, from sugar to Clabber Girl to Christmas ornaments, were swept up in a wave of regulation and restriction that spread over the country like an immense, tangled Victory Garden.[16]

The War Production Board's problem was not a shortage of factories or tools, but a lack of raw materials. Rubber, for instance, was especially tight. As Hopkins had predicted in 1940, Japan's occupation of Indochina and the Dutch East Indies sent American stocks of crude rubber tumbling to dangerous levels. The Army was forced to replace rubber tank treads with less-efficient steel ones, and it issued gas masks made of a new synthetic substance called neoprene. None of these substitutes were entirely satisfactory to the men whose lives depended upon them, and some items, like aircraft tires and intravenous bags, required the real McCoy.[17]

The public would have to pitch in. Two weeks after Pearl Harbor, FDR authorized a ban on the sale of new automobile tires. He directed the Office of Price Administration to draft a list of essential professionals whose work required them to buy tires for their cars. From his cramped government office at the corner of Second and D streets, assistant pricing chief John Kenneth Galbraith drew up a list that he limited to classes of professionals whose work was deemed essential to the national interest, such as doctors and public officials.

Galbraith, not a religious man, excluded ministers from his list. The omission outraged rank-and-file faithful, especially in the rural South, where clergymen often rode big circuits and needed cars to comfort their sick or grieving parishioners. When FDR learned of Galbraith's slight, he gave his pricing chiefs a fire-and-brimstone sermon of his own. "F.D.R. was outraged that anyone should be so casual about both fundamentalist religion and the fundamentals of American politics," Galbraith later recalled. Ministers were quickly added to the "tired" list.[18]

Other means to conserve rubber backfired. When production of women's girdles was banned, a backlash from the fairer sex over "Price Regulation 220" forced the War Production Board to find ways to weave rayon and synthetic rubber threads into civilian garments. For a public already weary from meat, sugar, and gasoline rationing, rubber was becoming a sore spot.[19]

After consulting with Wall Street investment guru Bernard Baruch and other informal advisers, FDR attacked the problem with a national rubber drive. He asked his fellow citizens to bring old tires, bicycle inner tubes, garden hoses, rubber overshoes, and anything else made of rubber to gas stations, which would pay a reimbursable penny a pound for used rubber.[20]

He opened his appeal with a radio address that began, "I want to talk to you about rubber, about rubber and the war—about rubber and the American people." He described the embryonic state of the synthetic rubber industry, Japan's control of rubber trees, and the importance of rubber to victory. He explained how the shortage of rubber on the home front affected the fighting man on the battlefront. Until synthetic rubber became available in quantity, he asked Americans to donate their old tires, raincoats, and worn-out overshoes to the war effort.[21]

The drive was a popular success. Thousands of ordinary citizens turned in dog chew toys, rubber-band balls, and galoshes in a roundup that garnered more than two hundred tons of rubber for the war effort.

For the Army's needs, the rubber drive was no more than a drop in a very large bucket. But the rubber drive, like FDR's fireside chat with the maps, raised the public's consciousness of the relationship between conservation and victory, and it helped ease some of the grumbling over draconian rationing laws. Roosevelt's goal was not merely to glean secondhand rubber, or even to educate the public about a specific problem, but to broadly enlist the public's help—to give all Americans a chance to feel they were hammering a nail into Tojo's coffin.[22]

. . .

Not everyone would be hammering the same nails. Naval regulations obliged officers to drop their anchors at age sixty-four, and as Ernie King approached his sixty-fourth birthday, he saw plenty of unfinished business on his paper-strewn desk. The foundations of victory had been laid, but the Solomons still hung in the balance. TORCH had not yet been launched, and there would be plenty of rough water ahead: U-boats to sink, leathernecks to land, and islands to conquer before the sails could be trimmed and the guns stoppered.

With the pounding he was taking in the press over Guadalcanal, the president just might feel it was time to make a change at the top, and King believed he should offer Roosevelt that chance. On October 23, one month before his sixty-fourth birthday, he sent the president a short letter calling attention to the date and its significance. He concluded, "I am, as always, at your service."[23]

Reading King's letter, FDR shook his head. "He didn't have to remind me of that," he told McCrea. With his fountain pen, he scribbled his endorsement across the letter's foot and sent it back to Main Navy:

E.J.K.,
So what, old top?
I may send you a birthday present!
F.D.R.

True to his quip, on King's birthday, Roosevelt sent Old Top an autographed photo.[24]

The president was in King's corner for now, but how long he would stay there was a question other Washingtonians began asking. As the struggle for the Solomons dragged on and casualty reports made the papers, rumors flew around press circles that the Navy wasn't giving the American people the full scoop on U.S. losses. Clouds began gathering in Congress over the Navy's credibility gap, and MacArthur's sycophants quietly accused the Navy of bungling the Solomons operation.[25]

Apart from the naval committee chairmen, King had few allies in Congress. He spent little time courting most legislators, and when he defended the Navy's budget requests, he made clear that if Congress didn't support him there would be a political cost.

"War inevitably results in waste," he told one committee. "We cannot afford to pussyfoot when it comes to appropriating money to carry on." Leaving nothing to congressional imagination, he added, "It is our belief that if the matter of appropriations is approached from a peacetime point of view, instead of from the war or realistic point of view, we shall be unable to prosecute the war effectively. In other words, Congress automatically shares the responsibility for the conduct of the war in respect to the funds appropriated."[26]

The growing tide of criticism worried King's friends. Secretary Knox took public relations out of the admiral's fumbling hands, and Undersecretary Forrestal began biweekly lunches with congressmen, where he introduced them to war heroes who had just returned from the front, to underscore the difficulty of the task the Navy was undertaking in the Solomons.[27]

Lobbying efforts by Knox and Forrestal generally shored up the Navy's rigging, but they did little to insulate its top sailor. It was up to two of King's most unlikely friends, a lawyer and a reporter, to save the admiral from himself.

Cornelius Bull, King's personal attorney, was having drinks one afternoon at the National Press Club with his friend Glen Perry of the *New York Sun*'s Washington bureau when the subject of Admiral King came up. As their glasses sweated and they swapped Washington gossip, the two men hatched a plot to break the ice between Ernie King and the Fourth Estate.[28]

The admiral, they acknowledged, was a man "reputed to hate newspaper men worse than anyone but Japs and Germans." But they figured they could persuade him to meet informally with a few carefully selected reporters at Bull's house in Alexandria. King could give journalists background information and reasons for military decisions that might seem inexplicable to those on the outside. No notes would be taken, and everything would be off the record. If word of the meetings leaked, they would end—as would, most likely, King's career. But if the reporters didn't muck it up, they would get two things every scribbler craves: access and information.[29]

As Bull broached the subject to King, the wary admiral listened. While he disliked newspapermen, he realized he was drifting into dangerous shoals; to get back to safe, deep water, he had to win over some allies among the press. He agreed to give it a try.[30]

Bull and Perry called six of their closest colleagues—Scripps-Howard columnist Raymond Clapper and editors from five major dailies. The group gathered at Bull's home on the Sunday before Election Day, and waited patiently for the legendary admiral to arrive.

King knocked on the door promptly at eight o'clock. Bull took his greatcoat and scrambled-eggs cap, and he seated himself in an easy chair. Sipping a glass of beer, a cigarette between his fingers, he appeared at ease among his natural enemies. "He spoke with great facility, without notes of any kind, and—I might add—without any expletives," recalled journalist Phelps Adams of the *New York*

Sun. "He told us what had happened, what was happening, and often, what was likely to happen next. He reported the bad news as fully as the good news, and in many cases, explained what went wrong on the one hand, and what strategy, weapon, or combination of both, had brought success on the other. Throughout, he was completely at ease and always patient, welcoming the questions that frequently interrupted his narrative, and answering them frankly and easily."[31]

The secret conclaves, which continued for the next two years, narrowed a credibility gap that had yawned during the uncertain months of 1942. The roaring sea kraken the newsmen had expected turned out to have an approachable, human side. Said Perry afterward, "A lot of firm friendships grew out of the correspondents' discovery that King was not only a great officer but a great guy, and King's discovery that the correspondents were smart, closed-mouthed when it counted, and also nice guys."[32]

.   .   .

For King's commander-in-chief, however, the political picture was a mixed bag. Polls suggested the public was behind him on rationing, compulsory work, and even drafting eighteen-year-olds. Solidarity behind the troops was as robust as ever; Frank Loesser's hymn "Praise the Lord and Pass the Ammunition" became the war's early hit song, and Spike Jones rode to fame on "Der Fuehrer's Face," a novelty song written for a Walt Disney short film titled *Donald Duck in Nutzi-Land*.

But hardships at home and the absence of victory abroad—or even a second front—exacted a steady political toll. Transportation and food shortages caused the frequent stirs, and on October 26, the Office of Pricing Administration announced a one-cup-per-day coffee-rationing measure, an unpopular move designed to cut down on cargo ships hauling the sacred bean from Central and South America. On hit song charts, "Pass the Ammunition" would soon give way to Louis Jordan's "Ration Blues," and the demands of a war economy breathed new life into Republican prospects for the midterm elections.[33]

Shuffling to voting booths through the specter of caffeine withdrawal, the public arrived at their precincts on November 3 in a predictably cranky mood. In an election with unusually low voter turnout, Republicans picked up forty-seven seats in the House, leaving Democrats with a four-vote majority. They harvested another nine in the Senate, cutting the Democratic lead to ten. In FDR's home state, voters in the gubernatorial election chose a Republican, prosecutor Thomas E. Dewey, to end a two-decade run of Democratic rule in Albany.

Many Democrats who kept their seats took the election as a warning sign that the public was growing tired of the traditional Roosevelt slogans, at least on domestic issues. As *U.S. News* concluded, "An unofficial coalition was in the making between anti–New Deal Democrats and Republicans to pluck all budding social reforms from future war legislation."[34]

Beneath the poplars of Shangri-La, the president's wooded retreat in Maryland's Catoctin Hills, FDR could breathe only a shallow sigh of relief when the returns were counted.* The public hadn't turned out his party, but that was about all the solace he could take. With conservative Democrats liable to defect on any number of issues, he was now, in effect, the head of a coalition government—a coalition in which Republicans would be almost as important as the New Deal cadre had been ten years earlier.

To remain in power, FDR needed a victory, and his next opportunity would dramatically alter the lives of thousands of men who, at that moment, were being drawn toward a storm of fire.[35]

---

* Shangri-La, a rustic Civilian Conservation Corps project, was later renamed Camp David by President Eisenhower, after his grandson.

TWENTY-SIX

THE DEVIL'S BRIDGE

Roosevelt was a master of deception, but secluded among the rustic furniture, martini shakers, and worn-down rugs of Shangri-La, he could not conceal the cocktail of tension stirring within him. He made little small talk, grinned less, and seemed distant, as if he were staring out from the back of a long, dark cavern. At one point, his secretary Grace Tully gave him a worried look. Noticing her concern, Roosevelt brushed her off, muttering something about an important message he was expecting.[1]

It was Saturday evening, November 7. Morning had broken on Africa's shores.

As the clock patiently ticked off the minutes, Roosevelt waited for his phone to ring. Ring it did, shortly before nine p.m. Grace took the call and announced that the War Department was on the line.

Roosevelt's hand shook as he reached for the receiver. He sat in silence as the voice on the other end told him of events unfolding four thousand miles away.

"Thank God. Thank God," he said at last. "That sounds grand. Congratulations. . . . Thank God."

With the blissful face of the reprieved, he returned the handset to its cradle, then emerged from his dark cavern. "We have landed in North Africa," he announced. "Casualties are below expectations. We are striking back."[2]

Over the next three days, Signals men at the new Pentagon building decrypted cables from Eisenhower's headquarters on Gibraltar. French and colonial troops had resisted the landings, but casualties on D-day had been light. At the harbor city of Oran, the defenders had fought back fiercely, but the surrounding beaches and a key airfield were secured by twilight. Algiers fell at noon on D-day, and

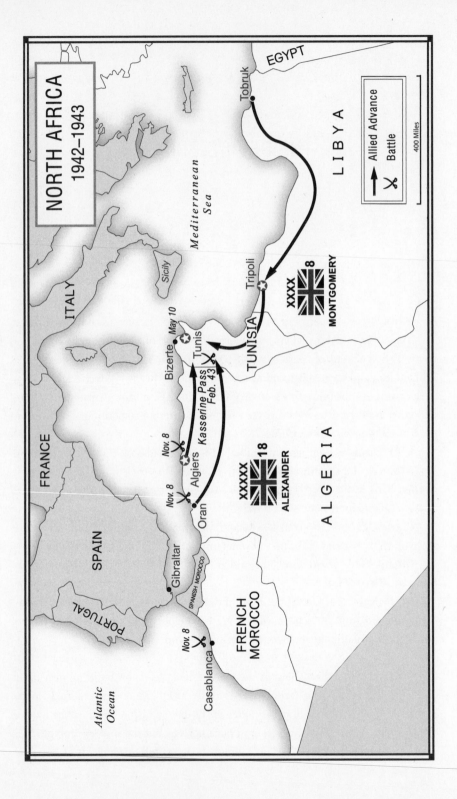

NORTH AFRICA
1942–1943

EGYPT

Tobruk

LIBYA

Mediterranean
Sea

Sicily

ITALY

Tripoli

XXXX
8
MONTGOMERY

TUNISIA

Bizerte *May 10*

Tunis

*Kasserine Pass
Feb. 43*

Nov. 8

Algiers

Nov. 8

Oran

XXXXX
18
ALEXANDER

ALGERIA

FRANCE

Gibraltar

SPANISH MOROCCO

SPAIN

PORTUGAL

Nov. 8

Casablanca

FRENCH
MOROCCO

Atlantic
Ocean

Allied Advance
Battle

400 Miles

Eisenhower's deputy, Major General Mark Clark, began moving Ike's advance headquarters to an Algiers hotel. And in an incredible stroke of luck, Admiral Darlan, Vichy's number two leader, had been captured in Algiers while visiting his polio-stricken son.[3]

On Africa's treacherous Atlantic coast, General Patton landed at three points in French Morocco, and by mid-morning he was rolling toward the port city of Casablanca. Casablanca was defended by a strong garrison of regulars, Foreign Legionnaires, and the French battleship *Jean Bart*. Patton's men converged on Casa from three sides, and based on early reports, Marshall expected the city to fall into Allied hands soon.

Three days after the landing, all was quiet. Under a cease-fire deal brokered by General Clark and Robert Murphy, Admiral Darlan was appointed High Commissioner of French North Africa, and General Giraud was made commander-in-chief of all French forces there. Eisenhower flew to Algiers on November 13 to ratify the arrangement, and the truce became permanent. The Allies were ashore in Africa.*[4]

Eisenhower's decision to place a pro-German Vichyite in charge of liberated French Africa set off a political firestorm. From London, American correspondent Edward R. Murrow bellowed, "What the hell is this all about? Are we fighting Nazis or sleeping with them?" Referring to Darlan, the liberal magazine *Nation* took the metaphor a step further: "Prostitutes are used. They are seldom loved. Even less frequently are they honored." State Department mandarins—"starry-eyed circles," Hull called them—were outraged, and Secretary Morgenthau ranted to Stimson at length about the shame of dealing with "a most ruthless person who had sold many thousands of people into slavery."[5]

FDR understood that the cease-fire order had saved U.S., British and French lives, so he held his nose and backed the "Darlan Deal" as a matter of military necessity. He cabled Eisenhower, "You may be sure of my complete support for this and any other action you are required to take in carrying out your duties. You are on the ground and we here intend to support you fully in your difficult problems."

With the heat turning up at home, however, he would not commit himself publicly on Eisenhower's behalf unless he had to. And for the moment, he didn't have to. He told Eisenhower not to make his letter public, and cautioned him, "We do not trust Darlan."[6]

---

* Though not part of the deal, Roosevelt invited Darlan's wife and son, Alain, to come to Warm Springs, where Alain received polio treatment until the war's end. A grateful Alain Darlan returned to France in 1946.

*   *   *

"Do not worry about this. Leave the worries to us, and go ahead with your campaign."

So Marshall wrote to Eisenhower on November 20. He and Stimson wanted their top general focusing on the battle for Tunisia without looking over his shoulder at daggers flying back home. To shield Ike from political flak, he held a secret meeting with a group of congressmen and read them a long cable from Eisenhower laying out the military necessity behind the Darlan Deal. He then held a press conference where he told reporters the deal had saved 16,000 American casualties, and that criticism of Eisenhower over his pact was "incredibly stupid."[7]

To win over a few influential liberals, Stimson and his wife, Mabel, hosted a tea at Woodley for Morgenthau, speechwriter Archibald MacLeish, and Justice Felix Frankfurter. Over china cups, Stimson explained the rationale for the Darlan arrangement. His explanation, given and received in good faith, blunted their wrath. On the Republican side, when Stimson learned that Wendell Willkie would be making a speech lambasting the Darlan Deal, he rang up Willkie and warned his former client in blunt, Long Islander terms that his speech "would run the risk of jeopardizing the success of the United States Army in North Africa." Willkie hung up in a huff, but he modified his speech.[8]

Willkie, Morgenthau, and a few other opinion leaders had been defused, but visceral reaction—from journalists and activists who had no understanding of the complexities of Eisenhower's war—stung Roosevelt. Furious with the sanctimony of liberals who formed his political base, FDR fumed with indignation from within the White House, but could say nothing publicly.[9]

As political winds howled, FDR shut the cellar doors and waited for the storm to pass. But after enduring a week of editorial assaults, FDR felt he had to make an announcement of some kind. He issued a clarifying statement that, while not renouncing the Darlan Deal, put Darlan and Eisenhower on notice that the decisions made in Algiers were not set in stone. "I have accepted General Eisenhower's political arrangements for the time being," the statement read. "We are opposed to Frenchmen who support Hitler and the Axis." The arrangement, he emphasized, "is only a temporary expedient justified solely by the stress of battle."[10]

After releasing his statement, FDR met with reporters in the Oval Office, who pressed him further. Why were the Allies working with Darlan, an anti-Semitic French fascist? In his typically whimsical manner, Roosevelt quoted them "a nice old proverb of the Balkans that has, I understand, the full sanction of the

Orthodox Church: 'My children, you are permitted in time of great danger to walk with the Devil until you have crossed the bridge.'"*[11]

For Roosevelt, that bridge stretched out further than anyone's eyes could see. And Roosevelt had a hunch they would be walking with the Devil many more times before they were safely on the other side.

. . .

As 1942 drew to a close, Franklin Roosevelt could look over a year of shock and convulsion with a certain justifiable pride. He had pushed the conversion of a third of America's industry to military production. Landing craft from Louisiana, bombers from Seattle, and small-arms cartridges from Minnesota were flowing into Army and Navy warehouses. The automotive industry had produced nearly 25,000 tanks during 1942—dwarfing the world's previous leader, Germany, which had turned out panzers at an annual rate of 4,000. Antiaircraft gun production jumped from 696 in 1941 to a predicted 24,000 for 1943. The twenty-one billion rounds of small-arms ammunition scheduled for the next year was ten times that produced in 1941. Combat ship production had increased sixfold by the end of 1942, while merchantman tonnage leaped sixteenfold.[12]

The war was also changing American society. With the support of Congress—not without dissent from both parties—FDR had built an administrative apparatus that regulated American life from paychecks to pleasure drives. Women were going into the workforce for the first time in large numbers, the population began migrating toward the industrialized coasts, and Negro citizens had a vague promise of rough equality in war industries. The Army and Air Forces ballooned from 1.7 million men to 5.4 million, and by the end of 1943 that number would reach seven million. The Navy would grow to 2.4 million sailors by the middle of the coming year, of whom half would be sent overseas.[13]

There were mistakes, obstacles, and shortages, of course. Although Congress extended the draft to eighteen-year-olds, there were still not enough men to feed the industrial, agricultural, and military machines. Donald Nelson's War Production Board couldn't decide whether to give first priority to synthetic rubber, high-octane fuel, destroyer escorts, or merchant vessels—all of which the president had thrown onto the "Number One Priority" list. And beating plowshares into swords incurred a political cost. New Deal stalwarts complained that the pro-

---

* Stalin's thinking was nearly identical to Roosevelt's on this subject. Two weeks later, he borrowed an emphatic Russian idiom when he told Churchill that the Allies should feel free to use "even the Devil himself and his grandma" to defeat Hitler.

gressive agenda was being sacrificed to the war effort; Eisenhower's Darlan Deal enraged liberals, and conservatives rallied against a three-term monarchy cloaked as wartime necessity.[14]

Most of all, the picture beyond America's shores remained dark. In the Atlantic, U-boats were sending ships to the bottom in terrifying numbers. In Russia, Germans were fighting from house to house in Stalingrad, and in the Pacific, marines, soldiers, and sailors were still dying in the Solomon Islands. In Africa, Eisenhower's November offensive bogged down in the mud of Tunisia's rainy season, and Marshall and Stimson warned FDR that America would be mired strategically if it didn't get out of the Mediterranean, and soon.[15]

To fix this last problem, Roosevelt decided it was time to see Churchill again. But as 1942 drifted into the history books on a subdued New Year's Eve, he gave himself a little time to relax. He worked late into the afternoon, but spent part of the evening watching a new Warner Bros. hit starring Humphrey Bogart and Ingrid Bergman. It was called *Casablanca*.[16]

Roosevelt must have smiled to himself as he watched Bogey's intrigues against the fascists, for few people knew the home of Rick's Café Américain was about to become history's next turning point.

TWENTY-SEVEN

"HOLLYWOOD AND THE BIBLE"

∽

E RNIE KING ENJOYED WOMEN, LIQUOR, AND A GOOD BOXING MATCH. HE
knew the technical details of sea power and global logistics like few other
men, but the old sailor in him thought of grand strategy in earthy terms that a
forecastle tar would appreciate.

During a secret meeting at Nellie Bull's house in early November, King explained to his circle of journalists that the Pacific war would go through four
phases, which he likened to a reeling prizefighter's comeback. In the first, defensive
phase, the boxer regained his footing and covered himself against his opponent's
blows. In the second, "defensive-offensive" phase, the fighter covered himself
while looking for an opening to land a blow of his own. In the third, "offensive-
defensive" phase, he used one glove to block while he jabbed and struck with the
other. And in the last, offensive phase, he threw haymakers with both fists.[1]

By the end of 1942, King felt the war with Japan was entering its third,
"offensive-defensive" phase. Coral Sea was a jab. Midway was a clean uppercut.
Guadalcanal, though messy, was a body blow. If Japan were denied enough time
to consolidate her defenses, the tide could be turned with as little as 30 percent
of the combined Allied strength. Admiral Nimitz could light across the Central
Pacific, and a blockaded Japan would be starved into submission.[2]

The war against Germany had not yet revealed its path, at least to King.
Churchill's strategy, to encircle Germany from the Mediterranean, seemed to him
a fool's errand, because the German reich was economically self-sufficient. With a
strong Italian fleet at Taranto and excellent defensive ground on the Italian peninsula, the Mediterranean, King said derisively, was the "soft underbelly of the
turtle."[3]

King figured Northern France, Marshall's obsession, would eventually become the war's focus. The wolf had to be killed in its den, and the entrance to that den was the French coast. But until enough landing craft, soldiers, and airplanes had been collected in England, the time for a cross-Channel invasion was not ripe.[4]

If there were one way the Allies would *not* defeat Germany, it was through airpower alone. As he observed over a beer, "The Germans are pretty much the same stock as the British and ourselves and are tough people. The blitz did not crack the British morale and there is no more reason to suppose that Germany can be blasted into surrender by planes than the British."[5]

That left Stalin's Russia, which was swallowing entire German armies around Leningrad to the north and Stalingrad to the south. As King saw it, the Red Army had the manpower, the land route to Germany, and if Lend-Lease continued, the materiel needed to break the Wehrmacht. "In the last analysis," King told his listeners, "Russia will do nine-tenths of the job of defeating Germany."[6]

The last tenth would be up to the American warlords.

· · · ·

Looking through the plane's round windows, the passengers could see a kaleidoscope of ancient homes and modern office buildings, of consulates and kasbahs, French architecture and fountain squares. The clash of modern European and old Moorish confirmed General Patton's lyrical description of Casablanca as a combination of "Hollywood and the Bible."[7]

Flying over a patchwork quilt of sand, orange groves, and turreted mosques, the big four-engine Skymaster descended to treetop level as it neared Casablanca. The plane touched down on the runway and gently rolled to a stop, and a company of men in dark suits surrounded the Skymaster. A door, tiny against the plane's fuselage, opened. Through the door and onto a long wooden ramp debouched a wheelchair-bound man traveling under the code name "Don Quixote." Behind Quixote shuffled his squire, a lanky, sickly former commerce secretary named "Sancho Panza."[8]

The Secret Service men drove Quixote and Panza through heavily guarded streets to Anfa, an upscale suburb of Casablanca. There the three-story Anfa Hotel—nicknamed "Hotel Patton"—awaited its distinguished guests. The agents installed Quixote in Villa No. 2, one of the hotel's large, modern lodgments, and before long the man in the wheelchair was welcomed to Casablanca by a plump, cigar-smoking gentleman registered on British travel manifests as "Air Commodore Frankland."[9]

• • •

Roosevelt and Churchill had been planning this rendezvous since early December. They had invited Stalin—"Uncle Joe," as they privately called him—but Uncle Joe refused to leave Russia while a battle raged around Stalingrad. Because decisions on post-African operations could be deferred no longer, in late December Churchill had proposed a two-power conference to take place in Morocco. He dubbed the conference "SYMBOL."*[10]

Roosevelt brought along his military advisers, but not his military secretaries. It was "the prima donna element," he explained to Bill Hassett. "If I sent Stimson to London, Knox would have thought he should go."

But with a bit more candor, he hinted at his real reason for excluding the cabinet: "Churchill and I were the only ones who could get together and settle things."[11]

As they arrived in Casablanca, FDR's advisers were not unanimous about what to do after Rommel had been run out of Africa. At the only pre-conference meeting between FDR and the Joint Chiefs, held at the White House on January 7, Marshall admitted his own planners were not of one mind. While the U.S. chiefs preferred a cross-Channel operation, they conceded many logistical obstacles.

Yet a British-led detour through, say, Sardinia, would draw fierce Luftwaffe and naval attacks from Axis-held France, Italy, and the Balkans. Given enough time, those attacks would inflict heavy casualties and cost the Allies irreplaceable landing craft and transport ships. "To state it cruelly," Marshall told Roosevelt, "we could replace troops whereas a heavy loss in shipping, which could result from the [Sardinia] operation, might completely destroy any opportunity for successful operations against the enemy in the near future."[12]

For the moment, King's eyes turned to the sea-lanes between Britain and Asia. The shortest route from England to India ran through the Mediterranean to the Suez Canal, the Arabian Sea, and the Indian Ocean. He agreed with Marshall that Sardinia was a bad idea, but told Roosevelt that Sicily, whose airfields covered southern Italy and the central Med, would make a worthwhile target.

In the Oval Office that cold January afternoon, Roosevelt had given no guidance to his lieutenants. He wanted to know what the half million Allied troops in Tunisia would do when the shooting stopped, and warned his warlords that in the upcoming conference in Morocco the British would have a firm plan and would stick to it.[13]

---

* King later remarked, "Every meeting had to have a special name which Mr. Churchill liked to use. In fact, he rather fancied himself as a 'namer.'"

When the meeting broke up, the men advising Roosevelt had no firm plan. In fact, they were no closer to a decision than they had been when they sat down.

FDR's near-magical ability to give each visitor the impression that he agreed with him once again worked on Marshall. As Stimson told his diary, *"Marshall said that thus far they have the backing of the President. In a word, it is that just as soon as the Germans are turned out of Tunisia and the north coast of Africa is safely in the hands of the Allies we shall accumulate our forces in the north and prepare for an attack this year on the north coast of France."*[14]

*   *   *

George Marshall arrived in Casablanca at the end of a miserable trip that took him through Miami, Puerto Rico, Brazil, Gambia, and Marrakesh. The first leg occasioned a snub between King and himself over landing priorities. When they left Miami, Marshall flew with General Arnold and Sir John Dill, while King flew in a second plane. King habitually badgered his pilots to go full throttle, and his plane arrived over Puerto Rico ahead of Marshall's. Marshall's pilot, technically flying the senior officer, ordered King's pilot to remain aloft until Marshall disembarked, and after some argument, King's pilot circled the airfield until Marshall's plane touched down.[15]

When they landed, a riled King told Marshall his performance "cost you about a hundred gallons of gas." Never one to let go a grudge, he recalled years later, "I didn't like spending one hour in the air just so that [Marshall] would be the first to land."*[16]

For the voyage, Marshall's industrious aide, Lieutenant Colonel Frank Mc-Carthy, had supplemented the plane's standard survival equipment—life rafts, flare guns, drinking water, and fishing tools—with a complete set of desert gear: tents, hats, dried food, and trinkets to trade with the natives. Hearing that Marshall might also be sent to Russia after the conference ended, he added snowshoes, parkas, mittens, fur-lined boots, fire-starting tools, and heating blankets. Then, warned by the Quartermaster's office of a malaria threat in Bathurst, Gambia, McCarthy added mosquito boots, mosquito gloves, and floppy veiled hats that resembled beekeeper bonnets.

As the plane touched down on Bathurst's runway, the lieutenant colonel solemnly addressed Marshall's entourage about the dangers of malaria and instructed everyone that before leaving the plane they must be protected against local mos-

---

* To keep peace among their chiefs, the pilots agreed that in the future, Marshall's plane would take off well enough ahead of King's plane to ensure that Marshall landed first.

quitoes. Marshall obediently donned the oversized boots, floppy hat, and big gloves, leaving no skin surface uncovered. With an indulgent smile that preserved some of his dignity, he climbed down the plane's ladder to the runway, where a delegation of British notables and an honor guard awaited the general's review.

In Bermuda shorts and short-sleeved shirts.

Marshall, looking like Buck Rogers in an interplanetary fashion show, glared at McCarthy. The bemused hosts stared at the Americans.

Flushed with embarrassment, Marshall yanked off the hat and gloves and thrust them into McCarthy's shaking hands. "You take these," he said coldly, and abruptly steered the conversation with the British governor to any subject other than mosquitoes.[17]

The Anfa conference began on January 14 with preliminary meetings among the war chiefs in the hotel's main conference room. Marshall, King, Arnold, and a few planning staffers sat on one side of a long, narrow table, while General Brooke, Admiral Pound, and Air Marshal Portal took up chairs on the opposite side.*

In a pattern that would hold for the rest of the war, the service chiefs had circulated papers outlining their positions before the conference began. If the two sides agreed on a concept—say, an invasion of Corsica or a bombing campaign in Romania—their planning staff would draft a memorandum setting forth the game plan. If they disagreed, they worked out their differences in face-to-face meetings by arguing, cajoling, reasoning, and outwaiting them until a consensus was reached. Once they were in agreement, they presented their final plan to Roosevelt and Churchill in the form of a final memorandum.[18]

For the conference's first hour General Brooke described the basic options as the British saw them: The Allies could build up for an invasion of France in late 1943, doing nothing in the meantime, or they could capitalize on the momentum they had gained by rolling up the Mediterranean. In his staccato cadence, Brooke declared that the Mediterranean course made the most sense. It would knock Italy out of the war, possibly draw Turkey into the Allied camp, and still allow for a small, opportunistic landing in France if German resistance there collapsed.[19]

As Roosevelt had warned, the British service chiefs had worked out their strategy before setting foot in Casablanca. On the voyage from London, British spokesmen rehearsed arguments they and the Americans would make when

---

* Admiral Leahy, usually serving as President Roosevelt's representative, was absent from the conference; he had made the first legs of the trip with Roosevelt but had come down sick en route to Trinidad. He was left behind on doctor's orders.

the conference began. Like Cecil B. DeMille on a big-budget film set, Winston Churchill set the role each chief would play, and coached his marshals on their delivery. He stressed tact and diplomacy. The British chiefs, he said, must not cut off discussion prematurely, they must let the Americans have their full say, and they must listen carefully. In particular, he said, Admiral King must be allowed to "shoot his line about the Pacific and really get it off his chest," while Marshall must be allowed to thunder about France.[20]

When they arrived, the Americans were taken aback by the size of the British delegation. Roosevelt wanted a lean party to accompany him, while Churchill brought a small battalion of planners, aides, technical specialists, and code clerks aboard HMS *Bulolo*, a specially fitted headquarters ship that linked Churchill to London almost as easily as if he were sitting in 10 Downing Street. Brigadier General Al Wedemeyer, Marshall's planning assistant, told a War Department friend afterward: *"They swarmed down upon us like locusts with a plentiful supply of planners and various other assistants with prepared plans to insure that they not only accomplished their purpose but did so in stride. . . . [I]f I were a Britisher I would feel very proud. However, as an American I wish that we might be more glib and better organized to cope with these super negotiators."*[21]

Against the locusts, Marshall, Arnold, and King stood nearly naked. They had pared down the military contingent to the Army's head of supply, Lieutenant General Brehon Somervell; Savvy Cooke, King's planner; Wedemeyer; and a few others. They had little information at their disposal, other than what they and their assistants carried in their briefcases or in their heads. Worst of all, while Churchill and his warlords had a settled plan of action, President Roosevelt had not firmly backed any strategy at the January 7 White House meeting, principally because the Joint Chiefs had offered him none.[22]

Since the British held most of the cards, the only question was how much, if anything, they would concede toward a future cross-Channel invasion. They had been enthusiastic about ROUNDUP in 1943 when Marshall and Hopkins were in London, but now they were ready to sidetrack ROUNDUP for another Mediterranean jaunt. As Air Marshal Portal privately quipped to his fellow Britons, "We are in the position of a testator who wishes to leave the bulk of his fortune to his mistress. He must, however, leave something to his wife, and his problem is to decide how little he can in decency leave apart for her."[23]

Roosevelt's Three Musketeers held their line against the Mediterranean strategy. Marshall argued vehemently that the Soviet Union needed a true second front, as Roosevelt had promised Molotov the year before. Mediterranean actions, he warned, would become a "suction pump" pulling landing craft and men from their primary mission: the conquest of Germany. Operations in the Pacific and Burma would also suffer if the Allies committed themselves to "interminable" operations on the Continent's periphery.[24]

General Brooke thought he glimpsed military politics behind Marshall's insistence on ROUNDUP. He told his diary Marshall *"has found that King, the American Admiral, is proving more and more of a drain on his military resources, continually calling for land forces to capture and hold bases. On the other hand MacArthur in Australia constitutes another threat by asking for forces to develop an offensive from Australia. To counter these moves Marshall has started the European offensive plan and is going 100% all out on it! It is a clever move."*[25]

Brooke launched a vigorous counterattack against ROUNDUP. An invasion could not be mounted until late in the year, leaving the Allies fighting in France during the wet and winter seasons, when air cover would be limited and armored mobility impaired. The number of landing craft then available would not support a large invasion force, and no more than twenty-three divisions would be available by September 15—hardly a serious threat when Hitler had almost two hundred divisions arrayed against Stalin in the east.[26]

On the other hand, an invasion of Sicily would give the Allies air bases covering southern Italy. That alone might prompt the wobbly Italian government to sue for peace. If Italy surrendered, Hitler would be forced to pull large forces off the Russian Front and send them down the Italian boot. Additional bombers could be concentrated in England for the strategic bombing campaign over Germany—a carrot Brooke dangled for General Arnold—while Admiral Pound claimed that opening the Mediterranean would allow Allied shipping to India to go through the Suez Canal rather than around the Cape of Africa, an effective savings of 225 cargo ships. King found this angle enticing.[27]

Through a three-day campaign of repetition, argument, and attrition, Marshall saw the ground slipping beneath his feet. Eisenhower privately warned him that an invasion of France would require a mammoth commitment—possibly more than the Allies could spare for that year—and Roosevelt appeared lukewarm to any particular strategy. In sharp contrast to the year before, nightmares of early defeat had vanished; the combination of Midway, Guadalcanal, Stalingrad, and Tunisia appeared to have staved off German or Japanese victory. But recognition of the vast resources needed to parry the Axis on these far-flung fronts drove away hopes for a rapid victory on the Continent. The optimism of July had been replaced by January's stark realization that the war would not be won quickly or easily.

By the fourth day of battle, Marshall was prepared to give the British what they wanted. The ground war in the Mediterranean would continue after Tunisia was captured, and the next major target would be Sicily. ROUNDUP, the Americans conceded, was too great a stretch for 1943.[28]

•   •   •

During lulls between meetings, the American and British chiefs held social gatherings—dinners, luncheons, and cocktail hours where they mingled informally and steered clear of talk of war. Brooke's passion, they discovered, was birdwatching, while Marshall's was gardening. Ian Jacob, the British secretary, wrote in his diary, *"British and Americans met round the bar, went for walks down to the beach together, and sat around in each other's rooms in the evenings. Mutual respect and understanding ripen in such surroundings, especially when the weather is lovely, the accommodation is good, and food and drink and smokes are unlimited and free."*[29]

Social hours gave Admiral King a better look at the men he was dealing with. Portal, he thought, was the most intelligent of the group—certainly smarter than Arnold—while Admiral Pound seemed groggy except when naval matters cropped up. He disliked General Brooke. "He talked so damned fast that it was hard to understand what he was saying," King recalled. That was one of the nicer things King said about him.[30]

The meetings also reinforced British impressions of the Ohioan. Jacob told his diary, *"King is well over sixty, but active, tall and spare, with an alert and self-confident bearing. He seems to wear a protective covering of horn which it is hard to penetrate. He gives the impression of being exceedingly narrow-minded and to be always on the lookout for slights or attempts to 'put something over' on him. He is se-cretive, and I should say he treats his staff stiffly and at times tyrannically. . . . His manners are good as a rule, but he is angular and stiff and finds it difficult, if not impossible, really to unbend."*[31]

While King was a man of few words and fewer hobbies, the British learned he could provide entertainment value when he let himself splice the main brace too handsomely. *"King became nicely lit towards the end of the evening,"* Brooke told his diary. *"As a result he got more and more pompous and, with a thick voice and many gesticulations, he explained to the President the best way to set up the political French organisation for control of North Africa. This led to many arguments with the P.M. who failed to appreciate fully the condition King was in! Most amusing to watch."*[32]

At the negotiating table, Churchill's lieutenants found King less amusing. He was the British bugbear, the monster that comes out of the closet at night and carries off little children's landing craft. "Unless you tied him down, he felt it was quite all right to run off to the Pacific with all his craft," remarked Portal. One British admiral mused, "King is said to have his eye on the Pacific. That is his Eastern policy. Occasionally he throws a rock over his shoulder. That is his Western policy."[33]

At Casablanca, King's Pacific eye fixed on two strategic issues. The first was the allocation of ships, planes, and fighting men between the Asian and European theaters. As if announcing a mathematical fact, King claimed a mere 15 percent of all Allied resources were going to the Pacific, and he wanted that number raised to 30 percent. A 30 percent allocation, he said, would be enough to get the Navy west of Truk, the Japanese stronghold in the Central Pacific. Thirty percent might even get them to the Philippines. Once their ships were that far west, the Navy could slice through Japan's supply lines and eventually hit the Home Islands.*[34]

Second, to weaken the Japanese "front door" in the Pacific, King asked the British to attack Japan's "back door," Burma. The proposed operation to open this door, code-named ANAKIM, was an amphibious landing near Rangoon on the Burmese coast.

The logic behind ANAKIM was as winding as the Burma Road itself. Chiang Kai-shek's Chinese nationalist army was too poorly equipped to threaten the Japanese forces in mainland China. Its supply lines from the United States ran through the southern port of Rangoon, up Burma's Irrawaddy River valley to the Burma Road, and east along the Burma Road to Chiang's camp at Kunming. When the Japanese invaded Burma, they severed that vital artery; now, the only way to get anything into China was by flying from India over the treacherous "Hump" of Nepal—land of the Himalayan mountains whose 25,000-foot peaks and sudden downdrafts were as deadly to aircrews as German 88s were to bombers over Berlin.

The capture of Rangoon, combined with a Chinese drive to the south, would reopen Chiang's Burma Road supply line. Liberally supplied with American weapons trucked up the Burma Road, the generalissimo's 3.8 million levies could tie down the huge Japanese Army occupying eastern China—keeping them out of the "American" Pacific.[35]

Brooke, Portal, and Pound wanted nothing to do with Burma, and they shifted in their chairs uneasily as King held forth on ANAKIM. The immediate obstacles, in their view, were a lack of naval power in the Bay of Bengal to protect the landing fleet and the ever-present shortage of landing craft.

ANAKIM became a triangular standoff: Chiang refused to attack Japan's army in Burma until the Royal Navy controlled the Bay of Bengal; the demands of the Atlantic and Mediterranean campaigns ruled out naval control of the Bay of Bengal; and the British refused to mount ANAKIM unless Chiang agreed to attack Japan's army in Burma. ANAKIM, concluded Brooke, was a dead issue until late

---

* King's 15 percent estimate was pure guesswork, but he presented his "fact" so forcefully—as the alleged product of U.S. naval studies—that none of the chiefs challenged him.

1943, at the very earliest. The war against Germany was and must remain everyone's top priority.[36]

Never a patient man in conference, King saw perfidy in Albion's reluctance to launch ANAKIM. The British chiefs seemed more interested in preserving their empire in the Mediterranean than in defeating Japan, and he was getting tired of his Pacific being shortchanged. His face reddening, he opened the door to a question no one had been rude enough to ask: How much help did the British intend to give the United States in the Pacific after Hitler was defeated?[37]

King's question was a slap in the face of British honor, and the implication that the British would abandon the Americans once Germany surrendered could not go unanswered. Brooke and Portal indignantly replied that His Majesty's Government was fully committed to the defeat of all Axis powers, including Japan, and the meeting ended on a sour, divisive note.

Worried lest doubts linger in Roosevelt's mind, Churchill vehemently assured the president, "If and when Hitler breaks down, all of the British resources and effort will be turned toward the defeat of Japan. Not only are British interests involved, but her honor is engaged." He offered to sign a treaty to this effect, but FDR shook his head. Churchill's word was good enough.[38]

The accumulated stress of four days spent sitting around a crowded, narrow table began fraying the men's nerves. On the morning of January 18, the sixth meeting of the Combined Chiefs began unraveling over King's proposed Pacific offensive until Portal stepped into the role of peacemaker. He tactfully suggested that the British had perhaps misunderstood the Americans, who were not trying to shift the war's focus from Germany to Japan.

Marshall heartily agreed with Portal, but added that he was anxious not to get stuck in the Mediterranean while MacArthur's men were short of ammunition, aircraft, and ships. King, in a conciliatory mood, assured the British chiefs that he was only aiming to lay a foundation for the eventual defeat of Japan, not steam into Tokyo Bay in 1943. The United States was not going to throw too much at Hirohito while Hitler and Mussolini were going concerns.[39]

As Eisenhower reflected ten days later, "One of the constant sources of danger to us in this war is the temptation to regard as our first enemy the partner that must work with us in defeating the real enemy." Brooke eyed King and Marshall with suspicion, and he demanded an explicit agreement that the Americans would stop moving west in the Pacific once MacArthur captured Rabaul. Neither King nor Marshall would agree to this limitation, which the Americans considered drastic and unnecessary. The two sides remained deadlocked, and the morning meeting ended in a fog of distrust and animosity.[40]

During an early-afternoon break, Field Marshal Dill visited Brooke in his hotel room and had a candid talk with him about the consequences of stubbornness. Dill was a close personal friend of Marshall, a frequent dinner guest at the general's home, and he held the respect of both sides. He provided the British perspective to Marshall, and he could articulate the American view to London without being perceived as selling out to the colonists.[41]

Sitting in Brooke's room, Dill convinced Brooke that the Americans, for political reasons, would not budge on the Pacific. They felt they had to give MacArthur and Nimitz some flexibility to move west if the opportunity arose, and they could not afford to let anyone accuse them of shortchanging the soldiers and marines in the Pacific theater. And if the Combined Chiefs remained deadlocked on the subject, the whole lot of them would have no choice but to refer the matter to Roosevelt and Churchill. "You know as well as I do what a mess they would make of it," Dill said.[42]

Brooke had worked with Churchill long enough to imagine the Frankenstein monsters he would create if left to his own devices. He later reflected, *"Winston never had the slightest doubt that he had inherited all the military genius from his great ancestor Marlborough! His military plans and ideas varied from the most brilliant conceptions at one end to the wildest and most dangerous ideas at the other."* While British policy was remarkably realistic—often much more so than the plans of their cousins—the prime minister's strategic attention deficit sometimes sent the Empire into places best left alone. One British staffer joked, "Some Americans are curiously liable to suspect that they are going to be 'outsmarted' by the subtle British— perhaps that is because we sometimes do such stupid things that they cannot take them at face value but suspect them of being part of some dark design."[43]

Faced with the horror of letting politicians decide military strategy, Brooke relented. Portal prepared a draft memorandum permitting the Americans to "maintain pressure" on Japan and "retain the initiative" in the Pacific. They would be free to push westward through Rabaul, the Marshalls, and the Carolines, so long as they could do it with forces already allocated to the theater. In return, the British insisted that the Americans stop at Rabaul and the Carolines for 1943, so as not to jeopardize their ability to land in France if opportunity knocked. Operations in Burma would remain "on the books," as Brooke put it, but a final decision on ANAKIM would be deferred until summer.[44]

The British chiefs presented their proposal to Marshall and King during the afternoon session. The U.S. chiefs huddled over the document like corporate lawyers reviewing a contract—which, in a sense, it was. After carefully reading the fine print, they agreed to Portal's compromise.[45]

At four o'clock, the Combined Chiefs gathered with their leaders in Roosevelt's villa to outline their agreement to invade Sicily, bomb Germany, sink the U-boats, study Burma, and take Rabaul, the Marshalls, and the Carolines. To their relief, neither Roosevelt nor Churchill offered any changes. The Allied course for 1943 was finally settled.[46]

So they thought.

As the Combined Chiefs were drafting their final memorandum, Roosevelt took an inspection tour to Rabat, eight-five miles up Morocco's Atlantic coast. He had told Marshall and Eisenhower that he wanted to visit the troops. In the last war, he said, he had been up to the front as assistant navy secretary, and he would visit this one, too.

Marshall and Eisenhower told him a trip to the front—central Tunisia—was out of the question. Even Eisenhower was an unwelcome guest at any place the Germans could reach. The Army would have to allocate so much fighter, armor, and infantry protection for the visiting V.I.P. that it would jeopardize operations against the Boche elsewhere. Roosevelt tried to argue the point, but Marshall shut down the discussion by telling Roosevelt that if he issued those orders, no one in a U.S. Army uniform could take responsibility for his safety.[47]

Roosevelt backed down sullenly and settled for a consolation prize, a trip to General Clark's encampment near Rabat. Accompanied by General Patton and a small entourage, FDR rode in the back of a command jeep and reviewed soldiers of Clark's Fifth Army. Men standing at attention broke into huge grins when they saw the "old man" nodding at endless lines of olive drab powdered in road dust. The sights and sounds of the hard, boisterous youths of the Ninth Infantry Division, with whom he lunched, infused Franklin Roosevelt with a fresh ray of optimism—that spark he and Churchill threw off that first night at the White House. "Those troops, they really look as if they're rarin' to go," he told his son Elliott. "Tough, and brown and grinning."[48]

.   .   .

Soldiers in Tunisia, however, were not grinning. Lieutenant General Eisenhower's disorganized rush to the critical ports of Tunis and Bizerte had sputtered out in late November, and Hitler had poured men, artillery, tanks, and aircraft into "Fortress Tunisia." Nestled behind the forbidding Dorsal Mountains and protected by a moat of gooey rainy-season mud, Fifth Panzer Army and Rommel's *Deutsch-Italienische Panzerarmee* fended off blows from Eisenhower to the west, and from General Bernard Montgomery to the east.[49]

Marshall had loyally supported Eisenhower through his education as a supreme commander, but the situation in Tunisia required a shake-up in the command structure. General Montgomery's British Eighth Army was driving from Libya toward Eisenhower's force. The command structure had become too complex for Eisenhower to handle everything by himself, so the Combined Chiefs agreed to create a new army group under British General Sir Harold Alexander. Eisenhower would remain "supreme commander," coordinating land, sea, and air forces, and Alexander would run the land battle.

The deal, like so many others, involved compromises. It placed Alexander, Montgomery, and Anderson, all four-star British generals, under a three-star American with no previous combat experience. It also moved Eisenhower further from the battlefield. "We were carrying out a move which could not help flattering and pleasing the Americans," Brooke wrote long afterward. "They did not fully appreciate the underlying intentions. We were pushing Eisenhower up into the stratosphere and rarefied atmosphere of a Supreme Commander, whilst we inserted under him one of our own commanders to deal with the military situations and to restore the necessary drive and coordination which had been so seriously lacking of late!"[50]

During a private lunch with Roosevelt near the end of the conference, Marshall brought up the subject of Eisenhower's rank. Ordinarily, he admitted, it would be hard to justify Eisenhower's promotion to full general when Allied armies were stuck in the mud. But under the new command structure, Eisenhower would be commanding generals who outranked him. In those circumstances, it would be difficult for Eisenhower to keep headstrong four-stars like Alexander and Montgomery in check while wearing only three stars.

Harry Hopkins popped into Roosevelt's room just as he was telling Marshall "he would not promote Eisenhower until there was some damn good reason for doing it." Promotions, he declared, would go to people who had done the fighting, and "while Eisenhower had done a good job he hadn't knocked the Germans out of Tunisia."*[51]

Marshall dropped the subject. He could do nothing for Eisenhower's rank, but he did help the beleaguered lieutenant general by moving the Army's best and brightest into the African theater. He arranged for the State Department to send him diplomats, to lift some of the political load, and he urged Eisenhower to consider appointing Patton as his deputy for ground forces. He also sent Eisenhower's West Point classmate Major General Omar Bradley to Algeria, where he would serve as the supreme commander's "eyes and ears" at the front, diagnose problems, and recommend solutions.[52]

---

* Roosevelt did not disregard Marshall's point, however. Three weeks later, he promoted Eisenhower to full four-star general.

. . .

Early in the conference, Roosevelt met with General Henri Giraud to discuss the French Army. Admiral Darlan's assassination by a radical monarchist on Christmas Eve left a power vacuum in French North Africa, and Roosevelt told Giraud it would be a "very splendid thing" if he and de Gaulle could work out an acceptable division of power. Giraud, who had once been de Gaulle's commanding officer, nodded agreeably and told Roosevelt he expected the resistance leader to fall into line.[53]

De Gaulle not only refused to fall into line; he refused even to show up. He told British foreign secretary Anthony Eden he would neither attend nor parlay with Giraud, and he even bristled at Roosevelt's invitation to come to Casablanca, huffing that a foreign leader could not invite a French leader to meet on French soil. It was a matter of honor; if Roosevelt wanted to invite de Gaulle to a meeting, de Gaulle would meet him in Washington.[54]

Roosevelt was furious with de Gaulle's impertinence. He managed to restrain his anger only by alloying it with a little humor, writing to Cordell Hull:

> We delivered our bridegroom, General Giraud, who was most cooperative on the impending marriage, and I am quite sure was ready to go through with it on our terms. However, our friends could not produce the bride, the temperamental lady de Gaulle. She has got quite snooty about the whole idea and does not want to see either of us, and is showing no intention of getting into bed with Giraud.[55]

FDR's son Franklin, serving as his father's aide for the conference, recalled Churchill arriving for a luncheon with a scowl wrapped around a fat Havana cigar. With a peevish look, he muttered that he had been unable to convince de Gaulle to come to Casablanca.

"Winston, I don't quite understand that," replied Roosevelt. "Payday is coming up on Friday and you issue checks to the Free French. I think you ought to just send a message to de Gaulle saying if the bride won't come to Casablanca, there will be no payday on Friday."[56]

Payday was not interrupted, for two days later the six-foot, four-inch Frenchman was standing over Roosevelt, arguing with him. Secret Service agent Mike Reilly, who watched the meeting from behind the villa's curtains, remembered de Gaulle marching up to Roosevelt with the "unmistakable attitude of a man toting a large chip on each shoulder."

The general towered over the man in the wheelchair and the two launched into a rancorous argument, FDR in his passable French, de Gaulle prattling on about *"ma dignité"* and his status among the Allies. Reilly, hidden behind the curtain with his pistol drawn, grew tense as the conversation heated.[57]

Roosevelt's appeal to the cause of French liberation eventually overcame—just barely—de Gaulle's ego. On January 24, Roosevelt persuaded him to sign a lukewarm statement of French unity with Giraud. De Gaulle's signature was hardly dry when Roosevelt and Churchill ushered bride and groom to four chairs set up on the lawn behind Roosevelt's villa. In his seat, the long-legged de Gaulle sulked like a French maiden betrothed to a hideous troll in some bad fairy tale.

As Roosevelt recalled afterward, before the two Frenchmen left, Army Signal Corps photographer Sammy Schulman, an old press friend of FDR's, asked the leaders if he could snap a picture of the two Frenchmen shaking hands.

When Roosevelt translated the request, Giraud immediately stood up and extended his hand with an affable *"Mais oui!"*

De Gaulle eyed Schulman's camera warily and refused to budge through five full minutes of cajoling by the Anglo-American leaders. Finally he stood up and shook hands with his fellow liberator.

Roosevelt grinned as Sammy snapped pictures of the shotgun wedding. The photograph went round the world, proving, as Mike Reilly concluded, "Cameras do lie."[58]

Like many of his generals, FDR's mental picture of a militant Germany was heavily influenced by the events of 1918. The Kaiser's empire had capitulated before Allied troops had overrun the *Vaterland*, a circumstance that had allowed Hitler's thugs to claim that Germany had never really been defeated. According to the Nazi line, the Second Reich's death was not due to military defeat, but to a stab in the back by Jews, communists, and the West.[59]

The State Department had discussed the idea of "unconditional surrender" the previous May, and at his pre-Casablanca meeting with the Joint Chiefs, FDR told them he would talk it over with Churchill. At the time, he did not ask his military men for their opinions.

Had Roosevelt asked, Marshall and King would have advised him not to draw a line that would spur the enemy to fight harder. "As the war carried on," King later wrote, "I, myself, became more and more convinced that this slogan was a mistaken one. Of course, the people of the United States like slogans because they are terse and sometimes fit the situation; at least, they don't have to <u>think</u>!!"[60]

After photographs of Giraud and de Gaulle were safely part of the historical record, Roosevelt shifted over to a wooden chair next to Churchill, where he casually addressed some fifty newsmen sitting cross-legged on the lawn. As the journalists took notes, he explained the proceedings of the eventful conference. His eyes sparking, he trod upon a subject he had not fully discussed with anyone:

> I think we have all had it in our hearts and our heads before, but I don't think that it has ever been put down on paper by the Prime Minister and myself, and that is the determination that peace can come to the world only by the total elimination of German and Japanese war power.
>
> Some of you Britishers know the old story—we had a General called U.S. Grant. His name was Ulysses Simpson Grant, but in my, and the Prime Minister's, early days he was called "Unconditional Surrender" Grant. The elimination of German, Japanese and Italian war power means the unconditional surrender by Germany, Italy and Japan. That means a reasonable assurance of future world peace. It does not mean the destruction of the population of Germany, Italy or Japan, but it does mean the destruction of the philosophies in those countries which are based on conquest and the subjugation of other people.[61]

FDR later claimed that his announcement was an impromptu decision. The effort to get Giraud and de Gaulle together was so difficult, he said, it reminded him of getting Grant and Lee together at Appomattox. "And then suddenly the press conference was on," he said, "and Winston and I had no time to prepare for it, and the thought popped into my mind that they had called Grant 'Old Unconditional Surrender' and the next thing I knew I had said it."*[62]

But a demand for subjugation, without any terms, would only make Germany fight harder, and Roosevelt's chiefs knew the decision would bind together the fates of Mussolini and Hitler. There would be no divide and conquer anymore, just conquer. It was a monumental strategic choice that the president neglected to examine with his advisers before committing himself in public.[63]

But for better or worse, as Eleanor had once remarked, "The president does not think. He decides." During the next few months, there would be plenty of agonizing decisions to make.

---

* In 1948 Churchill claimed he first heard the demand for unconditional surrender from Roosevelt's lips that day. Neither Churchill nor Roosevelt was accurate in this regard. The concept had been raised before Casablanca by FDR on January 7 at the White House, and during the conference Churchill informed the British War Cabinet of the issue, telling the War Cabinet that he and FDR wanted to omit Italy from the "unconditional surrender" demand.

TWENTY-EIGHT

"A WAR OF PERSONALITIES"

⌇

THE CASABLANCA CONFERENCE HAD SET THE WAR'S STRATEGIC DIRECTION for the next six months, but for Marshall, Stimson, Arnold, and King, that direction was a decidedly mixed bag.

Marshall returned home with the least to show for his troubles. He lost his second effort to revive ROUNDUP, and his only Moroccan trophy was an iffy British promise to think about invading Burma by the end of the year. As a frustrated Al Wedemeyer wrote, *"We lost our shirts and are now committed to a subterranean umbilicus in mid-summer. . . . We came, we listened and we were conquered."*[1]

The sister services fared better. King pried loose permission for a Pacific offensive. Arnold did even better: He came away with a commitment to concentrate heavy bombers in England for a strategic air campaign, and when Sicily was taken, his bomber bases would sit within easy range of Rome.

Having outlined grand strategy at Casablanca, Roosevelt's lieutenants were free to return to parochial problems. As in the troublesome year 1942, the biggest problem was identifying the man who would conduct the slow, bloody steps across the Pacific Ocean. The question of command once again dumped Marshall into the middle of another power struggle among MacArthur, Arnold, and King.

Tamping down interservice rivalry was never easy, and the attitudes of MacArthur, the Army Air Forces, and the Navy did not help. From Brisbane, MacArthur worked himself into a frenzy over conspiracies festering against him at Main Navy, and his xenophobia filtered down to his staffers planning the nuts and bolts of operations with their naval counterparts. Hap Arnold, who had

237

toured the Pacific theater the previous year, came home believing there was a lobbying campaign by the Navy against his high-altitude bombing strategy.[2]

Like everyone else, the admirals harbored their own prejudices. MacArthur had rubbed King the wrong way ever since 1942, when the general had carped about Admiral Hart not doing enough to help him in the Philippines, and King didn't think much of MacArthur as a strategist. "King said MacArthur wasn't using what he had, that he didn't know how to use naval power, and in his opinion never will know how," one of King's journalist friends wrote his editor. "MacArthur, [King] remarked, had just discovered the airplane."[3]

At one meeting of the Joint Chiefs, King began harping, bitterly and at length, on MacArthur's arrogant personality. Fed up with King's harangues, Marshall rose to his feet, pounded his fist on the table, and bellowed, "I will not have the meetings of the Joint Chiefs of Staff dominated by a policy of hatred. I will not have any meetings carried on with this hatred!"[4]

As Marshall later told one interviewer, "On the matter of the respective attitudes of the Navy and MacArthur, the feeling was so bitter and the prejudice was so great that the main thing was to get an agreement. . . . Because you were in a war of personalities—a very vicious war." A partnership between Army and Navy "was arrangeable if the two commanders wanted to get together. But their approaches, particularly on MacArthur's side, were so filled with deep prejudices that it was very hard to go about it."[5]

Through a thick stew of suspicion, King and Marshall understood that a fracture among the services would only benefit Japan; they would have to put a stop to the rancor their men were making. In March, King and Marshall hosted a meeting of the Pacific staff leadership to clear the air. MacArthur sent a delegation led by his chief of staff, Major General Richard Sutherland, and his air force commander, Lieutenant General George Kenney. Nimitz sent his deputy, Rear Admiral Raymond Spruance, and Halsey sent his chief of staff, Captain Miles Browning. Day-to-day meetings were refereed by the irascible Admiral "Savvy" Cooke and the suspicious General Wedemeyer.[6]

The conference was held ostensibly to decide what the Pacific theater needed and who would command the various operations. But nearly everyone at the conference—Army, Air Forces, and Navy—spent their time trying to grab men, ships, and planes for their own fiefdoms. In a case of politics and strange bedfellows, the Pacific delegations representing MacArthur and Nimitz lined up with King's men against staffers representing Arnold and Marshall. Admiral Cooke harangued Arnold's staff over air allocation to the Pacific. Knitting his thick black eyebrows, a scowling Savvy Cooke bellowed that the Pacific was being shortchanged by the War Department. Arnold's chief planner, Brigadier General Orvil Anderson, countered that the buildup of a concentrated bomber force in England was mandated by the Casablanca agreements; the Allies could not shut down the air war over Germany just so MacArthur and Nimitz could bomb Rabaul into the Stone Age from which it had so recently emerged.[7]

The conference degenerated into a squabble among khaki squires that did their overlords little credit. One witness recalled, "The admiral and the general quarreled, as was not uncommon in days when neither had an intelligent comprehension of the other's business." Another, Captain Charles Moore, was less understated: "I was really shocked to hear Cooke lay Anderson out in the most violent and vicious terms at this meeting, in public. . . . He didn't spare a word in cracking Anderson down."[8]

The delegates deadlocked on what to give the Pacific, and how to divide the Pacific's allotment of ships, transports, planes, and fuel between MacArthur and Halsey. Unable to agree, they dumped the logistical spaghetti bowl into the laps of the Joint Chiefs.

The Chiefs assembled in a closed-door session on March 19. King wanted every bomber coming off the line sent to the Pacific, and he was willing to give up a division of Army ground troops to get them. Major General George Stratemeyer, sitting in for Arnold, wanted no weakening of the bomber force in England; theater commanders, he argued, should be told what they could have and should make their plans accordingly, not the other way around. Marshall wanted to go slow. He felt a better inventory of the Pacific's resources was in order; once the Joint Chiefs knew exactly what the theater had, they could better decide what, if anything, to take from other places.[9]

Leahy weighed in on the Navy's side. Looking at the problem through Roosevelt's political lens, the old admiral knew that strategic bombing, no matter how important from a military perspective, was a concept the public would not condone if it meant American sons and husbands were being bombed and strafed on the ground. "Whether Germany is bombed or not," he concluded, "the American forces in Africa and the South Pacific must be adequately protected. If those troops are neglected, the Joint Chiefs of Staff could not face the people of the United States."[10]

Speaking for Roosevelt, Leahy resolved the "where" question the same way Roosevelt had resolved it before the midterm elections: More resources would go to the Pacific, even at the expense of a bomber buildup in England.

But there was still a "who" question that defied solution: Would the Pacific's next phase be led by MacArthur, or by Nimitz?

. . .

In July of '42, the Joint Chiefs had given Guadalcanal to Admiral Ghormley by shaving off one degree from MacArthur's Southwest Pacific Area. Now King's men suggested lopping off the rest of the Solomons and giving them to Admiral William Halsey, Ghormley's replacement. When Marshall refused, King suggested putting the entirety of the Pacific under a single Navy man—Nimitz—just as everything in Europe and the Mediterranean was under a single Army man, Eisenhower. Again Marshall refused.[11]

Marshall countered by nominating MacArthur as supreme commander of all operations in and around New Guinea. This time King refused, though not just for reasons of service pride. The fleet battles around Guadalcanal had taught Ernie King a healthy respect for Japan's striking power. The Imperial Navy could show up anywhere in the South or Central Pacific, and King didn't want his carriers tied to MacArthur's schemes in New Guinea if Yamamoto decided to launch another attack. Nimitz needed the flexibility to fight another Midway without asking permission from MacArthur, so MacArthur could not monopolize the Navy.[12]

Leahy broke the volley of refusals and counterrefusals by pushing King into a compromise. They could find a way to support MacArthur without leaving Nimitz in the lurch. Knowing MacArthur's propensity to blame the Navy for not doing enough to support him, Leahy assured King that Nimitz would remain in command of naval forces everywhere, but he made King promise that Nimitz would not shortchange MacArthur during the Solomons-Rabaul campaign.[13]

On March 28, the Joint Chiefs finally had an order they could send to MacArthur, Nimitz, and Halsey. Four months of studies, discussions, and arguments—sometimes vicious arguments—were distilled into six sterile paragraphs. The Pacific would receive two more infantry divisions and substantial air support. Under MacArthur's general direction, the Army and Navy would capture the Solomons and Cape Gloucester on New Britain, a stepping-stone to Rabaul. Heavily defended Rabaul would be crossed off MacArthur's target list for 1943, though it would be picked up the next year, and Nimitz would command all naval forces not assigned temporarily to MacArthur.[14]

For all their disagreements, King and Marshall had formed a good working partnership. Marshall did not ask for too much on MacArthur's behalf, and King refused to squeeze every drop of blood from the Army at the cost of his relationship with Marshall. The hard-swearing, hard-drinking womanizer from Ohio understood, better than most Army strategists, the symbiotic relationship among air, land, and sea forces. It would be a long war, and though they would lock horns time after time, he and Marshall knew they would have to "get along" a while longer.[15]

.   .   .

If Savo Island had been King's Garden of Gethsemane, Marshall found his in the Dorsal Mountains of Tunisia. As Rommel's force retreated from Libya before Montgomery's Eighth British Army, it came within striking distance of the American II Corps, under command of Major General Lloyd Fredendall. On February 14, two panzer divisions smashed into the American corps at the crossroads town of Sidi bou Zid in eastern Tunisia. Rommel's panzers drove the apple-green Americans back through Kasserine Pass, killing, wounding and capturing more than six thousand soldiers and destroying hundreds of tanks, trucks, and artillery pieces. Seventy miles behind the front lines, General Fredendall prepared for a siege in his fortresslike bunker while his scattered, leaderless units retreated in disorder.

Watching from Algiers as the American line crumbled, a depressed Eisenhower kept Marshall dutifully informed of the fiasco—so much so that Marshall felt Eisenhower was spending too much time thinking about Washington and not enough thinking about the battle. Marshall brusquely replied, "I am disturbed that you feel under the necessity in such a trying situation to give so much personal time to us." He promised Eisenhower, "You can concentrate on this battle with the feeling that it is our business to support you and not harass you and that I'll use all my influence to see that you are supported."[16]

The crisis passed after four nerve-racking days. Rommel, lacking fuel to follow up his initial victories, grudgingly pulled back his panzers, leaving a charred wake of American corpses. But the Kasserine blitz brought into stark focus the problem of poor American generalship, and it was Marshall's job to ensure that the problem was fixed immediately.

As the Americans were reeling, Eisenhower sent observers to the front. He listened carefully to reports from Major Generals Omar Bradley, Lucian Truscott, and Ernest Harmon, whose conclusions were remarkably consistent: air-ground cooperation was weak, antitank doctrine was flawed, and General Fredendall's handling of his corps was terrible.

Eisenhower asked Marshall to recall General Fredendall. Because he believed the top man in each theater—Eisenhower, MacArthur, Stilwell in China—should be free to promote and fire whomever he chose, Marshall approved Fredendall's relief.

Marshall knew that relieving Fredendall and demoting him to his lower, "permanent" rank would confirm what some newspapers were saying: the war was being bungled. Fredendall could still be useful as a training commander, imparting practical lessons to trainees while not risking men in another Kasserine fiasco. He had Fredendall shipped to the States, gave him a third star, and appointed him deputy commander of the Second U.S. Army, a training outfit. But he kept Fredendall at home for the rest of the war.

The change paid off. Eisenhower brought in General Patton to take Fredendall's place, and Patton appointed Bradley, another Marshall protégé, as his corps deputy. After a month of heavy fighting, Eisenhower sent Patton back to Morocco to plan the invasion of Sicily, and moved Bradley up to II Corps command. Under Bradley's careful leadership, the II Corps fought its way to the sea.

On May 8, the Axis armies in Africa surrendered, exactly six months after the TORCH landings. A quarter million Italian and German soldiers checked into Eisenhower's prisoner-of-war cages, there to await an all-expenses-paid cruise to their new homes in Aliceville, Alabama and other fine towns of the American South and West.

During the Tunisian campaign, Marshall kept a close eye on the Army's public relations. He appreciated the need to create American heroes. While Army doctrine stressed teamwork in military operations, the public needed vibrant personalities to rally behind. Marshall encouraged reporters to give Eisenhower his share of credit for the victory in Tunisia, while Eisenhower liberally spread the laurels among Bradley and his British partners.

When the influential British press threw the largest and greenest laurels at Montgomery's feet, however, Marshall grew irritated. He did not want to see American contributors overshadowed by the larger-than-life personality of Montgomery, whose reputation in America soared with the U.S. release of the British documentary *Desert Victory* in April. Marshall wrote his press relations officer, "You can tell some of these newsmen from me that I think it is a damned outrage that because [Eisenhower] is self-effacing and not self-advertising that they ignore him completely."[17]

• • •

At Casablanca, one of the few points of wholehearted agreement had been the U-boat peril. During 1942, the Allies lost nearly two million shipping tons and thousands of sailors to German submarine, surface raider and air attacks. Successive waves of Liberty ships, tankers, and colliers coming out of American and British shipyards could not keep up with German sinkings. Worse yet, Admiral

King estimated that the Kriegsmarine had more serviceable U-boats in 1943 than when it started the war.

On any given day, the Allies had about 1,500 ships plying the waves, and by mid-1943, the United States would be producing about 1.5 million tons of transport vessels each month. Admiral King, calculating losses against production, felt that if the Allies could reduce U-boat sinkings to half a million tons per month, the problem would be licked by American production power.

But for now, ships went down and King knew the Navy's fortunes would get worse before they got better.[18]

Henry Stimson had never taken his mind off the idea of using radar-equipped bombers to sink German submarines—an idea for which the Navy, it seemed to him, had no real interest. As Stimson saw it, land-based bombers with radar could hunt down U-boats on the open waters while Navy planes provided close convoy protection. But to his disappointment, in early 1943 Secretary Knox, Undersecretary James Forrestal, and the Navy's admirals insisted that the problem was a simple lack of escort vessels. They refused to adopt an Army air group to do their job.[19]

Stimson renewed his efforts to get Army planes into the U-boat war after the Battle of the Bismarck Sea in early March. There, off the New Guinea coast, U.S. and Australian bombers wiped out an entire Japanese troop convoy. *"The thing that cheers me about it,"* Stimson wrote, *"is that it was accomplished without any help of the Navy and only by land based planes—the kind of planes Knox is always saying are no good."*[20]

Stimson went back to see FDR with a proposal to use land-based Army bombers as submarine hunters. He claimed the Army could deploy an effective antisubmarine force six months before Admiral King could perfect the Navy's convoy system. To Stimson's delight, Roosevelt agreed with him—enough, at least, to suggest a trial run over the Bay of Biscay on the occupied French coast. Then, considering the explosion once word of Stimson's plan reached Main Navy, Roosevelt added, "I don't want to go over Knox's head." He asked Stimson to talk to Knox about it first.[21]

Knox and King had considered Army sub hunters back in 1941, when King headed the Atlantic Fleet. At the time, King insisted that the Navy needed its own bomber fleet. "The medium bombers are primarily to help the Navy do its job at sea," he had argued. "They are best manned by Navy personnel who know the ways of the sea and of sea-going people." Long-range bombers were, to King, just another weapon the Navy should have in its arsenal whenever it could be useful.[22]

To King, the job, not the weapon, determined who would command. In this case, the job was hunting maritime threats. That was the Navy's area of expertise, and that meant the Army should butt out. "There was only one solution," recalled Captain "Frog" Low, King's antisubmarine head. "Get the Army out of the antisubmarine business."[23]

On the first of April, a grimly determined Henry Stimson approached Knox with his proposal for a "Special Anti-Submarine Task Force" under command of an Army officer. Speaking as one old soldier to another, Knox told Stimson he worried that the admirals would "think that we were setting up an independent command right in the middle of their submarine work in the Atlantic." By endorsing the proposal, he would incite a mutiny on his unstable quarterdeck.[24]

Four days later, Stimson received the Navy's reply, written by Admiral King. Specially equipped aircraft flying from land, King allowed, would be an effective supplement to the convoy system. In time, the Navy would have sufficient bombers of its own to take over the functions Stimson proposed giving to the Army. But unity of command, the Great Commission preached by both services, dictated that when antisubmarine bombers were launched, they would—they must—fall under the jurisdiction of the Commander-in-Chief, U.S. Navy.[25]

In the wake of his encouraging meeting with Roosevelt, Stimson felt he had the upper hand, even without Knox's support. He pressed his view upon Harry Hopkins and ordered Marshall to take the matter up with the Joint Chiefs.[26]

But Henry Stimson underestimated Ernie King. Learning that the president, Stimson, Knox, and Hopkins were leaning against him, King retreated to his inner keep and prepared for a long siege. He threw together a separate command, designated the Tenth Fleet, to "control allocation of anti-submarine forces to all commands in the Atlantic." The Tenth Fleet, as King envisioned it, would absorb all functions that Stimson seemed bent on giving the Army.[27]

King's "Tenth Fleet" wasn't really a fleet, but a coordinating office for a small group of scientists, engineers, and communications workers, and its commander was one Admiral Ernest J. King. When Stimson, Marshall, and Arnold realized the Tenth Fleet was only a paper organization, they renewed their efforts to send Army bombers on ocean patrols.[28]

To Stimson, the antisubmarine war "provided an almost perfect example of the destructive effect of the traditional mutual mistrust of the two services. . . . The Navy and the Air Forces had a mutual grudge of over twenty years' standing— the Navy feared that the Air Forces wished to gain control of all naval aviation, while the Air Forces saw in the Navy's rising interest in land-based planes a clear invasion of their prescriptive rights. The Air Forces considered the Navy a backward service with no proper understanding of air power; the Navy considered the Air Forces a loud-mouthed and ignorant branch which had not even mastered its own element."[29]

The deployment of antisubmarine bombers remained stalled among Stimson, Marshall, Arnold, and King until early July. In a deal brokered by Rear Admiral John McCain and Lieutenant General Joseph McNarney, the Army Air Forces agreed to provide more than two hundred bombers to the Navy. In return, General Arnold asked only that bombers ceded to the Navy would be used exclusively for antisubmarine warfare and reconnaissance of America's coasts; strategic bombing missions would remain the responsibility of the Army Air Forces.[30]

King again refused. He wanted those bombers, and he would do with them what he pleased. He would agree that their primary function would be antisubmarine warfare, but he would not restrict himself or the Navy as to how those bombers would be used or what they could carry. He might need to bomb a Japanese air base or a U-boat pen somewhere, someday, and he refused to handcuff himself just to keep Arnold's staff happy.[31]

Now it was Marshall's turn to be furious. Tired of King's empire building, he warned King that Secretary Stimson would never agree to give away bombers that the Navy could send to bomb Berlin. "If the matter is taken to the President," he told King, "the Secretary desires to be heard by him on the subject."[32]

Eventually, King, ever sensitive to the power of words, agreed to language emphasizing that the "primary" function of the Navy's bombers would be for antisubmarine warfare. But he stipulated that no commander, Army or Navy, would be forbidden to hit the enemy if some unexpected opportunity presented itself.

King's wordsmithing was comfort enough for Marshall and Stimson, and on July 9 they accepted his conditions. American bombers, under King's Tenth Fleet, would take to the skies against Hitler's U-boats. The Army disbanded its antisubmarine unit, and the organization of the bombers was finally settled, more than a year and a half after Pearl Harbor.

"Boy, what a fight!" King remembered. "But we won."[33]

.   .   .

King didn't win every battle, though, and now and then he picked awkward moments to run to cable's length and engage with broadsides. He was supposed to be a big-picture admiral, but as the Navy grew under his watchful eyes, his restless mind drifted back to his experiments with uniforms on the *Texas*.

Though he had no business thinking about what clothes his officers wore, King decided to create a new uniform, something different from the ubiquitous khaki that, he sneered, looked too much like the Army's garb. "In my view, khaki is not to be considered as other than a stop-gap," he had written the previous June, and by April 1943 he had conjured up a dark gray uniform suitable for both ship and shore.[34]

Gray, thought King, was perfect for its "simplicity and utility." Unlike whites, it would not attract the attention of strafing enemy pilots against a steel superstructure. It did not show small stains, making it suitable for engineer and ordnance officers. And the color gray, King thought, was "more nearly consonant with democratic principles."[35]

While his *Texas* experiments had been a flop, King had better reason to hope his new uniform would stick. In April, Secretary Knox approved the new design, and with Knox's signature, "King Gray" became the Navy's regulation uniform for commissioned and petty officers. King's staff obediently paid for new uniforms from their modest government salaries, and a supportive Admiral Leahy showed up for meetings of the Joint Chiefs wearing the new regulation color, even while King wore the traditional Navy blue.[36]

King was not the only commander to indulge sartorial whims. Marshall and Eisenhower tinkered with the infantryman's uniform, and before the war Patton had become infamous for an outlandish green-and-gold suit he personally designed for his tankers.

But King had not considered the sailor's fondness for tradition, and the new uniform was a dud. At sea, young officers wearing "King Gray" were hooted at as bellhops and doormen by old salts. King's chief of staff, Admiral Willson, was asked by two Washington teenagers if he was a bus driver or a train conductor. One of King's longtime friends grumbled, "None of us ever felt that the gray uniform he insisted on during the Second World War ever made any contribution to us winning the war. It cost a lot of money for those who had to buy them and also put an extra load on the textile industry, all for no real purpose."[37]

Some commanders paid lip service to the new uniform regulations, but most benignly neglected it, especially in the Pacific. Nimitz, King heard, went so far as to ban it. When he got wind of the revolt in the Pacific, King cabled Nimitz to ask whether it was true that Pacific officers were not permitted to wear the new gray. Summoning his powers of military obfuscation, Nimitz replied, "Inquiry discloses some local commands, while not prohibiting gray, have expressed preference for khaki. Am fully aware gray is authorized and will issue appropriate clarifying directive."[38]

Any "appropriate clarifying directive" from Nimitz would be a far cry from what King intended.

· · · ·

King's wrangling with the Army, the Air Forces, the naval bureaus, Knox, and a thousand other foes ground away at his leather-clad soul. "Life here is strenuous, of course," he confessed to a Naval Academy classmate in a rare moment of openness, adding quickly, "So far I have managed to keep well and, on the whole, to remain as cheerful as circumstances allow."[39]

While King did his exercises in the morning, and managed to snatch some pleasure reading in his cabin at night, he had no horseback rides, vigorous exercise regimen, or evening social hour, as Marshall, Stimson, and Roosevelt did. At home, Marshall had a supportive wife in Katherine. Stimson had the faithful Mabel, and FDR had women, official and unofficial, to complete his domestic life. But King, living several miles from his uninspiring wife, Mattie, had no regular intimates. He had to find them outside the great Washington bubble.

For King, one placid harbor was the Shenandoah Valley farm of Paul and Charlotte Pihl. Paul, a Navy captain and old friend, was a member of King's logistical team, and Charlotte, sister of Wendell Willkie, was a charming, vivacious woman whom King had first met at a party when her husband happened to be out of town.

King would arrive at the Pihl house alone on weekends and stay with Paul and Charlotte in their farm's guesthouse. There he could relax, take in the quiet country air, muse about becoming a gentleman farmer, and most importantly, not think about war. "The farm was such a great relaxation for him from the tensions of his work," Charlotte told an interviewer years later.[40]

When Captain Pihl was transferred to California, Charlotte stayed on the farm and King continued his weekend visits. His long-standing reputation as a rake had been an article of faith in naval circles for decades, and rumors of an affair between Captain Pihl's wife and the "boudoir athlete" were natural. Mattie knew of her husband's visits and resented them, but did not try to stop him. She would call the Pihl farm and ask if the admiral was there; Charlotte would offer to have him come to the phone, but Mattie always refused, begging Charlotte not to tell King she called. When Charlotte invited her to come to the farm, she always declined. "Ernest doesn't want me to come up there," she would say before hanging up.[41]

The Pihl farm was not King's only port of call. Another naval wife, Abby Dunlap—a longtime friend married to one of King's Atlantic Fleet officers—lived on a farm in Cockeysville, Maryland, and King was among her frequent visitors. Abby and her sister, Betsy Matter, offered King a place to unwind, where he could sip a glass of Southern Comfort, engage in light banter, and think of nothing strenuous. Betsy was the subject of at least some of King's affections, and his letters to her were mildly flirtatious.[42]

"There would be times," said Betsy, "when he would come in and sit in the tiny drawing room and we would leave him alone. He would stay in there for perhaps fifteen minutes to a half hour and never say a word to anyone. He would just sit there on the Victorian sofa with his feet stretched out and his head back." She remembered, "Just before King would arrive we would kill chickens and fix corn and we would all sit out on the terrace and eat. Everyone would put their feet up, completely relaxed, and eat."[43]

Though marital fidelity was not among King's strong points, the admiral sought out the company of women for something beyond mere carnal diversion. In his line of work, there were not many men he could trust completely. He could let his few hairs down more easily around women, and with a wife and six daughters, he was accustomed to a house filled with feminine banter.

King's women friends saw it as their contribution to the war effort. "When King was with Abby he could be amused and relaxed," Betsy said later. "He could return to work refreshed and ready for anything."[44]

. . . .

That anything might present itself at any time. In the spring of 1943, a bright spot flickered when Navy cryptologists stumbled upon the travel itinerary of Admiral Isoroku Yamamoto, Japan's naval genius behind the offensive of 1941.

Before the war, Yamamoto had spent a great deal of time in the United States. He loved American sports, read American magazines, and respected America's production capacity. He harbored no illusions about its latent power, and as Japan looked ahead to war with the United States, he had warned his government, "To make victory certain, we would have to march into Washington and dictate the terms of peace in the White House."

By April of 1943, the prospect of making it to Washington looked dimmer than ever, and to boost the morale of his soldiers and sailors, Japan's most famous admiral decided to make a personal inspection to the front lines near New Guinea.[45]

Learning from Nimitz's codebreakers that Yamamoto would be flying from Rabaul to the Solomon Islands, Roosevelt and Knox arranged for a welcoming party of P-38 Lightnings to greet him over Bougainville. On April 18, the interceptors zeroed in on the bomber carrying the admiral and sent Yamamoto into the next life with a burst of machine-gun fire. His body was recovered by Japanese soldiers near his plane's wreckage.[46]

The killing, on the first anniversary of the Doolittle raid, was a shot in the arm for the American public, which still burned to avenge the dead of Pearl

Harbor. When the Japanese government released the news of Yamamoto's death, Roosevelt, in a jocular mood, sent Leahy a letter for personal delivery to "Mrs. Admiral Yamamoto, Tokyo, Japan." Borrowing a line from Mark Twain, the letter read:

> *Dear Widow Yamamoto:*
>
> *Time is a great leveler and I never expected the old boy at the White House anyway. Sorry I can't attend the funeral because I approve of it.*
>
> *Hoping he is where we know he ain't.*
>
> *Very sincerely yours,*
>
> *Franklin D. Roosevelt*

He scribbled a postscript to Leahy: Ask her to visit you at the Wilson House this summer.[47]

TWENTY-NINE

BLIND SPOTS

∽

I N HIS OLD CHAIR, AT HIS OLD DESK IN THE NEW Pentagon building, Henry Stimson shook his aged head in disbelief. Marshall had come to him with a lengthy memorandum from Giraud's staff listing the items the French general wanted for his army. Giraud brusquely informed Marshall that the president himself had approved the requests, and demanded to know when his equipment would arrive.[1]

Caught off guard, Marshall went to Stimson. What was this all about? he asked.

Stimson frowned. The president hadn't told him of any specific aid list for the French.

But neither was he surprised. Stimson knew how infuriating it could be to work for a man of Roosevelt's improvisations. During the Casablanca conference, he had complained to Justice Frankfurter about FDR's disorganized method of running the world's largest government. It was far worse, he said, than anything under Hoover, Taft, or even Warren G. Harding.

Nearly everyone drawn into Roosevelt's circle became a willing accomplice to the death of orderly procedure, and Frankfurter could offer his mentor neither reassurance nor remedy. He told the secretary that "he had better make up his mind that orderly procedure is not and never has been characteristic of this Administration—it has other virtues, but not that." The jurist advised Stimson to "reconcile himself to looseness of administration and the inevitable frictions and conflicts resulting therefrom which naturally go against the grain of an orderly, systemic brain like his."[2]

When FDR returned from Casablanca, Stimson marched over to the White

House and dressed FDR down for failing to ask his military advisers if what he had agreed to give General Giraud was even possible. Roosevelt shrugged off Stimson's point, but Stimson, a bulldog in argument, refused to let go. Tightening his jaws around the bone, Stimson sputtered that Hull had complained that the military aid agreements with Giraud might have been signed over a drink, for all the thought that went into it.

Throwing his head back, Roosevelt laughed and said other agreements he had signed might well have been reached in the same way.[3]

But Roosevelt wasn't laughing in June, when the tenuous peace between Giraud and de Gaulle fell apart. In a blitzkrieg power play, de Gaulle seized control of the Comité Français de la Libération Nationale, a body claiming leadership over freed French territories. By the end of the month, de Gaulle's men had effectively shut out the hapless Giraud.

Roosevelt, who still bore a grudge against the snooty bride of Casablanca, was incensed. He wrote Churchill, "I am fed up with de Gaulle. I am absolutely convinced that he has been and is now injuring our war effort and that he is a very dangerous threat to us. . . . I agree with you that he likes neither the British nor the Americans and that he would double-cross both of us at the first opportunity."[4]

Much as Roosevelt would have liked to jettison de Gaulle, the uncomfortable truth was that the Gaullists were more fanatical—and more willing to trade French unity for power—than any other faction. They had outshouted their rivals and had the relentless staying power that other resistance groups lacked.

There was little Roosevelt could do, though that didn't stop him from trying. When he suggested that Marshall send troops into Dakar to keep the country free of Gaullist influence, Marshall recommended against it. The move, he said, would legitimize Axis propaganda claims that the Allies were trying to dominate France's political affairs. Until an alternative French leader emerged, FDR would have to live with Charles de Gaulle.[5]

* * *

FDR refused to bet on de Gaulle's horse, but in Asia he found another filly that seemed like a sure thing: Chiang Kai-shek.

As American strategists saw it, the great mass of Chinese troops would keep the Japanese Army pinned down on the Asian mainland if they were given enough tools to do the job. But Japan had cut the Burma Road in early 1942, and without that carotid artery, supplies could only be airlifted over Himalayan peaks that might as well have been guarded by Shiva and Bhairab. The "Hump" route over the Himalayas was too dangerous and too prone to weather close-outs to allow

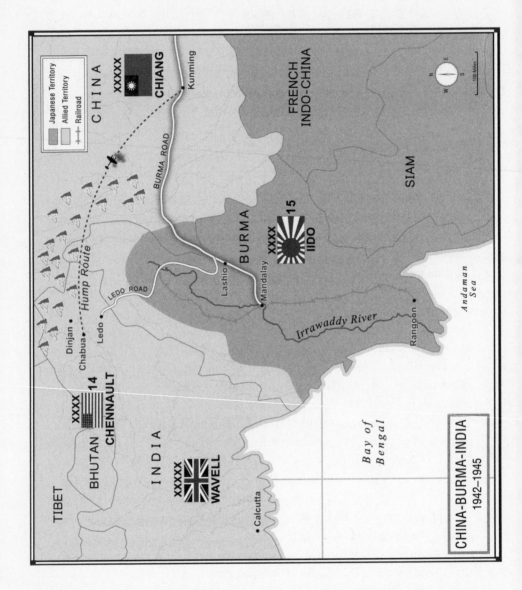

CHINA-BURMA-INDIA
1942–1945

Chennault's supply planes to move more than the barest trickle of fuel, weapons, and equipment. If Chiang's government collapsed from lack of sustenance, King feared the war against Japan might be prolonged by ten to fifteen years. As with Stalin the year before, FDR felt the Allies could not afford to let Chiang drop out of the war.[6]

Churchill had never understood Roosevelt's attachment to Generalissimo Chiang. The British Foreign Service, which had dealt with the Chinese for one and a half centuries, had no faith in Chiang's ability to weld together a nation convulsed by civil war since 1927. Chiang's struggle against the communists would claim the lives of twenty-five million of his countrymen before it was over, and Chiang and his lieutenants seemed more interested in re-arming for a re-newed civil war—or lining their own pockets—than defeating the Japanese.*[7]

But the picture from 10 Downing Street looked very different from the picture at 1600 Pennsylvania Avenue. Roosevelt, who had not traveled beyond the Western Hemisphere in a quarter century, was a political genius with a tre-mendous grasp of America's economic and social issues. Foreign policy was often his blind spot. He had a habit of cutting out experts like Cordell Hull and Sumner Welles, and he formed snap judgments from reports by a handful of ambassadors, cronies, spymasters, and White House visitors.

Having spent so much time canvassing all strata of American society—New York financiers, Georgia sharecroppers, California factory workers—Roosevelt assumed he also understood the mind-set of other world leaders and their people. People are the same everywhere, the egalitarian in him proclaimed. He couldn't bring himself to believe that some people, at least the ones making the decisions, were not animated by common human motives.

In late February, Generalissimo Chiang Kai-shek came to Washington to present his case for more military assistance, and this time he brought one of the Kuomintang regime's most important assets: his wife. Mei-ling Soong, an artic-ulate, Wellesley-educated power broker, was an Oriental blend of Eleanor Roo-sevelt and Marlene Dietrich, a rare woman whose sharp mind and vague sexual charm mesmerized westerners in equal measure.

In her formfitting black dresses, a lit cigarette nestled between her glossy red lips, Madame Chiang electrified the Washington press corps and politicians on both sides of the aisle. After her speeches to the Senate and House urging support

---

* Stilwell agreed with the British assessment. In July 1944 he told his diary that Chiang *"hates the Reds and will not take any chances in giving them a toehold in the government. The result is that each side watches the other and neither gives a damn about the war. If this condition persists, China will have civil war immediately after Japan is out."*

for Chiang's people, Washington gossips buzzed over the tigress of the Orient. "I never saw anything like it," gushed one congressman after her speech to the House. "Madame Chiang had me on the verge of bursting into tears."[8]

Pushing as many buttons as her thin fingers could reach, the first lady of China met privately with FDR, Wendell Willkie, and even the aging Henry Stimson, who found her a "most attractive and beguiling little lady." A bemused Eleanor Roosevelt wrote to her daughter, Anna Boettiger: "In a queer way I think the men (including FDR) are afraid of her. She is keen & drives her point & wants to nail them down and they squirm."[9]

While the Chiangs pitched for aid from Washington, the Joint Chiefs struggled over how to split that aid between their two American field commanders—Major General Claire Chennault, the tough, lantern-jawed commander of the Fourteenth Air Force, and Lieutenant General Joseph Stilwell, who wore the hats of commander of U.S. ground forces in China, chief of staff to Generalissimo Chiang, and commander of U.S. and Chinese ground forces in India.

The two American officers, both thin and bronzed from months in the Asian countryside, were natural rivals for the trickle of supplies flowing over the Hump. Like hungry lion cubs, they scratched and clawed over every scrap of food, fuel and equipment the Joint Chiefs would give them.[10]

Meeting with the Joint Chiefs on May 4, Chennault argued that all supplies should be earmarked for his air forces, which could increase the tempo of the bombing campaign against Japan. Stilwell proposed that the Allies divide supplies evenly between Chennault's airmen and the Chinese infantry, who would spearhead an offensive to open the Burma Road.[11]

The China problem split the Joint Chiefs along unfamiliar lines. Stilwell, one of Marshall's men at Fort Benning, enjoyed the wholehearted support of the chief of staff and secretary of war. Arnold just as strongly backed Chennault, and the Navy tended to support the air solution. Roosevelt would have to make this call.[12]

Stilwell had served in China for years in military and quasi-diplomatic roles. He traveled the countryside, mingled with its people, and spoke the language. But Roosevelt thought his brash approach was the wrong tack with the sensitive Chiang. FDR's grandfather Warren Delano II had once been a merchant in China, and his late mother, Sarah Delano, had lived in Hong Kong. Believing he knew more about the situation in China than his Mandarin-speaking general did, he wrote to Marshall, "My first thought is that Stilwell has exactly the wrong approach in dealing with Generalissimo Chiang who, after all, cannot be expected, as a Chinese, to use the same methods that we do."[13]

FDR's second thought was that the dashing Chennault should receive the lion's share of supplies. But after listening to Stilwell, Chennault, the Combined Chiefs, and Chiang's foreign minister, Dr. T. V. Soong, Roosevelt split the *bǎobǎo*: Chinese troops under Stilwell would receive enough supplies to defend Kunming, and Chennault would have most of the rest for his air offensive.[14]

It wasn't a lot, but it was all America could spare for the moment. Roosevelt was getting ready to make an all-out push for the invasion of northern Europe, and he was finding himself thinly spread.

THIRTY

STICKPINS

〰

Two weeks after his return from Casablanca, FDR held his 879th press conference since becoming president. Sitting back in his worn office chair as a crowd of notepad-wielding journalists pressed against the desk's edge, he explained his method of managing his war chiefs:

> You can't leave things to the military, otherwise nothing gets done. Now that's a dreadful thing to say, but the fact is that if you get almost all admirals and generals from different nations, or even one nation, talking over future plans, they spend a month or two in talking about why each plan or suggestion won't work—get just a series of "No's."
>
> On the other hand, if you get certain laymen to stick pins into them all the time—prod them, if you like—and say you have got to have an answer to this, that, and the other thing within so many days, you get an answer.[1]

To get those answers, FDR would stick a few more pins. In the spring of 1943 Army intelligence analysts concluded that Hitler's big summer blow against Russia would put renewed pressure on the Allies to open the long-promised second front. Where that front should be opened—and what character it should take—were questions that would stare the Allies in the face when the upcoming Sicilian campaign came to an end. It was time for another meeting of the western warlords.[2]

As Churchill prepared to embark on the ocean liner *Queen Mary* for his third transatlantic crossing to meet Roosevelt, Roosevelt called Marshall to the White

House to talk about the next invasion. Churchill's roving eye, they both feared, would be directed toward the Italian mainland, the Balkans, or some ancient Greek island that might make a fine USO stop but had no strategic value.[3]

At Casablanca the Americans had been outmanned, out-thought and outmaneuvered. But for the next round, they would be fighting on far better terrain. At the Washington conference, which Churchill code-named TRIDENT, the American leaders would have the full resources of the War and Navy Departments close to hand. Their limber chests would be filled with shipping timetables, production forecasts, and training schedules. Staffers, defending the flanks, would be ready to attack obscure questions of logistics and air allocation.

And most of all, they would have the unwavering support of their president.

To deafen Roosevelt to Churchill's Mediterranean siren song, the Joint Chiefs prepared a flurry of studies outlining the American game plan. Marshall and King acknowledged that post-Sicily operations in the Italian theater might garner some benefits, and they agreed to discuss them with the British as a possible compromise. But they insisted that any new operations there must reduce the Allied forces in the Mediterranean, not add to them. Furthermore, they wanted any new operations there to be launched in the western half of the great sea, such as on Corsica, Sardinia, or the French Riviera. "The United States will not become involved in operations east of Sicily except possibly for special air operations," they declared. "If the British insist on doing so, they do it alone."[4]

Roosevelt agreed. It had been a year since he had sent Marshall and Hopkins to London to win British approval for ROUNDUP, and in that year the coalition had followed a decidedly un-American path. That path had done little to alter the titanic struggle between Russia and Germany, it hadn't opened the Mediterranean to Allied shipping, and it helped neither MacArthur nor Nimitz. It was time to lunge onto the Continent.[5]

The day after discussing these points with Roosevelt, the Joint Chiefs suffered an unexpected blow. A heart attack sent General Arnold into Walter Reed Hospital for the second time in ten weeks. The strain of interservice conflicts, the impossible demands on his overstretched forces, and the physical stress of spending thousands of hours in unheated bombers had overloaded his ticker.

Hap's absence would deprive the Joint Chiefs of a booming voice on air matters at a critical point in the war. *"It is a bad blow to the strength of our smoothly working military machinery,"* Stimson wrote. *"He is not as cautious and diplomatic as Marshall, who is a good counterpoise to him, but on the other hand he does not hesitate to espouse the unpopular side of a discussion and make it clear even in the face of his chief."*[6]

Marshall, worried about his friend's health, grew alarmed when he heard that Arnold intended to leave Walter Reed early to deliver a commencement address at West Point in June. *"Please don't do it,"* he pleaded. *"Your Army future is at stake and I don't think you should hazard it with a matter of such trivial importance."* He ended his letter with a rare personal closing: *"Please be careful. Affectionately, G.C.M."*[7]

Roosevelt understood failing health. He suffered from bronchitis, sinusitis, high blood pressure, and fatigue far more than he would admit. He worked in sustained bursts of energy, but his mainspring would run down so low he needed long periods to rest at Hyde Park, Warm Springs, and other places of seclusion.

Sympathy aside, Roosevelt wasn't sure that a man with a bad heart ought to be running the war's air fleet. "If he continues as commanding general," Roosevelt asked Arnold's cardiologist, "is it likely to endanger his life?" The doctor said it could—though at Arnold's request, he reminded the president that when combat crews go on missions they too put their lives in danger; there was no reason their commanding general should shrink from the same risk.

FDR, who had used that same argument against friends who advised him not to run for a third term, agreed with Arnold. Before the war, Arnold had been in Roosevelt's doghouse, but he was playing ball now, and Roosevelt liked the man's fighting spirit. He would keep Hap in the pilot's seat. For now, at least.[8]

The *Queen Mary* steamed into New York's crowded harbor on May 11, her hold crammed with 5,000 German prisoners bagged in Tunisia. Her upper-deck state-rooms, crammed with papers, baggage, and wartime luxuries, catered to Winston Churchill and one hundred military advisers, political staffers, valets, secretaries, and personal assistants. Harry Hopkins greeted the prime minister at the Staten Island pier and whisked him to the White House for one of Mrs. Nesbitt's bland dinners.[9]

Formal meetings of the Allied high command began the next afternoon at the White House with a broad overview of global strategy by the two political leaders. Churchill argued that Hitler's Atlantic Wall was still impregnable. Better, he argued, to invade the Italian mainland. A successful landing near Rome might knock Italy out of the war, which would eliminate the Italian Navy as a naval threat. The surrender of Italy would, in turn, encourage Turkey to declare for the Allies, and it would loosen the Nazi grip on the Balkans. "The collapse of Italy would cause a chill of loneliness over the German people, and might be the beginning of their doom," the minutes record Churchill prognosticating. "Even if not immediately fatal to Germany, the effects of Italy coming out of the war would be very great, first of all on Turkey, who had always measured herself with Italy in the Mediterranean."[10]

FDR's focus, understandably, embraced both Europe and Asia. He said he expected a commitment to invade the Continent during 1944. On the Asian mainland, he said the Chinese were not "crying wolf" when they demanded material support. They were on the brink of collapse, and Roosevelt wanted western China preserved as a fortress from which attacks against Japan could be launched. It was imperative that the United States and Britain sustain the Kuomintang Army.[11]

With those remarks, the battle lines were drawn.

While Churchill and FDR continued their talks over tobacco and drinks in the Rose Suite, the Combined Chiefs met in the Board of Governors room at the Federal Reserve Building to put meat on the strategic bones. Chaired by Admiral Leahy, the morning session saw Marshall and Brooke plunge into the first and most intractable problem: where to go after Sicily.

Surrounded by a silent audience of staffers, the two generals sparred like gladiators over the next big invasion. Brooke wanted a leap from Sicily to the Italian mainland; Italy's surrender, he claimed, would force Hitler to pull at least twenty German divisions off the Russian front to secure the Balkans—and several more to block the exits around the Italian Alps. If they did not invade Italy, then three Allied armies would sit idle until the spring of 1944. "It is unthinkable that we should be inactive during these critical months when Russia is engaging about 185 German divisions," the British chiefs declared.[12]

Marshall replied that ROUNDUP could go forward only if all detours were, for once, ignored. The Allies would "deeply regret not being ready to make the final blow against Germany, if the opportunity presented itself, by reason of having dissipated ground forces in the Mediterranean area." Raising the specter of the Pacific again, he declared that a burgeoning Mediterranean strategy would "mean a prolongation of the war in Europe, and thus a delay in the ultimate defeat of Japan, which the people of the U.S. would not tolerate."[13]

Brooke felt that Marshall was missing the point, though in an unusual display of tact, he didn't say so outright. He did say that an Italian landing to draw off German troops was a prerequisite to ROUNDUP. Without that diversion, even if the Allies could land fifteen or twenty divisions of green troops on the French coast, they would be quickly bottled up by battle-hardened Germans moving unhindered from the Russian Front. Brooke warned the Americans, "No operations would be possible until 1945 or 1946, since it must be remembered that in previous wars there had always been some 80 French divisions available on our side."[14]

Thunderstruck by Brooke's suggestion that Europe might not be invaded until 1945 or 1946, Marshall dug in his heels and refused to budge. Churchill wouldn't let his chiefs make a tactical retreat from Italy, even had they been so inclined, so progress quickly ground to a halt.

In their meetings at Casablanca and Washington, the Combined Chiefs struggled to reach compromises that cut into their deeply held beliefs. But smaller annoyances added to the basic problems of strategy and national war aims.

The English language was one of them. Things necessary for operations, to the Americans, were "requirements," while in British vernacular they were "demands." The word "demands" had an imperious ring to the American ear, and when the British presented "demands," it sounded like an edict from King George III to his colonists. Similarly, the British might suggest that the group "table" a difficult matter—meaning lay it on the table for discussion—while to the Americans, to "table" something meant to set it aside for the future.[15]

It took time to overcome these small but irritating differences, particularly when they were shot from General Brooke, whose rapid-fire delivery, French-Ulster accent, and easy excitability grated on American nerves. Alan Brooke, for all his strategic acumen, lacked tact. The British chiefs, recognizing Brooke's limitations, sometimes tried to channel the discussion through the smoother Air Marshal Portal, but that was not always an option.

Then there was the "cloud of witnesses," as Dill called the staffers, aides, and assistants who watched the Combined Chiefs like spectators at Rome's Circus Maximus. For every British staff officer who thought the Americans were strategic rubes, there was an American staffer who believed the British were devious agents of imperialism. In front of staffers with whom they worked every day, it proved difficult for both groups of chiefs to back off positions they had worked up with their staffers before the meetings.[16]

Through three days of negotiations, a deadlock set in over the interconnected problems of the Mediterranean, France, and the Pacific. Tempers became short, veiled threats grew less veiled, and the meetings lurched forward without consensus. Distrust penetrated the Anglo-American relationship the way a vine's tendrils reach into a wall and separate brick from mortar over time. Something had to be done.

To rebuild their working relationships, Marshall arranged for the Combined Chiefs to take a break. On Saturday, May 15, he brought the British chiefs and their aides to Colonial Williamsburg for a weekend of sightseeing, fine dining and relaxation. Chaperoned by General McNarney and Marshall's aide, Colonel Frank McCarthy, the party lodged at the Williamsburg Inn, a magnificent period hotel built by millionaire John D. Rockefeller Jr.

As they stepped back in time along wooden shops and cobblestone streets, the masters of war withdrew from the horrors of industrial-scale death. Strolling along Duke of Gloucester Street in the evening, walking past the Capitol, Raleigh Tavern, and the old brick magazine, the men shared a blissful moment when they could mingle as friends and ponder an era when their forefathers had carved out an empire together. Isolated from the cares of office, Marshall talked of duck hunting with Wavell and the frail Admiral Pound, while Brooke spoke of his passion for birdwatching and Portal discoursed on falconry.[17]

To ensure the camaraderie flowed into the evening, Marshall's aide brought a case of bourbon, a case of scotch, two bottles of gin, two of brandy, and a bottle of vermouth. The men ate like kings, a few swam in the inn's pool—Air Chief Marshal Portal lost his borrowed swimming trunks after one audacious dive—and General Marshall plucked out "Poor Butterfly" on an antique harpsichord as the warlords sipped their drinks.

Next morning they attended services at a local church. Admiral Pound, the senior British guest, read the Scripture from the sixth chapter of Matthew: "Take therefore no thought for the morrow, for the morrow shall take thought of the things of itself."

As they drove back to Langley on Sunday afternoon, Marshall hoped cooler minds would prevail when they returned to work.[18]

To Marshall's disappointment, the British chiefs had hardly sat down at that long, remorseless table before they locked shields and argued their plans for Italy as if the weekend had never existed. Brooke watching red-breasted robins, Portal swimming bare-assed in the pool, Pound reading, "Take no thought for the morrow." Back in Washington, those men were taking thoughts of far too many morrows for Marshall's liking. Morrows of 1946, or even later. Marshall was personally prepared to violate the biblical injunction into, say, the morrows of 1944, but no longer.

Discussions again grew strained as the chiefs replowed old ground: Mediterranean against France, Europe against the Pacific, Burma against everywhere. But on Wednesday, May 19, the logjam began to shake loose as the two sides began working from points of common agreement. Both agreed that ROUNDUP was ultimately necessary to defeat Germany, and the only questions were when ROUNDUP should take place, whether operations against Italy should be put on ice until then, and how an invasion of Italy would affect the timing of ROUNDUP.

Brooke said he felt ROUNDUP could go forward in the spring of 1944, *if* operations designed to knock Italy out of the war were carried out in 1943. Italy, he claimed, could be purchased on the cheap—as few as three or four divisions— and its surrender would reap huge rewards by tying down dozens of Hitler's divisions around the Italian, French, and Austrian Alps. Those troops would not be waiting for the Allies on the beaches of France, making the landings infinitely easier.

Warming to a settlement, Marshall asked for time to study the British idea. The next day he and Brooke worked out a compromise they called ROUNDHAMMER: something less than a full-blown ROUNDUP, but more than SLEDGEHAMMER. ROUNDHAMMER could go forward by May 1, 1944, but its reduced size would still allow a knockout blow against Italy in the fall of '43.[19]

The next region they bulled through was China, a vast strategic swamp in which there was no discernible Allied policy. Like an old ecclesiastical court, the chiefs sat behind their table for an afternoon and listened to opposing sides argue strategy. As in February, the debonair Claire Chennault pressed for the lion's share of supplies in China, while "Vinegar Joe" Stilwell, still looking tired and irritable, advocated a ground campaign to reopen the Burma Road.

The British chiefs saw little to be gained in either Burma or China. The terrain was terrible, the Japanese had better supply lines, and a ground operation to open the Burma Road would have little immediate effect, as the decrepit, winding road could not be improved to carry large-scale supplies until 1945 or even later. They agreed with Chennault that air supplies could get over the Hump in sufficient quantities to keep China in the war. ANAKIM, Brooke and his colleagues argued, should not be attempted that year.[20]

Siding with Stilwell, Marshall argued that the air route was too weak to sustain the Chinese field army. Without ground operations, China might fall, and without China, the Pacific war might drag on for years.[21]

On the afternoon of May 20, the chiefs met without their staffers and hammered out another compromise. The Hump capacity would be expanded to ten thousand tons of supplies per month. Ten thousand tons, the Americans calculated, would be enough to feed the air war over China and leave a few crumbs for Stilwell's ground troops. Beyond that, the Allied chiefs agreed to work with the Chinese on some sort of "vigorous and aggressive" operations. Because the sense of urgency of the Pacific and Europe was not mirrored in the China-Burma-India theater, everyone sitting around the table knew nothing would be done there on the ground.[22]

Admiral King's Pacific bell clanged loudly in Brooke's ears. In their discussions, King had asked the Allies to maintain and extend "unremitting pressure" on Japan while the Germans were still fighting in Europe.

King's calls for offensive action were anathema to the British, who saw King as a wolf baying at the Germany First flock. But even those shoals were crossed when King stopped pressing for abstract, open-ended authority in the Pacific—in words whose flexibility frightened the British—and told Brooke, Pound and Portal what he had in mind.[23]

King said his service had been studying the problem of Japan for more than three decades. His own ideas on the Pacific, he said, were neither original nor novel. They represented orthodoxy laid down at the Naval War College by the Navy's best planners.[24]

As the Navy saw it, there were three routes to Japan. One was the Central Pacific route, through the Mariana Islands. The others were through the southwest, MacArthur's area, or the far north, near Alaska. The tall, churlish admiral became animated, specific, and eloquent as he described a limited drive across the Marshall, Gilbert, and Caroline island chains during 1943. He assured the chiefs he meant only to threaten Japanese supply lines to Truk and Rabaul during 1943, setting the stage for larger steps in 1944. With air and naval bases on a few key islands of the Central Pacific, he said, the Allies could move against the Philippines, and perhaps even take the big Chinese island of Formosa.[25]

Stripped of its troublesome abstractions, King's explanation satisfied the British. They agreed to keep "unremitting pressure" on Japan by pushing as far as the Caroline Islands, so long as the Combined Chiefs were able to give "consideration" to each specific operation before it was launched. King agreed, because to him, "consideration" was a damned long swim from "authorization."[26]

With an agreement on the Pacific, the last of the great blocks fell into place. The path to victory by 1946 had been set, and all the chiefs had to do was get the approval of their national leaders.

Selling war strategy to two energetic meddlers like Churchill and FDR required military men to hold their breath. Like two octopi groping at a school of fish, the long tentacles of president and premier darted into every economic, political, military, and social crevice within reach. The Combined Chiefs feared that their bargain, reached after so many hours of wrangling, would be entangled and shredded by the leaders to whom they reported.[27]

Roosevelt was the easy one. Content to accept most unanimous decisions of his military experts, his interference in grand strategy was usually limited to political matters, like putting American troops into the field in Africa in 1942. For this round, he wouldn't be the problem.

Churchill's penchant for meddling ran deeper. Like Bulgakov's cat, his notions leaped from nowhere, wreaked destruction around the room, then, self-satisfied, sauntered off on two legs to await the next opportunity to pounce. As an exasperated Sir John Dill remarked, "It is a thousand pities that Winston should be so confident that his knowledge of the military art is profound when it is so lacking in strategical and logistical understanding and judgment."[28]

When the Combined Chiefs presented their agreement to the two leaders on May 24, they described a cross-Channel invasion in May 1944, rechristened "OVERLORD." The Allies would commit twenty-nine divisions to OVERLORD, nine of which would form the initial assault. As collateral for the British promise, seven divisions would be transferred from the Mediterranean theater to the United Kingdom in November 1943. The Allies would expand the air corridor over the Hump, and American forces would launch campaigns to take the Marshall and Caroline islands, the western Solomons, the Bismarck Archipelago, and the rest of New Guinea.[29]

In return for British concessions, the Americans agreed to a blow designed to knock Italy out of the war. General Eisenhower would be tasked with coming up with a recommendation for what that blow might be—the Riviera, Corsica, Sardinia, Italy proper, or someplace else. Putting Eisenhower in the driver's seat, the Americans hoped, would squelch any wild Churchillian ideas about running off to obscure Aegean islands or the Balkans.[30]

After they presented the report, Churchill expressed dismay that the Allies might take seriously a recommendation from Eisenhower to assault something west of Italy—say, Sardinia—rather than strike the Italian boot near Rome. Jutting out his pudgy chin, he declared that a Sardinian operation would be unacceptable to His Majesty's Government. Operations should be commenced toward the Balkans, where perhaps thirty-four German divisions might be worn down in the mountains and ravines of Yugoslavia.

Marshall and King did not need to attack, cajole, hector, or circumvent Churchill; the British chiefs would do that for them. To the British chiefs, an ill-conceived Balkans adventure would wreck the Anglo-American relationship built over two painstaking years. Brooke was personally mortified by Churchill's recalcitrance, for he knew the Americans would think he had double-crossed them by allowing Churchill a second bite at an apple he had personally selected.

There was plenty of suppressed anger on the American side over Churchill's objections. In his diary, Admiral Leahy complained that Churchill was advocating a *"permanent British policy of controlling the Mediterranean Sea regardless of what may be the result of the war."* Even the normally indulgent FDR told Stimson that Churchill was acting "like a spoiled boy" over the Balkans.[31]

The spoiled boy came around the next day, when Churchill announced his agreement to go along with the group. The exhausted warlords breathed a collective sigh of relief, and went back to draft the necessary orders for their theater commanders.

The Yankees had, for once, obtained a better bargain than their cousins. The great prize was British commitment to a cross-Channel assault in May 1944, but the Americans also won British assent to sending Nimitz as far as the Caroline Islands. While agreements on Burma and China did not give Roosevelt everything he wanted, as he told Leahy, it was "the best I could get at this time."[32]

When Churchill left for England, Roosevelt knew he had leaped over the TRIDENT and landed on both feet. He took the *Ferdinand Magellan* home to Hyde Park and slept in his own bed for thirty hours over the next three days.[33]

No one who knew him begrudged Roosevelt his rest. The man had been balancing Allied relations, arbitrating Army-Navy disputes, battling old foes at home, and fighting Hitler, Mussolini and Hirohito abroad. He had survived another visit from Churchill, and he had wrestled with a nasty moral dilemma that began when a hungry wolf went scrounging for food in a faraway forest.

THIRTY-ONE

THE FIRST CASUALTY

〰

T RUTH IS NOT, AS THE ADAGE CLAIMS, THE FIRST CASUALTY OF WAR. IN THE war for Asia, the first casualties were the unfortunate Chinese of Manchuria. In Europe, they were the Poles of the German frontier.

But truth was not long in catching up, for like everyone else, in wartime honesty had to make its fair share of sacrifices.

To dictatorships, truth and lies are tools of power, just like secret police, rigged elections, and state-run media. Democracies, built upon the dissemination of truth, have to be more circumspect. Yet even democracies depend on the suppression of information—and the occasional lie—as the price of survival.

Duplicity was second nature to FDR. His mind was not a windowless house, but a house with mirrors, prisms and false doors, easy to enter but hard to navigate. Visitors on opposite sides of an issue left the White House convinced that Roosevelt was in their corner, and FDR's compulsion to tell visitors what they wanted to hear sometimes required him to affirm untruths. He rarely placed himself in the position of telling a bald-faced lie he'd have to ask forgiveness for later, but he would go to extraordinary lengths to stretch facts and conceal the truth when the truth became uncomfortable.

Roosevelt's relationship to public candor was ambiguous. As president, he had to maintain a basic level of trust with the newsmen who made up the Fourth Estate. For their part, most journalists put country ahead of scoop, and accepted that in war there are secrets that must be kept, truths that must not be told. While the American high command sometimes manufactured a falsehood, such as King's explanation of the Midway ambush, more often it simply withheld small but critical facts.

One of the more difficult feats for an idealistic democracy is collaborating with a dictatorship on an abject lie, and in the winter of 1943, a wolf roaming through western Russia dumped just such a problem on to FDR's lap.

The wolf, scavenging near Smolensk, sniffed out a pile of human bones buried near the bivouac of the German 537th Signal Regiment. The animal tracks, bones, and stories from local peasants led German troops to a massive burial trench in the nearby Katyn Forest. As they excavated, the cairn rendered up some three thousand rotting corpses clad in uniforms of Polish officers.

Radio Berlin, in its broadcast of April 13, announced the find. It claimed that the hands of the victims had been tied and each had been shot in the back of the head. The graves at Katyn and other sites, said the Nazis, were evidence of a massacre of some ten thousand captured Polish officers by the Soviet NKVD after Stalin occupied eastern Poland in 1939. Goebbels and his propagandists offered the bodies as Exhibit A to their indictment that the communists were every bit as murderous as the Allies claimed the Germans to be.[1]

The response from both Poles and Russians was immediate and bitter. London-based Poles, led by General Wladyslaw Sikorski, publicly called for a Red Cross investigation. Stalin claimed the murders were the handiwork of Germans in the summer of 1941, and accused Berlin of waging a propaganda war with the help "of certain pro-fascist Polish elements picked up in occupied Poland." *Pravda* branded the Polish exiles in London "Nazi collaborators," and the USSR broke off diplomatic relations with the London Poles. Stalin's apparatchiks began building a government-in-waiting among communist partisans of the Polish underground.[2]

American reaction to German accusations was mixed. Most newspaper editorials focused on the larger war aims rather than the specific crime. *Newsweek* ran the story under the headline "Poles vs. Reds: Allied Unity Put to the Test Over Officer Dead." *Time* remarked: "The U.S. State Department and No. 10 Downing Street were in complete accord: nothing must be allowed to create a final schism between Russia and the Anglo-American coalition."

On the other hand, the Polish-American and anti-Roosevelt communities were outraged. McCormick's *Tribune* published an article headlined, "American and Polish Leaders Brand Russia a Nation of Liars and Old Conspirators." Two senior Democrats, Congressman John Lesinski and Senator Burton Wheeler, saw the bad old Russia of the 1920s coming back, and called on Roosevelt to respond to Soviet persecution of Poles. Russia's popular image in the United States as a champion of the anti-Nazi cause lost some of the luster of 1941.[3]

Caught off guard, FDR did his best to prevent the atrocity from widening into a full-fledged rift between himself and Stalin. Without passing judgment on the Red Army's guilt or innocence, he drafted a cable to Stalin accusing Sikorski of making a "stupid mistake in bringing this issue to the international Red Cross." But he cautioned that with a great number of Polish Americans serving in uniform, it would not help him politically if Stalin broke relations with Sikorski.

Washington and London pressured the exiled Poles to drop their demand for an investigation, and tried to downplay the story at home. As evidence of German innocence quietly worked its way through the Army's intelligence branch, Roosevelt had Elmer Davis, head of the Office of War Information, urge reporters not to give credence to "phony propaganda stories," including a "very fishy statement" about an unproven Soviet massacre of Polish prisoners.[4]

Churchill set Anthony Eden, his foreign secretary, to work on the breach. Eden discreetly asked Owen O'Malley, his representative to the Polish exiles, to find out what really happened, and a month later O'Malley reported that Soviet denials were not credible. The victims wore winter coats, while the Germans had occupied Katyn in the summer. The ropes binding their hands were of typical Russian manufacture. The victims belonged to units that retreated east from the German invasion, toward the Red Army, and their letters home and pocket diaries all ceased in April 1940, when the Red Army occupied the area. Small, concealing pine trees planted atop the graves were old enough to have been introduced to the area in 1940.[5]

In light of the evidence, O'Malley was troubled by His Majesty's Government's inclination to deny Soviet guilt. "We have in fact perforce used England's good name like the murderers used the little conifers to cover up the massacre," he told Eden. Yet he concluded that in light of the overwhelming need to defeat Germany, "few will think that any other course would have been wise or right."[6]

Churchill told Sikorski, "If they are dead, nothing you can do can bring them back." In August he forwarded O'Malley's report to Roosevelt, though if FDR read it, he kept his mouth closed. There was a war to be won, and spring 1943 was not the time for moralizing to one's ally. The year would bring enough problems of its own.*[7]

· · · ·

As a cabinet member, Henry Stimson was focused on the home front. He supported Marshall's military strategy, but like FDR, most of his thoughts centered

---

* In 1945, Captain George Earle, a former attaché to Bulgaria and an old Roosevelt family friend, threatened to publicize information implicating the Soviets in the massacre. Roosevelt insisted the incident was manufactured by the Germans, and when Earle persisted, he was abruptly transferred to the Samoan Islands.

on domestic matters like production, funding, conscription, transportation, and weapons development. A car's engine will shut down if an air bubble gets into the gas line; if America's farms, factories, or railroads faltered, it wouldn't be long before its great military engine would grind to a halt.

One of those air bubbles formed in April, when half a million coal miners led by CIO head John L. Lewis threatened a strike to force management to boost wages by two dollars a day. Wildcat strikes by another 100,000 miners shut down some 3,000 mines and produced a daily loss of two million tons of coal, the stuff that fired every steel, automotive, and textile plant in the nation.[8]

Stimson struck back. Livid over labor agitation, he asked Roosevelt to use his emergency powers to reopen the mines, and he urged the president to order the War Manpower Commission to draft any miner who refused to work.[9]

In a national emergency, President Roosevelt could simply order the Army to take over the mines. He and Stimson hoped the strikers would back down first, however. *"We hate to have the Army do it,"* Stimson told his diary when a strike threatened in 1941. *"It will be misrepresented all over the world and be treated by the Nazis as a revolution."*

But German propagandists were already playing up the strike as evidence of a collapse of national will, and coal production was so critical that Stimson told Roosevelt he saw no other option. FDR agreed, and on April 30 he ordered the Army to seize and reopen the mines. Lewis declared a temporary truce, and the miners went back to work.[10]

Labor was one of Roosevelt's basic constituencies, and to Stimson's dismay, he refused to take punitive action against the unions. When Congress passed a bill imposing drastic penalties on anyone urging a strike in government-owned plants, Roosevelt vetoed it. He told Congress that during 1942, "99 and 95/100 of the work went forward without strikes, [and] only 5 one-hundredths of 1 percent of the work was delayed by strikes. That record has never been equaled in this country. It is as good or better than the record of any of our Allies in wartime."[11]

An unimpressed Senate overrode Roosevelt's veto in fifteen minutes. Two hours later, the House followed suit. The antistrike bill became law, and Stimson told his diary, *"The President met with a bad rebuff and an unnecessary rebuff. His administration really is beginning to shake a little and throughout the country there is evident feeling that he has made a mistake in regard to labor."*[12]

In the hot summer of 1943, the nation's collective nerves frayed as America struggled to swallow domestic sacrifices and population upheaval. Food prices had risen nearly 25 percent since 1941, and race riots scarred Detroit and Mobile.

In Southern California, sailors, marines, and white mobs waged a two-week running battle with Latino youths wearing "zoot suits," long, baggy, flamboyant outfits openly defying federal cloth-rationing laws.

Everyone seemed to be complaining about someone else's lack of shared sacrifice, and everyone seemed unhappy. Sketching out the tattered landscape that was the Home Front of 1943, the *New York Times* lamented, "Our President has said, and often, that we would show the world that a democracy could be made to work at war as well as any totalitarian state. . . . Therefore, much depends on how we run our democracy at war. And we are not running it any too well."[13]

With the war stuck in a doldrums and the public likely to take its frustrations out on Roosevelt in 1944, members of the Democratic National Committee quietly proposed offering up Stimson as a sacrificial lamb. The old Republican was becoming a political liability, they suggested. Dumping him would rid the administration of excess baggage without signaling a Republican purge, since Frank Knox, another Republican, would remain in the cabinet.

Roosevelt considered the idea. Morgenthau, Ickes, and Hopkins would have jumped at the chance to head the War Department, and bleak times make a dramatic shake-up an attractive option. But FDR stood by his man. He told the party elders he believed Stimson was "a man of courage, fairness, and great capacity." He had a high reputation throughout the country, said Roosevelt, and he was an asset to the nation at war.[14]

However duplicitous, however disorganized Roosevelt might be, he was loyal to those who were loyal to him. Henry Stimson, the establishment Republican, had set aside party loyalty for the good of the country, and he would remain part of Roosevelt's war team.

THIRTY-TWO

LANDINGS, LUZON, AND *LADY LEX*

〰

N THEIR PENTAGON SUITES, MARSHALL AND STIMSON BUTTED HEADS OVER the unconditional surrender of Italy.

In his "unconditional surrender" announcement at Casablanca, FDR had mentioned Italy in the same breath with Germany and Japan, but Stimson felt Italy might warrant an exception to the rule. He had traveled there as secretary of state under Herbert Hoover, and had met with Mussolini and other government officials. In their hearts, he said, the Italian people were not genuine Axis material. If they ejected *Il Duce* and his thugs, he said, it would make sense to treat them with a light hand. Instead of humiliating the Italians, the Allies should come as liberators, bearing the torch of freedom that Garibaldi had carried during the previous century.[1]

Marshall, who was studying an amphibious leap onto Italy's bony knee, told Stimson he was surprised to hear him side with mushy sentimentalists who wanted to go easy on the cradle of fascism. It was no easy job keeping the public mad enough to make the sacrifices needed to defeat the Axis, and he thought it a terrible idea to water down the cask of hate by invoking romantic images of Garibaldi and good Italians.[2]

Stimson's mustache twitched and he grew hot under his starched collar. He wasn't trying to go easy on the Italians, he said. "Divide and conquer" was a better strategy. By telling the Italians they were in for the same treatment Germany and Japan had coming, the Italian soldier—or at least his officers and leaders—would fight that much harder.[3]

As the invasion of Sicily drew near, they were about to find out which man was right.

271

• • •

Japan's surrender, unconditional or otherwise, remained a landfall far below the horizon. The TRIDENT agreement had authorized Admiral King to drive through the Carolines, and shortly after the conference ended he flew to San Francisco for another conference with Nimitz. There the two men mapped out an aggressive campaign for the Marshall and Gilbert islands, stepping-stones to the Carolines and Marianas.

They also decided on the all-important command assignments. The main fleet would go to Rear Admiral Raymond Spruance, the hero of Midway, whom King would bump to vice admiral. The amphibious commander would be Rear Admiral Kelly Turner, a veteran of Guadalcanal. The landing force commander, necessarily a marine, would be Major General Holland M. Smith—one of the best amphibious experts alive, but, like King, a difficult person to get along with. Nimitz, as theater head, would select the islands to assault, and King would ensure that Nimitz had every landing craft, warship, and marine that the country could spare. It was a good working team, and it was King's job to see that no dumb bastard in Washington blew them off their course.[4]

In early June, King told Marshall he wanted to put Nimitz in charge of the Pacific war's scheduling and allocation of resources. He also proposed that the Joint Chiefs approve an invasion of the Marshall and Gilbert islands by November 1. He wanted the Joint Chiefs to order MacArthur to give them a timetable for his upcoming operations against the Western Solomons and New Britain, home of Fortress Rabaul, so they could coordinate blows. Finally, he proposed transferring the veteran First and Second Marine Divisions from MacArthur to Nimitz.[5]

Marshall usually deferred to King on details of Pacific strategy. The general had one idea for Europe—to invade northern France—but he was out of his depth when evaluating the strategic merits of operations in the Dutch East Indies, the Marianas or the western Carolines. Those details were best left to MacArthur, Nimitz, Arnold, and King.

Command was another matter, however, and he pushed back when King proposed placing Nimitz in charge of "scheduling." Control of scheduling meant control of the Pacific war, for Nimitz would be able to postpone MacArthur's campaign until the Marines had driven far enough to make MacArthur and the Army irrelevant.[6]

Marshall kept MacArthur informed of his deliberations with King, and told him of King's request for the two Marine divisions. MacArthur shot back an indignant message, blasting the Joint Chiefs for even thinking of taking away his only trained amphibious fighters. The loss of the First and Second Marine Divisions would doom operations against Rabaul, he declared. He also warned of dire political repercussions that would follow the emasculation of his Pacific strategy.[7]

But King was not finished with MacArthur's Pacific strategy. One week after MacArthur's reply, the Joint Strategic Survey Committee, the senior Army-Navy strategy think tank, released a memorandum cautioning that MacArthur's planned drive to the Philippines was not necessarily a sacred cow. In fact, the committee concluded, the Allies would be better served by taking Nimitz's Central Pacific route to Japan, through the Marianas, than by crashing through the southern door at Rabaul and Luzon.

The study was a godsend to King, who had never liked MacArthur's Rabaul operation. He crowed that the JSSC had done a "great service" to the nation by presenting its paper. King's Central Pacific strategy had been vindicated.[8]

As he planned his advance through the Pacific, King decided that Marshall needed some basic education in the Navy's way of war. It had taken him three months to school Marshall on the importance of the Marianas, and now, he decided, the general needed a graduate course in carrier procedures. In early July, he invited Marshall to accompany him to inspect USS *Lexington*, a brand-new *Essex*-class fleet carrier and the namesake of his late, lamented flagship.

"The reason for the trip," King wrote in his postwar notes, "was for me to look over the new carrier (Lady Lex), but foremost to get Gen. Marshall educated, a little, about carriers since he had never been on board one and didn't even know how they were managed." Marshall, he wrote, "had a <u>little</u> idea about naval aviation, which it was my idea to get across to the Army people (and the Army Air Corps people), so they might understand at least a little about 'sea power' itself."*[9]

Marshall listened politely as King's men regaled him with the power of their new carrier. But it was wasted effort, for carrier procedure meant nothing to the strategic debate among the Joint Chiefs. With an independent committee advising them to go through the Central Pacific, Marshall could hardly dissent from King's plan. He was, however, able to keep the First Marine Division under MacArthur's wing for the Rabaul operation. To pay back the Navy for the loan of the marine division, Marshall found an infantry outfit undergoing amphibious training in Hawaii, the 27th Division, which he gave to King.[10]

With that shaky bargain, King and Marshall had set the stage for the next hard fighting in the Pacific. MacArthur's borrowed marines would pry open the

---

* King continued to refer to air service as the "Army Air Corps" long after its redesignation as the "United States Army Air Forces," and even after it was reorganized by Congress as a separate service named the "United States Air Force."

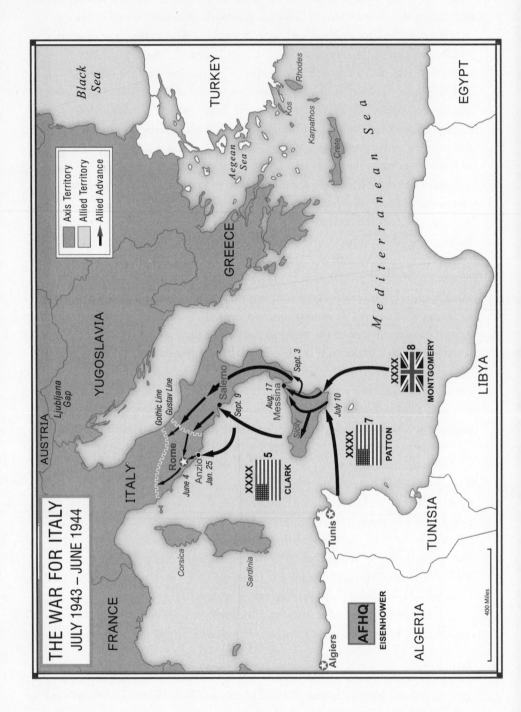

THE WAR FOR ITALY
JULY 1943 – JUNE 1944

Axis Territory
Allied Territory
Allied Advance

FRANCE
AUSTRIA
YUGOSLAVIA
Black Sea
TURKEY

ITALY
Ljubljana Gap
Gothic Line
Gustav Line
Rome
June 4
Anzio
Jan. 25
Salerno
Sept. 9
Sept. 3

GREECE
Aegean Sea
Kos
Karpathos
Rhodes
Crete

Mediterranean Sea

Corsica
Sardinia

XXXX
5
CLARK

Aug. 17
Messina
Sicily
July 10

XXXX
7
PATTON

XXXX
8
MONTGOMERY

Tunis

Algiers

AFHQ
EISENHOWER

ALGERIA
TUNISIA
LIBYA
EGYPT

400 Miles

road to the Admiralty and the Bismarck islands near New Guinea, keeping the campaign for Rabaul and the Philippines alive. At the same time Nimitz, supported by Army aircraft and ground troops, would have his shot at the Gilberts and Marshalls, the next steps on the bloody path to the Marianas.[11]

* * *

On the morning of July 10, landings on a scale that MacArthur and Nimitz could scarcely fathom commenced halfway across the globe. The greatest armada in history gathered off the southeastern corner of Sicily for the long-anticipated Operation HUSKY.

That morning, eight divisions from two armies fought their way ashore and locked up a thin slice of the Sicilian coast. General Montgomery's Eighth British Army smothered a stretch of the island's rocky east side, and General Patton's Seventh U.S. Army covered the south. Ahead of Patton's troops, the 82nd Airborne Division made a daring night drop and seized the high ground beyond the beaches.

On D-day and the next, men of Patton's army slugged their way inland. On Patton's right, Lieutenant General Omar Bradley's II Corps, veterans of the Tunisian campaign, drove back a vicious panzer counterattack by the Hermann Göring Division. On Bradley's left, Major General Lucian Truscott's Third Infantry Division captured coastal ports and fought its way toward the island's capital. Montgomery's army ground to a halt against stiff German resistance on the road to Messina, the island's key port. After a violent three days, the Allies were fixtures on Mussolini's island, and they were not leaving.

Yet the question weighing on the minds of the American warlords was "Where are those troops going next?"

THIRTY-THREE

"A VITAL DIFFERENCE OF FAITH"

WHILE PATTON'S FOOT SOLDIERS TRAMPED DOWN CENTRAL SICILY'S DUSTY roads, Henry Stimson sat for dinner at London's No. 10 Downing Street.

As a cabinet member, Stimson rarely accompanied Roosevelt to meetings with America's allies, so his inspection trip of U.S. troops in Iceland, the U.K., and North Africa afforded him his first opportunity to come to London. There he planned to sound out the prime minister on a matter of utmost concern to him.[1]

From the beginning, Stimson had been a disciple of OVERLORD, the operation's Peter, James, and John. Marshall even claimed the old man literally included the invasion in his nightly prayers. Over china plates and glasses of sherry, Stimson stressed to Churchill the political operation's importance in America. Roosevelt would face another election in 1944, and if the Allies could not pull off an invasion of the Continent by then, Republicans—Stimson's own party—would tap into anger at Japan as a campaign issue. The American voter, Stimson explained, was intellectually persuaded that Germany was the greater foe, but "the enemy whom the American people really hated, if they hated anyone, was Japan, which had dealt them a foul blow."[2]

Churchill kept his distance when Stimson pressed him about the political importance of OVERLORD. He blithely replied that Allied victories anywhere would suffice, but did not explain how the capture of some unknown Mediterranean island or patch of desert would give Roosevelt a selling point with voters in 1944.

Churchill's nonchalance toward France worried Stimson. It worried him still further when the prime minister spoke of operations on Italy's mainland and future landings in the Balkans. With a troubled mind he told his diary, *"The PM renewed his pledge of loyalty to Bolero with conditions it seemed to me less destructful than before, unless his military advisors could present him with some better opportunity."*[3]

It was this last bit, "unless his military advisors could present him with some better opportunity," that worried him most of all.

Five days later, while Stimson was riding with Churchill to Dover on an inspection tour, the hair on his neck stood up when Churchill began reading aloud a Combined Chiefs memorandum supporting a landing at Naples, halfway up the Italian shin. Because the chiefs had previously considered a more cautious effort against Italy's toe, Churchill inferred that Marshall and the Americans now supported a far more aggressive effort in the Mediterranean. Something more opportunistic that would take them farther up the Italian boot—and perhaps open new fronts leading to the Alps, Austria, and the Balkans. Energized by the possibilities, he waxed eloquent to a rattled Stimson about the prospects of stirring up the Balkans against the Hun.[4]

Stimson was certain that the prime minister had the wrong impression about Marshall's intentions. When they arrived back in London, he rang up Marshall on a new transatlantic phone line set up in an American bunker. Through the squelch of the phone scrambler, Marshall told Stimson what Stimson already knew—the Joint Chiefs had suggested landings in Naples only to hasten the fall of Rome and clear the way for OVERLORD. The operation, called AVALANCHE, was Marshall's way of avoiding a slow, debilitating crawl through the toe of Italy, up the foot, ankle, and lower shin, so everyone could get back to OVERLORD.[5]

Stimson told Churchill he was mistaken about Marshall; he had it straight from the horse's mouth that the general's eye was fixed on France, not the Balkans. A deflated Churchill glumly remarked that fifty thousand Allied troops would have a terrible time fighting their way into France. *"On direct questioning,"* Stimson wrote, *"he admitted that if he was C-in-C, he would not figure the Roundhammer operation; but being as it was, he having made his pledge, he would go through with it loyally."*[6]

Stimson spoke bluntly to Churchill. The Americans, he said, were looking for the quickest end to the war. He accused the prime minister of "hitting us in the eye" by stonewalling OVERLORD with his unending Italian schemes.

Churchill backed off. He said he did not propose going north of Rome—unless northern Italy were relatively free for the taking, that is. He had no desire

---

* By "Bolero," Stimson probably meant OVERLORD. It was another instance of the code names for the buildup (BOLERO) and the invasion (OVERLORD) tripping up the men who approved them.

to send troops into the Balkans. "If we start anything," Churchill said, referring to OVERLORD, "we will go through it with the utmost effort."[7]

It was hardly the ringing endorsement Stimson was looking for.

A year before, when FDR ordered the invasion of North Africa, a bitter Stimson had railed in his diary against Great Britain, *"a fatigued and defeatist government which had lost its initiative, blocking the help of a young and vigorous nation whose strength had not yet been tapped."* He returned to Washington convinced that the British wanted out of OVERLORD. As in London, and Casablanca, they would find a way to strangle OVERLORD if the Americans were not vigilant and inflexible. The only way to safeguard the operation, he felt, was to appoint an American commander. Someone of stature. Someone who would not be blinded by British arguments. Someone like George Marshall.[8]

.   .   .

As he ate lunch in the Oval Office with Stimson on August 10, Roosevelt carefully studied Stimson's report of his visit to London. The report laid out Stimson's case for an American commander of the OVERLORD operation. "We cannot now rationally hope to be able to cross the Channel and come to grips with our German enemy under a British commander," he wrote. "His Prime Minister and his Chief of the Imperial Staff are frankly at variance with such a proposal. The shadows of Passchendaele and Dunkerque still hang too heavily over the imagination of these leaders of his government. Though they have rendered lip service to the operation, their hearts are not in it." He concluded, "The difference between us is a vital difference of faith."[9]

The Allies were approaching their supreme test: Did they believe they could beat Germany in a head-to-head battle on the Continent? Stimson knew the British did not. He believed, and so did Marshall.

For OVERLORD, the Americans would need to field their most capable leader. Stimson, the Athos of FDR's Musketeers, saw only one choice for the job. His report to FDR concluded:

> You are far more fortunate than was Mr. Lincoln or Mr. Wilson. Mr. Lincoln had to fumble through a process of trial and error with dreadful losses until he was able to discover the right choice. . . . General Marshall already has a towering eminence of reputation as a tried soldier and as a broad-minded and skillful administrator. I believe that he is the man who most surely can now by his character and skill furnish the military

leadership which is necessary to bring our two nations together in confident joint action in this great operation.[10]

When he finished reading the report, Roosevelt looked up at Stimson and said he emphatically agreed. Fired up about OVERLORD, Roosevelt called in the Joint Chiefs and had Stimson describe the apathy he witnessed from Churchill in London. *"The president went whole hog on the subject of Roundhammer,"* Stimson wrote, using the operation's old TRIDENT name. *"He was more clear and definite than I have ever seen him since we have been in this war."*[11]

Roosevelt told his chiefs he wanted a large American force in Britain before the invasion—a force larger than the British force. "Frankly, his reason for desiring American preponderance in force was to have the basis for insisting on an American commander," the minutes recorded.[12]

Marshall seemed to be the only one in the room with no opinion on the identity of the invasion's commander. Taking a spartan pride in his stoicism on personal matters, Marshall refused to express any preference on the subject of his own appointment to lead OVERLORD.[13]

But everyone crowding the president's office that day knew who OVERLORD's commander would be.

THIRTY-FOUR

PLAINS OF ABRAHAM

〰

I N MID-AUGUST THE JOINT CHIEFS FLEW TO QUEBEC FOR QUADRANT, THE fifth major conference with Churchill and the Imperial General Staff. On the banks of the stately St. Lawrence River, they would fight another round over the fates of Italy, the Balkans, the Far East, and OVERLORD.[1]

Flying under dark skies, the American chiefs were comfortably billeted at the Château Frontenac, a palatial hotel whose spires, towers, and European charm were ornamented by red-coated Canadian Mounties who patrolled the grounds and stiffly saluted anyone in uniform. Inside the hotel, black-jacketed servants provided every accommodation to the generals, admirals, and staffers who would decide where great masses of men would die.[2]

Canada's governor-general lodged the stars of the show, Roosevelt and Churchill, in the Citadel, an imposing fortress dating back to an age when French and British kings fought over an empire in the west. Perched atop the Plains of Abraham, an ancient battleground for the New World, the two men would chart the fate of the Old.[3]

Roosevelt and Churchill enjoyed two different kinds of football. The football of FDR's America—an obsession for men like Eisenhower and MacArthur—was based on set-piece plays worked out between coach and quarterback. At the snap, linemen would hurl themselves forward, opening a hole, and the plunging back would charge through the hole for a touchdown if everything went as planned. If it did not, the team would line up and try a different play. A fixed plan and vigorous execution were the keys to the football played on green American fields.

The football of Churchill's Britain was more fluid and less choreographed than the gridiron battles of Knute Rockne or the Chicago Bears. An English attacker might send the ball to the wings, slow the tempo of the drive, or pass back to midfielders as he looked for an opening to exploit. The path to the net could play out in many different directions, and an unplanned opportunity to score would vanish if not seized quickly.

Churchill's strategic heart lay closer to the English style of play than the American variant. To Winston, OVERLORD, the Mediterranean, the Balkans and the air war were simultaneous plays on a monstrous, moving pitch, any of which might put the Allies in Berlin. He later explained the different modes of thinking:

> The American mind runs naturally to broad, sweeping, logical conclusions on the largest scale. It is on these that they build their practical thought and action. They feel that once the foundation has been planned on true and comprehensive lines all other stages will follow naturally and almost inevitably. The British mind does not work quite in this way. We do not think that logic and clear-cut principles are necessarily the sole keys to what ought to be done in swiftly changing and indefinable situations. In war particularly we assign a larger importance to opportunism and improvisation, seeking rather to live and conquer in accordance with the unfolding event than to aspire to dominate it often by fundamental decisions.[4]

At Quebec, the planners and the opportunists clashed in a fog of suspicion and disdain. Sitting across a long table in the Frontenac's Salon Rose, Marshall vehemently objected to further adventures in the Mediterranean. Brooke then sallied forth with a proposal to move through Italy as far north as the airfields around Milan. In response, Marshall and King unsheathed the old Pacific threat and laid it on the table like a rusty yet still-lethal sword.

"The conference at Quebec was the scene of two 'show-downs,'" King told an admiral a few days later. "When I say 'show-downs,' I mean just that!"

King dropped all pretense of courtesy. Using "very undiplomatic language," in Leahy's words, he fought like a shark for drives in the Pacific and a very limited Mediterranean effort, pushing the British to the end of their patience.[5]

"Come on, Ernie, you know you are talking bullshit!" said one exasperated

British admiral.* But for King the Pacific drive was anything but bullshit. OVERLORD and a push to the Marianas were the two most important thrusts of the war for 1944. As King saw it, the Americans were making the largest human contribution to the war in the west; the British could no longer pretend they were entitled to direct it from Whitehall.[6]

An exasperated Brooke told his diary, *"This is the sixth of these meetings with the American chiefs that I have run, and I do not feel that I can possibly stand any more!"* But Brooke knew that the time had come to concede the primacy of OVERLORD. It would be the decisive hammer-blow against Hitler, and like Marshall, Brooke had dreamed of commanding the cross-Channel invasion. He genuinely believed the Mediterranean could pin German troops and give OVERLORD a better chance of success, but he knew the time for open-ended adventures in Italy was drawing to a close.[7]

Feeling that matters could be better resolved on a personal level, without a large, high-level audience watching their every utterance, Brooke had the conference room emptied of planners, aides, and secretaries. When they had left, he told Marshall, King, Leahy, and Arnold the problem was lack of trust. The Americans doubted British intentions to put their full energies into OVERLORD, while the British were not confident the Americans would seize other opportunities once a contractlike memorandum had been signed.[8]

Brooke's candor melted the ice. After two closed-door sessions, the Combined Chiefs emerged with yet another compromise: OVERLORD's target date was confirmed as May 1, 1944, and its commander would be selected by President Roosevelt. OVERLORD would take priority over the Mediterranean, but the Allies would continue to exert "unremitting pressure" on the Germans in Italy, to borrow King's favorite Pacific phrase. The only new amphibious landing in the Mediterranean would be along the southern French coast, near Toulon or Marseilles, as an adjunct to the landings in northern France.[9]

As the last closed-door session let out, Lord Mountbatten approached a worn-out General Brooke. Blocking the door, Mountbatten begged a few minutes of the Chiefs' time to show them Project HABAKKUK, a plan worked up by his scientists

---

* The admiral, Sir James Somerville, was one of the few men who stood up to King to his face, and King respected him for it. After the meeting, King approached Somerville and told him, "James, if ever you wish to see me about something, the latch is always on the door." One astonished aide remarked, "Admiral King had never been stood up to quite like that before and it worked like a miracle."

to create floating airfields from immense blocks of pykrete, a frozen mixture of wood pulp and water.

An exhausted Brooke relented, probably against his better judgment. With the earnestness of a Fuller Brush salesman, Lord Mountbatten told the puzzled chiefs the new substance was many times stronger than ice, and would bear the weight of aircraft. If towed as a giant sheet and anchored off the French coast, the pykrete blocks would give the Allies new airfields from which to cover the invasion.[10]

To demonstrate the hardness of pykrete, Mountbatten had two large blocks—one of ice, one of pykrete—wheeled into the Salon Rose by his scientists. An assistant drew a pistol from his pocket with a flourish and announced that he would demonstrate the substance's resistance to gunfire. Nervous chiefs stepped back as the man leveled his arm and aligned the sights. With the first shot, the ice block shattered. At the second, the pykrete deflected the bullet, flinging it back toward its audience. The lead slug buried itself in a wall behind the Combined Chiefs of Staff.

"The damn fool!" an incredulous King said later. "One of those 'nuts' brought into the room a revolver and made a shot at the 'mush' which shot at least passed to nearby one of my own shins." Outside the room, a staffer joked, "My God! Now they're shooting at each other!!"[11]

The chiefs unanimously agreed they didn't need to see any more. When Mountbatten proposed a similar presentation to the president and PM, the chiefs suggested it would be best to leave out the pistol part.

After seeing the proposal, the normally inquisitive FDR told King, "We better leave HABAKKUK to the British." King heartily agreed, and he had the project quietly scuttled beneath a wave of staff papers and feasibility studies.[12]

Another scientific oddity, swallowed with more conviction than HABAKKUK, was TUBE ALLOYS, or "S-1," as the Americans called it. The year before, at Hyde Park, Churchill had agreed to pool British atomic fission research with the Americans, and he believed FDR had agreed to reciprocate by sharing the fruits of American labors. Since then, however, British military and scientific leaders complained, with good reason, that American scientists were keeping them in the dark.[13]

Roosevelt and Churchill saw the weapon as a possible war-winning breakthrough, and U.S. reluctance to share information became a discordant note in the partners' duet. "There is no question of breach of agreement," fumed Churchill in one of several irate letters to Harry Hopkins on the subject. American reticence, he said, "entirely destroys the original conception of 'A coordinated or even jointly conducted effort between the two countries.'"[14]

Churchill raised the issue privately with Roosevelt, who agreed that his team needed to be more forthcoming. He signed an aide-mémoire promising the British full access to American research, and the two men further agreed that neither side would use the bomb without the other's consent. Churchill left reassured, and before long British scientists began arriving at the heart of America's most secret program. Information began flowing in two directions, not one.*[15]

As the Quebec conference wrapped up on August 25, the only political meddling in military plans was Roosevelt's insistence on a SLEDGEHAMMER-like invasion if the Russians overran the Wehrmacht and were racing toward Berlin. Marshall deduced that Roosevelt wanted to maintain a U.S. military presence during the reconstitution of Europe.[16]

Otherwise, QUADRANT ended exactly as the Combined Chiefs recommended. Top priority would go to an invasion of France around the first of May, 1944; tides, moon, and cloud cover would dictate the exact date. Mediterranean operations would go forward, so long as they didn't restrict men, ships, or planes needed for OVERLORD. An invasion of the French Mediterranean coast, code-named ANVIL, would be launched at the same time as OVERLORD to draw Germans in southern France away from the OVERLORD landing site.

In the Pacific, King's "unremitting pressure" would be aimed at Japan's defeat within one year of Germany's surrender. In the Central Pacific, the Navy would take the Gilberts, Marshalls, Carolines, Palaus, and Marianas islands during 1944, while to the south MacArthur would bypass Rabaul and push through New Guinea toward the Philippines. Strategic bombing would be provided by a fleet of heavy long-range bombers still under development, and the Navy's job would be to take one or more islands that would put those monsters within range of Japan.[17]

Burma had consumed more time than anything else on the agenda, but even that problem was, more or less, peaceably resolved. A new headquarters, called South-East Asia Command, or SEAC, would be formed under Lord Mountbatten.† The Allies would open the Burma Road by attacking the Burmese coast, provided there were enough resources to meet all commitments in France, Italy, and the Pacific. The British insisted, as usual, that "priorities cannot be rigid," and Mountbatten's Burma Road work would be considered "a longer-term development." In other words, China could be ignored for a little while longer.[18]

---

* Actually, three directions. Klaus Fuchs, a British scientist, was a communist spy who would secretly forward to Moscow thousands of pages of documents describing Allied atomic bomb research.

† Cynical Pentagon staffers claimed "SEAC" stood for "Save England's Asian Colonies."

The plan satisfied no one entirely, but it was achievable, and that was the important thing. It was also the best deal each side could get, given the limitations of the moment and the lingering distrust. When they packed their bags and left Quebec, the Allied war leaders knew the next big job was to prepare for the invasion of Europe.

And that job started by selecting the man who would lead the invasion.

THIRTY-FIVE

THE INDISPENSABLE MAN

〜

S EPTEMBER 1, 1943, MARKED THE FOURTH ANNIVERSARY OF MARSHALL'S AP-
pointment as Army Chief of Staff—a post lasting, by tradition, four years.
But FDR, who had already run past his traditional term limit, didn't care about
either tradition or the calendar. George Marshall was a steady hand in war
councils, congressional hearings, and press conferences. He was a gifted speaker,
he had good political sense, and he harbored no ambitions, at least not the un-
healthy kind Greek playwrights warned against. Army dogma decreed that there
was no such thing as an indispensable man, but to Roosevelt, General George
Catlett Marshall came pretty damned close. He would keep Marshall on his in-
field team.

Though FDR agreed with Stimson that Marshall should command the
OVERLORD invasion, he believed the general's guiding hand would be needed in
Washington at least a little longer, and an unexpected political spat over the
Army's logistics branch underscored the need to keep Marshall near the throne.

While Marshall was negotiating at Quebec, General Brehon Somervell, commander of the Army Service Forces, unveiled a proposal to reorganize the Army's
sprawling supply network. Somervell's plan would bring the disparate elements of
the Army's rambling, unruly hydra under the control of one general, and that
general, naturally, would be Somervell.[1]

Somervell had made a long list of enemies. The ambitious, sharp-tongued
bureaucrat had been head of New York City's Works Progress Administration
under Harry Hopkins before the war, and many Washingtonians, including
Harold Ickes, saw in Somervell's plan an empire grab by a Hopkins protégé. Conservatives in both parties grew rabid at the thought of a New Dealer eventually

succeeding General Marshall as chief of staff, either formally—by sending Marshall to Europe—or, as seemed more likely, through a murky reorganization scheme that gutted Marshall's authority.[2]

Ickes, who loathed Hopkins, warned Roosevelt that MacArthur and Somervell were the two greatest threats to democracy among men in uniform. *"He agreed as to MacArthur but said that he had not thought of Somervell,"* Ickes wrote in his diary. *"I told him that [Somervell] would bear watching, pointing out that he was ambitious, ruthless, and vindictive."*[3]

But in preparing his bureaucratic coup, General Somervell made two fatal mistakes: He took an around-the-world inspection tour of Army posts after announcing the plan, and he failed to win the advance backing of Stimson and Undersecretary Robert Patterson, Stimson's logistics chief. He would be a long way from Washington when opposition to the Somervell plan found its voice.[4]

·   ·   ·

Marshall's services, thought King, were invaluable as Army chief of staff, and it was a dumb idea to send him off to Europe as a field commander.

King had butted heads with Marshall since early 1942, but he respected the general and felt they could see the war to victory if they stuck together. They were hardly friends—most of their meetings were conducted with the warmth and cheer of a real estate closing—but since that long-ago day in Ernie King's office, the two men had learned to "get along." It bothered King when Knox privately told him the president wanted to send Marshall to Europe, for the move would require someone else—possibly Somervell, maybe Eisenhower, perhaps even MacArthur—to step into Marshall's large brogans.

"This was no time to 'swap horses in mid-stream,'" King told Knox, and he made it his personal mission to see that George Marshall remained in Washington. He went to Roosevelt and urged him to keep Marshall in Washington. Knox, Leahy, and Arnold did, too. But Roosevelt, as usual, nodded along, refusing to commit himself.[5]

King had no illusions about his influence over FDR. He knew the president wouldn't listen to him about Marshall, and he probably wouldn't listen to Knox, Leahy, or Arnold, either.

But Roosevelt did read the papers, and he listened to their editors. So when King returned from Quebec, he told his newspaper friends the British "were raising Heaven and Earth to get Marshall removed from the Combined Chiefs to command the invasion." Stressing Marshall's value to the Allied team, he asked his confidants to publish editorials extolling Marshall's virtues as Army chief of staff and stressing the need to keep him home.[6]

Before long, Marshall's virtues as chief of staff were a matter of public debate, and it was clear his ticket to Europe had not been punched. At least not yet.

In a year when politicians were sharpening their knives for 1944, Marshall's upcoming transfer to London set off a firestorm of conspiracy theories. The *Army and Navy Journal* speculated that Marshall may have "come into conflict with powerful interests which would like to eliminate him from the Washington picture and place in his stead an officer more amenable to their will." Frank Waldrop of the conservative *Washington Times-Herald* wrote, "General Marshall right today is out as Chief of Staff, because he won't further subordinate his 'technical' views on global strategy to Messrs. Roosevelt and Churchill." Representative Jessie Sumner of Illinois had another theory: Marshall, Leahy, and MacArthur were targeted by the British for standing up for "American rights." Arthur Krock of the *New York Times* warned, "If the Army's Chief of Staff is replaced, those who have urged it will more easily be able to displace Admiral King."

The story of Marshall's "Dutch promotion" became such an article of faith that Senator Harry Truman warned Marshall of a palace coup in the making. Somervell, he told Marshall, was out to get his job.[7]

White House fumbling made things worse. Denying stories of a demotion in the works, press secretary Steve Early privately told a reporter that Marshall would become a sort of global supreme commander.

Early's off-the-cuff remark was a preposterous notion—the sort of rot, Stimson feared, that would upset the delicate game FDR and Churchill were playing with their voters. *"I can well understand that Churchill has got a terrific difficulty in getting his constituents to consent to putting in another general, an American general, in a key position over in the European theater,"* he told his diary. *"This has been made a heap worse by all this gossip that has gone on in America about Marshall getting not only a total European position but a global position over the world."* Early denied making any such announcement, but as Stimson noted, *"the White House has not got a great reputation for truth in these matters of publicity."* The storm lashed without letup.[8]

Stimson had seen the thunderheads forming, and he spoke with Roosevelt about naming Marshall's successor as chief of staff. *"We both agreed Eisenhower was probably the best selection in his place,"* said Stimson. *"He mentioned MacArthur, but I told him strongly that I thought MacArthur was not up to Eisenhower and he agreed. MacArthur is so lacking in the fine disinterestedness that Marshall has shown that I could not possibly feel the same confidence in him."*[9]

Stimson was not the only one worried about changes in the American command. Three ranking Republicans on the Senate Military Affairs Committee drove out to Woodley to tell Stimson they needed Marshall's prestige in Congress to win over fellow Republicans on controversial issues. Stimson assured the senators that while Marshall deeply desired command of the invasion, the president had not yet ordered him to England. If the president did send Marshall to Europe, he suggested, Marshall's influence might still be felt in Washington if he were promoted to the five-star rank of General of the Armies—a title previously held only by General Pershing.[10]

That suggestion resonated with the senators, and the next day Stimson asked Roosevelt to urge Congress to pass a bill promoting Marshall. He suggested that both Congress and Marshall would accept the new rank as long as it had the blessing of Pershing, Marshall's revered mentor. Stimson then asked FDR to push for Marshall's command of both the European and the Mediterranean theaters, and the two men discussed having General Brooke lead the cross-Channel attack and General Alexander, Eisenhower's ground commander, run the Mediterranean war.[11]

It was a trying time personally for George Marshall, who hated to see his name tossed about like a political football for politicians and newspaper editors. But he understood Washington's mercurial ways and took the teacup tempest in stride. When German-controlled Radio Paris announced that Roosevelt had fired Marshall and personally assumed the position of Army chief of staff, Marshall sent a report of the announcement to Hopkins with a note:

Dear Harry,
  Are you responsible for pulling this fast one on me?
  GCM

Hopkins passed the note to Roosevelt, who returned it to Marshall with his penciled endorsement,

Dear George—
  Only true in part—I am now Chief of Staff but you are the President.
  FDR[12]

As the autumn days grew short, Marshall contemplated the commanders he would lead in Europe. For field command, he saw Omar Bradley, Jacob Devers, Mark Clark, or Dwight Eisenhower leading the American army group, and he told his aide Frank McCarthy to be ready to fly to England to secure quarters and offices. His wife, Katherine, began quietly moving their furniture from Fort Myer to their home in Leesburg. She planned to live there until her husband's return from Europe.[13]

In late October Churchill cabled Roosevelt to urge him to make an official announcement. But now that the moment was upon him, Roosevelt was strangely indecisive. He wrote Churchill, "The newspapers here, beginning with the Hearst, McCormick crowd, had a field day over General Marshall's duties. . . . It seems to me that if we are forced into making public statements about our military commands we will find ourselves with the newspapers running the war. I therefore hope nothing will be said about the business until it is accomplished."[14]

But the business would not be accomplished until Roosevelt made a decision. In the midst of the newspaper field day, Admiral Leahy told his diary that the president *"was stalling for time. I definitely had the impression that he was being influenced more by the adverse public reaction than by anything that took place within the military groups."*[15]

Though not prone to revealing his inner thoughts, Roosevelt expressed himself most directly on the subject in a letter to General Pershing: *"I think it is only a fair thing to give George a chance in the field. . . . The best way I can express it is to tell you that I want George to be the Pershing of the second World War—and he cannot be that if we keep him here."*[16]

Eleanor once said that her husband "doesn't think. He decides." But this time he had a lot more thinking to do before he made a decision about George Marshall.

THIRTY-SIX

"DIRTY BASEBALL"

〜

T HE LINE THAT DIVIDES POLITICS AND STRATEGY IS NEVER AS SHARP AS politicians and strategists pretend, and those lines all but disappeared in the summer of 1943. As rhythmic waves of Allied bombers blasted factories and rail-yards in Rome, Pope Pius XII and U.S. Catholic groups called on both sides to declare Rome an open city, neither attacking it nor using it for military purposes. Rome, they emoted, was a cultural treasure and the seat of one of the world's great religions. It would be a crime to turn the home of the Forum, the Sistine Chapel, and Michelangelo's *Pietà* into a field of rubble.

But Rome was also home to Italian military offices and communications hubs. Its railyards funneled thousands of Axis soldiers to points south, and these soldiers had sent plenty of young Americans to places painted on Vatican chapel ceilings. So portions of Rome would fall under British and American bombs.[1]

Roosevelt, eyeing Catholic voters, could not ignore calls to spare the Eternal City, and he took the matter up at his regular press conference on July 23. He as-sured everyone he hoped the Axis would declare the city off-limits to both armies, but until they did, Allied bombers must strike military targets wherever they were found. "I don't believe in destruction merely for retaliation," he emphasized. "It's the wrong basis to put it on. But destruction for saving the lives of our men in a great war sometimes is an inevitable necessity."[2]

On July 25, three days after Patton's tanks clattered into Palermo, Sicily's capital, Mussolini's tottering government fell. King Vittorio Emanuele III placed the dictator under arrest and appointed Field Marshal Pietro Badoglio as his re-placement. Badoglio, desperate to end the war without drawing Hitler's armies into Rome, secretly contacted Eisenhower's diplomatic agents to arrange an armistice.

When news of *Il Duce*'s fall reached the Allied capitals, an elated FDR wrote Churchill, "It is my thought that we should come as close as possible to unconditional surrender followed by good treatment of the Italian populace." Sticking to his Casablanca announcement, however, the next day Roosevelt announced that the unconditional surrender terms would stand.[3]

Badoglio's emissaries secretly agreed to an armistice on September 3, but when rumors of Italo-Allied negotiations reached German ears, Hitler's Mediterranean commander, Field Marshal Albrecht Kesselring, funneled troops down the Italian boot and prepared to turn the ancient battlefield into a modern one.[4]

On September 8 Eisenhower announced the armistice and Badoglio belatedly ordered his forces to cease resistance to United Nations forces. The next morning the Fifth U.S. Army under Lieutenant General Mark Clark splashed ashore at Salerno, south of Rome on the Neapolitan coast. Churchill's dream of an invasion of Italy was the war's new reality.

By the time the Higgins boats dropped their ramps, however, Kesselring had captured Rome and sent five divisions lurching toward Salerno Bay. His panzers crashed into Clark's VI Corps at the Sele River, and with a bloody counterattack, they threatened to drive the Americans into the sea.[5]

· · ·

As Clark's men clawed their way onto Salerno's beaches, Roosevelt and Churchill puzzled over Italy's new status. Should Italy be treated as an ally? A vanquished enemy? A neutral?

Like most things connected with Italian government, the answer was messy and ambiguous. Messy and ambiguous had never bothered FDR, however, so long as he got what he wanted. What he got in Italy was, to him, a self-evident good: the long, murderous stretch of Mussolini's wars, from his invasion of Ethiopia to his last gasps in Sicily, was finally over. *Finito Benito.* In Roosevelt's mind, it didn't matter how his war chiefs got there, as long as they got there.[6]

On September 19, General Eisenhower wired Marshall to ask for authority to recognize the Badoglio government on a provisional basis. In Ike's view, recognition was necessary to enlist active Italian support in driving the German Army out of Italy. Marshall forwarded Eisenhower's request to Roosevelt and the State Department. After consulting with Churchill, FDR instructed Eisenhower to recognize the Badoglio government as a co-belligerent—not an ally—provided Italy immediately declared war on Germany.[7]

After tiresome, Italian-style negotiations concluded the following month, Italy formally declared war on Germany. No promises were made by either Roosevelt or Churchill. When the war was over, debts would be settled, and surrender would be, more or less, unconditional. The liberation of Italy had been won, on paper.

But it would cost the lives of thousands of Clark's men to turn that paper victory into a real one.[8]

. . .

As the Italian campaign slowed to a bloody *minuetto* south of Rome, a restless Churchill proposed opening yet another front, this time in the Aegean. He wanted to invade the Dodecanese Islands—Rhodes, Samos, Kos, and a few other specks of Greek mythology—to see if he could draw Turkey into the Allied camp.[9]

The Eastern Mediterranean had always been a hot button for Marshall. Churchill's designs there smacked of a grand scheme for Britain's postwar empire. An attack on the Dodecanese, he feared, was Churchill's down payment on a much larger investment of ships and men. It was an old military trick to commit men to battle on the assumption that where lives and victory hung in the balance, the brass would have to back the commander's check no matter how much he wrote it for. The Allies would have to double down if they got into trouble—and perhaps triple down if they found a success to exploit.

Reading Churchill's proposal, Marshall's every suspicion bubbled to the surface. He would have to kill this new diversion quickly and without mercy. After conferring with Leahy, he and King drafted a reply for Roosevelt's signature refusing any assistance in the Dodecanese.[10]

Roosevelt, safely in Marshall's corner on this subject, agreeably signed the cable. To Marshall's draft he added, "Strategically, if we get the Aegean Islands, I ask myself, where do we go from there and vice versa where would the Germans go if for some time they retain possession of the islands?"[11]

Churchill, who had no answer, backed down. *"He did it with a bad grace and with almost a childish squawk, but he yielded and the lesson will prove salutary,"* remarked Stimson.[12]

He did not yield for long, however. General Harold Alexander, Churchill's senior commander in the Mediterranean, began peppering the Ministry of Defence with pessimistic reports describing setbacks in Italy. Alex's messages, which Churchill circulated to the Americans, had the ring of a cry for more ships, men, and planes for the Mediterranean theater.[13]

Stimson watched from the Pentagon as events confirmed his fears. In September Churchill sent a small British force to occupy three key Aegean islands.

He didn't ask for any American help, at least not directly, but in late October he warned FDR that the transfer of divisions from the Mediterranean to England would cripple the progress of the Italian campaign. He complained that two of his best British divisions were sitting idle in Sicily, and would not be put into action for seven more months if they had to await OVERLORD's launch.[14]

To Stimson, the PM was blowing the old Mediterranean trumpet to renege on the OVERLORD deal. *"Jerusalem!"* he exclaimed, using one of his stronger expletives. *"This shows how determined Churchill is with all his lip service to stick a knife in the back of Overlord and I feel more bitterly about it than I have ever done before."*[15]

In a White House meeting at October's end, Roosevelt assured Stimson he would not touch the Balkans unless Stalin asked for help there, and both knew Stalin wouldn't. Much relieved, Stimson asked Roosevelt never even to mention the Balkans again. The mere word, he said, threw his OVERLORD planners into fits.

As Stimson was leaving the Oval Office, he turned to Roosevelt, held up his hand, and said with a smile, "Remember, no more Balkans!"

Roosevelt took Stimson's injunction in good humor and agreed to avoid the infernal word. But Stimson returned to the Pentagon rattled by the close call. *"It was an unpleasant incident,"* he wrote. *"What I would call dirty baseball on the part of Churchill."*[16]

· · ·

Throughout November Roosevelt kept his cards close to his vest. He wanted Marshall in Europe, but it would pose a political risk to defy Pershing's judgment and a chorus of criticism from Republicans and vulnerable Democrats. And in the fall of 1943, criticism of his war management was a dicey problem, because of a certain general in Australia.[17]

Though he had not lived in America for more than seven years, Douglas MacArthur's political star refused to die. In early 1942, two Republicans, Senator Arthur Vandenberg of Michigan and Representative Clare Boothe Luce, of the *Time-Life* publishing empire, began seeing in MacArthur a possible Republican opponent to Roosevelt in 1944. Responding the way men do when they hear words they wish to hear, MacArthur began quietly flirting with Vandenberg. In April 1943, rumors of a MacArthur ticket reached the ears of Stimson, Marshall, and Leahy, and as speculation swirled through the media, MacArthur coyly announced his intention to do his duty and retire. That announcement, as he well knew, was chum in the shark tank.[18]

Marshall was aware of MacArthur's ambition. As a bipartisan Army booster, he kept close ties with Senator Vandenberg—even if, to avoid stoking suspicions among Roosevelt loyalists, he kept his relationship so quiet he had to weather occasional press accusations that he was a New Deal lackey. ("Actually," Marshall said, "we couldn't have gotten much closer together unless I sat in Vandenberg's lap."[19])

Marshall watched the fall drama unfold but saw no reason to weigh in on MacArthur's theoretical candidacy. Stimson, however, did. Seeing a monster growing in Brisbane, Stimson enraged the Vandenberg clique by reinstating an old Army regulation banning political activities by Regular Army officers. At Stimson's weekly press conference, journalists pointedly asked whether he had targeted MacArthur when he reinstated the old rule.

Stimson denied it, but few believed him. Vandenberg attacked Stimson for squelching the constitutional and God-given right of a man to run for president. Representative Hamilton Fish, Roosevelt's nemesis from Dutchess County, introduced a bill to repeal the ban on candidacies for men in active service, and proposed drafting MacArthur on a "win-the-war platform and on a one-term plank, as opposed to a fourth term and military dictatorship." Colonel McCormick gleefully told one interviewer, "Roosevelt's in a hell of a position. If MacArthur wins a great victory, he will be president. If he doesn't win one, it will be because Roosevelt has not given him sufficient support."

Vandenberg summed up Republican hopes in his diary: *These people can easily martyrize [MacArthur] into an irresistible figure. . . . It is obvious on every hand that the movement is making solid headway in all directions.*[20]

MacArthur's polling numbers lagged well behind Wendell Willkie and New York Governor Thomas Dewey, and a spring poll pitting Roosevelt and Wallace against a Dewey-MacArthur ticket had the Democrats winning 54 to 46 percent. But a Gallup survey released in September 1943 showed MacArthur outpolling Roosevelt among Midwestern farmers, a key swing region. Roosevelt's global war strategy, and his choice of commanders, would be wrapped up in the politics of 1944.[21]

THIRTY-SEVEN

VINEGAR JOE AND PEANUT'S WIFE

C HINA'S WAR WITH JAPAN NEVER WENT ANYWHERE. WHEN THE ALLIES poured in supplies, Generalissimo Chiang Kai-shek said he needed trained men to attack the Japanese. When they cut supplies back, he said he couldn't move without more equipment.* Having spent a generation fighting warlords and communists, he fought like an old man set in his ways, waiting for his new enemy to leave so he could go back to battling the old one he knew better.

Chiang had long complained of General Joseph Stilwell, Marshall's honest but tactless commander in China. "Vinegar Joe" had the sharp tongue of a cranky country parson, and he blamed Chiang's corrupt cast of bureaucrats for the ills plaguing China's war effort. He called Chiang "Peanut" behind his back and refused to kowtow to Chiang's lackeys. Unsurprisingly, Chiang and his wife liked the dashing General Chennault better, and tended to support Chennault's air war at Stilwell's expense.[1]

The bad blood between Chiang and Stilwell grew more septic when Chiang's foreign minister, Dr. T. V. Soong, asked President Roosevelt to slice off chunks of Stilwell's authority as U.S. ground commander. To sideline the general further, Soong also asked Roosevelt's blessing for the appointment of a supreme allied commander—Chinese, of course—to be placed in charge of the theater.

Roosevelt liked Chennault better, and his distant cousin, Joe Alsop, was Chennault's publicity adviser. Hearing Soong's complaints, Roosevelt concluded

---

* OSS China hands derisively nicknamed the nationalist leader "Cash My-Check."

that Stilwell lacked the patience and diplomacy to get along with the touchy generalissimo.[2]

Normally unsentimental, Marshall was a deep shade of blue over the hatchet job being done on his old friend—a job for which Marshall would swing the hatchet. Stilwell had been an exemplary leader at Fort Benning before the war, and corruption and civil strife were hardly problems that could be laid at his muddy boots. Marshall deeply regretted having given Stilwell the assignment in the first place.

But the theater was stalled, and there was a job to be done. Perhaps someone else could make a better show of it. With a heavy heart, Marshall drafted the order recalling Stilwell to Washington.[3]

In one of those odd twists of fate that dictate so much of China's history, Madame Chiang leaped into the fray. In shrill tones, she told Soong, her brother, to back off Stilwell. She ordered her husband, politely but firmly, to give him one last chance. Then she summoned Stilwell to her home and ordered him to seek an audience with her husband. Stilwell, she commanded, would show Chiang the respect due his office, and would recant any discourtesy he had shown his Chinese commander.

Stilwell's initial reaction was to tell the Chiangs and Soong where to go. But after thinking over Madame Chiang's proposal, he gave in. He submitted himself to a sharp lecture on respect and decorum from the Chinese leader, and the interview cleared the air between Vinegar Joe and Peanut. At least, for a while.

With a better understanding of where each man stood, Stilwell won a reprieve. Marshall laid his draft order aside.[4]

THIRTY-EIGHT

A RUSSIAN UNCLE

〜

Admiral King's *Dauntless* rarely piped visitors aboard, much less important ones. Most evenings, King would emerge from his Cadillac, stride up the gangplank and salute the colors as he made his way to the dining cabin. That was about all the ceremony the converted yacht's crew ever had to observe.

Armistice Day 1943 was different. That morning, the ship's hundred-man crew welcomed Generals Marshall and Arnold, together with sixteen senior military and naval staffers. Though none of the crew knew it, they were about to carry the Joint Chiefs on the first leg of what would be a 17,000-mile odyssey to the far side of the world.[1]

Roosevelt had been trying to entice Josef Stalin to a three-power meeting since shortly after Pearl Harbor. To FDR, Hitler's one advantage over the Allies, unity of command, could be offset by careful coordination among the three great "United Nations" leaders.

Roosevelt, a man for whom politics was personal, believed a direct, one-on-one connection with the Soviet dictator would bridge decades of long-distance distrust. As Churchill's physician, Lord Moran, told his diary, *"The President is convinced that even if he cannot convert Stalin into a good democrat he will be able to come to a working agreement with him. After all, he had spent his life managing men, and Stalin at bottom could not be so very different from other people."*[2]

For weeks Stalin resisted Roosevelt's advances. He said the demands of a 260-division front, stretching from Leningrad to the Caucasus, ruled out meetings in Alaska, Scotland, Cairo, Iraq, Beirut, and every other place Roosevelt suggested. Finally, on September 8, he agreed to a three-power meeting in Tehran, a city nominally under Soviet control. The western leaders accepted, and Churchill, ready as always with a new code word, dubbed the summit EUREKA. Before meeting with the Russians, Churchill and FDR agreed to hold an Anglo-American conference in Cairo, code-named SEXTANT. To give the impression that the British and Americans were not conspiring against their eastern ally at Cairo, Roosevelt invited the Chinese and Soviets to send representatives to the SEXTANT conference.[3]

None of the Combined Chiefs relished the next round of discussions. Mutual distrust over means to an end, tamped down momentarily at Williamsburg and then at Quebec, kept bubbling over each time the delegations returned home. *"I wish our conference was over,"* a gloomy Brooke told his diary two days before the meetings even started. *"It will be an unpleasant one, the most unpleasant one we have had yet, and that is saying a good deal."*[4]

The crux of the unpleasantness was the incestuous ménage à trois of OVERLORD, Italy, and the Eastern Mediterranean. On the American side, Marshall and King had little faith that Churchill and his chiefs would keep their end of the OVERLORD bargain. Churchill's doubts, expressed to Stimson in London, rang loudly in their ears, and the PM's sputtered references to Rhodes, Crete, and the Balkans gave the Americans ample reason to doubt British fealty to OVERLORD.[5]

The British chiefs still saw strategic incompetence among their Yankee counterparts. Their prejudice was, however, neither an echo of 1776 nor the dread of corpses washing up on Dover's beach. They distrusted the American mindset. Fixated on northern France, the Americans viewed military strategy as an unalterable blueprint, and strategic agreements as binding contracts among lawyers. Marshall and King could not see the value in extemporizing in northern Italy or the Aegean—extemporization that would aid OVERLORD while lopping off key morsels of the German empire.[6]

To soften their suspicions, in October Sir John Dill approached Marshall and King with a proposal to allow the British and Americans to post junior officers on each other's planning staffs. Marshall, who had great faith in Sir John, agreed to the arrangement.

He then tried to persuade King to sign on to the experiment. "We have to work with these people and the closer the better, with fewer misunderstandings I am certain," Marshall stressed. "We are fighting battles all the time, notably in regard to the Balkans, and other places, and the more frankness there is in the business on the lower level the better off I believe we are."[7]

King didn't give a damn about frankness and cooperation. "Frankness" went against his grain, and he forwarded Marshall a memo from his staff that predictably concluded, "It would mean the injection of a low level group into our

Joint War Plans Committees which would permit us no privacy in the consideration of problems which are purely those of the United States." King would let no
subject of the Crown near his Pacific theater plans.[8]

With a heave, *Dauntless* slipped her moorings and steamed down the Potomac
River to Point Lookout on Chesapeake Bay. There she spoke the battleship *Iowa*,
a 58,000-ton behemoth boasting sixteen-inch guns that could throw a one-ton
explosive shell nearly a mile in two seconds. Welcoming the four-star guests to his
new ship was Captain John McCrea, the presidential naval aide who had survived
his encounters with Admiral King and had gone on to skipper the largest ship the
Navy had ever put to sea.[9]

Iowa and her consorts waited patiently at anchor in Chesapeake Bay until
early the next morning, when USS *Potomac*, the presidential yacht, hove to and
quietly transferred her passengers. Roosevelt, Hopkins, Leahy, Pa Watson, Dr.
Ross McIntire, Rear Admiral Wilson Brown (the president's new naval aide), and
a small group of staffers moved onto the battleship.[10]

Roosevelt loved to travel on naval warships. He especially enjoyed cruisers,
and four years earlier he'd had a grand time watching fleet war games in the
Caribbean from the bridge of the cruiser *Houston*. At one point in the war game
Admiral Leahy, then chief of naval operations, had to break some bad news
flashed in by naval umpires: "Mr. President, we have just been sunk by an enemy
submarine."*[11]

On this Friday evening, *Iowa* topped off with fuel and was ready to weigh
anchor by ten o'clock. But Roosevelt shared an old sailor's aversion to beginning a
voyage on a Friday, and he asked Captain McCrea to remain at anchor until 12:01
on the morning of Saturday, November 13. So at one minute past midnight, *Iowa*
fired her engines and stood out to sea.[12]

McCrea, intimately familiar with the president's routine, attended to every
detail for a VIP unable to walk. The Iowans built a special elevator to give the
president access to the bridge, and McCrea even had a bathtub installed—said to
be the only bathtub in the entire United States Navy.† To these structural accommodations Secret Service Chief Mike Reilly added a cache of presidential neces-
sities: Saratoga Springs mineral water, cigarettes, wooden matches, and enough
reels of musicals and comedies to run a different film every night.[13]

---

* In early 1942 Roosevelt's beloved *Houston* would be sunk, this time physically, by a Japanese
squadron at the Battle of the Java Sea.

† It was the only U.S. *battleship* of its day to boast a bathtub. The presidential yacht *Potomac*,
a commissioned vessel, also had one.

Welcome cards prepared by McCrea's staff told each guest where he would be living for the duration of his stay. Roosevelt's read:

WELCOME

The Captain, officers, and men of the Iowa are happy to have you on board.

Your room number is—Captain's Cabin

You will mess in—Flag Mess

Your Abandon Ship Station is—Lee Motor Whaleboat

Your Action Station is—Conning Tower (Flag Level)

and on the list went. In typical Navy fashion, some items seemed overdone:

The General Alarm and Gas Attack Alarm will be tested daily at 1200. They are not sounded for drill. When sounded at any time other than 1200, enemy action is expected.[14]

Enemy action was apparently expected the next day when, during an antiaircraft exercise, the General Alarm broke out. The siren was followed by the frantic call, "Torpedo defense! Torpedo defense! This is not a drill!"

The massive battleship listed as her rudder turned hard and *Iowa*'s screws revved to flank speed. The thump of hundreds of feet boomed as sailors scrambled to their battle stations like a herd of stampeding buffalo. *Iowa* executed a surprisingly nimble change of direction, and every antiaircraft gun on her starboard side lowered its muzzle and began shooting into the churning water.[15]

Two of those thumping feet belonged to Admiral King, who thundered up to the flag bridge as a concussive thump and geyser abaft announced a torpedo's death. King's face was turning that deep shade of crimson again, and he was working up to a three-day blow as he howled for an explanation from the bewildered Captain McCrea. In seconds, *Iowa*'s signal flags burst out, demanding answers from all ships in the convoy.[16]

The reluctant truth emerged in a flashing signal from a destroyer off the starboard bow. The destroyer was USS *William D. Porter*. While tracking *Iowa* as if she were an enemy during the drill, the destroyer had accidentally let loose a torpedo in *Iowa*'s direction. Fortunately, *Iowa*'s lookouts had spotted the torpedo's wake and Captain McCrea had time to swing clear of the deadly fish. The torpedo did not run "hot and straight," one passenger remembered, "which was indeed a

302 ★ AMERICAN WARLORDS

lucky break, as it would have been most embarrassing for the U.S. Navy to have torpedoed their Commander in Chief."[17]

Leahy didn't have to tell his commander a second time that they had been sunk by a torpedo, but King was purple with rage. Had there been planks to walk in the modern Navy, he would have sent the destroyer's captain to Davy Jones's locker without a second thought. But Roosevelt, who had followed the ruckus with interest but no evident alarm, shrugged it off. He had survived an assassination attempt in 1935, and an anarchist's bomb years earlier, and showed no sign of fear during the torpedo warning. He told King not to lop off any heads.

"It seemed to me," King recalled, "that it was F.D.R.'s idea to cover up the error, especially since there were some few reporters on board who apparently were told to forget the incident because it wouldn't look well that all hands were in danger by our own people."[18]

William D. Porter was sent off for the obligatory investigation under the humiliating track of *Iowa*'s guns. At her next few ports of call, her mortified crew would be welcomed with the flashing signal "Don't shoot! We're Republicans!"*[19]

After seven days at sea, *Iowa*'s passengers disembarked at Mers-el-Kébir, the Algerian port captured during the TORCH landings. FDR, unable to descend to the ship's whaleboat by gangway or Jacob's ladder, was sent over the side in a bosun's chair, evidently unconcerned as his chair hovered above the chopping waves. As ship's surgeon Robert Coffee recalled, "It was a very dramatic sight to see this single, lonely figure lowered precariously into this 'rowboat.'"[20]

Once safely ashore, Roosevelt and his entourage were met by General Eisenhower, two Roosevelt sons—Lieutenant Franklin D. Roosevelt Jr. and Colonel Elliott Roosevelt—and a retinue of British and U.S. officials. The core group boarded a trio of C-54 Skymasters—Roosevelt's plane was nicknamed the *Sacred Cow*—and the American warlords flew along the African coastline to Carthage. They would spend the night in Tunisia before pushing on to Cairo for the SEXTANT conference.[21]

In Carthage that evening, as Eisenhower prepared to leave his home for dinner with the president, he spent a few moments with King and Marshall in his villa's living room. As the men nursed cocktails in the ornately tiled salon, King casually opened the subject of the OVERLORD command.

"The time has come," King said, "for the President and Churchill to decide who the OVERLORD commander should be."

---

* The incident's cause, originally ascribed to an electrical short, was more likely carelessness in leaving the torpedo system armed during the tracking exercise.

Marshall sat, silent as the Sphinx, as King explained to Eisenhower that the other chiefs wanted Marshall to remain in Washington, but Roosevelt had decided to give command of OVERLORD to Marshall. Walking Ike to the door, he remarked, "I hate to lose General Marshall as chief of staff, but my loss is consoled by the knowledge that I will have you to work with in his job."[22]

As Eisenhower recalled, "General Marshall remained completely silent; he seemed embarrassed." Harry Butcher, Eisenhower's naval aide, told his diary, *"General Marshall had not mentioned or indicated anything about Ike's probable assignment, and Ike was embarrassed, not only by the warmth of the Admiral's statement, but by the spontaneity of his comment in General Marshall's silent presence."*[23]

Caught off-guard in an awkward moment, Eisenhower mumbled something about the president having to make his own decisions. Marshall, annoyed with talk of his own prospects, growled, "I don't see why any of us are worrying about this. President Roosevelt will have to decide on his own, and all of us will obey." The conversation abruptly ended.[24]

Outside Tunis the next day, Eisenhower's drab-colored Cadillac gunned its big engine as it rumbled over the dusty battlefields of Medjez-el-Bab and Tebourba. Travel security required the president to travel by night, so during his extra day in Tunisia, he asked for a tour of some nearby battlefields.

He and Eisenhower rode in the car's spacious main seat; Ike's attractive driver, Kay Summersby, and his dog, a scrappy Scottie named Telek, accompanied the two commanders. As they rumbled over unpaved roads, the Cadillac and its protective convoy of motorcycles, half-tracks, and armored trucks passed burned-out tank carcasses—American and German—flagged minefields, Arab farms, Red Cross tents, and the ruins of an old Roman aqueduct.[25]

At lunchtime, they pulled over for a roadside picnic, where Roosevelt's thoughts turned to command of OVERLORD. Acutely conscious that Americans would soon outnumber Commonwealth troops, he had asked Marshall for an estimate of the two armies by January 1944. Marshall told him Britain and her dominions would have 4.92 million men, while the United States would have 10.53 million in uniform, of whom 3.78 million would be serving overseas. Because the numbers were tilting steadily toward the Americans, at Quebec Churchill had agreed that the man who led OVERLORD should be an American. That American, Roosevelt knew, would go down as one of history's greatest generals.[26]

As he and Eisenhower rode back to Carthage, Roosevelt brought up the subject of the invasion. He mused, "Ike, you and I know who was chief of staff during the last years of the Civil War. But practically no one else knows, although the names of the field generals—Grant, of course, and Lee, and Jackson, Sherman, Sheridan and the others—every schoolboy knows them. I hate to think that fifty years from now practically nobody will know who George Marshall was. That is one of the reasons I want George to have the big command—he is entitled to establish his place in history as a great general."[27]

The balding Kansan sat beside Roosevelt, nodding, listening. Saying nothing, except that he would do his duty wherever his president sent him.

Sacred Cow and her consorts touched down at Cairo West airfield the next morning, November 22, and Marshall, King, Arnold, and their staffs spent the next five days at Giza's Mena House Hotel, in the shadow of the Great Pyramid.

The hotel, whose manager had been told of the conference only three days earlier, had been crash-converted from a luxury resort to a heavily guarded military camp. From its hill overlooking the Pyramids, Mena House was ringed with barbed wire, antiaircraft guns, and battalions of MPs demanding identification cards from hotel staff wherever they went. Hotel guests were moved out, with the abject apologies of the manager. Bedroom furniture was pushed aside for desks, radios, and file cabinets, and Mena House joined a long list of Egyptian headquarters from which the campaigns of Pharaoh's chariots, Caesar's legions, Selim's Turks and Napoleon's Frenchmen had been directed.[28]

As the war chiefs debated operations for 1944, FDR's overriding concern was keeping Chiang in the Allied camp. He felt the Kuomintang regime could make a material contribution to the war against Japan, but even if it if could not, China would play a major role in the postwar world. Roosevelt wanted to ensure that after the war China would remain firmly in the western fold, and for that reason he had invited Chiang and his entourage to come to Cairo.

The price of Chiang's enthusiasm for battle would be steep. The generalissimo demanded that the Allies launch an amphibious assault on the Burmese coast and drastically increase the level of supplies over the Hump. If they would take those first great steps, he promised to launch a drive into Burma. If the two efforts were successful, the Allies could reopen the Burma Road, and that would enable Chiang to ask for more supplies from the United States.

Roosevelt assured Generalissimo and Madame Chiang that the Allies would mount an invasion somewhere along the Bay of Bengal. Until the land route to Kunming was reopened, he pledged to increase supplies coming over the Hump to ten thousand tons of equipment per month, promising supplies at a level his air chiefs found dangerously optimistic.[29]

Marshall did not share Roosevelt's confidence in Chiang. From the detailed

A rare moment of levity for Admiral Ernest J. King.

WAR'S FIRST DAY. General George Marshall is sworn in as U.S. Army Chief of Staff on September 1, 1939—the day Hitler launched his war in Europe.

Churchill and FDR meet to discuss the Nazi threat aboard Admiral King's flagship, USS *Augusta,* in August 1941. To steady himself, FDR leans on the arm of his son Elliott.

FDR'S CABINET. Roosevelt (*rear center*) with his cabinet in the fall of 1941.

THE FIRST CHIEFS. General Marshall, Admiral Harold Stark, Admiral King, and Major General Henry "Hap" Arnold formed the first Joint Chiefs of Staff.

THE DOOR THAT NEVER CLOSED. Stimson and Marshall worked in adjoining offices at the Pentagon, where they could consult freely with each other. Throughout the war, they were an effective, harmonious team.

As an approving Navy Secretary Frank Knox looks on, Admiral King takes the oath of office of U.S. Navy Chief of Naval Operations.

BROTHERS IN ARMS. FDR and Churchill talk after church services aboard the battleship *Prince of Wales* in August 1941. An emotion-charged FDR told his son, "If nothing else had happened while we were here, that would have cemented us." Next to Churchill are General Marshall and his new friend, Britain's General Sir John Dill.

UNEASY PARTNERS. Admiral King tolerated Navy Secretary Knox—so long as he kept out of King's naval war plans.

GLOBAL STRATEGIST. FDR looks at a globe made for him by General Marshall.

THE CHAIRMAN. Admiral William D. Leahy (*center*) served as chairman of the U.S. Joint Chiefs. His official title was "Chief of Staff to the Commander-in-Chief."

THE CHIEFS AT WAR. King, Marshall, and Arnold after a White House meeting with FDR in 1942.

Paratroopers in England train for combat in October 1942. Marshall was instrumental in the development of the Army's airborne divisions.

AMERICAN WARLORDS. Marshall, FDR, and King at Casablanca, January 1943. Standing behind them are Harry Hopkins, General Arnold, Lieutenant General Brehon Somervell, and Lend-Lease coordinator W. Averell Harriman.

THE COUSINS. Air Chief Marshall Charles Portal, Fleet Admiral Sir Dudley Pound, and General Alan Brooke meet with Churchill aboard the liner *Queen Mary* in May 1943 to map out strategy before meeting with the Americans.

Marshall confers with Churchill and an unhappy-looking General Bernard Montgomery in May 1943.

PROTÉGÉ AND PATRON. Generals Eisenhower and Marshall in a light moment before a press conference in Algiers, May 1943.

Marshall gives a speech at West Point, 1942.

THE ARSENAL OF DEMOCRACY. Ford Motor Company's Willow Run plant turns out waves of B-24 Liberator bombers. Roosevelt's production goal of fifty thousand planes per year was ridiculed by Axis leaders when he announced it in 1940. They didn't laugh long.

MAIN NAVY. King and Knox attend a press conference in Knox's office. King hated talking to the press—but learned to use journalists to influence decisions.

Chinese Generalissimo Chiang Kai-shek, FDR and Churchill meet with their military staffs in Cairo in November 1943.

THE BIG THREE. Stalin, Roosevelt and Churchill pose for history at Tehran, December 1943.

INTO THE MED. U.S. soldiers lie crammed into a troopship hold on their way to Africa. The American warlords fought bitterly to limit the Allied commitment in the Mediterranean, a "British" theater.

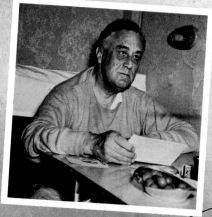

COMING HOME. A tired FDR catches up on paperwork during his return trip from Casablanca in January 1943. His "unconditional surrender" demand, announced at the end of the Casablanca conference, gave his military advisers fits.

NEW HEROES. Fighter pilots of the all-black Tuskegee Airmen (332nd Fighter Group) discuss German tactics over Italy in early 1944. Roosevelt, Stimson, Marshall, Knox and Eleanor Roosevelt struggled over the role non-white servicemen would play on the front lines.

An M-4 tank halts in Kasserine Pass, February 1943.

U.S. Marines scramble over the wreckage of Tarawa in 1943.

OVERLORD IN ACTION. Riding over Omaha Beach, Eisenhower and King discuss the progress of the Normandy landings.

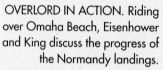

BIG PRIZE. Admiral King, Major General Holland Smith (*with carbine*) and Admiral Nimitz tour Saipan, a vital step in King's plan to take back the Pacific.

Troops of the 34th Infantry Division advance in Tunisia.

THE PATH TO VICTORY. In July 1944, MacArthur, FDR, and Leahy listen as Admiral Nimitz advocates the "King Plan" for the conquest of Japan.

A NATION TURNED UPSIDE DOWN. A Frenchwoman passes a German antitank gun abandoned in the wreckage of her town.

DEATH'S DOOR. Photographs of FDR accepting the 1944 Democratic nomination showed him looking gaunt, tired, and old, in stark contrast to his energetic looks of 1940.

An enemy goes down during the Battle of the Philippine Sea.

First Lady Eleanor Roosevelt dines with American troops in South America. Eleanor traveled widely in her unofficial role as FDR's emissary and U.S. goodwill ambassador.

Stimson and Marshall discuss the military picture in Asia and the Pacific.

One of FDR's closest advisers was Treasury Secretary Henry Morgenthau Jr., an old friend from Hyde Park.

VISITING THE FRONT. Marshall confers in France in October 1944 with Major General Walton Walker and Patton.

WAR CHIEFS AT SEA. King, Leahy, Roosevelt, Marshall, and Major General Laurence Kuter confer aboard the cruiser USS *Quincy* on the journey to Yalta to meet with Churchill and Stalin.

LORDS OF THE OCEAN. Admirals King and Nimitz flank Navy Secretary James Forrestal in late 1945. "I hated his guts," King said of his Navy Department boss.

A LION IN WINTER. FDR sits with (*clockwise*) Churchill, Anna Roosevelt Boettiger, and Sarah Churchill Oliver aboard the cruiser USS *Quincy* in Malta Harbor, January 1945.

Henry Stimson and his aide, Colonel William Kyle.

DEATH FROM ABOVE. A mushroom plume rises over Nagasaki, Japan. Nagasaki was bombed because Stimson couldn't bear to destroy the cultural sites of Kyoto.

WELCOME TO THE ATOMIC AGE. Stimson guided the development of the atomic bomb from its early stages through its use to destroy Nagasaki. Here technicians give the "Fat Man" plutonium bomb its final paint job.

Patton, Stimson (*with hat*), and Assistant War Secretary John J. McCloy salute the American flag in Germany, 1945.

VICTORY! A beaming Harry Truman points to the signature of Emperor Hirohito on the instrument of surrender, as Henry Stimson and General George Marshall look on.

INSTRUMENT OF SURRENDER

"THESE PROCEEDINGS ARE CLOSED." General MacArthur watches as the belligerents sign the instrument of surrender.

Allied war prisoners liberated from Aomori Prison, August 1945.

reports of General Stilwell, he held the firm conviction that whatever Chiang's personal intentions, he was being "sold down the river" by his advisers.

But it was the president's decision, not his, and Marshall would have to toe the line when he met with the skeptical British. In the Combined Chiefs meetings on Asian operations, he and King made a pitch for Operation BUCCANEER, the capture of the Andaman Islands near Rangoon's coast. In Allied hands, the Andamans would give the Allies a naval base to cut Japan's supply line to Burma, which would be the first step toward reopening the Burma Road.[30]

General Brooke, Britain's Doubting Thomas on all things Burmese, believed Chiang was a broken reed whose only proven skill was talking the Americans into giving him weapons. Madame Chiang, he concluded, was the "leading spirit" of the Chinese nationalists, but he did not trust her. As an afternoon meeting devoted to Burma wore on, Brooke moved to postpone discussion of BUCCANEER, and he later suggested that landing craft slated for Burma might instead be sent to the Aegean. If BUCCANEER went forward, he said, then OVERLORD must be postponed.[31]

The Aegean had always been a lightning rod for the Americans. By late 1943, the mere suggestion of another foray into the Eastern Mediterranean sent the U.S. chiefs into a rage. They had given in on the Mediterranean a thousand times more than they had intended, and after months of compromise and ultimatum, they had finally stuffed Churchill's Balkans cat back into the gunnysack. Now Churchill's lieutenants were trying to weasel their way back to the Aegean—out of China and, most likely, out of France.[32]

"Things got so hot," wrote Army staffer Charles Donnelly, "Admiral King and General Brooke traded insults." The RAF's Air Marshal Sholto Douglas called it "the father and mother of a row," while a bemused Stilwell scribbled in his diary, *"Brooke got nasty & King got good & sore. Brooke is an arrogant bastard."*[*][33]

Arrogant, perhaps, but Brooke's demands came from a higher source. Aware that Britannia was being eclipsed as the senior partner, Winston Churchill had become so obsessed over the Dodecanese Islands that Brooke wasn't sure he could rein in the prime minister. He fretted to his diary, *"[Churchill] is inclined to say to*

---

* The venom between Brooke and King is, like all other heated disagreements, nowhere hinted at in the meeting's sanitized official minutes. Lord Ismay later remarked, "One can read the official minutes of these meetings without suspecting that a single harsh word had been exchanged." Secretaries, including Generals Smith and Handy, were not above editing quotes before committing them to paper, partly as a matter of sound office politics. "If you will always quote people with what they should have said, you will have no trouble," Handy explained after the war.

the Americans, all right if you won't play with us in the Mediterranean we won't play with you in the English Channel. And if they say all right well then we shall direct our main effort in the Pacific, to reply you are welcome to do so if you wish!"[34]

Churchill's Dodecanese dream was a sepsis poisoning Allied blood. The evening after the row over BUCCANEER, Churchill invited Marshall to dine with him. Surrounded by a few personal retainers and Pug Ismay, he discoursed at length on Balkan partisans, the Dodecanese Islands, and Aegean air operations. After dinner, he carried the conversation into the courtyard and was still holding forth on the Aegean well after midnight.

Churchill was at his persuasive best in these late-night discussions, while Marshall, an early riser, was tired, ornery, and short. In the wee hours of the morning, Churchill pontificated as if he were giving a speech to the House of Commons. The war, he thundered, could not wait until May 1944 for a new offensive. "His Majesty's Government can't have its troops standing idle. Muskets must flame!"

Marshall gave Churchill a look that, for what it lacked in eloquence, made it up in brute force. "God forbid if I should try to dictate," he growled to Churchill, "but not one American soldier is going to die on that goddamned beach."

Aides stared at Marshall as if he had slandered Churchill's wife, Clemmie, and a disheveled Pug Ismay had to stay up late that morning calming down the irate PM. Marshall didn't care. Churchill had made a mess of the Mediterranean in 1915, and nobody, including Churchill's own generals, wanted to see that happen again.

"The others were horrified," Marshall chuckled afterward, "but they didn't want the operation and were willing for me to say it."[35]

. . .

The flight from Cairo to Tehran took the American delegation over the ancient cities of the Holy Land, the Dead Sea, the Transjordan, Babylonia, and Persia. After many a barren stretch, their wheels touched down five hours later at a Red Air Force base outside Tehran.

Emerging from their Skymasters, the travelers were surrounded by ranks of American, British, and Soviet sentinels armed with rifles and stern looks. At Stalin's request, Roosevelt agreed to stay at the Soviet embassy compound, where he could rest safely from the prying eyes and pistols of the assassins Molotov assured him were skulking about Tehran.[36]

The Soviet embassy was reasonably comfortable and—as some in his party suspected—infested with listening devices. But, as Molotov had promised, it was secure. At least it seemed to be, judging from the valets with holsters bulging under their uniforms, and the sprinkling of GRU agents whose clumsy efforts to hide behind lawn trees proved more comical than insulting.[37]

"I am glad to see you. I have tried for a long time to bring this about."

Josef Stalin came to the opening meeting early to pay his respects to the man he had heard so much about. From his wheelchair, Roosevelt extended his strong hand to Stalin. Smiling broadly, the dictator shook it and patiently waited for his translator to finish Roosevelt's greeting before replying in kind.[38]

The short, clubfooted man who ruled the empire of the tsars blended the political fire of St. Petersburg with the pitiless fatalism of the Caucasus. Georgian, not Russian—his family name was Dzhugashvili—Stalin's human empathy and compassion traced their roots to Ivan the Terrible. As Marshall later remarked, "He was a rough SOB who made his way by murder and everything else." Brooke's first impression was not far from Marshall's: *He has got an unpleasantly cold, crafty dead face, and whenever I look at him I can imagine him sending people off to their doom without turning a hair.*[39]

Yet Stalin had not climbed the Kremlin's bloodstained ladder merely by pounding fists and murdering rivals. Quiet and laconic, he wielded the blunt comment as his usual weapon, though the occasional humorous quip would waft through his lips like smoke from his briar pipe. Leahy appreciated Stalin's direct methods and plain speaking, while Brooke, a military snob if there ever was one, considered Stalin an abler strategist than either Roosevelt or Churchill. Stalin, concluded Brooke, was a politician with "a military brain of the very highest calibre."[40]

In their first conversation in Roosevelt's sitting room, Stalin and Roosevelt touched on many subjects they would delve into later in detail. But Roosevelt knew the smiling man in the mustard suit was looking for one thing above all others: a true second front. When it would open, how large it would be, where it would take place. Those were the questions the Georgian sphinx would ask. And soon.[41]

Formal discussions of the "Big Three" began in the embassy boardroom a few minutes after Stalin left Roosevelt's sitting room. The conference had been scheduled with some ambiguity as to its starting time, but FDR and Stalin, not wanting to waste hours in Tehran, spontaneously agreed to move the first plenary meeting to that afternoon and rang up Churchill.[42]

By mutual agreement, FDR presided over the meetings. He delivered a lengthy and eloquent opening speech in which he thanked the other members for the privilege of chairing the conference—a meeting on which so much of the world's future would turn.

Picking up Roosevelt's theme, a hoarse Winston Churchill, suffering from laryngitis, stressed the momentous nature of the decisions before them. "In our hands we have perhaps the responsibility for shortening this war," he croaked. "In our hands we have, too, the future of mankind. I pray that we may be worthy of this God-given opportunity."[43]

Stalin sat silently through the perorations, chain-smoking Latakia cigarettes, betraying no thoughts. When Roosevelt and Churchill concluded their flowery remarks, Stalin roused himself just long enough to say, "I think the great opportunity which we have and the power which our people have invested in us can be used to take full advantage within the frame of our potential collaboration. Now let us get down to business."[44]

Roosevelt sketched out American concerns in China, Burma, and the Pacific before turning to what he called the most important theater of the war: Europe. A contested landing in France, he cautioned, would be no easy task. The English Channel is a disagreeable body of water, and operations would be impossible before May 1944.[45]

He then asked Stalin and Marshal Kliment Voroshilov, the Soviet defense minister, how best the western allies could help the Soviet Union. Voroshilov knew better than to move a muscle, and Stalin commenced with a thorough overview of the Soviet battlefront. Approximately 260 Axis divisions were facing the Red Army, he said. He needed a genuine second front, not a diversion. Italy was good, and the Balkans were good, but neither was the decisive theater. The only way to destroy Hitler was to invade through northwestern France. The Germans would fight like devils there, but a landing in France was the only way to win the war.[46]

All eyes turned to Churchill. Speaking through the twin handicaps of raspy voice and Russian interpreter, he told the conferees he fully supported an operation in France. In Italy, he offered no plans to move north of Pisa and Rimini, just above the Italian knee, and he said he had always considered Mediterranean operations of secondary importance, compared to OVERLORD.[47]

Yet, he continued, there was much to be gained in the Aegean. Partisans in Yugoslavia could strike a blow at the German war machine. The entry of Turkey into the war would open the sea to a prodigious flow of Allied supplies to the Soviet Union. Soldiers standing idle in the Middle East, and a very small part of the Allied fleet, could provide great assistance to the Red Army.[48]

As Churchill expounded on the Mediterranean, Stalin sat in silence, watching mechanically, looking down every now and then to doodle on a small pad of paper, which he would fold, then fold again, and when he could fold it no further, he would stuff it into his pocket and reach for another piece of paper.*[49]

When Churchill finished, Stalin said the Anglo-American situation was clear to him. It was useless to scatter British and American forces. The presentation seemed to indicate plans to send some soldiers to Turkey, some to northern France, and some to the Adriatic. Everywhere, which meant nowhere. The Allies, he said, should consider everything besides OVERLORD to be diversionary and focus on France. If other operations were considered, let them go forward in the Western Mediterranean, toward Corsica and the southern French coast as a supplement to OVERLORD.[50]

As his words carried around the room, first in Russian, then in English, the Americans smiled inwardly. Uncle Joe had come down hard for OVERLORD, and Churchill, a shorn Samson, had little choice but to back down. *"I thank the Lord Stalin was there,"* a grateful Henry Stimson wrote when he read the dispatch from Tehran. *"In my opinion, he saved the day. He brushed away the diversionary attempts of the PM with a vigor which rejoiced my soul."*[51]

Two and a half years into the war, Stalin's blunt decision meant the battle for the west would begin, finally and irrevocably, with OVERLORD.

Marshall's face reddened when he learned the first plenary session had started while he and General Arnold were sightseeing in Tehran. Roosevelt had said nothing of a meeting that afternoon, but after he moved into the Soviet compound, the Big Three decided to hold the general session at four o'clock that day. While Marshall and Hap were gawking at Persian architecture and Tehran's squalor, their colleagues were making history.[52]

King gave the two generals a rundown of the meeting. Stalin had cut through the rhetoric and put a stop to Churchill's diversions. As King remarked later, "The British had to be pitchforked into the coming invasion, and it was our pal Joey that applied the pitchfork."[53]

The next few days were filled with more conferences, more discussions, and more estimates. It was also filled with dinners, luncheons, and the inevitable lengthy toasts. But Stalin's cold pronouncements on that first day in Tehran set the tone for the military advisers. Nothing would be permitted to drain men or machines from OVERLORD or delay its commencement beyond May 1944.[54]

---

* When Stalin left one of his doodles on the table at the end of a meeting, King made a beeline for the souvenir. On his way, Roosevelt intercepted him with a question, and while King was answering the president, out of the corner of his eye he saw a British staffer get to the sketch and make off with it.

• • •

Between courses of vodka, beef, and sturgeon, Roosevelt considered how to build a personal relationship with a tyrant. FDR had built his career by forging bonds with antagonists, but at home he had months, even years, to win over men like Sam Rayburn, Joe Kennedy, or Wendell Willkie. In Tehran for only a few days, and handicapped by translators, there seemed no way he could crack the armor of a Georgian whose soft-spoken words masked a supernatural relentlessness.

"I had come here to accommodate Stalin," he told Frances Perkins afterward. "I felt pretty discouraged because I thought I was making no personal headway. I couldn't stay in Teheran forever. I had to cut through this icy surface so that later I could talk by telephone or letter in a personal way."[55]

Roosevelt believed Stalin feared the two western powers "ganging up" on him, and at Cairo he had taken pains to limit both U.S.-British military talks and his personal meetings with the prime minister. Spooling out that line a bit, on the second day of the conference he experimented by ribbing Churchill in front of Stalin. Before the meeting began, he joked to Stalin, "Winston is cranky this morning. He got up on the wrong side of bed." The remark survived translation into Russian, and though Stalin seemed puzzled by the display of disunity between the world's capitalist leaders, his face betrayed a slight smile.[56]

Using Churchill as bait, FDR set the hook over the next two days. "I began to tease Churchill about his Britishness, about John Bull, about his cigars, about his habits," he told Perkins afterward. "Finally Stalin broke out into a deep, hearty guffaw, and for the first time in three days I saw light. I kept it up until Stalin was laughing with me, and it was then that I called him 'Uncle Joe.' . . . That day he laughed and came over and shook my hand."[57]

Whether Stalin's laughter was contrived, or genuine delight at finding a rift between the two westerners, no one but Stalin knew. It made little difference, for strategic objectives were the coin of the realm; the dictator and the Democrat stood united on grand strategy. In Italy, Eisenhower could advance as far north as Pisa, but go no farther. The Allies would invade southern France, and OVERLORD would go forward in May 1944. In return for a second front, Stalin promised that the Soviet Union would join the war against Japan once Germany was defeated.[58]

The agreement left Churchill as the odd man out. The one big British-dominated theater, the Mediterranean, was being straitjacketed, and further efforts to open another front in the Balkans proved futile. As Churchill later told a diplomat, "There I sat with the great Russian bear on one side of me, with paws outstretched, and on the other side the great American buffalo, and between the two sat the poor little English donkey who was the only one . . . who knew the right way home."[59]

One question Stalin refused to drop was command of OVERLORD. The western Allies were disunited, and as Stalin saw it, putting one man in charge—a man figuratively standing between victory and the firing squad (figuratively in the West, that is)—was a practical necessity. His brusque comments to Roosevelt and Churchill on the summit's second day told them both that Soviet faith in the West was still lacking.

"Who will command OVERLORD?" he asked abruptly.

Roosevelt, caught off guard, replied that no decision had been made.

"Then nothing will come out of these operations," said Stalin. Until a commander is appointed, he said, "the OVERLORD operation cannot be considered as really in progress."

Roosevelt had no satisfactory answer for Stalin, and he knew it. Leahy recalled Roosevelt leaning to him and whispering, "That old Bolshevik is trying to force me to give him the name of our supreme commander. I just can't tell him, because I have not made up my mind."

Playing for time, Roosevelt said decisions made at the conference would affect his choice of commander. Stalin shrugged, unconvinced. The Soviets did not desire a voice in the selection of a commander, he said. They just wanted the commander appointed as soon as possible, and they wanted to know who it would be. Until that happened, he repeated, nothing would happen.[60]

Over lunch with Stalin and Churchill the next day, Roosevelt promised to name a commander soon. He would release the name in a few days, shortly after his return to Cairo.[61]

⋆   ⋆   ⋆

As the Big Three shifted their discussion to political matters, the Combined Chiefs flew back to Cairo to figure out how to implement the sweeping military decisions. Brooke, backed by Air Chief Marshal Portal and Fleet Admiral Andrew B. "A.B.C." Cunningham, Pound's successor, insisted that an invasion of the French Riviera required two divisions, not one. This meant finding more of those precious landing ships—or "lift," to use the British term—and the only place to find more lift was the Burma theater.[62]

The mundane question of lift would dominate Allied war councils for the next ten months, for it was a deeply rooted problem whose seeds were planted shortly after Pearl Harbor. In the war's first months, a large landing craft construction program had been planned around ROUNDUP in 1942. When ROUNDUP was canceled in favor of TORCH, the Navy and the War Production Board put their resources into destroyers, cargo ships, and convoy escorts. As a result, the Allies found themselves some two dozen landing craft short of the requirements of BUCCANEER, OVERLORD, an assault toward Rome, and an invasion of southern France.[63]

For all his expertise in logistics, Marshall had been slow to grasp the importance of landing craft. When he went to Europe with Pershing's army in 1917, "lift" simply meant renting enough ocean liners to carry two million doughboys to Calais. "Prior to the present war I never heard of any landing craft except a rubber boat," he had joked at Tehran. "Now I think about little else."[64]

Portal suggested they face the math and cancel BUCCANEER. But canceling BUCCANEER was anathema to the Americans, since FDR had promised Chiang that the Andamans invasion would go forward. King strongly believed canceling BUCCANEER would make the Chinese "feel that they had been 'sold out,'" and until Roosevelt returned from Tehran, he would not budge on Burma.[65]

Roosevelt returned on December 2 and met with the Combined Chiefs and Churchill two days later. Churchill pressed for a two-division ANVIL in southern France, and argued that the Allies could find the necessary lift in Burma. King tried to salvage BUCCANEER by suggesting a slight delay in OVERLORD; if OVERLORD could be moved from May 1 to the end of May, the Allies would have an extra four weeks to make up the landing craft deficit from new production. BUCCANEER need not be canceled.

King's logic strained under the weight of OVERLORD's requirements. As originally written, OVERLORD contemplated only a three-and-a-half-division assault. That scale now seemed too puny to ensure the success of a second front. If OVERLORD needed more muscle, BUCCANEER was the only place to find the extra lift.[66]

Roosevelt's first impulse was to keep his promise: he could not, in good conscience, tell Chiang the Allies were going back on BUCCANEER. But after listening to his chiefs and running through the math, he concluded that operations in Burma left too few ships to mount OVERLORD with the margin of safety it needed.

BUCCANEER was an American commitment, so it would fall to Roosevelt to break the news to Chiang. Perhaps, he mused, he could instead offer Chiang a renewed bombing campaign when the B-29 "Superfortress" was deployed in the spring of 1944.[67]

King was profoundly disappointed to lose BUCCANEER, for he believed passionately that China could pin down a large part of Japan's main army, keeping it out of his Pacific theater. As he saw it, the nation was also breaking faith with an important ally; FDR, King felt, had been too worn down from weeks of travel and banquets and meetings to make the right decision every time.

General Stilwell, who had waited in Cairo for final instructions on BUC-CANEER, was floored when FDR told him the Andamans project was dead. "I've been stubborn as a mule for four days, but we can't get anywhere," Roosevelt told Stilwell, stretching the truth again. "The British just won't do the operation, and I can't get them to agree to it."[68]

Stilwell wasn't buying it. The president had made an improvident promise, and now he was going back on his word. Topping it off, Stilwell would have to break the bad news to Chiang. *"God-awful is no word for it,"* a blue Stilwell told his diary after emerging from his meeting with FDR. *"The man is a flighty fool. Christ, but he's terrible."*[69]

.   .   .

While men in uniform wrangled over landing craft and divisions, Roosevelt faced a question he could no longer defer: whether to spare Marshall for OVERLORD.

His intuition told him Marshall was the best man for handling MacArthur, Chiang, Republicans, the British chiefs, the press, and Congress. Eisenhower, for his part, was doing an admirable job of keeping the war machine together in the Mediterranean. On the other hand, Stimson insisted that only Marshall had the stature to pull off OVERLORD in the face of British resistance.[70]

Then there was Marshall's place in history. Roosevelt had told both Eisenhower and Pershing that Marshall had earned his place in the Pantheon of American generals. While Marshall professed to have no such ambitions, his denial flew in the face of human nature. FDR didn't believe him.

Perhaps, thought Roosevelt, there was another way to decide the question. He called for Harry Hopkins.

Hopkins had been George Marshall's protector and patron since 1938. Marshall had become one of Harry's personal heroes, and Harry used his influence as Roosevelt's closest adviser to prevent Marshall's opponents from gaining access to the president.[71]

Hopkins pulled Marshall aside before dinner on December 4. The boss, he said, was "in some concern of mind" over Marshall's appointment as supreme commander. Without letting on which way Roosevelt was leaning, he asked Marshall to express his preference.

Command of OVERLORD would be the greatest opportunity any American soldier had ever been given. To Marshall, it would be the culmination of his life's work, his chance to make history on a grand stage. The Great War, St. Mihiel, Meuse-Argonne—those battles were small potatoes compared to what the Pershing of World War II would do in France, the Low Countries, and Germany. That general would be known as the liberator of Europe.

To be that general, all Marshall had to do was ask.

But that was what bothered Marshall. He hated asking. He took a perverse pride in refusing to let anyone pull strings on his behalf, and he had never hinted at his personal ambition. Characteristically, he told Hopkins he would go along, wholeheartedly and cheerfully, with whatever the president decided. The issues were too great to turn on personal preferences. "He need have no fears of my personal reaction," Marshall assured him.[72]

Roosevelt was puzzled when Hopkins shuffled in and told him what Marshall had said. After nearly five years of working with George Marshall, Roosevelt still didn't know what to make of him. Now that the job of a lifetime—perhaps ten thousand lifetimes—was his for the taking, Marshall didn't seem to want it. Or if he did, he wanted Roosevelt to reach across the table and stuff it into his pocket.[73]

But Roosevelt also knew Marshall was a man who could handle anything the war could throw at him. He had equipped an army, convinced Congress to extend the draft, guided strategy, and handled MacArthur, the press, the Navy, and his field commanders masterfully.

Around lunchtime on December 5, after a meeting with Churchill and the Combined Chiefs ended, Roosevelt took Marshall aside. When Marshall sat down next to him, Roosevelt, in his usual indirect way, made a weak attempt at small talk before broaching the subject burning in his mind: What were Marshall's feelings about commanding OVERLORD?[74]

"I was determined," Marshall said, "that I should not embarrass the president one way or another—that he must be able to deal in this matter with a perfectly free hand in whatever he felt was the [nation's] best interests."

Marshall drew a breath and told Roosevelt the same thing he had told Hopkins. "I just repeated again in as convincing language as I could that I wanted him to feel free to act in whatever way he felt was to the best interest of the country and to his satisfaction and not in any way to consider my feelings," he said later. "I would cheerfully go whatever way he wanted me to go."[75]

Much relieved, Roosevelt gave Marshall his answer. "I didn't feel I could sleep at night with you out of the country," he said. General Dwight Eisenhower would command OVERLORD, and Marshall would remain U.S. Army chief of staff.[76]

If Marshall was deflated by Roosevelt's decision, he never let on. After the last meeting, he pulled out a piece of paper and drafted a message to Stalin for the president's signature. It read: "The appointment of General Eisenhower to command of OVERLORD operation has been decided upon." Roosevelt interlineated the word "immediate" at the beginning, signed the note, and handed it back to Marshall, who sent it to Signals for transmission to Moscow.

The next day Marshall had Colonel McCarthy retrieve his handwritten message from the coding clerks, and he forwarded it to Ike with a personal note:

> Dear Eisenhower:
>
> I thought you might like to have this as a memento. It was written very hurriedly by me as the final meeting broke up yesterday, the President signing it immediately.
>
> G.C.M.[77]

THIRTY-NINE

RENO AND GRANITE

∿

T HE SHORTEST DISTANCE BETWEEN TWO POINTS IS USUALLY A STRAIGHT LINE. On Admiral King's charts, the straightest line from Pearl Harbor to downtown Tokyo ran through the Central Pacific. *Nimitz's* Central Pacific.[1]

By the fall of 1943, Nimitz's marines were closing in on Makin and Tarawa, the last holdouts of the Gilbert Islands. Their capture would pave the way to the Marshall and Caroline islands, which included the formidable naval base on Truk. After the Carolines, the Navy's next ports of call would be the Palaus, to the south, and the Mariana Islands—Saipan, Tinian, and Guam—which Nimitz hoped to reach by November 1944.

As King saw it, with the Marianas in American hands, the Navy could push through to the Philippines and coastal China. Bases in China would complete the encirclement of Japan. The Home Islands would starve, and their cities would fall under the bombsights of Hap's new B-29s. The war would be as good as over.[2]

GRANITE, Nimitz's Central Pacific plan, made the most of the mobility and striking power of the carrier fleet that King had husbanded since Pearl Harbor. It required comparatively little "lift," and allowed American airpower to do much of the heavy demolition work.[3]

But the realist in King knew Japan would not sit still as Nimitz hopped from Makin to Honshu. In his cabin late at night, that realist kept asking, "Where would the Japanese turn and fight?"

Picturing the great ocean from Japan's point of view, King felt the Imperial Navy might let the United States reach the end of its supply tether before delivering a knockout punch. For that reason, he did not think the enemy fleet would sally too far east of the Philippines, much as he would love to see them try.

"Because we would like them to come out is no military reason why they should do as we wish," he reminded his journalist friends one evening. "Also—and maybe more important—the Japs have had a taste of what a wallop the American Navy packs and does not like the taste."[4]

The American Navy packed a wallop, but so did Japan's. While King, Marshall, and FDR were crossing North Africa to meet Stalin, the Second Marine Division landed on Tarawa, an equatorial atoll with one airstrip not much larger than an oversized carrier deck. Japanese marines fought like demons on the beaches, and the bitter fighting for this obscure strip of coral produced the bloodiest battle in the long annals of the Marine Corps. In two days, the Marines suffered 3,149 casualties, and the final toll would include more than a thousand American dead. Japanese losses were 4,609 killed, seventeen captured.[5]

Before Roosevelt returned from Tehran, word of the butchery at Tarawa made the front pages of U.S. newspapers. "It has been the bitterest, costliest, most sustained fighting on any front," a war correspondent told the *New York Times*. "Something suddenly appeared to have gone wrong." The press and Congress raised a ruckus over the island's steep cost, and Secretary Knox had to hold a press conference defending the Navy's performance from "armchair strategists" on Capitol Hill.[6]

Admiral King, seeing the larger picture, took both carnage and criticism in stride. "Nobody should be surprised if we get a bloody nose in the Pacific," he told his journalists two months later. "The way we are throwing our weight around, it is natural to suppose that the Japanese will react very violently, and a fight is sure to follow."[7]

• • •

As the Army's senior soldier, George Marshall made an effort to visit troops whenever he left Washington. He had shaken hands with soldiers of every stripe at Casablanca, Algeria, Iran, and throughout the United States, but two years into the war, there was one soldier he hadn't seen—a soldier who refused to be ignored.

Taking leave from Cairo at the end of the SEXTANT meetings, Marshall flew east to MacArthur's advance headquarters on Goodenough Island, off New Guinea, for a conference on strategy and supply. There, outside a small shack MacArthur's staff had fixed up as a war room, the two generals shook hands for the first time since the war began.

Pacing in his rumpled khaki uniform, MacArthur had what he called a "long and frank discussion" with Marshall about the theater's many shortages: aircraft, ammunition, reinforcements, landing craft, and food, for starters. But the biggest problem of all was lack of "unity of command," by which MacArthur meant that the Allies needed a supreme commander over all naval, air, and ground forces. MacArthur had more than a faint idea who that supreme commander should be.[8]

For all his flaws—and he had many—MacArthur's eloquence flowed like a mountain river when making his case in person. One of Marshall's staffers described his performance that day:

> He stood at various maps, strode back and forth, and talked for about two hours without notes of any sort. He had at his fingertips all the dispositions and recent actions of his troops. He seemed equally well acquainted with his enemy. He named Japanese organizations and their commanders everywhere and seemed well informed of their competence. . . . Throughout the presentation he employed wit and charm with devastating persuasiveness. Although I had from the first been an advocate of a "Europe first" strategy, with attendant delay against Japan, I simply melted under the persuasive logic and the delightful charm of the great MacArthur. By the time he had finished, I was anxious to find some way to give him what he had asked for.[9]

So was Marshall, but only to a point. The chief of staff's job was to keep one eye focused on the global picture, regardless of momentary pressures afflicting any theater, even MacArthur's. He would push King and the other chiefs to support the Southwest Pacific, but he reminded MacArthur of the "Germany First" strategy and told him many requests would not be fulfilled because of decisions made in Tehran and Cairo.

MacArthur accepted Marshall's response politely—outwardly, at least. The meeting ended on a courteous and correct note.[10]

When courtesy and logic failed, MacArthur did what came natural to him; he went over Marshall's head. He asked Brigadier General Frederick Osborn, a War Department staffer, to give Secretary Stimson a personal report on his battle plan, code-named RENO III. The Philippines had some 1,500 islands, but from a military standpoint the commonwealth, like Caesar's Gaul, was divided into three parts: Mindanao in the south, Leyte in the center, and Luzon, home to Manila, in the north. If the Navy would lend him a healthy portion of its Central Pacific forces, Sutherland said MacArthur could sweep through New Guinea, the Dutch East Indies, and Mindanao by the end of the year.

If he could control the Navy's ships, men, and aircraft, he told Osborn, he could be in Mindanao by the end of 1944. "I do not want to command the Navy," he added, "but I must control their strategy, be able to call on what little of the

Navy is needed for the trek to the Philippines. The Navy's turn will come after that."

To Osborn, MacArthur explained the true obstacle to American victory. "Mr. Roosevelt is Navy minded," he emoted. "Mr. Stimson must speak to him, must persuade him. Give me central direction of the war in the Pacific, and I will be in the Philippines in ten months. . . . Don't let the Navy's pride of position and ignorance continue this great tragedy to our country."[11]

To King, pride of position was MacArthur's credo; for all he knew, it was probably inscribed in Latin on the MacArthur family crest. The real tragedy was that MacArthur was scaring the War Department, and probably scaring the White House, too. MacArthur was a good soldier, but having been run out of the Philippines, he had become obsessed with returning to Luzon. That obsession warped his strategic vision, because Luzon was a hell of a long way from Tokyo. Just to get there, he'd have to slog through New Guinea, finish off the Admiralties, and capture some of the Dutch East Indies. Only then could he move against Formosa or Japan.[12]

To King, the dogs of war were barking in one direction: across the Central Pacific, to the Marianas, then Formosa, then Japan's Home Islands. At Cairo, the Combined Chiefs had approved a two-pronged advance across the Pacific by Nimitz and MacArthur, but if the Joint Chiefs had to choose between the two, they leaned toward Nimitz's Central Pacific route.

MacArthur knew he would have to make an exceptionally persuasive sales pitch to move Allied strategy from the Marianas to Manila. He started by sending his chief of staff, Major General Richard Sutherland, to Pearl Harbor to argue the case for RENO III to the Navy's Pacific planners.

Nimitz's staff generally agreed with MacArthur. Unlike King's planners in Washington, Nimitz's men leaned toward slowing the Central Pacific route in favor of a New Guinea–Mindanao advance under MacArthur. They also never seriously questioned the importance of Luzon. Vice Admiral Thomas Kinkaid, commander of the Seventh Fleet coordinating with MacArthur, and Vice Admiral John Towers, commander of Pacific naval air forces, thought the Marianas too far from Japan to serve as an effective B-29 base. MacArthur's Philippines route through Luzon seemed preferable.[13]

MacArthur lost no time in capitalizing on the momentum. On February 2 he wrote Marshall, "All available ground, air and assault forces should be combined in a drive along the New Guinea–Mindanao axis. . . . This axis provides the shortest and most direct route to the strategic objective." He warned that a two-pronged approach would result in "two weak thrusts which cannot attain the major strategic objective until several months later."[14]

• • •

King blew his stack when he read the minutes of the Pearl Harbor conference. He could not believe Nimitz would be so unreliable, or so dumb, as to question the Marianas-Formosa-China strategy. He fired off a blistering cable to Nimitz that began, "I have read your conference notes with much interest and I must add with indignant dismay." Reminding Nimitz of the importance of the Marianas, he concluded, "The idea of rolling up the Japanese along the New Guinea coast, throughout Halmahera and Mindanao, and up through the Philippines to Luzon, as our major strategic concept, to the exclusion of clearing our Central Pacific line of communications in the Philippines, is to me absurd." Formosa and the China coast, he thundered, were the proper targets for the Pacific war. Those targets must be attacked through the Central Pacific.[15]

To let some of the wind out of MacArthur's sails, King wrote Marshall to remind him that at Cairo the Combined Chiefs had agreed to make two drives through the Pacific, not one, with the Central Pacific taking priority if there weren't enough resources to go around. He admitted, for the moment, that B-29s could be used more effectively from the Philippines—if and when they were recaptured—but he insisted that the Marianas were the proper base for Hap's bombers. Lest there be any doubt about the Navy's position, King refused to allow MacArthur command of any Central Pacific naval assets. MacArthur already had Kinkaid's Seventh Fleet working for him, and King would not give him Halsey's Third Fleet, too.[16]

On February 17, Admiral Spruance's Fifth Fleet and a Marine-Army assault force captured Eniwetok, rolling up the Marshall Islands. Worried that King would accelerate the Central Pacific drive, cutting him out, MacArthur raised the stakes. He offered to bypass intermediate steps along the eastern New Guinea coast and lunge directly at the port of Hollandia, leapfrogging 40,000 Japanese troops holding the island's eastern half. It was a dicey gamble, he admitted, but the flanking move, if it worked, would accelerate his drive against Mindanao. All he needed was the loan of Nimitz's carriers and transport ships.[17]

As King and Nimitz considered MacArthur's Hollandia plan, they stumbled into another political thornbush. With the capture of the Gilberts and Marshalls, the South Pacific Area, home of Bill Halsey's powerful Third Fleet, had become the Pacific's rear area. King and Nimitz sensibly decided to move Halsey's headquarters forward, so it could contribute to future campaigns, and that meant finding Bill a new home.

Looking over a map of MacArthur's upcoming conquests, in late February Nimitz and Halsey decided Manus Island, a backwater speck in the Admiralties, would make a fine forward base for the Third Fleet. At Nimitz's suggestion, Marshall asked MacArthur to put the Navy in charge of building and running the Manus base.[18]

MacArthur saw the Navy's move as an invasion of his theater, and he fought back like he was defending the Bataan Peninsula. Under international agreement, he protested, the Southwest Pacific Area was off-limits to Halsey's fleet except when MacArthur was using the fleet. He warned Marshall that any attempt to insert naval forces into his theater would "cause a serious reaction, not only with the soldiery but in public opinion, that would be extremely serious."

It was not just the fate of a small island at stake, he said. The entire Pacific campaign hinged on his having a free hand throughout the Southwest Pacific Area. His personal integrity—"indeed, my personal honor"—would be tarnished if the Joint Chiefs dropped a Navy enclave in the middle of his theater. If the Joint Chiefs trimmed his territory, he demanded "an early opportunity personally to present the case to the Secretary of War and the President before finally determining my own personal action in the matter."[19]

"Personal action" was MacArthurian for "resign," and Marshall patiently assured the general he would not let the Navy pull Manus Island out of Army jurisdiction. But he told MacArthur that while he might command the island, he could not limit the right of Halsey's ships to use Manus in the war against Japan. "Your professional integrity and personal honor are in no way questioned or, so far as I can see, involved," a weary Marshall concluded. "However, if you so desire I will arrange for you to see the Secretary of War and the President at any time on this or any other matter."[20]

Bill Halsey, an admiral well liked among Army men, tried to settle the matter in person. He got nowhere. MacArthur again called the Navy's plan an insult to his personal honor, and an exasperated Halsey told MacArthur he was putting his personal honor ahead of the welfare of the United States.

Afterward, Halsey sent a rear admiral to Nimitz with an oral report. MacArthur, he said, was "suffering from illusions of grandeur," while his staff officers "are afraid to oppose any of their general's plans, whether or not they believed in them." As Halsey later remembered, "MacArthur lumped me, Nimitz, King and the whole Navy in a vicious conspiracy to pare away his authority."[21]

To King, MacArthur was becoming a genuine stumbling block to victory in the Pacific. Stimson and Marshall, he believed, would do anything rather than tangle with their subordinate, so he went to Leahy on Friday, March 10. He raised hell, knowing the word would get back to Roosevelt.

Leahy did him one better. The next day he summoned King and Nimitz to the White House and gave them time to plead their case to the president.[22]

Though he often meddled in Army-Navy disputes, Roosevelt refused to umpire this latest MacArthur-Navy fight. Beset by labor unrest, manpower shortages, nervous Democrats, and poor health, he told his admirals that Manus and MacArthur were problems the Joint Chiefs would have to handle themselves. He would not add to his overloaded plate a squabble among his military chiefs over a speck of coral no one had ever heard of.[23]

The Joint Chiefs handled it. On Sunday, March 12, they met in a closed-door session and thrashed out another messy compromise: MacArthur would capture Hollandia in mid-April, Mindanao in November, and, if necessary, Luzon in February 1945. Nimitz would bypass the Japanese base at Truk. He would capture the Marianas in mid-June, the Palau Islands in mid-September, and Formosa the following February. MacArthur would support Nimitz with land-based aircraft, and Nimitz would provide ships and marines for MacArthur's conquests. The grand plan made MacArthur and the Navy collaborators, while setting both horses on a race to Tokyo.[24]

As he looked at the new instructions, MacArthur spotted another trap. Under the new timetable, Nimitz would capture the Marianas eight months before MacArthur had slugged his way to Luzon. He knew King wanted to hop from the Marianas to Formosa, perhaps even to Japan itself, and he feared the admiral would try to bypass the Philippines.

When Nimitz paid a visit to MacArthur's headquarters in late March, MacArthur was impeccably courteous until Nimitz referred to the portion of the JCS order requiring them to prepare contingency plans to bypass Luzon. Then Nimitz wrote King,

> [MacArthur] blew up and made an oration of some length—on the impossibility of bypassing the Philippines, his sacred obligation there—redemption of the 17 million people—blood on his soul—deserted by the American people—etc. etc, and then a criticism of "those gentlemen in Washington who—far from the scene—and having never heard the whistle of bullets, etc.—endeavor to set the strategy for the Pacific War," etc.[25]

Those men, far from the scene, would have their hands full as the year's strategy played out.

. . .

In early 1944, Admiral King saw the enemy gaining strength. Not in the Atlantic, where the U-boat threat was waning, or the Pacific—though the fighting at Tarawa signaled that the Japanese were anything but licked.

King's enemies were closer to home.

His patrols smoked them out when he learned of a plot to separate his COMINCH and CNO titles. A draft executive order being circulated around Knox's Navy Department would make King a five-star "Admiral of the Navy and Commander, United States Fleets," while Vice Admiral Frederick Horne would become a four-star "Chief of Naval Logistics and Material." Horne, the draft said, would report to the navy secretary, not King. In other words, Horne would become the chief of naval operations.[26]

King saw Forrestal's fingerprints on the quarterdeck mutiny. The New Yorker was a slick political operator who resented the breadth of King's authority. "Forrestal believed, but he never said to me, that I had too much power myself," King remembered later. "He hated like hell that I had both jobs."[27]

At first, King figured Forrestal couldn't trim his sails, even with Leahy's backing, because President Roosevelt would have to repeal his executive order combining the offices of COMINCH and CNO. It would be a public admission that the president had made a mistake in giving King both jobs. He didn't believe FDR would publicly admit to a mistake like this.

But his bureaucratic alarm sounded general quarters when he learned that the president and Secretary Knox had buttonholed House Naval Affairs Committee chairman Carl Vinson about the change. If Forrestal convinced Chairman Vinson, King's strongest ally, King's position would become untenable.[28]

As Forrestal gained the weather gage on King, the admiral wore around and spread more canvas. He quietly went to Vinson's Capitol Hill office and pleaded his case.

Vinson rewarded King's fealty with a letter withdrawing his support for the new proposal. In no uncertain terms, the powerful chairman told Knox that he wanted King running the whole show at Main Navy.[29]

Knox was furious. He knew where Vinson's letter had been drafted, and it wasn't in the Cannon House Office Building. The secretary sputtered to Roosevelt, "[Vinson's letter] bears internal evidence of having been prepared, I think, in the Navy Department. . . . You will not fail to observe . . . that there is the same effort to consolidate all authority in one person, and that not the Secretary of the Navy."[30]

King didn't care what the Secretary of the Navy thought. He saw military politics in terms of power and weakness, not fair and equitable. The Secretary of the Navy could bitch all he wanted, but with Chairman Vinson in King's corner, there wasn't a damned thing Knox could do about it. The twin titles of COMINCH and CNO would hang on Ernie King's door a while longer.

He would need every bit of his authority to deal with a storm brewing over the upcoming invasion in Europe.

FORTY

"CONSIDERABLE SOB STUFF"

⌇

OPERATION NEPTUNE, OVERLORD'S LANDING PHASE, ORIGINALLY CALLED for a three-division assault on Normandy's beaches. It had been planned by the capable Sir Frederick Morgan, and his three-division operation had been limited by the estimated number of landing craft available in the spring of 1944. Three divisions seemed a reasonable assumption going into the Tehran conference.

After Roosevelt appointed Eisenhower to command OVERLORD, Marshall ordered Ike to return home for a few days to rest, see Mamie, and forget about the war. "You will be under terrific strain from now on," he told Eisenhower. "I am interested that you are fully prepared to bear the strain and I am not interested in the usual rejoinder that you can take it. It is of vast importance that you be fresh mentally and you certainly will not be if you go straight from one great problem to another. Now come on home and see your wife and trust somebody else for 20 minutes in England."

Upon Eisenhower's arrival in London, that rested mind focused on OVER-LORD's three-division pillar and found it unstable. At the same time General Montgomery, OVERLORD's ground commander, decreed that three divisions would be inadequate to capture the beaches and build a sufficient bridgehead to widen the invasion.

Before January's end, Ike and Monty had rewritten the playbook. NEPTUNE would include five reinforced assault divisions, parachute and glider troops, armor, reinforcements, and support troops. Its projected load was 174,320 men and 20,018 vehicles, and except for a small number of airborne troops, every mother's son would be carried across the English Channel on a ship.[1]

As dramatic as the D-day landings would be, NEPTUNE was merely the tip of

OVERLORD's much larger blade. The real struggle would begin once Rommel and his boss, Field Marshal Gerd von Rundstedt, figured out that Normandy was no feint and launched their panzer counterattacks. With those attacks, the battle for France would be joined.

Eisenhower's planners at Supreme Headquarters, Allied Expeditionary Force, or "SHAEF," believed they could pull off the invasion if Rommel and von Rundstedt could not throw more than thirteen mobile divisions at them within three days of the landings—but only if the Allies shoved troops, ammunition, and fuel across the Channel faster than the Nazis could move them from central France. Fighting men would decide the fate of NEPTUNE, but OVERLORD's fate rested in the calloused hands of engineers driving bulldozers, sailors piloting landing ships, and beachmasters moving ammunition and fuel from sea to shore.

The weapons of this battle would be combat loaders—ships carrying complete units of infantry, tanks, and artillery over open water—and the amphibious craft that would cross the home stretch and disgorge men, vehicles, and supplies onto French soil. These obscure vessels spawned an alphabet soup that New Deal bureaucrats would envy. Vessels bearing acronyms like LST (Landing Ship, Tank), APA (assault troop transports), LCVP (Landing Craft, Vehicle/Personnel, the "Higgins boat"), LCM (Landing Craft, Medium), DUKW (two-and-a-half-ton amphibious truck), and LVT (amphibious tractor, or "Amtrack") confounded planners with their nomenclature and scarcity. At one point, Eisenhower groaned to an aide that when he died, "his coffin should be in the shape of a landing craft, as they are practically killing him with worry."[2]

When Eisenhower told the Combined Chiefs that OVERLORD needed more assault divisions, he presented the bill in terms of those coffin prototypes. In addition to the fleet already amassed for the invasion, the chiefs would have to find him seven large infantry landing ships, forty-seven LSTs, seventy-two infantry landing craft, 144 tank landing craft, and a headquarters ship, for a total of 271 new vessels. To escort the 271 new vessels, he would also need an additional two dozen destroyers, twenty-eight motor launches, four flotillas of minesweepers, and a bombardment force of five cruisers, a dozen standard destroyers, and one or two battleships.[3]

As his dark eyes tore through a bewildering tide of production tables, Admiral King began to realize just how many of those 271 new vessels he didn't have. In February, he had assured journalists that "the craft in the Channel will be so numerous that one could walk dry-shod to the beach." But there would be a lot of wet-shod soldiers if Eisenhower added two divisions and King couldn't find the ships to ferry them.[4]

Before Pearl Harbor, no one had grasped the immense scale of shipping required to land soldiers on a hostile beach, much less execute near simultaneous landings on beaches in the Marianas, Italy, southern France, and Normandy. A crash program of landing craft construction in early 1942 had put a deep dent in production of destroyers and escort carriers desperately needed in the Atlantic. Stung by appalling losses to U-boats, the Navy was reluctant to resume priority for assault craft production until OVERLORD seemed certain. But by early 1944, it was too late; there was not enough "lift" to go around.[5]

Something had to give, and the question was who would be giving. To strengthen NEPTUNE and keep the momentum in Italy, Montgomery, Churchill, and the British chiefs were prepared to cut ANVIL to a one-division feint. Marshall and King, believing ANVIL vital to OVERLORD's success, were prepared to curtail operations in Italy, but they refused to cut either NEPTUNE or ANVIL.[6]

In Italy, Churchill insisted on launching Operation SHINGLE, a corps-size landing at Anzio about thirty-five miles from the Italian capital. But to reassure the Americans, he suggested that if SHINGLE were launched around January 20, the LSTs used for SHINGLE could be sent back to England in plenty of time for OVERLORD. Roosevelt agreed, but he insisted that the ships must go to England as soon as they could be released from Italy, without further delay.[7]

Launched on the twenty-second of January, SHINGLE bogged down under German counterattacks, and the Anglo-American force dug in for a long and violent struggle around Anzio. Churchill told the British chiefs a week later, "I had hoped we were hurling a wildcat into the shore that would tear the bowels out of the Boche. Instead we have a stranded vast whale with its tail flopping about in the water."

To add fuel to the Anzio attack, Churchill and his chiefs wanted those LSTs to tarry in Italy a while longer. The conflicting demands of OVERLORD, ANVIL, and SHINGLE tossed the Allies onto the horns of a dilemma: Which front would be robbed to ensure that Eisenhower had enough ships to smash into Hitler's *Festung Europa*?[8]

To the British, the answer was spelled "ANVIL." A landing in southern France, five hundred miles from Normandy's beaches, would do nothing to help Eisenhower's men. The Riviera had about as much to do with OVERLORD as Italy did, so why shut down a big Italian show to open a smaller French one? Better, they felt, to keep pushing in Italy and give Eisenhower ANVIL's ships. At most, ANVIL should be reduced to a one-division feint.

The Joint Chiefs had no authority, or inclination, to rob ANVIL to beef up Italy. During OVERLORD's follow-up phases, the Allies would use Marseilles, one of ANVIL's target ports, to pour in massive reinforcements from the south and augment Allied supply lines, an advantage that appealed to the logistician in Marshall. Roosevelt had also promised Uncle Joe that the Allies would land in southern France, and it was a promise he intended to keep. Strategically, they saw the Italian campaign ending at a frozen, easily defended wall known as the Italian Alps, where German soldiers, holding the high ground, would fight among the snowdrifts and edelweiss until someone told them the war was over. As correspondent Hanson Baldwin put it, "All roads led to Rome, but Rome led nowhere."[9]

In London, Eisenhower tried to broker a compromise among the Combined Chiefs by enlarging his "lift" pie. He played dangerous number games by overloading ships, swapping LSTs for less capable vessels, and decreeing, like King Canute, an increase in the percentage of craft that SHAEF would deem "serviceable," or ready for duty, at any given time. He also pushed back the invasion from early May to early June, to allow another month's landing craft production to arrive in England.

But creative accounting and aggressive management would get Eisenhower only so far. On February 19, he told Marshall that, in light of the lift shortage, the needs of the Italian theater probably required the chiefs to cancel ANVIL.[10]

In a White House meeting two days later, the Joint Chiefs informed Roosevelt the British were trying to cut ANVIL to a pittance that wouldn't make it worth the effort. FDR wouldn't hear of it. At Tehran, he said, the Russians had been "tickled to death" over ANVIL. If ANVIL were canceled, they "would not be happy, even if we told them it would mean two or more divisions for OVERLORD."[11]

Yet the British would not budge. ANVIL would require ten or twelve divisions to follow up the initial landing, and with a major battle raging south of Rome, they wanted those divisions in Italy, not ambling through the Coté d'Azure.

Because Marshall and King had no authority to compromise on ANVIL, they felt the best solution was to shelve the decision until late March and see where things stood. The Combined Chiefs agreed. They would defer their argument and take a second look at ANVIL on March 20.[12]

·   ·   ·

As logistical problems ensnared the Allied high command, George Marshall thanked his stars for his friend Field Marshal Sir John Dill, the senior British military representative in Washington. Handsome, tactful, and only a year younger than Marshall, Sir John had been Marshall's British counterpart when they first met at Placentia Bay in 1941. Since the ARCADIA conference the following December, he and Marshall had developed a personal bond unique among the members of the Allied high command.

Dill had the unusual ability to see the picture from both sides of the Atlantic, and the American chiefs trusted him. When Churchill complained about British exclusion from the Manhattan Project, the Americans offered to put Dill on the project's top policy committee, and when bitter disputes erupted over the Mediterranean and France, Dill minimized collateral damage to the "very special relationship."[13]

Sir John was also Marshall's back channel to the FDR-Churchill conversations. Unlike the secretive Roosevelt, Churchill ensured that copies of important communications between him and the American president were regularly circulated among his military staff. Dill would bring his copies to Marshall's office so Marshall would know what his commander-in-chief was telling his military allies.

Churchill had never liked Dill, and in disclosing the top secret correspondence to Marshall, Dill was taking an enormous professional risk. "Dill would be destroyed in a minute if this was discovered," Marshall reflected. Yet Dill assumed that risk to keep the vital engine of cooperation running just a little smoother.[14]

Dill's destruction looked probable in mid-February 1944, when Marshall heard rumors that Churchill was planning to recall him to London. Genuinely worried about the repercussions for the alliance, Marshall and Stimson cooked up a scheme to give Dill enough stateside clout to discourage Churchill from firing him. At Stimson's suggestion, Marshall went to see Harvey Bundy, special assistant on the Manhattan Project, with a novel idea.

"Bundy, I need help," said Marshall. "I have word that Churchill is likely to throw out Sir John Dill and my relations with Dill are vital. . . . I wish you would go up to Cambridge, to Harvard, and see if they can't give him a quick honorary degree."[15]

Bundy, a Yale man, promised to do what he could. He rang up the university administration, but reported back that Harvard could confer a degree only during a convocation, and a convocation was not be arranged on the spur of the moment.

"Try Yale," Marshall said.

Bundy tried his alma mater, and this time it worked. Yale president Charles Seymour not only promised to give Sir John an honorary degree, but he would award him the prestigious Charles P. Howland Prize for international relations and roll out the reddest of carpets (which, in New Haven, were blue). The ceremony would be adorned with academic robes, a mace bearer, and most importantly, extensive publicity.

To ensure the story received the kind of press coverage that would register in London, Stimson, Marshall, McCloy, and Bundy attended the ceremony. They sat in rapt attention as Sir John received his doctorate, and Marshall encouraged cameramen and movie crews to take their fill of pictures with Dill surrounded by War Department luminaries.

Other colleges followed Yale's lead. William and Mary, Columbia, and their sister universities began rolling out awards for the field marshal, and each ceremony was attended by news coverage monitored at Downing Street.

It wasn't long before a relieved Marshall heard that Churchill had remarked, "Dill must be doing quite a job over there," and Dill remained at his post.[16]

Dill's diplomacy was a big glob of the grease that kept the motor running smoothly—or as smoothly as it could run on a pockmarked, bumpy road. But diplomacy could change neither basic arithmetic nor Churchill's strategy, and the question of how to serve up three halves of a single pie dogged Marshall, King, and the rest of the Allied chiefs in the spring of 1944.

On March 20, Eisenhower told an unhappy Marshall that ANVIL had become the war's latest casualty. While Eisenhower supported the ANVIL concept, to give ANVIL the lift it needed, he would have to short OVERLORD by fifteen LSTs. That left too thin a margin to give OVERLORD a favorable chance. He asked the Combined Chiefs to cancel the ANVIL landings and give him an additional twenty-six LSTs, plus an assortment of other landing ships and minesweepers.[17]

Marshall was deeply disappointed, but he took Eisenhower at his word, for Eisenhower was not the type to demand more than he needed. Looking for a middle ground, Marshall suggested rescheduling ANVIL for July 10, and he told Eisenhower landing craft coming off the shipyards could be diverted from the Pacific and would be available for use in southern France if the British agreed to launch ANVIL by mid-July. If they would not, he said, the landing craft would go to the Pacific as scheduled.[18]

They would not. As Churchill saw it, the rewards of a drive through Italy into the Po River valley, toward Istria or north to Austria, would be lost if divisions and ships were committed to ANVIL. The British refused to launch ANVIL in June, July, or any other time, and they wanted the Pacific landing craft sent to Italy.[19]

Dill—*Doctor* Dill now—explained the problem to his countrymen from the American point of view. The U.S. chiefs were fighting a war in the Pacific, where places like Kwajalein and Tarawa commanded the U.S. public's attention. They courted a heavy political risk by diverting landing ships slated for MacArthur and Nimitz to Europe.

"It is difficult to realize how hardly anything can be taken from the Pacific in view of the fact that the U.S. Chiefs of staff are constantly being abused for neglecting [the Pacific] theater," Dill wrote the British chiefs on April 1. "The U.S.

Chiefs of staff made the offer [of additional landing craft for ANVIL] with a feeling of broadminded generosity and were shocked and pained to find how little we appreciated their magnanimity and how gaily we proposed to accept their legacy while disregarding the terms of the will."[20]

Churchill pressed Marshall to back down. "Dill tells me that you had expected me to support 'Anvil' more rigorously in view of my enthusiasm for it when it was first proposed by you at Teheran," he wrote Marshall on April 16. "Please do me the justice to remember that the situation is vastly changed."[21]

But Marshall would not retreat. The stalemate continued, and as before, a decision would have to await further developments on Italy's blood-soaked soil. ANVIL was still alive, for the moment, but Marshall could see its pulse was faint, its breath shallow. And Churchill was waiting by its bedside with a pillow.[22]

* * *

ANVIL was not the only plan in acute distress. One of OVERLORD's many subsidiary operations was the Transportation Plan, a massive bombing campaign to wipe out railyards near Paris, bridges over the Seine River, and road networks linking Normandy's beaches with the rest of France.[23]

SHAEF estimated that Hitler had fifty-one divisions in France and the Low Countries. His best troops were concentrated around Calais, north of Normandy, and in the French interior. To keep Allied troops from being overwhelmed in OVERLORD's critical first days, SHAEF had to keep those Germans away from the Normandy coast. That required Allied bomber command to wreck the French transportation network.[24]

A mass bombing of French rail and road centers would take many Frenchmen with it—possibly eighty thousand casualties, Churchill estimated, including ten thousand or more killed. Having endured the horrors of the German blitz over London, Churchill was repelled by the thought of dismembered French bodies strewn about Paris. He had been forced to fire on the French Navy in 1940, and he did not want to kill more Frenchmen in 1944. He saw in Eisenhower's plan an act not only morally objectionable, but one that would poison Anglo-French relations for years to come.[25]

The military men were less sympathetic. *"Considerable sob stuff about children with legs blown off & blinded old ladies but nothing about the saving of risk to our young soldiers landing on a hostile shore,"* grumbled Cunningham to his diary. Ike tried to reassure Churchill that estimates of eighty thousand Frenchmen had been very much on the high end, and he pointed out that the raids would kill a great many Germans, too.

In any event, he said, the Transportation Plan was critical. "The 'OVERLORD' concept was based on the assumption that our overwhelming Air power would be able to prepare the way for the assault," he told Churchill on May 2. "If its hands are to be tied, the perils of an already hazardous undertaking will be greatly enhanced."[26]

But Churchill was willing to tie OVERLORD's hands, and less than a month before the invasion he wrote Roosevelt, "The War Cabinet share my apprehensions of the bad effect which will be produced upon the French civilian population by these slaughters all taking place so long before Overlord D-Day. They may easily bring about a great revulsion in French feeling towards their approaching United States and British liberators. They may leave a legacy of hate behind them."[27]

Roosevelt avoided interfering in military tactics. He might tell his generals *when* to do something, or even *where* to do it, but not *how* to do it. That was their job, not his, and he trusted them to know best. He replied to Churchill, "However regrettable the attendant loss of civilian lives is, I am not prepared to impose from this distance any restriction on military action by the responsible commanders that in their opinion might militate against the success of Overlord or cause additional loss of life to our Allied forces of invasion."[28]

"This was decisive," wrote Churchill later. "The price was paid." Roosevelt would let his airmen, generals, and admirals run this part of the war. The Transportation Plan went forward, and more than 76,000 tons of bombs fell on bridges, open lines, and railyards.

French casualties were lighter than SHAEF had predicted.[29]

.   .   .

As FDR pondered the invasion of France, an aide brought him a letter from Eisenhower's chief of staff to General Marshall complaining that SHAEF was having problems arranging French help for the invasion force. The obstacle was that Eisenhower had no authority to deal with Charles de Gaulle's French Committee of National Liberation.[30]

De Gaulle was still the same haughty, headstrong bride that Roosevelt and Churchill had dragged to the altar at Casablanca the year before. He had formed the FCLN as "co-chairman" with General Giraud, but de Gaulle's coterie never considered de Gaulle "co"-anything. *Le général*, they insisted, was the rightful provisional ruler of France.[31]

As the invasion of Normandy nosed up over the horizon, Roosevelt believed de Gaulle was not only an annoying, self-serving Frenchman, but an unnecessary one. Brigadier General William Donovan, head of the Office of Strategic Services, assured Roosevelt that de Gaulle was not critical to a revived French democracy; other parties, Donovan said, would spring up to fill the political vacuum in French government.[32]

Yet no one had worked harder to rally the French people against the fascists. From England to West Africa, de Gaulle's was the voice of French resistance, and he led the most viable alternative to a resurrected Vichy government. He could be ignored as long as battles were being fought in Africa or the Mediterranean, but once the war moved into France, his faction's cooperation would be vital.

Roosevelt, Eisenhower later concluded, was "almost an egomaniac in his belief in his own wisdom." De Gaulle was an egomaniac, too, but by poking him in the eye, the Allies would be left to govern a liberated France, with all its civil and logistical burdens, without a political infrastructure while they were fighting their way toward the Rhine.[33] Observing that the Allies had two choices of provisional government—the FCLN and the "Vichy gang"—in mid-May Eisenhower cabled Marshall for instructions.[34]

After letting Ike's message sit in his office for almost two weeks, Roosevelt sent Marshall a blunt reply complaining that Eisenhower "does not quite get the point." He told Marshall, "[Eisenhower] evidently believes the fool newspaper stories that I am anti–de Gaulle, even the kind of story that says I hate him, etc., etc. All this, of course, is utter nonsense. I am perfectly willing to have de Gaulle made President, or Emperor, or King or anything else so long as the action comes in an untrammeled and unforced way from the French people themselves."

He added: "It is awfully easy to be for de Gaulle . . . but I have a moral duty that transcends 'an easy way.' It is to see to it that the people of France have nothing foisted on them by outside powers. It must be a French choice—and that means, as far as possible, forty million people. Self-determination is not a word of expediency. It carries with it a very deep principle in human affairs."[35]

In human affairs, perhaps. But a deep principle in FDR's affairs was to shut de Gaulle out of power. Eisenhower would keep his dealings with the general and his French committee to a bare minimum.

But even a bare minimum, he would find, was more than Roosevelt wanted.

FORTY-ONE

SORROWS OF WAR

౸

"I was somewhat intrigued by your question regarding the method of selecting generals, 'What are they like that makes you know they will be good ones?' This probably is the most important of my duties, the most difficult."

MARSHALL PENNED THESE LINES TO MISS LILLIAN CRAIG'S CLASS AT VIRGINIA Heights School in Roanoke, Virginia. In March 1944 Miss Craig's students, diagnosed as "strephosymbolics" because they read words or letters backward, sent a letter to Marshall asking questions about his life, work, and personality. The general brushed aside most of their personal questions without comment, but to this one he offered a thoughtful answer.

"It is comparatively simple to select the generals after a display of their military qualities on the battlefield," he said. "The difficulty is when we must choose them prior to employment in active operations." He told the children that an officer's bearing, his dependability and speech, and the recommendations of his commanding officers were important considerations. The most important trait, he said, was the character of the man, meaning his integrity, selflessness, and "sturdiness of bearing when everything goes wrong and all are critical."[1]

Before the war, Marshall and Stimson had agreed on three principles governing their selection of these sturdy, selfless men. Marshall would select the very best officer for each theater. He would give that commander the widest possible authority, and would hold him accountable for the results. As one staffer summed up Marshall's method: "He'd give you responsibility for something and let you do it. If you didn't do it right, or like he thought it was right, you'd catch hell."[2]

After the Casablanca conference, Marshall told Eisenhower, "Retention under your command of any American officer means to me that you are satisfied with his performance. Any man you deem unsatisfactory you must re-assign or send home." He warned Eisenhower, "This principle will apply to the letter, because I have no intention of ever giving you an alibi for failure on the excuse that I forced unsatisfactory subordinates on you."[3]

In August 1943, when it looked as if Marshall would lead the invasion of France, Marshall asked Eisenhower to release Lieutenant General Omar Nelson Bradley, commander of the U.S. II Corps in Sicily, for assignment to OVERLORD. Bradley had won high marks from Eisenhower and Patton, and Marshall wanted a battle-proven commander leading the American spearhead, the First U.S. Army. Eisenhower felt the same way about his West Point classmate, and when he took the reins of OVERLORD, he appointed Bradley as First Army commander for D-Day, then as an army group commander when a second force, the Third U.S. Army, arrived from England a month later.[4]

Eisenhower next asked Marshall to banish Lieutenant General Jacob L. Devers, commander of American forces in the United Kingdom. Devers, a onetime classmate of Patton, was the Army's senior man in England until Eisenhower's arrival. Ike had never worked with him and did not trust him. The previous year Devers had earned Eisenhower's enmity by denying him four bomber groups Ike wanted for the Salerno landing, on the grounds that the bombers were needed for the strategic bombing campaign against Germany.

Eisenhower rarely held grudges when he won an argument, but Devers made the fatal mistake of winning the bomber group fight. Marshall left the decision to Eisenhower, so Devers went to the Mediterranean to take the reins of ANVIL.[5]

One by one, the other pieces of Eisenhower's command team dropped into place: Major General Walter Bedell "Beetle" Smith would remain Ike's chief of staff. Lieutenant General Courtney Hicks Hodges, a self-effacing Georgian who served under Marshall at Fort Benning, would take over First U.S. Army when Bradley moved up to army group command. Major General J. Lawton "Lightning Joe" Collins, a young, handsome Pacific commander, would lead one of Bradley's two corps, while Major General Leonard T. Gerow, Eisenhower's Leavenworth classmate, would command the other. About a month after D-Day, the Third Army under George Patton, one of Eisenhower's oldest friends, would come ashore on Bradley's right. As Eisenhower saw it, First Army would be Bradley's workhorse, punching a hole in the German line and pinning down the enemy, while Patton's Third Army would be the racehorse, slashing deep on the right and driving toward Paris and the Rhine.[6]

Marshall's liberality did not mean Eisenhower would be free to strip the Mediterranean of everyone, however. Always on the lookout for "localitis," a communicable disease causing theater commanders to break out in a rash of cables demanding ships, men, and air support, Marshall told Eisenhower and Beetle to be judicious in their requests. When the exiled Jake Devers asked Marshall if he could keep both Patton and Major General Lucian K. Truscott in the Mediterranean theater, Marshall gave Patton to Eisenhower but told him Devers would keep the able Truscott.[7]

*   *   *

As a Washington general, Marshall had politics in the back of his mind every day. So when he opened his newspaper on the morning of April 26 and read the latest story on General Patton, he saw a political sledgehammer swinging toward the Army's head.

Patton, the *New York Times* reported, had made some ill-conceived remarks to a Welcome Club at Knutsford, England. After tossing out a few extemporaneous comments that he considered charming, Patton, according to a British reporter, added, "It seems to be the evident destiny of the British and Americans to rule the world."[8]

Patton was no stranger to bad press. He had caused a scandal the previous summer when he slapped two enlisted men suffering from battle fatigue in field hospitals. Editors and politicians were now quick to pounce on his Knutsford gaffe; he had excluded the Russians from his new world order, and they expressed outrage that a general would suggest a world order of any sort, a matter for politicians, not generals. In a scathing editorial, the *Washington Post* opined, "It is more than fortunate that these [flaws] have become apparent before the Senate takes action to pass upon his recommended promotion in permanent rank from Colonel to Major General. All thought of such promotion should now be abandoned. That the War Department recommended it is one more evidence of the tendency on the part of members of the military to act as a clique or club."[9]

Marshall was beside himself. He had known Patton for many years, and had supported him when the press crucified him for slapping those enlisted men.* This time his offense was more serious. Republicans were using the story as an example of the administration's inability to control its own officers, and Marshall

* Like Eisenhower and Marshall, Roosevelt refused to sack Patton over the "slapping incidents." When asked about Patton on his return from Tehran, FDR worked into his answer an apocryphal story of Lincoln and Grant: "If you want to write a piece, stick in there the story of a former president who had a good deal of trouble finding a successful commander for the armies of the United States. And one of them turned up one day and he was very successful. And some very good citizens went to the president and protested: 'You can't keep him. He drinks.' 'It must be a good brand of liquor,' was the answer."

feared what the Senate would do with his list of permanent promotions. "We were about to get confirmation of the permanent makes," he told Eisenhower the day the story broke. "This I fear has killed them all."[10]

Marshall ruminated over what to do. While Patton was Eisenhower's man and a longtime friend of Secretary Stimson, Marshall held the ultimate responsibility for Army assignments. If a general detonated a bomb at home, it was Marshall's job to remove him.

But it was also Marshall's job to win the war as quickly and cheaply as possible. Where the needs of victory conflicted with the whims of the electorate—in the case of the "Germany First" policy, for instance—it was the chief of staff's duty to take the heat.

Reflecting on what was riding on OVERLORD, he sent another cable to Eisenhower.

"You carry the burden of responsibility as to the success of OVERLORD," he told Eisenhower. "If you feel that the operation can be carried out with the same assurance of success with Hodges in command, for example, instead of Patton, all well and good. If you doubt it, then between us we can bear the burden of the present unfortunate reaction."[11]

On April 29 Eisenhower wired Marshall about their mutual friend. "Frankly I am exceedingly weary of his habit of getting everybody into hot water through the immature character of his public actions and statements," he wrote. While Patton had apparently been misquoted by the reporter, Eisenhower said he had "grown so weary of the trouble he constantly causes you and the War Department, to say nothing of myself, that I am seriously contemplating the most drastic action."[12]

Both Marshall and Eisenhower wanted to keep Patton, for he possessed fighting qualities uncommon among army commanders. He believed that a fast, violent assault produced fewer casualties than a slow, methodical advance—a principle other generals paid lip service to but tended to shy away from when shells exploded and men died. Patton may have been theatrical and vulgar, but he produced victories. That was what generals were paid to do.

In a follow-up letter the next day, a deeply conflicted Eisenhower told Marshall, "On all of the evidence available I will relieve him from command and send him home unless some new and unforeseen information should be developed in the case." He did not specify what "new and unforeseen information" might develop, but as he talked himself through the problem, Eisenhower grew reluctant to send his old friend to the scaffold. He contrasted Patton's proven ability to conduct "a ruthless" drive with the cautious approach of General Hodges and concluded, "There is always the possibility that this war, possibly even this theater, might develop a situation where this admittedly unbalanced but nevertheless aggressive fighting man should be rushed into the breach."[13]

Marshall believed Eisenhower's military judgment, not his political view, should rule the day. He cautioned Eisenhower against overreacting to a few caustic editorials. "Do not consider War Department position in the matter," he counseled. "Consider only OVERLORD and your own heavy burden of responsibility for its success. Everything else is of minor importance."

A relieved Eisenhower replied, "Because your telegram leaves the decision entirely in my hands, to be decided solely upon my convictions as to the effect upon OVERLORD, I have decided to retain him in command." They would not put down the old warhorse. Not yet.[14]

.   .   .

In war, death stalks the land with many weapons. Its favorite tool is usually famine or disease, though sometimes, to keep up with the fashion of the times, the scythe takes the form of a Mauser rifle, a B-17 bomber, or a Long Lance torpedo.

Sometimes that scythe takes the shape of fatigue, and when it takes that form, death does not swing, but jabs and nicks as if it has all the time in the world: Hap Arnold's two heart attacks since Pearl Harbor. Hopkins wheeling in and out of Bethesda Naval, Walter Reed, and the Mayo Clinic. Cordell Hull and John Dill, declining under the stress of war. Death could even find a man sitting quietly in a chair, thousands of miles from the nearest front line.[15]

In late April, Navy Undersecretary James Forrestal rang up Henry Stimson to tell him that Frank Knox had suffered a heart attack while attending the funeral of a former business partner in New Hampshire. Not appreciating the seriousness of his condition, the secretary flew home, and his heart began shutting down and he soon lost consciousness. The next day Stimson asked Mabel to go see Knox's wife, Annie, and offer her the family's friendship and support.

That support comforted Annie Knox, but her husband was beyond help. A little after one o'clock on the twenty-eighth of April, Jim Forrestal called Stimson and told him their friend and colleague Frank Knox had died.[16]

Knox had been Stimson's comrade in politics and war. The two old-line Republicans had answered their nation's call at the same moment, and each had sacrificed his standing with his party for the good of the country. Stimson would miss the smiling, combative Frank, and his condolence visit to Annie Knox left him badly shaken. *"She was a pathetic sight and my heart ached for her,"* he wrote that night. *"The whole thing was a very hard blow to me too because Frank had been a close friend and we had been so similarly situated we had seemed like side partners here in Washington."*

Stimson and Mabel attended Knox's funeral at the Mount Pleasant Congregational Church on Columbia Road. FDR was recovering from another bout of what he called the "flu" in South Carolina, and Eleanor attended the service in her husband's place. Supreme Court justices, diplomats, and Navy men, including Admiral King, came to pay their respects as Knox's flag-draped casket, laid atop an Army caisson pulled by seven white horses, bore the Rough Rider to his final resting place.[17]

"Well done, Frank Knox!" read the press release from Admiral King's office. "The nation has lost a great patriot, the Navy a great leader. We dedicate ourselves, one and all, to what would have surely been his last order—'Carry on!'"[18]

The "Well done" part lacked perfect sincerity, and King would have been more comfortable with Knox's death if he'd had a hand in selecting his successor, for King believed the obvious choice, James Forrestal, was a micromanager who committed the sin of butting his nose into Navy business. He had previously tried to strip King of his CNO position, and in early 1942 he even had the impertinence to ask King for a summary of the Navy's war plans. "I think Forrestal was undercutting Knox even before his death," King later remarked. "Forrestal was as tricky as he could be," and King hoped to steer the post to someone less "tricky."[19]

But Forrestal was an old Roosevelt neighbor whose father had been active in Democratic politics in the Poughkeepsie area. Moreover, Chairman Vinson wanted Forrestal to have the job, and Vinson's wish was as good as law. Without asking King's advice, Roosevelt appointed Forrestal as the new naval secretary.[20]

A disappointed Ernie King gave Forrestal what he considered to be an education about his new department. He told an associate, "After due reflection, I went yesterday afternoon to the new incumbent and laid the situation on the line, as I saw it." He added, "It was an unpalatable job for me but I am bound to say that it was, all things considered, received amiably enough."[21]

While Forrestal at first deferred to his senior admiral, King knew the hard-driving New Yorker would be a pain in his stern. Forrestal had a bad habit of overintellectualizing problems when direct solutions were called for, and he clearly disliked King. "Although he seemed to needle me in many ways, I tried to tell him the exact truth as I saw it," King told his biographer years later. "He would make dirty cracks at me."[22]

. . .

It was an unwritten rule that no high-ranking official could shield his son from danger once he joined the military. The repercussions to the boy's career, and to family relations, would be long-felt and severe.

FDR's sons served in combat positions. James, a Marine Raider, won a Silver Star for gallantry in action on Makin, and he weathered the storm of bloody Tarawa. Elliott rose through the ranks of the Army Air Forces reconnaissance group in the Mediterranean and Southwest Pacific. John served aboard the carrier *Hornet* as a Navy supply officer, and Frank Jr. earned his Purple Heart rescuing a fellow sailor when his destroyer came under enemy fire.

Other war leaders also had loved ones in harm's way. Marshall's stepson Lieutenant Allen Brown was a tanker in Italy with the First Armored Division, and another stepson in the Mediterranean worked an AA gun. Eisenhower refused Mamie's plea to keep their son John out of danger once he graduated from West Point. King's son, studying at Annapolis, would become a naval officer upon graduation, and the sons of five other generals came under fire in the Marshall Islands.

With time, the lottery of loss would catch up to the high command. The world, as Hemingway wrote, "kills the very good and the very gentle and the very brave impartially." George Patton's son-in-law, a colonel in the First Armored Division, had been captured at Kasserine Pass and languished in a prison camp somewhere in Eastern Europe. Harry Hopkins's son Stephen, a marine who turned down a college deferment, was killed at Kwajalein. General Leslie McNair and his son would die on opposite sides of the globe within two weeks of each other—Marshall flew down to break the news to Mrs. McNair personally—and Ambassador Joseph Kennedy would lose his eldest son when his experimental bomber exploded. *"The sorrows of the war are coming in on us on every side, and more and more of our friends are being stricken by its blows,"* lamented Stimson.[23]

On May 29, as the Allies closed in around Rome, George Marshall came home to find his wife, Katherine, ebullient over letters from her two sons. Gaily chattering, she showed them to her husband. He was fond of both stepsons, and Allen, the youngest, had been like a natural son to him. He smiled at Allen's reference to a horseshoe he once nailed to a barn door at their Leesburg home. "The horseshoe has held my luck," Allen wrote. "I shall take it down this Christmas and keep it for the rest of my life."[24]

That same day, a little south of Rome's Alban Hills, Allen stood up in the turret of his tank, field glasses pressed to his eyes. In the distance, a German sniper caught sight of the glass and dialed up his scope. The trigger moved, the rifle cracked, and Lieutenant Allen Brown tumbled into his tank. He was buried in a roadside grave.[25]

The next day Marshall's driver, WAC Sergeant Marjorie Payne, drove Marshall to the Pentagon and took her usual place in the antechamber near the general's office. Moments later, the door opened and General Marshall stepped out. His secretary, Cora Thomas, took one look at him and knew something was very wrong. "You had better run along," she told Payne. "Something is up."[26]

Katherine Marshall had been up since before her husband had left, and didn't expect to see him until around lunchtime. She was startled to see the front door open in the middle of the morning as her husband stepped inside.

As she walked toward him, the expression on his face told her something terrible had happened.

Marshall quietly closed the door behind him.[27]

FORTY-TWO

"DR. WIN-THE-WAR"

∽

W HEN HE RETURNED FROM TEHRAN, ROOSEVELT KNEW THE NEXT TWELVE
months would be a year of American fatigue. Fatigue over losses, fatigue
with his long administration, fatigue even with the qualified blessings of a wartime
economy in full swing. Republicans would attack him for overstaying his welcome
at 1600 Pennsylvania, and liberals would attack him for rolling back the New
Deal in the rush to mobilize for war.

Speaking to reporters, FDR explained the role of the New Deal in the
world war:

How did the New Deal come into existence? It was because there was an
awfully sick patient called the United States of America, and it was suffering from a grave internal disorder—awfully sick—all kinds of things
had happened to this patient, all internal things. And they sent for the
doctor. And it was a long, long process—took several years before those
ills, in that particular illness of ten years ago, were remedied. But after a
while they were remedied. . . .

Two years ago, on the seventh of December, he was in a pretty bad
smashup—broke his hip, broke his leg in two or three places, broke a
wrist and an arm, and some ribs; and they didn't think he would live, for
a while.

And then he began to "come to"; and he has been in charge of a
partner of the old doctor. Old Dr. New Deal didn't know "nothing"
about legs and arms. He knew a great deal about internal medicine, but

nothing about surgery. So he got his partner, who was an orthopedic surgeon, Dr. Win-the-War, to take care of this fellow who had been in this bad accident.

And the result is that the patient is back on his feet. He has given up his crutches. He isn't wholly well yet, and he won't be until he wins the war.

When FDR finished, a reporter asked a follow-up question: "Does all that add up to a fourth term declaration?"

A round of polite laughter went up from the newsmen and Roosevelt. As any politician would, FDR dismissed the question, saying, "Oh now, we are not talking about things like that now."[1]

Dr. Win-the-War spent most of his time peering into the dim future. From his Oval Office desk, a president could pull many levers of power, but most of those levers wouldn't change events for months, sometimes years. He could replace the man in charge of ship production, or approve a new bomber program, but those ships and bombers would not appear in equipment tables for many months—and it would be many more before they could be taken into battle by trained men. Like a billiard player who calculates his second and third shots when leveling his cue, FDR's sight had to remain fixed on front lines that were one to three years distant.[2]

The war's effect on the nation would last much longer. The butcher's bill would not be tabulated for many years, but everyone knew the final tally would be a big one. By February 1944, neatly tabulated columns in War Department files totaled 112,030 killed, wounded, or missing, while the Navy Department's books put its losses at 38,448. While small compared to other belligerents, these losses had no precedent in U.S. history save the carnage of the Civil War. And the heavy ground fighting to conquer "Fortress Europe" had not even begun.

For FDR, killed, wounded, and missing were only part of the picture. Men and women returning to civilian life physically unscathed—the vast majority of the eleven million who would don a uniform—would stow their duffel bags in attics and begin looking for work, an education, and a way to put their lives back together.[3]

Returning veterans have plagued democracies since the days of ancient Greece. When the crisis that binds the nation has passed, democracies tend to move on quickly. Hoplites and musketeers who defended their lands were sometimes honored, but rarely given material assistance. Poet Rudyard Kipling lamented Victorian disdain for the Empire's lower-class redcoats:

For it's Tommy this, and Tommy that, an' "chuck him out, the brute!"
But it's "Saviour of 'is country" when the guns begin to shoot.

America's attitudes toward its veterans—"bums," if they were poor—had not been, on the whole, much different. Revolutionary War claims had vexed Congress, Civil War pension laws were passed decades after the war ended, and the plight of the Depression-era Bonus Army marchers had helped bring down Herbert Hoover.

Eleanor Roosevelt had seen the American fighting man, thousands of them. She saw them in field hospitals and mess halls and rest camps at Guadalcanal and Christmas Island, San Diego and Australia. She had comforted them, heard their gripes, and won their admiration. And they won hers.

In her syndicated newspaper column "My Day," Eleanor quoted from a letter she received from a young officer in Europe: "There is one great fear in the heart of every serviceman, and it is not that he will be killed or maimed but that when he is finally allowed to go home and piece together what he can of life, he will be made to feel he has been a sucker for the sacrifice he has made." She asked her readers, "Will we see that they have a better job, a better chance when they come home, for health, education, working conditions, professional standards and above all, for a peaceful world in the future?"[4]

In his New Deal heyday, FDR had championed social security for the poor and work programs for the unemployed. He agreed with Eleanor that the nation had an obligation to care for citizens who answered the call to arms, regardless of whether they had lost limbs or earned medals. In a land where fewer than 5 percent of college-age Americans attended college, he understood that returning veterans would become either vibrant, productive contributors to American society or despondent, bitter men filling the ranks of the next Bonus Army.

In the fall of 1943, Roosevelt asked Congress to draft a bill providing vocational and educational benefits for returning veterans. He proposed government funding of college education for able veterans, and special rehabilitation benefits for veterans with service-connected disabilities.

"I believe this nation is morally obligated to provide this training and education," he told Congress on October 27. "While the successful conclusion of this great war is by no means in sight, yet it may well be said that the time to prepare for peace is at the height of the war." He returned to this theme in a Christmas Eve fireside chat: "We here in the United States had better be sure that when our soldiers and sailors do come home they will find an America in which they are given full opportunities for education, and rehabilitation, social security, and employment and business enterprise under the free American system."[5]

Over reflexive opposition to another New Deal program, the Servicemen's Readjustment Act, as the bill was called, gained legislative support. In time, it was broadened to include discharge pay, unemployment insurance for veterans who could not find work, medical and rehabilitation benefits, and credits for time served toward social security. The bill that took shape over the summer of 1944 provided soldiers and sailors with sweeping benefits, including home and business loan guarantees of $2,000 per veteran, a year's worth of relocation insurance, and $500 per year for college tuition. Dr. Win-the-War had launched the last, great New Deal program.[6]

The "G.I. Bill of Rights," as veterans groups dubbed the law, changed the way America lived, worked, and thought. Within seven years of the war's end, eight million Americans would claim educational benefits under the law. More than 2.3 million veterans attended colleges or universities on the G.I. Bill and accounted for nearly half of all college students. This newly educated class established a tradition of giving their children a college education. The percentage of Americans with college degrees rose from 4.6 percent in 1945 to 25 percent by the end of the original program, and the resulting income tax revenues repaid Congress many times over its $14.5 billion investment.

America's real estate engine would also rev into high gear, as the G.I. Bill provided low-interest, government-backed housing loans to returning veterans. Those veterans bought 20 percent of all new homes built in the war's wake, boosting the U.S. economy and making home ownership widely available to ordinary citizens. Better housing, education, and job prospects fueled a baby boom, which begat a generation that grew up believing a college education and home ownership lay within reach. As *Time* declared in retrospect, the G.I. Bill "effectively created the American middle class."[7]

•   •   •

Reshaping the way nations resolved their differences was a much more complex matter. Through two convulsive wars, it had become painfully obvious to Roosevelt that ad hoc alliances and antiaggressor coalitions were too weak, and formed too late, to keep dictatorships like Nazi Germany or Imperial Japan from shattering the world order. Tyrannies tended to get the jump on democracies.

The League of Nations, Woodrow Wilson's dream, had been a profound disappointment. Yet FDR, a Wilsonian at heart, felt a permanent world peace organization could impart a stabilizing force on an out-of-kilter world. In the Atlantic Charter, he included an oblique reference to a peacekeeping mechanism, and since that inspired week aboard *Prince of Wales*, he had never stopped thinking about how such an organization might be established by the thirty-four countries fighting against the Axis—belligerents he referred to as the "United Nations."[8]

The Cairo and Tehran meetings brought home to FDR a mathematical reality: The United States, the Soviet Union, China, and the British Commonwealth accounted for nearly three-quarters of the world's population. The arithmetic impelled one conclusion: "As long as these four nations with great military power stick together, there will be no possibility of an aggressor nation arising to start another world war."[9]

A bestselling book written by Wendell Willkie, titled *One World*, blew fresh wind into FDR's sails. Based on Willkie's travels as Roosevelt's roving emissary, *One World* predicted the global hegemony of the United States and the Soviet Union. He argued that, whether Americans admitted it or not, the United States would be the postwar era's democratic leader. As Willkie's book became a part of Washington conversation, Roosevelt decided the time was ripe to put the State Department to work on an international peacekeeping institution.[10]

During a meeting of foreign ministers from China, Britain, the United States, and the Soviet Union in the fall of 1943, Secretary Hull extracted a pledge of cooperation in forming a global organization. In November, a beaming Roosevelt presided over an East Room ceremony where representatives of forty-four countries pledged to support a United Nations Relief and Rehabilitation Administration charged with feeding, clothing, and housing the world's poor.

It was a first step on a longer path. Eleanor Roosevelt, who was present at the ceremony, wrote, "I watched each man go up to represent his country and thought how interesting it was that, before the end of the war, we have the vision this time to realize that there is much work to do and preparation by the peoples of the United Nations is necessary."[11]

In the summer of 1944, Roosevelt turned his rough sketch over to the diplomats, whose brushes and palette knives would add color, lines, and shading to the picture. In the fall, representatives of the great powers would meet at Harvard's Dumbarton Oaks mansion in Georgetown to plan an international organization, a security council with permanent and rotating members, and an initial roster of its general assembly. A United Nations peacekeeping institution, the holy grail of the Wilsonian idealist, was taking shape.

● ● ●

The United Nations held out the promise of a more secure future. But for the moment, it was nothing more than a promise, and one community in need of immediate help would continue to suffer until the Allies could find a way to end the war. In late 1942, Rabbi Stephen Wise, head of the American Jewish Congress and an old friend of Roosevelt, approached the president with a detailed dossier of Nazi atrocities committed against European Jews. His report described mass murders on a horrific scale—"the most overwhelming disaster of Jewish history," said Wise.[12]

FDR invited Wise and four other Jewish leaders to the White House on December 8 and asked them to draft a statement for his signature denouncing the slaughter. He also persuaded Churchill and Stalin to approve the Declaration on Jewish Massacres, which condemned "in the strongest possible terms this bestial policy of cold-blooded extermination," and indicted Hitler's "often-repeated intention to exterminate the Jewish people in Europe."[13]

Roosevelt was pragmatic on moral questions in wartime. When Hitler threatened to execute Allied bomber pilots in 1943, for instance, he said nothing publicly but quietly told Marshall to be ready to retaliate. He chose to believe, in spite of the evidence, that the Katyn Forest massacre was a Nazi crime, and although he promised that America would not be the first nation to use poison gas on the battlefield, he warned that it would be the second.[14]

Roosevelt saw he could do little beyond offer sympathetic words of support to Europe's Jews. When he asked Rabbi Wise's delegation about concrete actions the government might take, the group had no practical suggestions. FDR concluded that reprisals against German war criminals would have to await the war's end.

In January 1944, Morgenthau, who had been quietly working to fund rescue efforts for European Jews, handed Roosevelt a memorandum prepared by his staff. Morgenthau toned down the original title, "Report to the Secretary on the Acquiescence of This Government in the Murder of the Jews," to the more prosaic "A Personal Report to the President." The report concluded that Hull's State Department had become a serious impediment to Treasury's efforts to rescue Jews caught in the jaws of the Nazi killing machine. It quoted an October 1943 speech by Senator William Langer, a Republican from North Dakota: "We should remember the Jewish slaughterhouse of Europe and ask what is being done—and I emphasize the word 'done'—to get some of these suffering human beings out of the slaughter while yet alive."*[15]

Morgenthau urged Roosevelt to cut out the State Department from Jewish rescue efforts. FDR agreed, though he wanted State to have some voice in the problem. He signed an executive order establishing a War Refugee Board, which

---

* There were, of course, millions of victims in other groups slaughtered during the Holocaust in Europe—Poles, Slavs, communists, homosexuals, Catholics, Gypsies, German dissidents, and the insane, to name a few—and perhaps a comparable number of Chinese, Filipinos, Koreans, Australians, Indonesians, Vietnamese, and Burmese murdered in the Holocaust in Asia.

would act under the direction of Morgenthau, Hull, and Stimson. In classic Roosevelt style, the board was a mix of men whose opposing philosophies would keep the group from doing anything too precipitous without his personal approval.[16]

Hitler's occupation of Hungary in March 1944, and his demand that the Hungarian government deport or liquidate the country's 700,000 Jews, underscored the urgency of the crisis. The War Refugee Board funneled bribe funds to diplomats like Raoul Wallenberg, who saved thousands of Hungarian Jews, and from the Oval Office FDR denounced "the wholesale systematic murder of the Jews in Europe [that] goes on unabated every hour." Vowing to punish both Nazi and Japanese executioners, Roosevelt called upon the Allies "temporarily to open their frontiers to all victims of oppression."[17]

But until those executioners were caught, until those frontiers could be pried open, Roosevelt's words remained unfulfilled platitudes. Military options, such as bombing death camps in the Auschwitz-Birkenau complex, were given low priority by the Allied command. The War Department's Jack McCloy turned down several requests for bomb strikes against death camps on grounds that strategic air missions were better directed toward defeating Hitler—to stop the slaughter everywhere—than at a single camp or rail line. As long as refineries were brewing gasoline and factories were assembling Messerschmitts, neither the War Department nor the Army Air Forces had any interest in diverting bomb groups to concentration camps.[18]

FDR wished to alleviate some of the suffering he knew was being felt across Eastern Europe, but he would not alter military plans, not even to give the doomed a brief respite. Nor, does it appear, was he ever directly asked to do so.* Unless his captains told him that destroying death camps would hasten the war's end, he would not edit Hap Arnold's target lists. He might meddle in military strategy, but he would not elevate humanitarian operations over wartime necessity.[19]

◦   ◦   ◦

As Roosevelt wrestled with the G.I. Bill, a United Nations organization, and the plight of Europe's Jews, another watershed was slowly breaking in arid New Mexico.

Since early 1942, when the Army took over the "S-1" project, the Manhattan

---

* Jack McCloy contradicted himself on whether he laid the proposal to bomb German death camps before Roosevelt. In his early postwar years, he denied having done so. Then, at age ninety-three, he told historian Michael Bechloss that he took the proposal to Roosevelt, who nixed the idea as futile and objected that the administration would be accused of slaughtering innocent prisoners.

Engineering District and its sinuous web of affiliates had grown to 42,000 civilian workers, 84,500 construction contractors, and 1,800 military personnel working on half a million acres of land. The $85 million covertly funneled to the project in early 1942 turned into a $400 million estimate by the following December. By the time General Somervell secretly pulled $300 million from engineering funds in May 1943, he was told an additional $400 million would be needed just to get the project through 1944. By early 1944, the Manhattan Project had become the second most expensive weapon program in the Arsenal of Democracy, exceeded only by the snafu-plagued B-29 bomber.[20]

Though the number of people who knew small or isolated details of the project had grown to the size of a two-corps army, Manhattan remained one of the war's best-kept secrets. Within the administration, only a tiny, carefully guarded circle of senior officials, scientists, and administrators were privy to the full story. Few others in the American high command knew the big picture; Marshall did not even tell Admiral King about it until late 1943.[21]

As head of the project's "Policy Committee," Henry Stimson struggled to keep nuggets of technical information from spilling out, and first and foremost, that meant keeping Congress in the dark. To legislators whose investigators stumbled upon mysterious offices and fence-shrouded factories, Stimson would simply say the matter was too secret to discuss, he was acting at the direction of the president, and they should not ask any more questions.[22]

But democracy is a noisy, unruly affair, and by 1944 the project's cost had grown so prodigious that money could no longer be quietly pulled from harbor construction and land mine appropriations. Over an uninspired White House lunch on February 15 (Mrs. Nesbitt, despite Roosevelt's best efforts, was still on the job), Stimson and Roosevelt agreed that Congress would have to earmark specific funds for the project, though it could not know why it was shelling out all those greenbacks. Roosevelt suggested that Stimson and Marshall take Speaker Sam Rayburn and ranking members of the House into their confidence. The Senate, he cautioned, would be a tougher nut to crack, so he advised waiting until momentum had built in the lower house before walking the request to the other side of the Rotunda.[23]

Three days later, Stimson, Marshall, and Dr. Vannevar Bush, head of the Office of Scientific Research and Development, entered the plush office of Speaker Rayburn, who had invited House Majority Leader John McCormack and Minority Leader Joseph Martin Jr., the ranking Republican, to join them. In Rayburn's office, Stimson informed the three congressional leaders that the Army had been working on a uranium-powered explosive using funds appropriated for general military research.

Marshall and Dr. Bush explained the military and scientific fundamentals of the new weapon. Emphasizing that the project was so important, and so secret, that Congress would have to accept certain aspects of it on faith, Marshall told the congressmen the atomic weapon required massive congressional appropriations without the usual public disclosure.

Rayburn understood the delicacy and importance of the Manhattan Project. He liked to say, "Any jackass can kick down a barn. It takes a carpenter to build one." Rayburn would not let any jackass kick down Stimson's barn, and he promised to get the money from Appropriations and see to it that nobody talked.[24]

Sam Rayburn ran the House the way Admiral King once ran *Lady Lex*, and no one breathed a word. The Senate was another matter, however. When staffers began hearing rumors of mysterious military outposts in Oak Ridge in Tennessee and Hanford in Washington State, several upper-chamber members demanded to know what was going on.[25]

The most prominent of them was Senator Harry Truman of Missouri, the same gadfly who had backed off his investigation of undisclosed war plants the year before. He had a mandate to clamp down on War Department overspending, and this time he asked the War Department to allow his staffers to investigate rumors of financial waste at a Washington State defense plant.[26]

For a second time, Stimson asked Truman to call off his investigation. Now Truman objected. He had two uniformed staffers, both brigadier generals, on his staff. Surely, he told Stimson, the Army could trust its own one-star generals with military information. At the very least, they should be allowed to look into the cost of housing and roads leading to the plant.

Stimson held his ground, and the Missouri senator shot back a burst of plain Midwestern speaking. Stimson grumbled to his diary that night, *"He threatened me with dire consequences. Truman is a nuisance and pretty untrustworthy man. He talks smoothly but acts meanly."*[27]

Eventually Harry Truman backed down. While the project's cost was enough to unnerve everyone who knew about it, Truman had second thoughts, given Stimson's strenuous objections, and he declined to push the matter further.

With a sigh of relief, Stimson went back to shepherding the Manhattan Project, grateful that he would never have to worry about Harry Truman sticking his nose into the atomic bomb project again.

⋆　　⋆　　⋆

In December 1943, George Marshall had returned from Cairo, Tehran, and the Pacific to an overflowing assembly line of paperwork. But before he could plunge into decisions requiring his signature, FDR dropped another domestic bombshell on him and Stimson. The four largest railway brotherhoods, unions representing 1.4 million railway workers, were about to go on strike.[28]

Roosevelt's heart was usually with organized labor. Back in '37, when strikers waged sit-ins at steel and auto plants, he refused to send federal troops to reopen the factory gates. His solution back then was to call Bill Knudsen of General Motors, ask him to sit down with the strikers, and let common sense and the Wagner Act solve the problem.[29]

But that was in time of peace, when the consequences of a strike were limited to bad press, short-term unrest, and a disruption in the supply of consumer goods. In wartime, a railroad strike could cripple America's ability to defend itself. The nation had more than 233,000 miles of track crisscrossing the continent, and reinforcements, ammunition, fuel, food, and weapons ran along those steel arteries. A strike could leave planes stranded for lack of engines, cities cut off from the farms that fed them, and troop-marshaling points with no troops to marshal.

FDR told Stimson to be ready to take over the railroads if no agreement were reached by December 30. He added, to Stimson's satisfaction, that he would ask Congress to enact a general service law requiring undrafted men to serve in war industries.[30]

Like Roosevelt, Henry Stimson saw the home front as a continuation of the battlefront. Total war wasn't total if men lucky enough not to be drafted were free to become insurance salesmen or musicians—a reality other nations at war had accepted long ago. When Mussolini invaded backward Ethiopia in 1935, Emperor Haile Selassie issued one simple order to his villagers: "Everyone will now be mobilized, and all boys old enough to carry a spear will be sent to Addis Ababa. Married men will take their wives to carry food and cook. Women with small babies need not go. The blind, those who cannot walk or for any reason cannot carry a spear, are exempt. Anyone found at home after receipt of this order will be hanged."[31]

While he might not go as far as the Lion of Judah, Stimson felt that every able-bodied American citizen had an obligation to support the war effort, either on the front lines or behind them. A "National Service Act" would require personal service in the form of industrial labor—similar to a military draft—and for business owners, a formal requisition of factories and shipyards for the war's duration. The threatened railway strike provided Stimson with a platform for a loud, deep blow on this bugle.[32]

On December 27, FDR signed an executive order seizing the railroads and placing them under the secretary of war. Between the threats of a national service bill and railroad nationalization, the brotherhoods backed down for the moment. On December 29, the holdout unions canceled their strike.[33]

352 ★ AMERICAN WARLORDS

The unions had entered into a truce, but a truce isn't the same as a peace treaty. On the last day of 1943, Stimson heard fresh rumors that the brotherhoods were planning another strike. When he told Marshall what the unions were up to, the general exploded.

The war would go on another six months if the unions struck, Marshall raged, his words getting hotter and louder by the syllable. Any hope that Germany would collapse by spring would be "gone with the wind" if Nazi propaganda painted U.S. strikes as evidence of a near-collapse. Cheeks growing red, Marshall said he would not let civilian labor leaders condemn men to death and disfigurement just so their workers could make an extra twenty-five cents an hour.

Though he hated being dragged into domestic politics, Marshall decided to use his clout with Congress and the public to avert a railroad shutdown. He told former senator James Byrnes he was "sleepless with worry" over the looming strike, and openly wondered "whether it was his duty to go on the radio, give his opinion, and then resign."[34]

Marshall's standing with the public in some ways outsized Roosevelt's. That very week, his face stared out from the cover of *Time*'s "Man of the Year" issue, whose editors emoted, "The American people do not, as a general rule, like or trust the military. But they like and trust George Marshall. This is nothing more paradoxical than the fact that General Marshall hates war. The secret is that American democracy is the stuff Marshall is made of." FDR gave Marshall and King the lion's share of credit for the previous year's victories in his Christmas fireside chat, and Colonel Eddie Rickenbacker, America's great air ace from the previous war, called Marshall the kind of soldier who would make a great president. To this last, the *Memphis Commercial Appeal* remarked, "That's about all that's needed to start one of those 'Marshall for President' movements."*[35]

Marshall had no interest in partisan politics, but he would do his best to kill the railway strike in its cradle. He called in a group of trusted journalists for a "not for attribution" interview and told them that any suggestion that the United States was disunited would hinder Allied efforts to break off Germany's satellites. "He banged his white-knuckled fist on the desk," *Time* reported, "and although not a blasphemous man, he swore bitterly. For he was brimming over with indignation that Americans in their ignorance should do anything so tragic."[36]

Though the interview was off the record, Marshall knew his identity would leak, exactly as he intended. The *St. Petersburg Times* identified the "highly

---

* Ernie King told his confidants that Marshall rarely swore, but when his name was mentioned as a possible presidential candidate, "He really showed what he could do when he was roused."

responsible source" as the Army's chief of staff, and labor spokesmen struck back at Marshall, angrily challenging him to prove his implied charge that the brotherhoods were aiding the enemy.[37]

While the American Federation of Labor heatedly denied giving aid and comfort to the enemy, the fact that the unions even had to answer the charge from America's most trusted man was enough to break the strike. Even the Communist Party's *Daily Worker*, labor's staunchest ally, did some bobbing and weaving, recognizing that lives lost if capitalist production faltered would include citizens of the USSR, not just citizens of the United States.[38]

The agonizing episode of the railroad strike came to a muddled end two weeks later, when the brotherhoods agreed to accept an increase of six cents an hour. The trains kept chugging, goods moved from factories to ports, and Roosevelt's warlords had weathered another storm.[39]

• • •

To FDR, Douglas MacArthur was an amusing political diversion, a sort of boa constrictor behind glass who looks intimidating but is quite harmless so long as you don't get too complacent. MacArthur lacked the political stature of Governor Dewey or Ohio's governor John Bricker, but he was the hero of Bataan and Corregidor. Headstrong, unquestionably brave, and obsessive about his publicity, MacArthur had butted heads with Roosevelt during the Depression years, when he was the Army's chief of staff.

"He has the most pretentious style of anyone I know," Roosevelt once said. "He talks in a voice that might come from the oracle's cave. He never doubts and he never argues or suggests; he makes pronouncements. What he thinks is final." He told Harold Ickes that MacArthur and Louisiana's governor Huey P. Long were the two most dangerous men in America.[40]

Yet to many Republicans, MacArthur was also the most dangerous man to Roosevelt.

As the 1944 election campaign lurched into first gear in late 1943, pundit typewriters turned to MacArthur's burgeoning support among Republican voters. A Gallup poll in September put MacArthur not far behind Dewey and Wendell Willkie, a respectable showing for an Army officer who had never held elected office and had lived overseas since 1935. In November, a War Department staffer wrote Marshall, "The *Times-Herald* in editorial yesterday demoted General MacArthur to Secretary of War, and proposes that if any other Republican candidate for President carried the promise that MacArthur would be appointed Secretary of War upon his election, Republicans would win."[41]

But MacArthur already had a job. Unlike General George McClellan, Lincoln's old Democratic nemesis, MacArthur was far from the nation's capital. He could hardly fight Roosevelt at home when he was supposed to be fighting Tojo in the Pacific. MacArthur's reputation in middle America hinged on his single-minded focus on victory over Japan. Any hint that he might abandon his command post for Washington—or plot against his commander-in-chief—would unravel his support like a cheap sweater.

To get around the problem of MacArthur running for president, the general's backers tried to engineer a draft movement before the summer's Republican convention in Chicago. Initial results were encouraging. A grassroots "Bricker-MacArthur" campaign gained steam in October; one convention delegate ran as a MacArthur man in New Hampshire, while in Wisconsin, the general's adopted home state, MacArthur finished second to Dewey and ahead of Willkie. "MacArthur for President" clubs sprang up in seven states, and in Illinois he picked up half a million votes. The MacArthur bandwagon seemed ready to go the distance.[42]

But as the skies seemed bright, the air honeysuckle-sweet, the MacArthur bandwagon lurched into a ditch. The previous fall, Representative Albert Miller of Nebraska had written a private letter to MacArthur describing turmoil on the home front. "The New Deal, including President Roosevelt, is scared to death of the movement in the country for you," Miller said. "Unless this New Deal can be stopped this time our American way of life is forever doomed. You owe it to civilization and the children yet unborn to accept the nomination. . . . You will be our next president."

MacArthur, busy with his New Guinea campaign, sent a short, vaguely encouraging reply to Miller that concluded, "I do not anticipate in any way your flattering predictions, but I unreservedly agree with the complete wisdom and statesmanship of your comments."[43]

In reply, Miller sent MacArthur another, more pointed letter. He spoke of "a tremendous revolution in this country" and wrote, "If this system of left-wingers and New Dealism is continued another four years, I am certain that this monarchy which is being established in America will destroy the rights of the common people."

Encouraged to the point of incaution, MacArthur wrote back, "I appreciate very much your scholarly letter. Your description of conditions in the United States is a sobering one indeed and is calculated to arouse the thoughtful consideration of every true patriot." He concluded, "We must not inadvertently slip into the same condition internally as the one which we fight externally."[44]

MacArthur had a long history of leaving behind compromising letters, and when Miller released his exchange to the press in April 1944, the backlash killed

MacArthur's non-candidacy. For twelve years, Americans had sent FDR to the White House, and while most voters had their complaints about the administration, MacArthur had made a serious mistake by taking sides against his commander on domestic political issues. As a furious Senator Vandenberg wrote in his diary, *"Miller, in one inane moment, crucified the whole MacArthur movement and MacArthur with it."*[45]

MacArthur's contemporaries saw him as an able commander, but voters would not confuse a great general with a great president. Governor Dewey pulled ahead of the pack, Willkie formally withdrew from the race, and MacArthur's candidacy began a death spiral from which it could never recover.

Stung fatally at home, on Sunday, May 30, MacArthur told the press, "I have on several occasions announced I was not a candidate for the position. . . . I do not covet it, nor would I accept it." His gold-braided field cap would stay out of the ring.

With that face-saving measure, the bugler blew taps, three volleys were fired, and the "MacArthur for President" movement was planted in the earth. Republicans, looking elsewhere, would send an incumbent New York governor to replace the former New York governor.[46]

FORTY-THREE

HALCYON PLUS FIVE

꙰

A s the first of June came and went, Roosevelt, Marshall, Stimson and King watched the dice arc slowly through the air—tumbling, turning, spinning as they sailed toward the square marked "Normandy." The roll on which thousands of lives had been wagered would fall, for good or evil, in a matter of days.

There was nothing left for them to do. The orders had been issued, the men assigned, and Eisenhower's legions, not their warlords, would decide the fate of Europe. Eleanor Roosevelt recalled that her husband's inner circle, those in the know, seemed "suspended in space, waiting for the invasion, dreading it and yet wishing it could begin successfully."[1]

They had promised Stalin the invasion would go forward in May, but landing-craft shortages forced the Allies a postponement to HALCYON, the code name for June 1. But a favorable combination of moonlight and tides would not coincide until June 5, 6, or 7, so the invasion's D-Day would await the weather's leave.[2]

As May sauntered into June, the Atlantic weather turned sour and violent storms lashed the Normandy coast. After several stomach-churning days at Portsmouth watching thermal bubbles float between Iceland and Ireland, Eisenhower's meteorologists thought they saw a break. On Sunday, June 4, Eisenhower cabled Marshall a terse message: "HALCYON plus 5 finally and definitely confirmed."[3]

The invasion of Europe would begin on Tuesday, June 6.

Roosevelt had dreamed of sitting on English soil when the invasion was launched, but his health made the trip a fatal impossibility. His blood pressure was rising; in

April it had run as high as 234 over 124, and in May his diastolic was holding at a dangerous 120.

Hobbled by his heart, he opted for a weekend at Pa Watson's home in Charlottesville, Virginia, not far from Jefferson's Monticello. He brought with him a small personal retinue, including his daughter Anna and her husband, John Boettiger, and the three of them worked on a draft speech that Roosevelt would read to the nation on D-Day.[4]

As in the hours leading up to TORCH, to his friends Roosevelt exuded a serenity that masked a gut-tightening tension. FDR sat quietly in Watson's living room, going about his business with neither fuss nor flair. It was an imperfect facade. Grace Tully recalled, "The Boss was keeping up a pretense of normal activity, but every movement of his face and hands reflected the tightly contained state of his nerves."[5]

Eleanor believed her husband was better equipped than most men to withstand the strain of waiting, in part because of the many hours he had spent struggling with his crippling illness. "He'd learned from polio that if there was nothing you could do about a situation, then you'd better try to put it out of your mind and go on with your work at hand," she once explained.[6]

But waiting for a doctor's diagnosis was a far cry from waiting to see how many men would die for an idea—and whether their deaths would be in vain. On HALCYON plus five, the single-minded stoicism Eleanor spoke of would be put to the test.

When Rome fell to the Allies on June 4, FDR returned to Washington and delivered a fireside chat extolling the heroism of the Fifth Army's soldiers fighting in Italy. In words that would have appalled Fifth Army's General Clark, he downplayed the significance of Rome and hinted at bigger things to come: "It would be unwise to inflate in our own minds the military importance of the capture of Rome. We shall have to push through a long period of greater effort and fiercer fighting before we get into Germany itself."[7]

As he spoke those words, he knew thousands of young men were far from their homes, packed into large, slow landing ships, rolling over waves toward Normandy's forbidding shore.

He finished his address, and having nothing else to do, Franklin Roosevelt went to bed.

He was in the dawn of consciousness when his mind perceived a soft, familiar voice in the distance. It was Eleanor, speaking gently to him, quietly prodding him to take a telephone call.

As light drifted through his blinking eyelids, Roosevelt's mind registered what she was saying: General Marshall was on the line from the Pentagon. It was news the world had been waiting to hear.

He sat up in his bed and pulled an old, worn sweater over his head and shoulders. Then he picked up the receiver and listened.

Eisenhower's first cable confirmed what Marshall knew it would say: "I have as yet no information concerning the actual landings nor of our progress through beach obstacles. Communiqué will not be issued until we have word that leading ground troops are actually ashore." There was so little information in the message that Marshall's wife had not even awoken her husband to read it.[8]

When the communiqués came, they reported heavy losses and bad weather. The First and Fourth Infantry Divisions had made it ashore, but the "hinge" beach, code-named OMAHA, was defended by a strong German infantry division that happened to be working on beach defenses at the time of the invasion.[9]

FDR met with Sam Rayburn that morning, then with Marshall, King, and Arnold. He was anxious for news, and between meetings, Dr. Howard Bruenn, his cardiologist, took the president's blood pressure: a murderous 210 over 122.*[10]

No one, from the private at the shingle to the supreme commander in Portsmouth, knew if the invasion had succeeded, or what awaited them beyond the high ground overlooking the beaches.

"My fellow Americans," the radio's familiar voice began for the second night in a row, "last night, when I spoke to you about the fall of Rome, I knew at that moment that troops of the United States and our allies were crossing the Channel in another and greater operation. It has come to pass with success thus far. In this poignant hour I ask you to join me in prayer."

Twelve years of selling hope had led to this moment, the last, supreme roll of the dice that would, in time, mean the end of Nazi tyranny. Roosevelt reached out to that hope, and to its brother, faith, and he rolled them into a prayer blending his secular and religious convictions:

> Almighty God, our sons, pride of our nation, this day have set upon a mighty endeavor, a struggle to preserve our Republic, our religion, and our civilization, and to set free a struggling humanity. Lead them straight

---

* On June 8, Dr. McIntire would tell press correspondents that the president was "in better physical condition than the average man of his age—in better shape today than he had been at any time for a year."

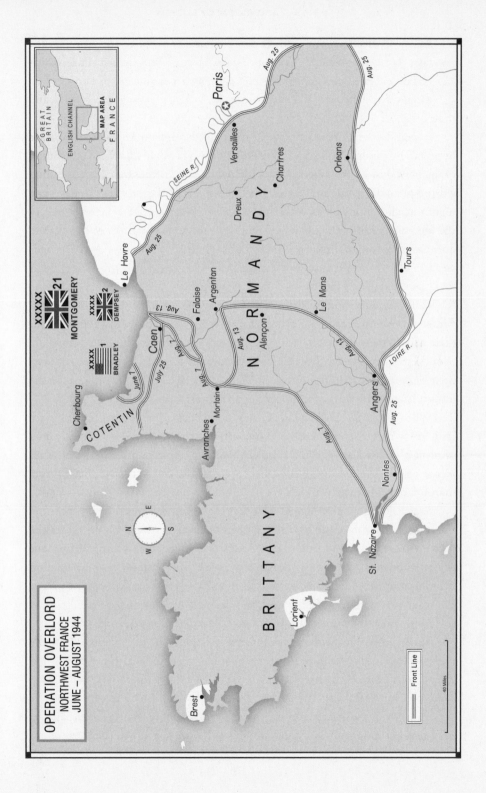

OPERATION OVERLORD
NORTHWEST FRANCE
JUNE – AUGUST 1944

GREAT BRITAIN

ENGLISH CHANNEL

MAP AREA

FRANCE

Paris

Versailles

Chartres

Orleans

Dreux

N O R M A N D Y

Le Havre

SEINE R.

Aug. 25

Aug. 25

Aug. 25

Aug. 25

Argentan

Falaise

Le Mans

Tours

Aug. 13

Caen

Aug. 7

Aug. 13

Aug. 13

Alençon

LOIRE R.

Angers

Aug. 25

July 25

Aug. 7

June 7

Aug. 7

Mortain

Avranches

COTENTIN

Cherbourg

Nantes

Aug. 7

St. Nazaire

BRITTANY

Lorient

Brest

XXXX 21
MONTGOMERY

XXXX 2
DEMPSEY

XXXX 1
BRADLEY

N
W   E
S

Front Line

40 Miles

and true; give strength to their arms, stoutness to their hearts, steadfastness in their faith. . . . Let our hearts be stout, to wait out the long travail, to bear sorrows that may come, to impart courage unto our sons wherever they may be.[11]

It would take more than stout hearts. Allied troops were fighting for their lives, and by the end of the day 2,499 American sons would lose that fight. But as twilight swallowed the Norman sky, the public joined Roosevelt in celebrating, worrying, and praying for the men on French soil. "The president's prayer last night was the nation's prayer," declared the *New York Times*. "We go forth to meet the supreme test of our arms and our souls, the test of the maturity of our faith in ourselves and in mankind."[12]

★ ★ ★

Marshall would learn more about that supreme test when he, King, and Arnold flew to Europe to see their handiwork. On June 8, the trio boarded Skymasters for their usual landing site, the Air Transportation Command hub in Prestwick, Scotland. When a heavy Scottish fog set in around Prestwick, the ATC diverted their plane to a base in northwestern Wales. Wiring ahead to London, the warlords hitched a ride to the capital aboard an express train named the *Irish Mail*.[13]

When *Irish Mail* clacked into Euston Station, the British chiefs of staff were on hand to greet them. "Marshall was as charming as ever, and Admiral King as saturnine," remarked "A.B.C." Cunningham. The Americans coiled themselves into a line of cars and were driven to their temporary quarters, a Tudor-style mansion in Middlesex.[14]

Marshall, King, and Arnold were itching to see the French coast, but first they had business to attend to with the British chiefs. So strong was the lure of the battlefield on everyone, the Combined Chiefs did virtually nothing except race through the various fronts at the double-quick. The minutes of their June 10 meeting describe discussions as "brief," "short," or "very brief" thirteen times in the scant two-page record. The next day they did little more than debate, then shelve, the lingering question of ANVIL, now renamed Operation DRAGOON.[15]

Their formal business concluded, they scrambled to join Churchill on the night of June 11 aboard his private train bound for Portsmouth, where they would board ships bound for the Normandy coast. As the train rattled south from London, Churchill threw an excellent dinner at his dining car's banquet table.

The evening warmed with the prospect of stepping onto the fulcrum on which the great war turned. After living for so long in the future, they were about to see the debate, planning, bitter argument, and nervous activity of the last two years come to life.[16]

King, more than the rest, had reason for the saturnine countenance that A.B.C. remarked on, for in addition to the Normandy landings he was thinking ahead to the invasion of Saipan, the Mariana island on the globe's far side. Churchill, noticing his sour expression, tried to lighten the atmosphere. When King didn't respond, a playful Churchill ribbed, "Don't look so glum. I am not trying to take anything away from the United States Navy just now."[17]

Eisenhower greeted the entourage at Portsmouth. The parties boarded two destroyers—USS *Thompson* for Ike and the American chiefs, HMS *Kelvin* for the British—and Marshall, King, Arnold, and Eisenhower took passage to Omaha Beach in the American sector. As *Thompson* cut her way through the Channel's choppy waters, the three Joint Chiefs gazed in awe at the floating city filling land and water from Portsmouth to Bayeux.

It was one thing to envision an operation as immense as OVERLORD; it was another to see that plan take physical form. Stapled memoranda had become ships. Tables had become bombers. Footnotes and charts had become fighters and supply dumps. Blueprints with far-fetched names like MULBERRY and GOOSE-BERRY had blossomed into artificial harbors warding off the Channel's blows.

The chaos that welcomed them was both wretched and breathtaking. It was as if a titanic bin holding the jumbled product of the capitalist world had been upended onto Normandy's shingle. Landing craft bellied up to beaches and spit out tanks, self-propelleds, ambulances, and bulldozers. Crates of medicines, grenades, bread ovens, spare tires, and howitzer tubes piled high. Beachmasters roared at longshoremen, while jeeps and scout cars scurried like Labrador retrievers from sand to grass, road to farm. Deuce-and-a-halfs, bunched like burros on the roads, carried their loads oblivious to the turmoil around them. The farms and factories of the New World had crashed into the Old with the fury of a democracy roused to anger.

The swarming anthill at Omaha Beach must have made sense to someone, but that someone was not among the chiefs. In his diary, an impressed Hap Arnold called it *"a regular mad house but a very orderly one, in which some 15,000 troops a day go from ship to shore and some 1500 to 3000 tons of supplies a day are landed."*[18]

Marshall, who had in 1939 been sworn in wearing a white single-breasted suit, now sported a short "Eisenhower" field jacket. In deference to combat regulations, he carried a modest, nearly invisible automatic pistol on his right hip. Like a child on Christmas Day, he could not suppress a broad smile as he rode in an amphibious truck over the newly conquered land. Admiral King—armed with a scowl and a lit cigarette—listened with interest as Eisenhower and his lieutenants described the fight for the beaches, the terrain ahead of them, and the supply situation for ten hard-pressed divisions ashore.[19]

Leading a small convoy of jeeps and scout cars away from sniper-infested villages to their front, General Bradley guided the chiefs to his headquarters, a small cluster of camouflaged tents near the hamlet of Saint Pierre-du-Mont. They washed up with water poured from a jerry can, ate C-rations and crackers, and savored what looked like the beginning of the conquest of Europe.[20]

Having seen what they came to see, after a few hours Marshall and King climbed onto an amphibious truck that rolled into the water and ferried them back to the *Thompson*. Before long, they were in Portsmouth sitting aboard Churchill's train.

Churchill was not there. He had ordered His Majesty's destroyer *Kelvin* to maneuver into firing position so he could have the pleasure of lobbing a shell at the enemy. "We arrived at the exact time," King later wrote, "but had to wait in the special train for 'W.C.' for almost an hour since he insisted on shooting at least one large gun at the Nazis in France, which was just like him!"[21]

While Churchill tarried off the Norman shore, King, in better spirits for the return trip, ordered up a glass of sherry. Then another. And another. By the time the prime minister arrived, King was well lit. Churchill, for whom wine was a breakfast drink, ordered several bottles of champagne for toasts, and King was obliged to join in.

"I managed it," King said charitably. But he could no longer handle grape or grain the way he did when he prowled the China coast with women and liquor on his mind. Two nights later, when a Churchillian session of cigars and brandy was cruising full sail as the clock struck one in the morning, a groggy King mumbled, "Don't anyone ever go to bed around here?"

With an impish grin, Churchill replied, "This is early yet. We have lots to talk about."[22]

·   ·   ·

France, for instance. Two days before D-Day, Charles de Gaulle refused Eisenhower's request to broadcast a speech urging his countrymen to obey the instructions of the supreme allied commander. Then he denounced the use of "invasion francs" printed for Allied troops to use in bargaining for goods and services with the locals.

Long before the invasion, Roosevelt and Morgenthau had foreseen that a large influx of pounds and dollars would destroy the value of the franc; local merchants would naturally prefer safer currencies, and would heavily discount anything French. So the two Hyde Park neighbors designed paper notes as a stopgap measure in the absence of a viable French government with viable French currency. Angry at the Anglo-American infringement on French sovereignty, de Gaulle ordered his faction to treat the Allied francs as counterfeit. When the Allies began circulating the invasion francs over his objection, he howled to Churchill's labor minister, *"Allez, faites la guerre avec votre fausse monnaie."**[23]

In another fit of pique, de Gaulle halted the shipment of nearly two hundred bilingual French liaison officers, whose job was to assist invading troops working with French locals. This infuriated the American commanders even more than his resistance to the francs, because it directly hampered their efforts to liberate de Gaulle's country and move east to fight the Nazis. "We had trained French officers for civil affairs, and he cancelled every damn thing," Marshall fumed. "They had things fixed up well and then, by God, de Gaulle cancelled it all."[24]

"The more I think of the whole French situation, the more I am convinced of the danger of de Gaulle," Stimson told his diary on June 12. *"He is in some ways a brilliant soldier but is a man of egocentric and unreliable nature and cannot be relied upon to be steadfast and to place the welfare of his country before that of himself."*[25]

But thinking it over at Woodley, he concluded the Allies needed to reach a workable arrangement. With no other viable leader on the French playing field, British Foreign Secretary Anthony Eden was pressing an angry Cordell Hull to recognize de Gaulle as the provisional leader of France.

Roosevelt was emphatic: American policy was to let the French people choose their own leader in a fair and free election. De Gaulle would have no leg up on domestic French opponents through Allied recognition.

FDR's policy was, however, blind to the reality of liberation. The people needed an interim government immediately. They were hungry, and winter would arrive before long. Cold, starving people don't have the luxury of waiting for fair and free elections. As Eisenhower's armies drove toward the Seine and beyond, FDR's reluctance to face this inconvenient truth left his commanding general governing a nation whirling in ecstatic disarray. This, Stimson knew, would have terrible consequences come winter.[26]

Reflecting on his experience with elections in Central and South America, Stimson concluded that without a provisional government under de Gaulle,

---

* "Go ahead and wage war with your false money."

America would end up arbitrating disputes among a kaleidoscope of bickering factions. Playing umpire, even temporarily, was a no-win proposition that would blacken relations between France and the United States for decades to come—no matter whom America backed or who won. The best Roosevelt could insist upon was de Gaulle's promise of free elections once the war was over.

With venom spewing from Hull, Marshall, and Roosevelt, it was up to Henry Stimson to broker a compromise. On June 14 he called Hull and told him that Roosevelt's position, while theoretically correct, was unrealistic. For the sake of the war effort, the United States and Britain must support an interim leader, even if that leader was the tall, obstreperous general.

Hearing this, Hull lapsed into a stream of Tennessee profanity—"almost incoherence," Stimson wrote. So Stimson gave up on Hull and called Roosevelt. He asked for an accommodation with de Gaulle.

FDR wouldn't budge. Relying on information from Bill Donovan's OSS sources, he said de Gaulle's popularity with the French people had been grossly exaggerated. While the French respected him as a symbol of resistance, that symbolism would not translate into political power. French parties were springing up all over France, and the Gaullists would crumble as new parties congealed into a governing coalition. "As the liberation goes on, de Gaulle will be a very little figure," FDR said confidently.[27]

Stimson wasn't so sure. *"This is contrary to everything that I hear,"* he wrote in his diary. *"I think de Gaulle is daily gaining strength as the invasion goes on and that is to be expected. He has become the symbol of deliverance to the French people."*[28]

Moving to the other side of the aisle, Stimson called on the British ambassador, Lord Halifax, and explained Eisenhower's dilemma. Eden's support for de Gaulle encouraged the Frenchman's antics, and those antics were hampering the war effort and making it hard to move Roosevelt and Hull. Was there something Halifax could do to bring Eden around?

Halifax agreed to look for a middle ground. He told Eden that the president was not dead set against de Gaulle's rise to power, so long as he was not installed by the Allied governments. If the French wanted to elect him as their leader, Roosevelt would abide by their decision.[29]

Returning home to Woodley after speaking with Halifax, Stimson found the secretary of state playing croquet on his mansion's neatly clipped grounds—a favorite pastime of Hull's. Stimson took up a mallet, joined Hull, and pressed his case for reconciliation. In this instance, he knocked no balls through the wicket. When the game ended, Hull left Stimson's croquet field, obstinate as ever.

A few days later Stimson again lobbied Hull, whose penchant for curse words and disgust with de Gaulle had not abated. Stimson told Hull he was being short-sighted. The Americans, he said, had the choice of *"telling [de Gaulle] he is a blank, blank, blank, or trying to get some working arrangement."*[30]

After more shuttle diplomacy, Roosevelt agreed to give de Gaulle a warm if unofficial reception in Washington. Because Roosevelt considered de Gaulle a short-term expedient—and because Hull had an endless supply of "blank, blank, blanks" to shoot at him—FDR told Stimson, rather than Hull, to make the arrangements. To avoid inflating *le général*'s troublesome ego, Stimson arranged for Marshall to keep the self-saturated general busy in military meetings, "so his visit would be filled up instead of having a parade through the country."[31]

● ● ●

As Stimson cauterized the oozing French sore, another wound broke its stitches in the Mediterranean. King and Marshall were adamant that the DRAGOON landings in southern France proceed on schedule. Churchill, however, was dead set against the operation. He favored using those ships and troops to push General Harold Alexander's armies in Italy north to Trieste, on the Adriatic coast, then north again through Slovenia's Ljubljana Gap, into Austria.

A drive into Austria—under a British commander—was a strategic vision that Churchill would not let go of, despite his staff's belief that any thrust into the Italian Alps was a fool's errand. A weary Brooke told his diary, *"We had a long and painful evening of it listening to Winston's strategic ravings! . . . In the main he was for supporting Alexander's advance on Vienna. I pointed out that even on Alex's optimistic reckoning the advance beyond the Pisa-Rimini line would not start till after September. Namely we should embark on a campaign through the Alps in the winter!"*[32]

Marshall agreed: "The 'soft underbelly' had chrome-steel sideboards." The Adriatic region would become a huge, parasitical drain as the defeated countries contributed nothing except demands for food, fuel, and money to support their populations.

Caught between Churchill and the Joint Chiefs, a frustrated General Eisenhower received a broadside of protests, declamations, and pleas from 10 Downing Street. But after talking over the difficulties with Field Marshal Henry "Jumbo" Wilson, Churchill's Mediterranean theater commander, Eisenhower recommended launching DRAGOON and scuttling stabs north of the Pisa-Rimini line.[33]

Churchill had once succeeded in persuading Roosevelt to overrule his warlords, and on June 28 he tried again. "Our first wish is to help General Eisenhower in the most speedy and effective manner," he wrote FDR. "But we do not think this necessarily involves the complete ruin of all our great affairs in the Mediterranean, and we take it hard that this should be demanded of us. I think the tone of the United States Chiefs of Staff is arbitrary and, certainly, I see no prospect of agreement on the present lines."[34]

Churchill's persuasive spell ebbed when communications were separated from the man. His plea to kill DRAGOON, encoded, wired to the Map Room and decrypted, then typed and retyped for FDR's perusal, lost much of the eloquence and force it had coming from Churchill's own lips.

In any event, Roosevelt was in no mood to break with his chiefs. "On balance I find I must completely concur in the stand of the U.S. Chiefs of Staff," he told Churchill. "General Wilson's proposal for continued use of practically all the Mediterranean resources to advance into northern Italy and from there to the northeast is not acceptable to me, and I really believe we should consolidate our operations and not scatter them. . . . ANVIL, mounted at the earliest possible date, is the only operation which will give OVERLORD the material and immediate support from Wilson's force."[35]

As a follow-up, Marshall, King, and Arnold drafted a second message from Roosevelt to Churchill stressing the superiority of the Rhône Valley approach to either the Italian mountains or the Ljubljana Gap. Roosevelt approved the message, but before sending it, he added a paragraph reflecting a more personal motive: "Finally, for purely political consideration over here I would never survive even a slight set-back in Normandy if it were known that fairly large forces had been diverted to the Balkans."[36]

The unwritten part of Roosevelt's message was just as clear: America was running the show now. After meeting with the prime minister, Brooke wrote, *"He looked like he wanted to fight the President. However in the end we got him to agree to our outlook, which is: 'All right, if you insist on being damned fools, sooner than falling out with you, which would be fatal, we shall be damned fools with you, and we shall see that we perform the role of damned fools damned well!'"*[37]

The next day a sullen Churchill wrote Roosevelt, "We are deeply grieved by your telegram. . . . The splitting up of the campaign in the Mediterranean into two operations, neither of which can do anything decisive, is, in my humble opinion, the first major strategic and political error for which we two have to be responsible." Looking for a hook that Roosevelt might respond to, he added that pouring French troops into the Rhône Valley "would no doubt make sure of de Gaulle having his talons pretty deeply dug into France."[38]

Roosevelt didn't take the bait. He suggested that the British and Americans make their cases to Stalin and ask what the dictator preferred. A shocked Churchill replied, "On a long-term political view, [Stalin] might prefer that the British and

Americans should do their share in France in the very hard fighting that is to come, and that east, middle and southern Europe should fall naturally into his control."

Churchill's rhetorical grenadiers—logic, pathos, geopolitics, even the threat of de Gaulle—failed to breach Roosevelt's defenses. With loud complaints to his cabinet and generals, Churchill gave ground to the senior partner. On July 1 he wrote Roosevelt, "It is with the greatest sorrow that I write to you in this sense. I am sure that if we could have met, as I so frequently proposed, we should have reached a happy agreement. I send you every personal good wish. However we may differ on the conduct of the war, my personal gratitude to you for your kindness to me and all you have done for the cause of freedom will never be diminished."[39]

But behind closed doors he ranted to Pug Ismay, "I hope you realize that an intense impression must be made upon the Americans that we have been ill-treated and are furious. Do not let any smoothings or smirchings cover up this fact. After a while we shall get together again; but if we take everything lying down there will be no end to what will be put upon us. The Arnold-King-Marshall combination is one of the stupidest strategic teams ever seen."*[40]

Histrionics were second nature to Winston Churchill. But when Jumbo Wilson told him the assault troops had embarked for the Riviera, Churchill gamely came around and took his station aboard the destroyer *Kimberley* to watch the landings.

After it was over, he sent Eisenhower a gracious note to compliment him on the operation. A relieved Ike replied, "I am delighted to note in your last telegram to me that you have personally and legally adopted the DRAGOON. I am sure that he will grow fat and prosperous under your watchfulness."[41]

---

* Churchill deleted this last sentence from his memoir *Triumph and Tragedy*, published in 1953.

FORTY-FOUR

HATFIELDS AND McCOYS

〜

F June 6, 1944, was the beginning of the end for Hitler, then June 15 was the beginning of the end for Tojo. On that day fifty of Arnold's new B-29 bombers winged their way from central China to southern Japan and hit the Imperial Iron and Steel Works at Yawata, on the island of Kyushu. And on that day, Nimitz's marines landed on the Mariana island of Saipan.

As the command ship USS *Rocky Mount* beamed encrypted reports to Pearl Harbor, a dim picture of the fighting began to take form. Eight thousand marines under Lieutenant General Holland M. Smith crashed ashore in twenty minutes, but their tracked landing vehicles and tanks bogged down in the rough, swampy ground of Saipan's shoreline. Blistering enemy shellfire and a Japanese night attack inflicted heavy casualties on the first day, and by D-plus-1, the Army's 27th Infantry Division, one of the assault units, was having tough going along the island's rocky central spine.[1]

Despite heavy casualties, the leathernecks made a good showing. The island would not be fully pacified until early July, and those twenty-five days of fighting would cost the Marines another eleven thousand men. But the casualties were well spent. Once Saipan, Tinian, and Guam were in U.S. hands, Japanese supplies to Truk would be cut, and the Emperor's bastion there would surrender or starve. The drive on Japan would begin in earnest.[2]

More than a thousand miles to the north, one geocultural fact had arrested Hap Arnold's attention: Japan was unusually vulnerable to air attack. Its houses were made of wood, its population was concentrated, and oil and steel, the lifeblood of modern war, were funneled through identifiable ports and refineries. To Arnold, Japan's main island, Honshu, and its southern neighbor, Kyushu, were ripe for saturation bombing.[3]

The problem was distance. Kyushu and Honshu lay beyond the thousand-mile radius of America's longest-range bomber, the B-24 Liberator. Until America produced a "very long range" bomber—an expensive, mistake-ridden process that took an agonizing four years to complete—strategic bombing of Japan remained a distant dream, and the Emperor's home remained inviolate.

The solution was the B-29 "Superfortress," a four-engine behemoth with a 1,500-mile radius and a ten-ton bomb load. The B-29 gave Arnold's pilots the ability to hit the Home Islands from either central China or the Marianas. Anticipating delivery of the weapon in the spring, at Cairo the Combined Chiefs had approved Operation MATTERHORN, a strategic bombing campaign against the Japanese homeland, to run from 1944 until the war's end.[*]

The Superforts that lifted off from China on June 15 caused little damage when they emptied their bomb bays and headed home. But it was only the beginning. With fuel, bombs, crews, and planes trickling into central China, and with new air bases under construction on Saipan and Tinian, the silver geese would return, and in greater numbers. The wind the Emperor sowed at Pearl Harbor was about to reap a typhoon, courtesy of the Twentieth Air Force.[4]

While the Marianas were the big prize, a second plum fell from the tree on June 19, when a five-carrier fleet under Admiral Jisaburo Ozawa attacked Admiral Spruance's Task Force 58 in the Philippine Sea, just west of the Marianas. American Hellcats, their pilots dead tired from defending the skies over Saipan, took to the air and intercepted Ozawa's torpedo and dive-bombers. Their adrenaline up, American fighter pilots slaughtered their poorly trained enemy. Some three hundred Japanese planes went into the ocean for a cost of twenty-nine Hellcats, while below the water's surface, two American submarines sank the carriers *Taiho* and *Shōkaku*.[5]

The next day, Admiral Marc Mitscher sent his own flattop bombers on a revenge raid. Flying to the edge of their fuel range into headwinds and fading daylight, the Americans found and pounced on Ozawa's fleet. They sank the carrier *Hiyo* and damaged two carriers, a battleship, and a cruiser. The full extent of the

---

[*] To keep MacArthur or Chiang from monopolizing the new weapon, Arnold concentrated all B-29s into one group, the Twentieth Air Force. Like King's Tenth Fleet, Arnold kept those bombers under his personal command.

enemy's loss would not be clear for some time, but it appeared the Japanese had lost between three and four hundred planes and three irreplaceable flattops. The Battle of the Philippine Sea had finally broken the Emperor's most feared weapon.[6]

. . . .

On the twenty-ninth of June, Admiral King learned of some three-star stupidity that threatened to sour his victory at Saipan. Lieutenant General Holland M. Smith, nicknamed "Howlin' Mad" Smith by fellow marines, had been giving the Army's 27th Division hell for its slow pace up Saipan's center.

The Marine way, grim as death, called for storming the enemy rapidly and violently wherever he could be found. After the main defense lines were ruptured came the cleanup: devil dogs would fire into tunnels and caves, or cauterize openings with flamethrowers, while the engineers moved up with detonation charges. Blow the mouth of the cave, and shoot or bayonet anyone left outside—dead or alive, since Japanese corpses had a habit of waking up and shooting passing troops in the back. That was how the Marines fight.[7]

The 27th Infantry Division's commander, Major General Ralph Smith, was trained in the Army method: Call down air and artillery support, move methodically, watch for gaps, and protect flanks. Replacing a few truckloads of howitzer shells cost a lot less than replacing a company of flesh-and-blood men. That was how Army soldiers fight, live, and fight some more.

To the bloodied Marines, the Army method might have worked fine in French dairy fields, but it sure as hell didn't work in the ravines of Saipan. As the Second and Fourth Marine Divisions advanced steadily across the island on the flanks, a U-shape formed in their line as the 27th Division, in the center, slowed to a crawl by heavy resistance.[8]

The two General Smiths had a history going back to the Tarawa campaign, where Holland Smith, then a two-star observer, excoriated the Army's Ralph Smith for taking so long to capture the nearby island of Makin. The Marine general raised hell over Army errors, like firing indiscriminately into jungle cover, while overlooking those same rookie mistakes by his younger leathernecks. Howlin' Mad's prejudice against the Army in general, and Ralph Smith in particular, deepened as he and the 27th were drawn toward Saipan's rocky center.[9]

On the tenth day of fighting, Marine Smith asked Admiral Spruance to relieve Army Smith, and the 27th's commander was promptly put on a boat for Hawaii. When news of Smith's relief reached Oahu, it unhinged Lieutenant General Robert Richardson Jr., an Army officer with a temper to match Howlin' Mad's. Richardson recommended to Nimitz that Ralph Smith receive a naval medal, to assuage the Army's hurt feelings, and Nimitz forwarded the recommendation to Washington with his approval.

King predictably turned down Richardson's medal request, so Richardson appointed a board of inquiry to investigate the matter. To head that board, he selected Lieutenant General Simon Bolivar Buckner Jr., whose own brawls with the Navy had already percolated to the desks of King and Marshall. To no one's surprise, Buckner's board found General Holland Smith at fault for mismanaging his Army units. Holland Smith howled about Richardson's investigation to Vice Admiral Richmond Kelly Turner, and Turner, his blood up, asked Nimitz to shut down Richardson.[10]

When the "Smith versus Smith" skirmish escalated into a "Richardson versus the Navy" battle, Nimitz and King tried to stop the fireworks. Nimitz deleted disparaging references to the 27th Division from Admiral Spruance's official report, and he and King created a new "Fleet Marine Force," with Holland Smith as its commander, ensuring that Howlin' Mad would never again lead Army troops.[11]

Word leaked out, as it always did. Hearst newspapers picked up the family feud and castigated the Marines for their casualty-filled methods. The Luce magazines, *Time* and *Life*, weighed in on the side of the leathernecks.[12]

Though he had approved the loan of the 27th Division to the Navy as part of a deal to support MacArthur, Marshall had never liked the idea of an Army division working under the Marines in general, and General Smith in particular. The year before, he had foreseen problems and hoped to nip them in the bud with a stern letter to General Richardson. "Under the circumstances," he wrote, "I want General [Ralph] Smith to be made aware of the critical importance of his training preparations for the operation and of the cooperative spirit of himself and his staff. There must be no weakness, no hesitations or reluctances in the action of units once they have landed. There must be no misunderstandings, jealousies, or critical attitudes."[13]

Now he had both critical attitudes and a very public feud. Marshall asked his operations man, Tom Handy, to figure out what the War Department should do, and Handy recommended pulling Ralph Smith out of the theater. Marshall agreed, and transferred Smith to Europe. Unless the Marines fought their way through Tokyo and kept going to Berlin, Ralph Smith would not encounter Holland Smith for the rest of the war.[14]

The Army-Marine problem on Saipan went far beyond "mere healthy rivalry," Marshall told King. He informed King that he had transferred General Ralph Smith from the theater, and suggested they send both Richardson and Turner a message ordering each to ensure that the problem didn't happen again.[15]

King had known of Howlin' Mad's temper long before the war, and King was one of the few men who could stop the rhino in his tracks. He was fond of Smith and his marines, but he saw the need to meet Marshall halfway. Besides, the Truman Committee had been inquiring into heavy casualties on Tarawa, and King didn't want disgruntled Army generals handing the press a fresh ladle to stir the pot.

King sent Marshall a reply promising to squelch any public statements from naval sources on the subject. He made it plain that he was still angry with Richardson for appointing an Army board to pass judgment on the Navy, but he said he would talk to Nimitz about smoothing relations between the Army and the Marines.[16]

Marshall let the matter die down. He was glad to sweep it into the dustbin, because a far more divisive matter was about to unfold, and it would drag in the entire American high command.

FORTY-FIVE

MR. CATCH

~

THREE MONTHS BEFORE THE CURTAIN WAS RAISED ON SAIPAN, MacARTHUR opened the strategic debate again, setting Army and Navy on another collision course. In March, the Joint Chiefs granted MacArthur permission to drive north through the Admiralties and Celebes Islands toward Mindanao, the big southern island in the Philippines group. By June the chiefs had made no decision on Luzon. Unsure of themselves, they asked MacArthur and Nimitz to weigh in on a proposal to bypass Luzon and move against the island of Formosa.[1]

The memory of Luzon, home to Manila, Bataan, and Corregidor, haunted MacArthur every day. His headquarters, code-named BATAAN, reminded him of the men he'd left behind, and that memory called him back to Luzon like a pilgrim to Mecca. His "I shall return" pledge rang with the clarity of "Remember the Alamo" or "Don't Give Up the Ship." The centerpiece of MacArthur's Pacific campaign had become a matter of personal honor.

Acutely conscious of his celebrity status back home, MacArthur pressed the case for Luzon with Marshall in sweeping moral terms. "We have a great national obligation to discharge," he declared with the voice of the oracle Roosevelt so despised. "If the United States should deliberately bypass the Philippines, leaving our prisoners, nationals and loyal Filipinos in enemy hands without an effort to retrieve them at the earliest moment, we would incur the greatest psychological reaction."[2]

Marshall quickly reassured MacArthur that the War Department wasn't selling out either American prisoners or the Filipinos. But when the Joint Chiefs debated the question of bypassing Luzon in favor of Formosa, Marshall saw the Navy's position as the more logical of the two. MacArthur's arguments, it seemed to Marshall, were grounded in emotion. And politics.

Formosa was three hundred miles closer to Tokyo than northern Luzon. Distance made a difference, for less fuel pumped into Hap's bombers meant more bombs carried to Japan. Mulling over the difficulties of an attack on either Formosa or Luzon, Marshall and King agreed that MacArthur was wrong to slog through the Philippines. "We should be going the slow way," Marshall explained to Stimson. "We should be butting into the large forces the Japanese have accumulated in the Philippines. . . . We should have to fight our way through them and it would take a very much longer time than to make the cut across."[3]

Marshall sent MacArthur a cable cautioning him that Formosa was not out of the question. Both options would remain under study, but he predicted, "A successful conclusion of the war against Japan will undoubtedly involve the use of a portion of the China coast." Knowing that MacArthur wouldn't like his message, he added, "We must be careful not to let our personal feelings and Philippine political considerations override our great objective, which is the early conclusion of the war with Japan."[4]

Though they hadn't made a firm decision on either Luzon or Formosa, in mid-June the Joint Chiefs asked MacArthur and Nimitz to submit proposals to accelerate the tempo of conquest. In early July MacArthur sent the Joint Chiefs a revised plan. He intended to hop to Mindanao in mid-November, then to Leyte the following month. Around April 1, 1945, he would launch a huge invasion—six assault divisions—at Lingayen Gulf on Luzon, and force the enemy off the island for good.[5]

King agreed that Mindanao needed to be taken, but he dismissed Luzon as strategically irrelevant. As he saw it, a base on Formosa would open a new supply route to Chiang, tighten the blockade of Japan, and starve the enemy into submission—a slow death, but a sure one. With Formosa in American hands, King doubted whether Luzon would even need to be taken before the war ended.[6]

King's lead admirals, Nimitz, Halsey, and Spruance, privately told King they would prefer to take Luzon, for it would give the Americans air bases and a fine anchorage for operations against Japan's small outer islands. But King didn't budge. Ships and planes operating from Luzon wouldn't cut the enemy lifeline, since Japanese tankers hugging the China coast would take shelter under a swarm of fighters based on Formosa. King would go along with an attack against Mindanao, then Leyte, Philippines midsection, but no farther.[7]

•   •   •

To satisfy his hunger for information about the Marianas, Admiral King visited Saipan while the embers of battle still smoldered. After a jeep tour of the island—

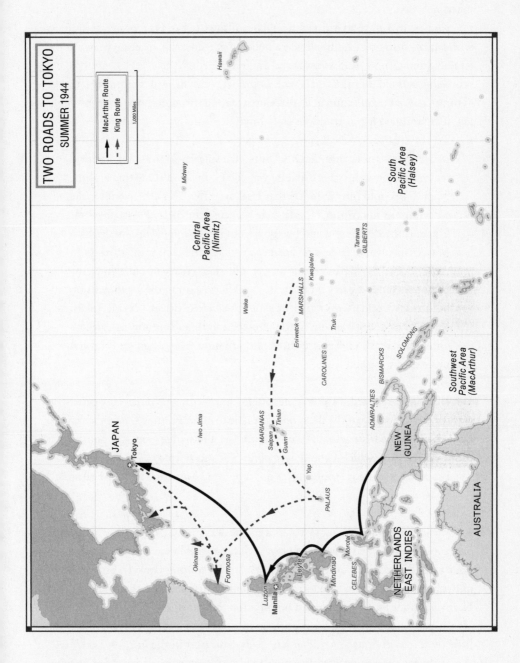

TWO ROADS TO TOKYO
SUMMER 1944

MacArthur Route
King Route

1,000 Miles

Hawaii

Midway

Central
Pacific Area
(Nimitz)

Wake

MARSHALLS
Eniwetok   Kwajalein

CAROLINES   Truk

Tarawa
GILBERTS

South
Pacific Area
(Halsey)

SOLOMONS

BISMARCKS

ADMIRALTIES

Southwest
Pacific Area
(MacArthur)

NEW
GUINEA

AUSTRALIA

NETHERLANDS
EAST INDIES

CELEBES   Morotai

Mindinao

Leyte

Luzon
Manila ☆

Formosa

Okinawa

MARIANAS
Saipan Tinian
Guam

Yap

PALAUS

• Iwo Jima

JAPAN

Tokyo

passing through a hot sector with Nimitz and a carbine-toting Holland Smith—King flew east to Hawaii, where he spoke with Nimitz about an upcoming strategy meeting.

Roosevelt, King had learned, would be sailing to Pearl Harbor to talk strategy with MacArthur and Nimitz. It was a political gesture in an election year, but the strategic question FDR raised was real and relevant.

King wanted Nimitz's personal assurance that he was committed to the "Formosa over Luzon" approach. Years later, leafing through the pages of his flight log, King recalled his instructions to Nimitz:

> "All hands" knew that "F.D.R." was en route to Hawaii for political reasons to show the voters that he was the C. in C. of the Army and the Navy. I also told him (Nimitz) that I believed that "F.D.R." would make a deal out of his trip to "The Islands" because the "set-up" was fixed for him and the decision would be about going in to the Philippines (which MacArthur wanted) or into Formosa which the Navy wanted to do in order to cut the Japanese out of gas, oil, rice, etc., without which the Japanese couldn't carry on the war. . . . After recalling the arguments on both sides I told Nimitz that he would be called before "F.D.R." with MacArthur and with Adm. Leahy present and that I was not making any orders but he should please think the situation over when he (Nimitz) was called to speak.[8]

King later recalled, "I was quite careful not to give [Nimitz] any orders on this occasion, because if the decision was made to go to Formosa he would have to do it. It would be his own job. I pointed out the alternatives, but did not think it proper to tell him, which I could have done, 'stick to Formosa.'"[9]

Proper or not, King would regret not ordering Nimitz to "stick to Formosa."

·   ·   ·

Before Roosevelt could "show voters that he was the C. in C.," he had some backroom business to transact at home. In June, Republicans meeting in Chicago nominated New York Governor Thomas E. Dewey for president. In a public letter on July 11, FDR told Democratic National Committee chairman Robert Hannegan that he would accept a nomination to run for office for a fourth term. "If the people command me to continue in this office and in this war, I have as little right to withdraw as a soldier has to leave his post in the line," he said.[10]

Whether the people would command him to stay in office depended on what they thought about a few key issues. The first was management of the war, of course. That favored Roosevelt, who could show tangible progress in Europe, the Pacific, and the Mediterranean since December 1941. It also played against Governor Dewey, a prosecutor with no military experience—not even as nominal commander of the New York National Guard, which had been federalized before Dewey became governor.[11]

The second issue was FDR's health. After his return from Tehran, he appeared fatigued. He suffered from bouts of what he and his doctors disingenuously called the "flu." He complained of headaches, he had trouble sleeping at night, and his jaw would hang open as he watched movies. He would even doze off occasionally while reading his mail or dictating long passages to secretaries. "It was evident that the grind was becoming too severe for him," remembered Grace Tully. "The next step might well be a real breakdown in his health or a dangerous decrease in the soundness and force of his judgment."[12]

By March FDR's famous voice had grown husky, his face pallid, his lips and nail beds bluish. He dropped hints that he wanted his cousin, Daisy Suckley, to have Fala should anything happen to him, and when the White House staff politely inquired how he was feeling, his routine answer was "Like hell."

FDR's daughter, Anna Roosevelt Boettiger, persuaded him to see a heart specialist at Bethesda Naval Hospital. "Appalled at what I found," that specialist wrote later. Dr. Howard Bruenn, an experienced cardiologist, summarized the results of his exam. The president's heart was enlarged, his blood pressure was dangerously high, and his left ventricle was on its last legs. He was a sick man whose hourglass sands were pouring fast. The only way he might prolong his life would be to rest, eat better, stop smoking, and avoid high-pressure activities. Like being president.[13]

FDR cut down on cigarettes and vowed to get more rest. He spent most of April in South Carolina at the 23,000-acre estate of his friend and adviser Bernard Baruch. He wore that broad, familiar smile, but reporters began asking Steve Early about rumors they heard of clandestine hospital visits. "Bunch of Goddamned ghouls," Roosevelt muttered.[14]

The commander-in-chief's health raised a third big election issue: the vice presidency. As always, the party, not the presidential candidate, would select the vice presidential nominee, though the headliner naturally had a big voice in his running mate's selection. Four years earlier, Roosevelt had supported the nomination of Henry Wallace, partly because he respected Wallace and partly as a sop to the party's liberal wing. But Wallace's eccentric beliefs in Eastern mysticism and his far-left views would probably drive moderate voters to Dewey. That would jeopardize the ticket's chances in big states like New York, Pennsylvania, Illinois, and California.

378 ★ AMERICAN WARLORDS

At the White House, Roosevelt caucused with a high-powered delegation of Democratic Party bosses. Chairman Robert Hannegan, Edward Flynn, the Bronx boss who helmed FDR's 1940 campaign, and Chicago mayor Ed Kelly reviewed the bullpen of possible running mates. Roosevelt felt the best man for the job was Jimmy Byrnes, but Byrnes, a segregationist from South Carolina, was unpopular with labor. He also was born Catholic but quit the Church, alienating many voters outside the Solid South. Age, demographics, or inexperience ruled out other candidates, like Sam Rayburn of Texas, Senator Alben Barkley of Kentucky and Supreme Court Justice William Douglas.[15]

That left Harry Truman, a moderate from Missouri who had made a name for himself investigating military waste. As a veteran of the First World War and a New Deal stalwart, Truman would appeal to both conservatives and organized labor, and he hadn't made any racial comments that might hurt the ticket with liberals or Negro voters. By default, the bosses agreed that Truman would be the least dangerous running mate.

Whenever potential running mates came to Roosevelt for support, he subtly reminded them that he could not select the nominee, a limitation he played up to avoid any personal responsibility to would-be candidates. But everyone knew his preference would be decisive, and he clouded the issue by assuring both Wallace and Byrnes that each had his personal support.

On the morning of July 11, he shook hands with Wallace and told him, "I hope it will be the same old team." He wrote a letter on Wallace's behalf to Democratic convention chairman Senator Samuel Jackson, saying of Wallace, "I like him and respect him and he is my personal friend. For these reasons I personally would vote for his nomination if I were a delegate to the convention."

Later that morning, he told Jimmy Byrnes, "You are the best qualified man in the whole outfit and you must not get out of the race. If you stay in you are sure to win."[16]

FDR knew the party elders would orchestrate the balloting to ensure Truman's selection, for the last thing they wanted at the Democratic convention was unbridled democracy. FDR could give Byrnes his assurances and pass off the vote as something beyond his control.

Similarly, his letter praising Wallace fulfilled a personal obligation wrung from him by an insistent friend. But the letter was so tepid by political standards that the *Philadelphia Inquirer* published a cartoon depicting Wallace as a barefoot farmboy standing before a locked convention door reading FDR's missive: *"Here's a nice boy! But don't hire him if you don't want to! FDR."*[17]

They didn't. When Democratic Party delegates met at Chicago Stadium on July 19, a Roosevelt-Truman ticket emerged the winner.

. . .

Voters thinking about Roosevelt's fourth bid wanted to know that he had a firm hand on the levers of power, and FDR obliged them. On July 13, he left Washington with Leahy, Pa Watson, Hopkins, Fala, and a bevy of aides, Secret Service agents, valets, and Filipino cooks. He boarded his armor-plated Pullman car, the *Ferdinand Magellan*, and arrived in California on the nineteenth. Stopping in San Diego to watch Marine landing exercises, he mingled with sailors and visited the wounded. Then he boarded the cruiser USS *Baltimore* and spent the next four and a half days at sea.[18]

The journey to California, and from California to Pearl Harbor, was, as King knew, FDR's way of dramatizing his role as commander-in-chief—and making sure the voters knew who was calling the shots. He could have easily summoned MacArthur and Nimitz to Washington at much less personal risk. But remaining in the capital would throw away the priceless political advantage of being seen as a true military leader in wartime, giving orders to men in uniform and making decisions other politicians could not.

It was an advantage FDR had every intention of pressing. Even the title "commander-in-chief" became a fixation with him. He asked Hull to refer to him as "commander-in-chief" rather than "president" at State Department dinners. At FDR's request, one day Admiral Leahy came to King's office—a rare excursion for Leahy—and told him the president wanted one of King's titles modified. Leahy said that "Commander-in-Chief of United States Fleets" seemed too expansive when Franklin Roosevelt, president of the United States, was the Navy's constitutional commander-in-chief.

King looked at Leahy. "Is that an order?" he asked.

"No," said Leahy sheepishly, "but he'd like to have it done."

King gave Leahy a hard stare. "When I get the orders I will do exactly that. Otherwise not."[19]

Unwilling to put that kind of order in writing, Roosevelt let the matter drop.

FDR's fourth presidential campaign stumbled on July 20, even before he left California. As he was about to leave his sleeping cabin on board the *Magellan* to watch landing exercises of the Fifth Marine Division, he turned white as a ghost and began to shake and shudder. As Pa Watson and Roosevelt's son James looked on in shock, he gasped, "Jimmy, I don't know if I can make it out. I have horrible pains."

Jimmy's first thought was to get a doctor, but Roosevelt refused. He and Pa helped FDR out of his bed and stretched him out on the cabin floor to ease the pressure on his heart and lungs. His torso convulsed and his brow furrowed as he shut his eyes and waited out the pain. After a few minutes of sharp spasms, he opened his eyes and said calmly, "Help me up now, Jimmy."

Jimmy and Pa managed to get Roosevelt upright and back on his bed. When he looked like he had recovered, the cabin door opened and unsuspecting valets entered. They wheeled the president into a waiting convertible.

Soon, he was on a bluff overlooking a slice of Oceanside, California, watching five thousand marines charge an "enemy" position, rifles cracking, mortars thumping, and flamethrowers billowing. Roosevelt leaned over the car's side to watch the show as photographers snapped pictures of the mighty Marines and their commander-in-chief.[20]

Though his bout aboard the *Ferdinand Magellan* was kept hidden from the public, a more damning sign of his health emerged that evening when he broadcast his acceptance speech for his party's nomination. Speaking from an observation car hooked up to the convention's loudspeakers, Roosevelt's speech buzzed over electrical lines to Chicago Stadium, where the party faithful listened to their three-time champion.

"I am now at this naval base in the performance of my duties under the Constitution," he said, his voice low and solemn. "The war waits for no elections. Decisions must be made—plans must be laid—strategy must be carried out." The job of the American people was, he said, "First, to win the war—to win the war fast, to win it overpoweringly. Second, to form worldwide international organizations. . . . And third, to build an economy for our returning veterans and for all Americans—which will provide employment and provide decent standards of living."[21]

The fifteen-minute speech was no better than workmanlike, and after it was over he sat at the table rereading portions of it as pool photographers snapped pictures for news services. From a batch of negatives dipped and dried in a local darkroom, the Associated Press editor in Los Angeles hurriedly picked out one of FDR with his mouth open, obviously reading his speech from a sheaf of papers spread on his desk.

As every politician knows, the millisecond a photo is snapped can make the difference between a glorious legacy and utter humiliation. Unfortunately for Roosevelt, the photo the AP editor selected became the first disaster in a campaign not three hours old. The camera's timing and angle revealed a gaunt FDR, his head bowed over his text. His mouth hung open, and glassy eyes stared blankly over a narrow face and oversize suit and a dark, outsize shadow hovered over his shoulder. The photo was a silent testament to a candidate whose running mate was probably the Grim Reaper.

The photographer—who had advised against the photo's publication—was promptly kicked out of the presidential retinue, but the damage was done. Anti-Roosevelt papers reprinted the photograph and Republican flyers showed side-by-side comparison photos with the full-faced, vibrant Franklin Roosevelt of 1940. Governor Dewey gleefully accepted the windfall and told audiences that a vote for Roosevelt was a vote for Harry Truman.[22]

The "death's door" photo of July 20 was an ironic setback on a day bearing sweet fruits of America's war leadership. In Tokyo that day, General Tojo's government collapsed, in large part due to the loss of Saipan to King's leathernecks. On the other side of Eurasia, a cabal of German officers tried to assassinate Hitler with a bomb—a reflection, in part, of the success of Marshall's army in France. Roosevelt's job as candidate would be to ensure that voters associated him with victory abroad, not the squabbles at home.

•   •   •

When *Baltimore* sailed into Pearl Harbor on July 26, Roosevelt was in better form. His cheeks had regained some of their color, his blue eyes sparkled in the tropical sun. The grin returned to his face. He beamed as he scanned Hawaii's great harbor, bustling with hundreds of ships, their rails lined with row upon row of white-suited officers and sailors. In open disdain for security protocol, the presidential flag snapped from *Baltimore*'s main, and for a brief, shining moment, Franklin Roosevelt looked like a lion in autumn, not winter.

Before they reached the pier, a motor launch pulled alongside the cruiser and sent aboard Admiral Nimitz with a delegation of white-uniformed flag officers. *Baltimore* proceeded majestically to Pearl's seawall, where the splash of the anchor as it hit the water announced the end of the voyage.[23]

As the crowd of vice admirals, rear admirals, generals, and commodores milled about *Baltimore*'s deck, Roosevelt looked over the group and realized one face was missing. "Where's Douglas?" he asked.

Nimitz had no answer. The party stood there in embarrassed silence.

Marshall had ordered MacArthur, whose code name for this occasion was MR. CATCH, to fly to Honolulu so as to arrive on July 26. Marshall told MacArthur to let as few people as possible know of his trip, and gave him no official instructions as to why he was going, other than "general strategical discussion." But Marshall did tell him he would see "Leahy etc.," and figured he probably knew who the "etc." in Leahy's party would be.

Marshall couldn't dictate security measures to the Secret Service or the Navy, but he could make certain there would be no leak from the Army's end. "I assume that there will be no publicity regarding the President's visit until after his return to the mainland and therefore there should be no reference to General Mac-Arthur's presence in Hawaii," Marshall instructed Hawaii's senior Army commander, General Richardson. "The restrictions regarding the President are not my affair, but I wish you to see that no reference is permitted regarding General MacArthur's presence in Hawaii except in strict accordance with the President's instructions."[24]

Forty minutes after *Baltimore*'s gangplank was lowered to the pier, the air was split by the shriek of a police siren. A motorcycle escort appeared, leading what Sam Rosenman remembered as "the longest open car I have ever seen. In front was a chauffeur in khaki, and in the back one lone figure." That figure wore a crushed general's hat and a brown leather jacket.

Mr. Catch had arrived.[25]

MacArthur's car drove to the gangplank, to the wild applause of the crowd. He bounded up the ramp—stopping halfway to acknowledge another round of applause—then strode onto the cruiser's deck. He saluted the commander-in-chief before shaking Roosevelt's outstretched hand.

"Hello, Douglas," said Roosevelt. "What are you doing with that leather jacket on? It's darn hot today."

"Well, I've just landed from Australia," MacArthur said with a smile. "It's pretty cold up there."[26]

He had not just landed from Australia, for *Bataan* had arrived an hour earlier; MacArthur had stopped at Richardson's home to shave and freshen up. But the jacket made a great prop, and FDR appreciated good political theater. On *Baltimore*'s sun-bathed deck, he sat between MacArthur and Nimitz for the obligatory photo session as Leahy, wearing his "King Gray" uniform, worked his way to the center of the group portrait.

The photo work done, the party set off for their quarters. They gathered the next morning for a day inspecting shipyards, airfields, hospitals, and training camps scattered across the island.

When Roosevelt decided to take Nimitz and MacArthur on his inspection, his advance men scoured Oahu for an appropriate car. The islanders had been squeezed by war rationing, so the garrison had only two choices to offer the president: a slightly cramped hardtop Packard, owned by the city's fire chief, or a larger, classier touring car owned by the proprietor of one of Honolulu's well-known brothels.

Sensibly concluding that the president of the United States could not be filmed wheeling about town in a car owned by a whorehouse madam, Richardson borrowed the fire chief's car, and Admiral Nimitz squeezed into the backseat with Roosevelt and MacArthur. Leahy rode shotgun as newsreels turned and cameras snapped.[27]

Flying from Australia to Hawaii, MacArthur had complained bitterly that he had been caught at a disadvantage—summoned from Brisbane on short notice to make a presentation with few notes and no staff support. But it would be Nimitz who labored under the greater disadvantage of keeping pace with two outsize personalities. During the ride, Nimitz recalled, the conversation was dominated by "Franklin" and "Douglas," as the two men called each other. MacArthur later reminisced that at one point he asked Roosevelt whether Dewey had a chance come November.

"I've been too busy to think about politics," Roosevelt replied.

MacArthur stared at him for a split second, then burst into laughter. FDR burst out laughing, too.[28]

As Waikiki's rhythmic waves lapped the shifting sands, Roosevelt brought Leahy and his two commanders over to his borrowed beach house for a private dinner. After a dessert of ice cream served in pineapples, the men retired to the living room. Roosevelt opened the discussion by asking, "Where do we go from here?"

The foursome batted around general considerations for three hours, but reached no definite consensus. Finishing late, they agreed to meet after breakfast for a more formal discussion.[29]

The next morning Roosevelt, Leahy, MacArthur, and Nimitz resumed their talk. Standing before a giant map of the Pacific theater mounted near one of the living room's large walls, for two and a half hours Nimitz and MacArthur took turns presenting their views.

Articulating Admiral King's position more than his own, Nimitz explained that Formosa was a prime location from which to cut off Japan's oil supplies by sea. Air forces based there could also strike Japanese armies in China and hit the Empire's home islands. Luzon was less vital, for if Leyte were in Allied hands, the Allies would have an adequate anchorage in the Philippines to springboard to Formosa.

Then came MacArthur's turn. Grasping the pointer in his right hand, his left casually shoved into his pants pocket, he made his military points in confident detail before hitting themes closer to Roosevelt's heart. Using every kilowatt of his charisma, MacArthur explained the military problem in political terms. America bore a moral obligation to liberate Filipinos and Americans abandoned at Corregidor and Bataan, he said. Those dependents were suffering untold hardships in Japanese concentration camps while the nations of Asia waited for their protector to liberate them. The Filipinos—indeed, much of Asia—might forgive America for abandoning them to overwhelming force in 1942, but they would never forgive America for bypassing them in 1945.[30]

To Leahy's surprise, MacArthur and Nimitz made their points professionally and respectfully. Nimitz even conceded that further developments might make Luzon an important objective. "I never heard two men expound their views more clearly, without deviating from the main issues, than did Nimitz and MacArthur," Roosevelt's naval aide recalled. Leahy agreed. "After so much loose talk in Washington, where the mere mention of the name MacArthur seemed to generate more heat than light," he wrote, "it was both pleasant and very informative to have these two men who had been pictured as antagonists calmly presenting their differing views to the Commander-in-Chief."[31]

MacArthur's political warnings were not entirely convincing, and after eleven years in power, Roosevelt knew the American voter far better than MacArthur did. But it was an election year, and FDR's sensitivities were acute. He could not chance a misstep now, especially when the warning came directly from the darling of Republican conservatives.

As with his endorsement of Harry Truman, Jimmy Byrnes, and Henry Wallace, FDR was coy about his decision on Formosa versus Luzon. But MacArthur came away with the impression that Roosevelt had committed himself to liberating the Philippines before moving against Japan. So, too, did reporters who gathered around the grassy terrace of the beach house the next day. FDR told them MacArthur would go back to the Philippines—though he cautioned that how or when he would go back "cannot be told now."[32]

•   •   •

Before the meeting in Honolulu, Admiral King had complained about Nimitz to his friend Charlotte Pihl: "I cannot make his back stiff. He'll give in, he won't even open his mouth." But the president had asked to see Nimitz, not King, so King's Pacific chief would have to carry the ball, even if Nimitz was not the orator MacArthur was. "Nimitz was a good sound man but he wasn't even in the same class to be pitted with MacArthur that way and, of course, Mr. Roosevelt was a 'past master,'" King conceded. But he nonetheless blamed his Pacific commander for the damage to his Formosa strategy. "He let me down," he said.[33]

"Of course," Admiral King later admitted, "MacArthur was appealing to political motives, to which 'F.D.R.' was always ready to listen."[34]

On further review of America's options in the Pacific, George Marshall grew lukewarm on King's Formosa plan. He did not dispute the advantages of taking the island; a glance at the map told him Formosa was a larger and better base than Luzon. But a glance at the map would not tell him how many dead and wounded Americans it would take to capture the island.

Throughout August, Marshall pondered the cost of an island defended by 145,000 fanatical soldiers. His Operations Division estimated Formosa would cost nearly 90,000 casualties, a number that Marshall noted "approximates our total U.S. ground force casualties in France during the first two and a half months of the present campaign."[35]

On September 1, the Joint Chiefs agreed to begin landings in December on the island of Leyte, the thoracic segment of the Philippines chain. They had not yet decided whether to leap from Leyte to Formosa, or take the short hop to Luzon. Admiral Halsey, General Richardson, and many of Nimitz's planners favored Luzon, and as the Combined Chiefs prepared for their upcoming conference with the British in Quebec, the pendulum seemed to be swinging MacArthur's way.[36]

FORTY-SIX

TRAMPLING OUT THE VINTAGE

"**W**HEN ARE WE GOING TO MEET AND WHERE?"

The war's most impatient man, Prime Minister Winston Churchill, was itching to decide political and economic matters, and he pestered Roosevelt for another summit with "Uncle Joe." Five weeks after D-Day, he wrote FDR, "That we must meet soon is certain. It would be better that U.J. came too. I am entirely in your hands. I would brave the reporters at Washington or the mosquitoes of Alaska!"[1]

As it turned out, Churchill didn't have to swat at buzzing reporters or give jostling mosquitoes a quote. After a few rounds of correspondence, the two men agreed to meet on September 12 at their old QUADRANT haunt, the Citadel in Quebec. As before, the Combined Chiefs would caucus at Château Frontenac.

The OCTAGON conference, as Churchill dubbed it, signaled a shift in the high command's focus. The big decisions had been made, and grand strategy had been set. Gone, too, was the desperation of earlier meetings. Russia was safe, England was safe, the second front was open, and the springboards to Japan were being graded and paved. Even the Citadel's antiaircraft batteries, which had watched over the leaders the year before, had been packed up and sent to places where they could fire at something with a swastika or a red meatball.[2]

The final campaign for Japan brought the only open blows of the military conference. At the first plenary meeting, Churchill delivered a typically grand soliloquy on British intentions to contribute to the war against Japan after Germany's defeat. As his perorations reached their climax, he offered Roosevelt the use of the Royal Navy in the Pacific. As Roosevelt's advisers sat around the conference table, poker-faced as always—and caught completely off guard by the proposal—Roosevelt graciously accepted.

Admiral "A.B.C." Cunningham later quipped that the meeting minutes for this point should have read, "Admiral King was carried out." King loathed the idea of giving the Royal Navy a place in the Pacific order of battle. In his mind, the British had broken their promises in Burma and tried to hobble his Pacific offensive when the chips were down. In 1941 he told a confidant that he considered his Admiralty counterparts "inefficient, blundering and stupid." On a more practical level, His Majesty's Ships were notoriously "short-legged" on logistics; his seaman's eye could picture U.S. tenders being sent hither and yon to refuel their British cousins.[3]

Arnold wrote in his diary that at the next Combined Chiefs meeting, *"Everything normal until British participation in Pacific came up. Then Hell broke loose."* When A.B.C. asked where the British fleet was to be used, King said the "practicability of employing these forces would be a matter for discussion from time to time."[4]

Leahy corrected King. "He did not feel that the question for discussion was the practicability of employment," the minutes reflect, "but rather the matter of where they should be employed." A.B.C. added that the prime minister had offered Roosevelt the Royal Navy for the main effort against Japan the day before.

King, the proverbial bull who carried his china shop with him, said the British could put forth their proposals if they cared to. Cunningham snapped back that the proposal was to use Britain's fleet against Japan.[5]

The discussion turned into a row. King denied recalling any mention of the *Central* Pacific the day before. Brooke, more peevish than usual, rejoined that the plenary meeting plainly decided the basic question: The Royal Navy would join the "main effort" against Japan. Piling on, the abrasive Cunningham reminded King that FDR and Churchill had agreed to include British forces. King said he did not remember that, leading an exasperated Brooke to interject that "the offer was no sooner made than accepted by the President."

Leahy, trying to defuse the quarrel, shut King down. He said the British offer was acceptable, though no one could say for certain where British ships would be deployed. If King didn't like it, he could take his objections up with the president himself.[6]

"We had great trouble with King, who lost his temper entirely and was opposed by the whole of his own committee!" Brooke wrote in his diary. *"King made an ass of himself,"* Cunningham told his. According to Cunningham, when King turned on Marshall over MacArthur's plans for the Pacific, he "was finally called to order by

388 ★ AMERICAN WARLORDS

Admiral Leahy, the President's Chief of Staff, with the remark: 'I don't think we should wash our linen in public.'"[7]

King lost this round, but he knew there would be more rounds before the final bell. He would order his intelligence chief to withhold from the British liaison anything about future U.S. operations, and he provided no logistical help for Britain's ships. The Brits might worm their way into MacArthur's Southwest Pacific, but he could probably keep them in the dark until it was too late to interfere with his Central Pacific plans.[8]

King forestalled British meddling in the Central Pacific, but stopping MacArthur wouldn't be so easy. His Formosa strategy took an unexpected blow during the Quebec conference when Halsey's Third Fleet launched carrier raids against Mindanao and Leyte and encountered no resistance worthy of the name. The enemy's oil supplies there were gone, there was "no shipping left to sink," and "the enemy's non-aggressive attitude was unbelievable and fantastic." One of Halsey's downed pilots had been rescued by Filipinos who told him there were no Japanese left on Leyte, he reported. "The area is wide open," Halsey assured King.*[9]

With Leyte allegedly free for the asking, Halsey recommended canceling operations against Mindanao, Yap, and the Palau Islands and going straight to Leyte Gulf. That would free up an amphibious force that MacArthur could use against Leyte. Nimitz radioed King that he would cancel Yap, but wished to proceed against Peleliu and its Palau Island neighbors.[10]

Since Leyte was in the Southwest Pacific Area, MacArthur's domain, the American chiefs asked MacArthur for his views. They left the decision to him, but hinted it was "highly to be desired and would advance the progress of the war in your theater by many months" if MacArthur accepted the gift.[11]

MacArthur's reply arrived as the Combined Chiefs were dining at a banquet thrown by their Canadian hosts. Arnold, Marshall, and King politely excused themselves to read MacArthur's response. MacArthur, through his chief of staff, proposed to move forward the attack on Leyte by two months, from December 20 to October 20. Within ninety minutes, MacArthur's headquarters was decoding formal orders from the Joint Chiefs directing him to abandon Mindanao and Yap and proceed directly against Leyte. They also ordered Nimitz to go ahead with his attack on Peleliu and the Palaus.[12]

---

* As MacArthur's staff informed the Joint Chiefs, the Japanese were actually still holding Leyte. Japan's 16th Division defended the island at the time of the invasion, and before the monthlong battle was over, Japan would feed more than 45,000 troops into the fight.

The next day King informed the British chiefs of the change in strategy. The minutes simply record that the British "took note with interest of Admiral King's remarks." Having no say in the matter, the British said nothing.[13]

The "Luzon versus Formosa" debate had been tilted in MacArthur's direction by some timely help from a downed Navy pilot. By moving up the attack on Leyte, Nimitz, King, and Marshall had shifted the critical element of time in favor of MacArthur. Luzon, Leyte's northern neighbor, would be within easy reach, while Formosa remained a shimmering vision on a distant horizon.

At the end of September, King's Formosa plan finally collapsed. After the U.S. chiefs left Quebec, King flew to San Francisco to bring Nimitz up to speed on the high command's thinking. There Nimitz, Spruance, and Rear Admiral Forrest Sherman, Nimitz's planning deputy, told King that Formosa was a strategic dead end. Without Luzon, air support for an attack on Formosa would be thin; with Luzon, Formosa would be irrelevant. The better solution, they argued, was to bypass Formosa and assault Okinawa and Iwo Jima.[14]

King frowned. "Why Iwo Jima?"

An airstrip, Nimitz replied. He and Sherman explained that Army air bases on Tinian could reach Japan with long-range bombers, but those bombers would have no escort fighters, which had a shorter fuel range. Taking Iwo would allow AAF fighters to protect the Superforts as they made their way to the Home Islands and back.[15]

King didn't like the idea of quitting on Formosa, for like Stimson and Churchill, he rarely gave up the ship without a fight. But Nimitz was adamant, and he knew Nimitz was probably right.

With a sullen expression King turned to Spruance and muttered, "Why are you so quiet?"

Spruance calmly replied, "Nimitz and Sherman are doing very well. I have nothing to add."[16]

King grudgingly struck his colors. On October 3 the Joint Chiefs ordered MacArthur to take Luzon beginning December 20. Nimitz would capture Iwo Jima.[17]

. . .

A curious feature of Rooseveltian government was that individuals meant everything and titles meant nothing. The treasury secretary had overseen the sale of arms to London, the secretary of war negotiated a destroyers-for-bases deal with the British Empire, the secretary of the interior regulated Japan's oil supply, and a sickly man with no title at all negotiated with Stalin and Churchill. When one department or agency didn't move fast enough, FDR found one with a better whip-cracker, which often left two competitors working on the same job. It made no sense when drawn on an organization chart, but to White House veterans, it was just how the Boss ran the shop.

As one of Roosevelt's oldest friends, Henry Morgenthau carried a roving license to insert himself into matters outside the Treasury Department's jurisdiction. His loyalty to FDR, not his financial expertise, was the reason he remained in Washington, and early in the war Roosevelt had quipped to him, "You and I will run this war together."

Britain's lifeline, the Lend-Lease program, had begun as a cash-and-carry arrangement. That gave the Treasury Department tremendous influence over British aid as Britain began picking up the pieces of its empire. When Roosevelt and Churchill agreed to refer postwar Lend-Lease issues to a committee chaired by Morgenthau, the apple farmer turned treasury secretary found himself with a booming voice in the postwar world.[18]

Henry Morgenthau had never been a very religious man, but as the highest-ranking Jew in Roosevelt's administration, he had been a natural focus of efforts to rescue Europe's victims of Hitler's abattoir. Morgenthau and his staff had been privy to information about the horrific slaughter in Eastern Europe—information, he believed, the State Department routinely ignored—and Morgenthau met with refugee groups, traveled to Western Europe, and gathered information on conditions behind enemy lines. Like others in the administration, he was shocked at what he heard. Unlike others, he was willing to upset a lot of applecarts to punish the Nazi architects and their followers.[19]

FDR quietly encouraged him. He liked to tell of bicycling through Germany as a boy during the days of the Kaiser's reign, and seeing firsthand militant Prussian attitudes that filtered down to the nation's roots. In mid-August he told Morgenthau, "We've got to be tough with Germany, and I mean the German people, not just the Nazis. We either have to castrate the German people or you have got to treat them in such a manner so they can't go on reproducing people who want to continue as they have in the past."*[20]

In a cabinet meeting in late August, a tired and petulant Roosevelt asserted

---

* While not agreeing that conquered Germans should be neutered, Morgenthau suggested to Stimson that German children might be taken from their parents and made "wards of the state, and have ex-US Army officers, English Army officers and Russian Army officers run these schools and have these children learn the true spirit of democracy." He did not explain how Red Army officers would teach the true spirit of democracy.

that the German people could live on soup kitchens if their heavy industry were dismantled. Warming up to the subject, he then pulled out a draft handbook on military government in Germany, written by Eisenhower's civil affairs men. He read with scorn passages describing the need to restore water, heat, and other basic services to the civilian population. Morgenthau, who had quietly passed the draft on to Roosevelt, beamed with approval as FDR blasted Stimson over Eisenhower's handbook and argued vehemently for a punitive peace with Germany.[21]

But for all his talk of soup kitchens and castrated Germans, Roosevelt was not about to give Morgenthau a free hand to reshape Germany. He wanted drive but not recklessness, and in a private sidebar conference, he quietly asked Henry Stimson to have a talk with Morgenthau.[22]

Stimson loathed Prussian arrogance, which he had seen up close during the First World War. He believed small, industrialized portions of Germany, such as Alsace and Silesia, ought to be lopped off and given to France, Poland, and Russia. German schoolchildren should be re-educated—a process he thought would take thirty years—and he favored prompt trials and severe punishment for Nazi leaders.

But he took a long view of Germany. Its population had numbered about forty million until 1870. With industrialization, it swelled to seventy million, and it was obvious that without industry and trade Germany would be unable to feed those thirty million extra mouths. Food shortages would set in motion the next stage of Adolf Hitler's campaign for *lebensraum*, and in the meantime, Central Europe would suffer economically from the loss of German markets and manufacturing.[23]

Henry Morgenthau spent the Labor Day weekend at his home in Hyde Park, near the Roosevelt place, where he had a full, uninterrupted day to talk with FDR about Germany. Buoyed by Roosevelt's enthusiasm for a harsh peace, Morgenthau returned to Washington the following Monday and invited Stimson and Assistant Secretary of War John J. McCloy over for dinner, at which he outlined his vision of a new Germany.[24]

He proposed to strip modern Germany down to its agrarian roots. The Ruhr and Saar regions, the industrial heart of Germany, would be dismantled, every mine flooded, every factory wrecked. Heavy industry would be banned, and the land would be converted to fields and pastures. The German people would be held to a subsistence level, kept in a perpetual noose of dependency on imports to sustain life.

If Morgenthau's vision meant mass dislocation, if it meant dispersing inhabitants to some faraway place like Africa, so be it. Never again would Krupp cannon or Porsche tanks spring from these troves of coal and iron. "Flood and dynamite," he told his Treasury Department lieutenants.[25]

Stimson and Morgenthau had exchanged frank views on Germany a few days earlier, but Stimson was still shocked when he heard Morgenthau outline the full extent of his plan. As the scion of a Wall Street investment banker, Stimson saw economic opportunity as a sanctified path to lift Germany above its violent habits. Instead of grinding Germany into a bitter, resentful poverty, Stimson thought the Allies should give Germans an economic incentive to stay out of future wars. Even if the German people were due for a good kicking—and they were—the Ruhr was too important to the rest of Europe to destroy. After going home that night, he told his diary, *"Morgenthau is, not unnaturally, very bitter and as he is not thoroughly trained in history or even economics it becomes very apparent that he would plunge out for a treatment of Germany which I feel sure would be unwise."*[26]

The next morning Hull, Hopkins, Morgenthau, and Stimson convened in Hull's office to work out the American position on postwar Germany. Over Stimson's heated objections, Hull, Morgenthau, and Hopkins pushed for the permanent destruction of the German economy. "This Naziism is down in the German people a thousand miles deep, and you have just got to uproot it," Hull declared. Hopkins and Morgenthau looked on approvingly. "You can't do it by just shooting a few people."[27]

The meeting sputtered into a hung jury, and Stimson shuffled back to his office in a blue mood. He groused to Marshall how ironic it was that he, the man in charge of the department that did the killing, was the only cabinet member with an ounce of mercy for the other side.[28]

To halt what he saw as the administration's runaway train, Stimson wrote a memorandum to Roosevelt warning of a second Treaty of Versailles in the making. He concluded, "By such economic mistakes I cannot but feel that you would also be poisoning the springs out of which we hope that the future peace of the world can be maintained." Stimson knew the memo probably would do no good, but as he told his diary, *"I feel I had to leave a record for history that the government of this Administration had not run amuck at this vital period in history."*[29]

The next day, September 6, the group met in the Oval Office to brief the president. Repeating his position, FDR opined that the German people could get along well enough with soup kitchens and an agrarian, subsistence-level economy. But when Stimson outlined his objections to the pastoralization of the Ruhr— "the destruction of a great gift of nature," he called it—Roosevelt waffled. Perhaps Britain would need German steel to help rebuild itself, he mused. In any event, he said he was in no hurry to dismantle German industry.[30]

Stimson felt he had at last made some headway with the president. Morgenthau, on the defensive, asked for two more hours of the president's time later in the week. He and Stimson made an appointment to come back to the White House on Saturday, to reach some definite answer.

For the rest of the week, Morgenthau set his Treasury staff to work dismantling Stimson's objections, and Stimson and McCloy did the same for Morgenthau. To cover his bases, Stimson consulted Dr. Isaiah Bowman, an economic adviser and member of the Council on Foreign Relations, and he buttonholed his old friend Justice Felix Frankfurter, "a Jew like Morgenthau" whose study of German saboteur trials gave him an expertise on "the common law of war crimes."[31]

Armed with studies and expert opinions, Stimson went back to the Oval Office on Saturday, where he found his chief in a wretched physical state. Roosevelt was suffering from a sinus cold. He was tired, he was thin, and the circles around his eyes were dark. He seemed in no shape to do much other than sleep, and deciding the future of a world power was probably not one of his doctor's recommendations for that day.[32]

Yet he listened quietly as Morgenthau spoke from a compendium of Treasury staff studies. The best way to help Britain, the secretary claimed, was to seal off the Ruhr and Saar so England could dominate Europe's coal and steel markets. The best way to help the world was to strip the Ruhr, Rhineland, and Kiel Canal areas of all industry. A list of "arch-criminals" should also be drawn up, and those on the list would be lined up before firing squads posthaste.

Stimson argued vehemently against the destruction of the Ruhr and summary execution of Nazis, but he knew he was in the minority. Through rheumy eyes, Roosevelt fell back on his old tenets about soup kitchens. His comments leaned toward Morgenthau's position, though he was, as usual, unclear exactly where he was going.[33]

Shortly after Roosevelt left for Quebec, he sent for Morgenthau to join him there—ostensibly to work out Lend-Lease questions with Churchill and Anthony Eden. Stimson, not invited to the conference, fretted to Justice Frankfurter about what FDR would do without a counterweight to Morgenthau. *"[Frankfurter] said I need not worry over it; it would not go anywhere and that the President himself would catch the errors and would see that the spirit was all wrong,"* Stimson told his diary. *"I wish I was as sure as he was."*[34]

· · ·

Churchill arrived at Quebec passionately opposed to Morgenthau's plan for Germany, for many of the same reasons that moved Stimson. "He turned loose on me the full flood of his rhetoric, sarcasm and violence," wrote Morgenthau, when the subject of Germany's future came up. Britain, Churchill declared, must not be "chained to a corpse." The PM called the plan "unnatural, un-Christian and un-necessary," and said the British people would not stand for it.*[35]

But two days later, he changed his tune. Morgenthau had been England's benefactor in the dark days of 1940. Morgenthau held America's Lend-Lease purse strings, and Morgenthau knew that one of Churchill's goals was to keep America from snapping those strings tight when Germany surrendered. After Morgenthau had a long talk with Lord Cherwell, one of Churchill's personal advisers, the PM claimed he had reconsidered what a crippled Germany would do for British exports. Churchill would play ball, and he and Roosevelt initialed a memorandum agreeing to the pastoralization of Germany. Morgenthau had won.[36]

Helpless to influence events in Quebec, Stimson was livid when he heard that Roosevelt and Churchill had accepted the destruction of the German economic engine. To Stimson, government should never become so personalized as to make revenge an object of national policy. *"I have yet to meet a man who is not horrified with the 'Carthaginian' attitude of the Treasury,"* he vented to his diary. *"It is Semitism gone wild for vengeance and, if it is ultimately carried out (and I cannot believe that it will be), it will sure as fate lay the seeds for another war in the next generation."*[37]

But those seeds would not fall onto Germany's bomb-furrowed soil. Days after the Quebec conference, the tides began flowing the other way. Word of the Morgenthau Plan was leaked by underlings in the War and Treasury Departments—for opposing reasons—igniting an election-year brushfire. "Mr. Morgenthau would destroy the economic utility of a continent to gain revenge on a few hooligans," a disgusted *New York Times* reader wrote. "Such a Carthaginian peace would leave a legacy of hate to poison international relations for generations to come," wrote another.[38]

Rumors of a rift within Roosevelt's team hit the newspapers. Editorial comments centered on the cabinet split, and on FDR summoning Morgenthau to discuss foreign policy at Quebec, but not Stimson or Hull. Sounding more like Oz the Great and Terrible than the self-assured president at Casablanca, Roosevelt heatedly denied the existence of a schism within his cabinet. Few believed him. Even the *New York Times* observed that the affair "underline[s] once more the President's vacillations and vagaries in the administrative field."[39]

---

* Morgenthau told Stimson afterward, "He was even more angry [at me] than you, Harry."

. . .

Overseas, German media seized on the Morgenthau Plan for its propaganda value. The *Berliner Morgenpost* called Morgenthau's proposal a "satanic plan of annihilation," while the *12 Uhr Blatt* declared, "the aim of these conditions, inspired by Jews, is the annihilation of the German people in the quickest way."

The press uproar signaled a stiffening of German resolve. Marshall complained to Morgenthau, "We have got loudspeakers on the German lines telling them to surrender and this doesn't help one bit." Senator Edwin Johnson of Colorado echoed Marshall's complaint: "Prior to the announcement of the Morgenthau Plan, the Germans were surrendering in droves," he told the press. "Now they're fighting like demons."[40]

Roosevelt's Republican opponent felt the same way. Governor Thomas Dewey claimed the Morgenthau Plan left the Germans with no choice but to fight to the bitter end. As he bellowed to a friendly crowd at New York's Waldorf-Astoria ballroom, "Almost overnight the morale of the German people seemed wholly changed. Now they are fighting with the frenzy of despair. We are paying with blood for our failure to have ready an intelligent program for dealing with invaded Germany." The Morgenthau Plan, Dewey claimed, was as valuable to Hitler as "ten fresh divisions."[41]

. . .

"I think you have to take whatever rap is coming during the next month," Treasury's senior adviser on Germany, Harry Dexter White, told Morgenthau. "The president will do whatever he thinks is of interest to him politically. After the election, it will be a different story."[42]

Politically, it was of interest to Roosevelt to distance himself from the Morgenthau Plan. In a radio address, he drew a distinction between the Nazi leaders and German citizens. "The German people are not going to be enslaved, because the United Nations do not traffic in human slavery. But it will be necessary for them to earn their way back into the fellowship of peace-loving and law-abiding nations. And, in their climb up that steep road, we shall certainly see to it that they are not encumbered by having to carry guns. We hope they will be relieved of that burden forever."[43]

FDR's praetorians scrambled to repair the damage. "The usual 'high administration sources' are passing out word that Mr. Roosevelt does not favor the plan, after all, and that he never really was ready to adopt its basic philosophy," wrote *New York Times* columnist Arthur Krock. Roosevelt ordered the Foreign Economic Administration, a State Department group, to study economic controls for postwar Germany, and the Morgenthau Plan was swept out to sea, never to return.[44]

When Roosevelt returned from Quebec, he had Henry Stimson join him for lunch at the White House. With a smile, he said, "Henry Morgenthau has pulled a boner." Morgenthau had come to Quebec looking to have Germany dismantled, but Roosevelt said he would never agree to turn Germany into an agrarian state.

Stimson was used to presidential dissembling. It was not the first time FDR had lied about what he and Churchill had agreed to, and in the past, Stimson refrained from correcting the president. But this time Roosevelt seemed to believe what he was saying; his manner was so emphatic, the old statesman worried Roosevelt might make the claim in public and get badly burned.

For the president's own good, he pulled out and read aloud a copy of the memorandum initialed "F.D.R." and "W.S.C." The *aide-mémoire* declared that the president and prime minister were "looking forward to converting Germany into a country primarily agricultural and pastoral in its character."

Roosevelt sputtered when he heard this. *"He was frankly staggered by this and said he had no idea how he could have initialed this; that he had evidently done it without much thought,"* Stimson wrote.[45]

When speaking in lies or half-truths, a younger Franklin Roosevelt had been skillful at covering his tracks. But back then he had been an energetic, healthy man. In the fall of 1944, FDR's mind was still remarkable, but at times it seemed the ancient mariner was losing some of his legendary ability to tack between shoal and shore.[46]

For all his bitterness over the Morgenthau Plan, Stimson took no pleasure in the backlash directed at the treasury secretary. He liked Morgenthau, and he attributed the Treasury Department's meddling to "youngsters around Morgenthau who are seeking to make Treasury the center of everything." After listening to Dewey lash out at Morgenthau in a radio address, Stimson told his diary, *"I am sorry for Morgenthau, for never has an indiscretion been so quickly or vigorously punished as his incursion into German and Army politics at Quebec."*[47]

But Morgenthau was not the only one Dewey intended to punish, as Marshall was about to learn.

FORTY-SEVEN

OLD WOUNDS

⟪﹏⟫

N O MATTER WHAT HE TOLD HIMSELF, GEORGE MARSHALL WAS A POLITICIAN. A different sort of politician, perhaps—one whose party, the Army, he subordinated to the needs of his country—but he was still a politician. He hosted dinners for congressmen, he monitored the *Post* and the *Times*, and he counted noses on Capitol Hill when military bills came up on a chamber floor.

But the professional soldier in him drew the line at partisan politics. He never voted, never registered to vote. He endorsed no candidate, and he never, under any circumstances, inserted himself into a presidential campaign.

Until 1944.

In the outpouring of fury following Pearl Harbor, Americans had, for the most part, drawn together as one. Republicans and conservative Democrats, isolationists and many pacifists fell in behind FDR. Nearly everyone loyally supported the president's approach to the war.

But 1941 was a long time ago. The Nazis hadn't bombed Times Square, California was still American, and it looked like Hitler might be finished by Christmas.

Besides that, 1944 was an election year. By the canons of politics, hallowed by solemn tradition and the sacred mud first slung by the Founding Fathers, every fourth year was open season on any man with his hat in the ring.

That year Republicans had few issues to trumpet and few advantages to press. Their candidate, Governor Dewey, was a short, humorless former prosecutor who vaguely resembled the mustachioed villain Spaldoni from *Dick Tracy*. Too stiff to connect with middle-class voters, he lacked Roosevelt's easygoing charm and mastery of the radio. Dewey and his campaign strategists would find it impossible to beat FDR on personal charisma, so they banked on moderate voters buying the right issues.[1]

Finding the right issues had always been the hard part. The big one, the war, seemed to be going well. Rome was liberated, Paris was liberated, the Germans were retreating to the Siegfried Line, and MacArthur announced, "I have returned!" when he landed on Leyte. The GOP could argue until it was blue in the face that the war was not going as well as it might, but Dewey, who had never worn a military uniform, knew he had no chance of matching military credentials with the nation's wartime commander-in-chief.[2]

"How can you challenge a will-o'-the-wisp?" Dewey groaned to Robert McCormick one day.

McCormick suggested calling for a debate. Dewey snorted, "Roosevelt won't debate anything with anybody, and he will laugh at the proposal from his positions at Pearl Harbor, Guadalcanal, or the White Cliffs of Dover."[3]

Dewey would have to find another issue.

The New Deal had become a dead horse to flog, especially since FDR had publicly retired old "Doctor New Deal" and referred the patient to the younger, better-looking "Doctor Win-the-War." That left Dewey, Vandenberg, McCormick, and the rest harping about the perils of communism—another old nag, if not quite dead yet—and Roosevelt's health.

But the charge with more latent electricity than anything else was Pearl Harbor.

Blame for the disaster had lain dormant as the nation rallied against Japan. General Walter Short and Admiral Husband Kimmel, the Army and Navy commanders at Pearl, had been sacked. Admiral Stark, Kimmel's boss, was also forced out of Washington, which seemed an appropriate sacrifice at the time. But Republican strategists figured the baskets might have room for a few more heads—including one head puffing a cigarette—and they climbed into the attic and unpacked an issue they had been saving for just this sort of rainy day.

In June 1944, Congress passed a law directing the Army and Navy to investigate what went wrong at Pearl Harbor. Over the summer, the Navy heard testimony from thirty-nine witnesses, the Army one hundred fifty-one, including Stimson and Marshall. As America wondered what the investigations would turn up, stories began circulating that U.S. codebreakers had given FDR advance warning of the attack. Fed by Bertie McCormick's gumshoes and his own well-placed sources, Dewey concluded that before Pearl Harbor, American cryptologists had unlocked the Japanese Navy's codes.*[4]

Whether FDR had been reading Tojo's mail was a question that began percolating into public discourse. On September 11, Representative Forest Harness, a Republican from Indiana, gave a speech on the House floor referring to a letter dated December 7, indicating that the Japanese would present an ultimatum at one p.m. that day, and that the Japanese embassy in Washington had been instructed to destroy its coding equipment.

It was a horrifying security breach, but it caused no damage. *Congressional Record* didn't exactly have the readership of the *Saturday Evening Post*, so few average voters knew about the congressman's claims. But three days later, Navy Secretary Forrestal handed Roosevelt more disturbing news. "Information has come to me that Dewey's [next] speech will deal with Pearl Harbor," Forrestal warned. An unidentified Army officer had told Dewey's operatives that the White House knew before the attack that Pearl Harbor would be a target.[5]

Marshall was appalled to learn that Dewey was going to discuss American codebreaking in public. The man evidently did not realize that the Japanese had never changed their diplomatic codes. Baron Hiroshi Oshima, Hirohito's ambassador to Germany, was still sending radio messages to Tokyo in those old codes, and Oshima's transmissions were a gold mine of information about Hitler's plans. Once word leaked out that the U.S. had broken Japan's diplomatic codes, Oshima's gold mine would shut down and Eisenhower's job would become much harder.[6]

"The matter was growing more pointed as the campaign was growing more violent," Marshall told an investigating congressional committee after the war. "Some action had to be taken in a hurry or we were going to lose our tremendous source of information in the Far East." Marshall reluctantly concluded that he must walk through a door he had never permitted himself to open.[7]

On September 25, after fretting over what he could do to stop the looming disaster, he drafted a letter to Governor Dewey, which he forwarded to King for his thoughts. "This letter of course puts [Dewey] on the spot," Marshall admitted. "I hate to do it, but see no other way of avoiding what might well be a catastrophe

---

* Dewey's belief that Roosevelt had read Japanese transmissions was accurate, but only to a point. By December 7, 1941, MAGIC cryptographers were decoding Japanese diplomatic messages as fast as Ambassador Nomura's staffers. But they made little headway against the more complex Japanese naval code, designated JN-25, and those naval operational codes were the keys to predicting specific military targets.

to us. Just what he can do in the matter without giving reasons I do not know, but at least he will understand what a deadly affair it really is."

By inserting himself into an election campaign, Marshall knew he was crossing a dangerous line. He confessed to King, "The whole thing is loaded with dynamite but I feel very much that something has to be done or the fat will be in the fire to our great loss in the Pacific, and possibly also in Europe."[8]

* * *

In Tulsa, Oklahoma, on a warm Tuesday afternoon, a nondescript man in civilian clothes stepped off a plane. In his briefcase he carried a sealed envelope addressed to Governor Thomas Dewey, who was stumping through the Sooner State as the presidential campaign galloped into its last full month.[9]

The governor had delivered a rousing speech the day before in Oklahoma City, and he stormed into Tulsa on the crest of a wave. Ten thousand supporters met his train at the station and roared with approval as he promised to clean out the communists in Washington, the Kelly machine in Chicago, the Pendergast machine in Kansas City, Big Labor everywhere and the "political satellites who have fastened themselves on your pocketbooks and mine for twelve years."[10]

Finishing his speech, he withdrew to his posh Pullman car. Once inside, the man with a briefcase handed him a letter from the Army's chief of staff:

TOP SECRET
FOR MR. DEWEY'S EYES ONLY

My Dear Governor:

I am writing to you without the knowledge of any other person except Admiral King (who concurs) because we are approaching a grave dilemma in the political reactions of Congress regarding Pearl Harbor.

What I have to tell you is of such a highly secret nature that I feel compelled to ask you whether you accept it on the basis of your not communicating its contents to any other person and returning this letter or not reading any further and returning this letter to the bearer.[11]

Dewey stopped. He glanced at "the bearer," Colonel Carter Clarke.

As an agile prosecutor, Dewey could smell a trap a mile away. If he read the letter, he might have to stand down from an attack that would win him thousands of votes—or lose votes for revealing the information and breaking faith with *Time*'s Man of the Year.

Marshall's letter was a grenade with the pin pulled, and Dewey politely returned the pineapple to sender. He told The Bearer he would not read the letter any further and handed it back. He said he already had some of the information he suspected was in Marshall's letter. Besides, the Japanese were not dumb enough to use the same naval codes that had been broken before Coral Sea and Midway.

Dewey added that he could not believe the letter was written without Roosevelt's knowledge. "Marshall does not do things like that," he told Clarke. "I am confident that Franklin Roosevelt is behind the whole thing."

With a flash of indignation, he added that Roosevelt "knew what was happening before Pearl Harbor, and instead of being elected, he ought to be impeached."[12]

Back in Albany, New York, two days later, Dewey was surprised to see The Bearer appear at the governor's mansion with the same letter attached to a new cover letter. Marshall's second cover letter told Dewey he would not ask him to keep secret any information he received from a source other than Marshall. He also assured Dewey that the only persons who knew of the letter's existence were himself, Admiral King, seven senior Army intelligence officers in charge of signals security, and his secretary. "You have my word that neither the Secretary of War nor the President has any intimation whatever that such a letter has been addressed to you or that the preparation or sending of such a communication was being considered," he emphasized. "I am persisting in the matter because the military hazards involved are so serious that I feel some action is necessary to protect the interests of our armed forces."[13]

Dewey frowned. He told Colonel Clarke he could not agree to the request to remain silent, even under Marshall's looser conditions, without making a trusted adviser aware of the circumstances of the letter. He said he would also need to keep a copy of the letter in his most secret files.

Clarke telephoned Marshall from Dewey's office and explained the governor's conditions. Marshall agreed, then asked to speak to Dewey directly. Dewey told Marshall he would keep the matter between himself and Elliott Bell, New York's superintendent of banks and Dewey's top campaign strategist. The letter would go into a safe. Marshall agreed, Clarke left, and the matter rested in Dewey's hands.[14]

"I should have preferred to talk with you in person," Marshall wrote, "but I could not devise a method that would not be subject to press and radio reactions as to why the Chief of Staff of the Army would be seeking an interview with you at this particular moment."

He outlined his dilemma: "The most vital evidence in the Pearl Harbor matter consists of our intercepts of the Japanese diplomatic communications. . . . [W]e possess other codes, German as well as Japanese, but our main basis of information regarding Hitler's intentions in Europe is obtained from Baron Oshima's messages from Berlin reporting his interviews with Hitler and other officials to the Japanese Government. These are still the codes involved in the Pearl Harbor events."

He explained that U.S. fleet movements before Coral Sea and Midway were guided by intercepts from encrypted Japanese sources. Submarine successes against Japanese convoys came, in part, from decoded signals. Conversely, when Bill Donovan's OSS spies once broke into Japanese embassy offices in Portugal, the Japanese changed their military attaché codes, depriving America of valuable information. He concluded, "You will understand the utterly tragic consequences if the present political debates surrounding Pearl Harbor disclose to the enemy, German or Jap, any suspicion of the vital sources of information we now possess."[15]

To Dewey, Pearl Harbor was a golden hammer to swing as the campaign reached its climax. The administration had been reading Japanese mail, and a credible charge that Roosevelt had fallen down on the job so tragically could blast a hole in Roosevelt's strongest defense—his war record.[16]

Marshall had no idea what Dewey would do with the information. He held an obstinate faith that the nation's civilian leaders—including opposition candidates like Willkie or Dewey—would put the country's interests above their own partisan ambitions. But a presidential election year was a funny thing. It was the nation's most bitter regularly scheduled conflict, and people do strange things in the heat of battle. Marshall would have to wait, perhaps until November, to know whether his appeal to Dewey's better nature would succeed.[17]

·   ·   ·

While Marshall bandaged the scars of Pearl Harbor, Admiral King was doing the same at Main Navy. As required by law, naval investigators had spent the summer investigating the disaster at Pearl Harbor. The Navy's report hit King's paper-strewn desk in October 1944, the last month before the general election.

The reports laid blame on General Short and Admiral Kimmel, a finding King endorsed.* Yet the principal admirals involved—Betty Stark and Husband Kimmel—were friends of his, and King saw Stark's exile to England as punishment for Pearl Harbor. He resented the way Marshall had escaped a similar humiliation. The Army had been in charge of Pearl's air defenses, and Marshall's

---

* After the war's end, King recanted his judgment on the Hawaiian commanders. Admiral Kimmel and General Short, he concluded, had been "sold down the river."

relationship to General Short was exactly the same as Stark's had been to Admiral Kimmel. King figured Marshall probably kept his job through the secretary of war's intervention.

"That's why I didn't like Stimson," King muttered, stewing on the subject years later. "I have never been able to understand how or why FDR could fire Admiral Stark without doing the same to General Marshall. In my opinion one could not possibly be more suspect than the other."[18]

Everyone in Washington, it seemed, wanted to know what was in the report on King's desk. Dewey's Republicans were certain it painted a damning picture of incompetence—though Dewey, to Marshall's relief, did not utter a peep about the Japanese codes. FDR's pitchmen, for their part, insisted that Roosevelt had nothing to fear because he and his men had done nothing wrong.[19]

To King, job number one was keeping American knowledge of Japanese codes out of the public eye. Four enemy carriers had been sunk because of those codes, Yamamoto had been sent to the hereafter through those codes, and there would be many American lives saved if the Japanese didn't wise up to what was going on. King recommended that Forrestal keep the Navy report under lock and key, and Forrestal suppressed the report on grounds that publication "would cause exceptionally grave damage to the nation."[20]

The outrage was predictable. *Time* remarked that the "damage to the nation" Forrestal spoke of would be "election damage," and Admiral Kimmel's lawyer complained that Forrestal was using "a specious pretext to keep the truth of Pearl Harbor hidden from December 7, 1941 to November 7, 1944."[21]

But Marshall and King knew that even in a democracy, there are some truths that cannot be told. Until the war's end, the secrets of Pearl Harbor would remain in Admiral King's cluttered office.

· · · ·

At 10:05 p.m. on Friday, November 3, Mike Reilly's Secret Service men wheeled Roosevelt aboard *Ferdinand Magellan* for his Election Day trip home to Hyde Park. Unlike his overseas voyages, which had been shrouded in cloak-and-dagger schemes worthy of a Dashiell Hammett novel, on this trip the presidential party included thirty reporters, a press secretary, and FDR's two best speechwriters, Sam Rosenman and Bob Sherwood.[22]

The Presidential Special chugged its way through short stops in Connecticut, Western Massachusetts, and Boston on its way to Hyde Park, where FDR would cast his vote. He was at his buoyant best along the way, and when Reilly's men helped him off the *Magellan*, his grin betrayed no doubts about the result.

"I can't talk about my opponent the way I would like to, sometimes, because

I try to think that I am a Christian," he told a platform crowd in Bridgeport, Connecticut. "I try to think that some day I will go to Heaven, and I don't believe there is anything to be gained in saying dreadful things about other people in any campaign. After next Tuesday, there are going to be a lot of sorry people in the United States."[23]

FDR believed he would not be one of them. But he was tired and had barely campaigned. He spoke at only five major events, three of them before large crowds. At each stop, his performance was exquisite—a taste of the old master—but his offstage indifference rattled his inner circle. "He doesn't seem to give a damn," fretted Pa Watson.

That attitude began showing in the polls. As summer gave way to fall, public opinion surveys found Dewey's stock rising steadily. By October it was even possible that Dewey might pull off an upset. In the campaign season's final Gallup poll, FDR led in states claiming 209 electoral votes, while Dewey held the lead in states with 255 votes. States holding another seventy were too close to call.[24]

The skies over Hyde Park were clear and cold on Tuesday, November 7. It was a fair mix of Republican and Democrat weather, but voter turnout was heavy in urban industrial centers, which favored Roosevelt. After voting at his home precinct—again, listing his occupation as "tree farmer"—FDR rode back to Springwood. There, in the comfort of his home, he, Eleanor, Hopkins, Pa, Leahy, and the rest gathered in the study at nine p.m. to hear returns on the radio.[25]

The news was little more than speculative banter until around eleven, when AP and UPI ticker machines in the smoking room began to chatter. As three times before, Dutchess County voters pulled their levers for a Republican. So did a majority of voters in a dozen states. But New York and thirty-five others went for Roosevelt, giving him a margin of 3,594,993 votes over Dewey, and a lopsided electoral count of 432 votes to Dewey's ninety-nine. FDR had won his fourth presidential election.[26]

Dewey would not concede the race until after three o'clock in the morning. Waiting up for Dewey's concession, FDR dictated a telegram to the governor thanking him, then went to bed at four.

"I still think he is a son-of-a-bitch," he grumbled.[27]

FORTY-EIGHT

VOLTAIRE'S BATTALIONS

∽

ROOSEVELT HAD THRASHED DEWEY, BUT THE "SON-OF-A-BITCH" HAD BEEN right about one thing: Below the surface, the war hadn't been going too well lately.

The bright hopes of August had ground to a screeching halt in early September, when Eisenhower's forces outran their supply lines and halted short of the Rhine River. German rockets—first the simple jet-engined V-1 "buzz bombs," then the infinitely more lethal V-2 ballistic missiles—began crashing down on London. Field Marshal Bernard Montgomery captured the vital port of Antwerp, a critical step in the Allied war plans, but he could not get the port working because the Germans doggedly held the estuary that linked Antwerp to the sea. Then First Airborne Army suffered a bloody repulse when it tried to capture a series of bridges across Holland. Operation MARKET-GARDEN, Montgomery's gambit to get across the Lower Rhine, cost the Allies fifteen thousand men and achieved nothing.

Montgomery's neighbor to the south, Lieutenant General Omar Bradley, fared no better in the Rhineland's rough terrain. His First Army under General Courtney Hodges lost five thousand men capturing the German city of Aachen, then another thirty thousand wriggling through the primordial Hürtgen Forest, a defender's dream where everything but numbers favored the Germans. Patton's Third Army, which had made headlines racing across France, now slammed ineffectually against the fortified city of Metz. At the southern end of the front, the Seventh U.S. and First French Armies made it to the banks of the Rhine, but would go no farther until Bradley's First and Third Armies pierced the Siegfried Line.[1]

• • •

A shortage of supplies, luck, and favorable weather swirled in the cauldron of Allied misfortune that fall. But the biggest problem—a problem Marshall wrestled with for nearly two years—was a shortage of war's most basic expendable commodity: soldiers.

Voltaire once quipped, "God is on the side of the big battalions," and George Marshall intended to keep American battalions big. During the first nine months of 1944, the Army moved two million men to England and France; Britons joked that their island would sink under the weight of all those Yanks if it weren't for the barrage balloons holding it up. On the other side of the globe, more than one million soldiers were arrayed throughout the Pacific, while another 150,000 worked the supply lines between India and China.[2]

Getting all those big battalions where they could fight Hitler or Hirohito was only the first problem. The second was keeping them big when the enemy began shooting back.

The volunteer wave that swept the nation after Pearl Harbor had broken long ago, and since 1943 the Navy, Air Forces, Marines, and Army had been competing for the same limited pool of draftees.* Even within the Army, there was intense manpower competition: Europe absorbed fighting men badly needed in the Pacific and Mediterranean, the Air Forces needed flight and maintenance crews, and the Army Ground Forces required over a million men to train, equip, transport, and process draftees who would be sent abroad.

Congress bowed to Stimson's calls to lower the draft age from twenty to eighteen, at considerable political risk to themselves. But even with eighteen-year-olds in uniform, the Army had only 6.7 million men by 1943, about 1.5 million fewer than the number FDR had approved in late 1942.[3]

As clouds darkened the Rhineland's rain-soaked battlefields, Allied casualties jumped. Besides the expected hazards—getting shot, blown to bits, or captured—cases of trench foot, self-inflicted wounds, combat fatigue, and the GI trots yanked a small army out of the line. AWOL cases jumped, and Eisenhower declined a clemency request by a deserter from Detroit named Eddie Slovik, who was tied to a post and shot. The Army needed an example, and Eddie was the example.[4]

Marshall understood the fine line America's war leaders had to walk. They

---

* Competition was heaviest to get into the prestigious Marine Corps, which soon reached its authorized quota. Early recruiting was so successful that in February 1942 Admiral King advised FDR, "Unless the strength of the Marine Corps is increased, it will be necessary to stop recruiting entirely within a few days."

could not simply draft another million men and hand out M-1 rifles. The Arsenal of Democracy required millions of miners, factory workers, oilmen, and steel refiners to produce the tools of war, and it needed millions more truckers, railyard workers, stevedores, and merchant mariners to move those tools overseas. Rosie the Riveter made a fine propaganda image, but throwing open factory doors to women still did not deepen the labor pool enough to solve the problem looming before Stimson and Marshall.

Assuming that Russia would be knocked out of the war no later than 1943, Marshall originally thought the United States would need to raise two hundred divisions. When it became obvious that the Red Army was in for the long haul, he revised his estimates down to ninety divisions. To him, it was better to have fewer divisions and an assembly line of trained replacements as the blades of war took their toll. Airpower, mobility, and firepower, he assumed, would make up for any imbalance on the ground.[5]

ROUNDUP's deferral in 1942 relieved immediate pressure to train teams of riflemen and tankers. Infantry units no longer needed for an invasion of France in 1943 were broken up to create specialist units like parachute, antiaircraft, and signal battalions. By the time OVERLORD was written back into the calendar, those fighting men had been scattered like millet seeds in the wind. New men had to be found to take their place.[6]

In late 1943, as OVERLORD and ANVIL grew into hungry, fussy babies, Marshall pillaged divisions in training and sent the gleanings to Europe. He scoured the Western Hemisphere for troops he could pull from Iceland, Panama, and California, and sent Japanese Americans to fight Germans in Italy.*

Marshall leaned on his personnel wizards to ship healthy conscripts overseas and refuse "limited service" ratings to draftees who were not really incapacitated. But those wizards made plenty of mistakes, and would make plenty more before the war's end. Marshall had bellowed to an aide about a story claiming that a major-league catcher had been put on limited duty by Army doctors because he had a couple of broken fingers. "It is ridiculous to place on limited service a man who can catch with his broken fingers a fast ball," Marshall fumed. "I have seen dozens of men with half a dozen serious complaints, in addition to their years, passed by Army doctors—and now to find great athletes, football and baseball, exempted, is not to be tolerated."[7]

---

* In Italy and France, soldiers of the 442nd Infantry Regiment, composed largely of Nisei, earned 21 Medals of Honor, 7 Presidential Unit Citations, 53 Distinguished Service Crosses, 560 Silver Stars, 4,000 Bronze Stars, 9,486 Purple Hearts, and 18,000 other individual citations, making the regiment, man for man, one of the most decorated units in U.S. history.

But cannibalizing regiments and drafting athletes wouldn't be enough. While Congress had authorized an Army of 7.7 million by the end of 1944, authorization was not the same thing as having 7.7 million fighting men. One Army survey showed an estimated 200,000 men living within the replacement pipeline; one general called them "an invisible horde of people going here and there but seemingly never arriving." Another 400,000 were casualties recovering in hospitals, while 150,000 were classified simply as "overhead," for lack of any better description.[8]

Of the men who actually made it overseas, only a relative handful would be doing the bloody work. Most would ride the Army's serpentine logistical tail— "one a-shootin', ten a-lootin'," the GI grumble went—and in early 1944 Marshall told Roosevelt that infantry, accounting for 11 percent of the Army and Air Forces, was taking 60 percent of all casualties in Italy. All told, European infantry and armored divisions were thought to be short about 100,000 men, though no one really knew for certain.[9]

To find replacements for rifle companies, Marshall began dismantling every non-essential program he could lay his hands on. He began by tackling the Army Specialized Training Program, a pet project of Stimson's that was educating 75,000 men for specialist jobs like engineering, dentistry, and foreign languages. He slashed furloughs, cut infantry training by two weeks, retrained 30,000 men from the Air and Service Forces, and told Eisenhower to cut his quartermaster count in Europe. He then stripped more infantry divisions training in the United States until he had moved another 78,000 replacements to Europe.[10]

Despite Marshall's cuts and cannibalization, the Army still lacked enough killers to drive the dagger home. Marshall and Stimson knew there would come a bitter season when the German will to resist would break, or the lines on both sides would stagnate and harden.

Back in May, Stimson had asked Marshall whether a stalemate would grip the western front in the autumn. At the time, Marshall could not say for certain, but by November 1944 the answer was plain to everyone. Heavy rains, swollen rivers, Rhineland mud, and desperate German resistance had left the Hun in possession of the forbidding land along the Rhine.[11]

∙ ∙ ∙

In times of peace, the death of an empire is brought on by old age, or by chronic diseases like complacency or debt or corruption. It comes so slowly its people do not even know their world is dying until the disease is irreversible. In times of war, the end comes like a wild boar from the woods, yellowed teeth bared, mud-spattered tusks lowered to gore. Those empires are overrun quickly and violently, and nearly every inhabitant, highborn or low, knows damn well what is happening to them—and what awaits them.

In the Third Reich, the few who did not know death was upon them lived like blind moles in their underground bunkers and Alpine redoubts. These men, to whom a far-off, indistinct gallows beckoned, saw no defeat, because defeat's implications were unthinkable. Defeat meant the end of their existence, the end of their ideology, the end of the world they had built.

Those men, the leaders of the master race, heard no Allied bombers and saw no despair in the faces of their famished, glassy-eyed followers. They put their faith in destiny, Hitler's uncanny luck, and the battered but unbroken Wehrmacht as it reared back for its most desperate gamble.

FORTY-NINE

COUNTING STARS

∽

A s General Bradley met with Eisenhower near Paris on December 16 to discuss the deteriorating manpower situation, word arrived of a German attack through the Ardennes Forest in Belgium and Luxembourg. Through softly falling snows and under cover of low, thick clouds, the Second SS Panzer Division led a quarter-million-man panzer army straight into the belly of Bradley's thin line. *"A complete surprise to our people,"* Jack McCloy told his diary when word reached the War Department. *"I could sense that General Marshall was very much upset over the developments."*[1]

Upset he was, but Marshall didn't want anyone in Washington bothering Eisenhower when he had his hands full. He forbade his staff from sending any message to SHAEF inquiring about operations in the Ardennes without his express approval, and as the German bulge deepened, Marshall refused to interfere, trusting that Eisenhower knew best what to do.[2]

Mulling over the attack in Stimson's office, Marshall and Stimson compared the battle raging in Belgium to von Hindenburg's last great offensive in the spring of 1918. When the Kaiserschlacht was broken, an exhausted Germany had no choice but to sue for peace. Perhaps, Stimson thought, history was ready to repeat itself. *"We know a good deal more today about the situation behind the German lines than we did twenty-six years ago,"* he wrote. *"So if this is stopped, particularly if stopped quickly and sharply and with big losses to the Germans, then we may have them crumbling sooner than expected."*[3]

But thoughts of a second Armistice Day were premature. Like a Boris Karloff monster, the Wehrmacht that had been destroyed at Stalingrad and Tunisia, at St. Lô, Belorussia and the Falaise Gap, kept shuffling back like an animated corpse, arms outstretched and claws spread, unaware of its own death. On December 19, Eisenhower sent word that he was throwing in his only reserves, the 82nd and 101st Airborne Divisions, and he placed the First and Ninth U.S. Armies, which had become separated from Bradley's headquarters, under command of Field Marshal Montgomery.[4]

Roosevelt spent the morning of the twentieth in the Map Room reviewing dispatches with Leahy and pinpointing obscure crossroad towns like St. Vith, Spa, and Bastogne. But throughout the Battle of the Bulge, as the papers began calling it, he maintained his usual routine, taking trips to Warm Springs and Hyde Park to celebrate Christmas, keeping one tired eye on the battle's progress.

"Roosevelt didn't send a word to Eisenhower nor ask a question," Marshall recalled. "In great stress Roosevelt was a strong man." Stimson agreed. *"The anxiety on [Roosevelt's] part must have been heavy,"* he wrote in his diary at year's end. *"He has been extremely considerate and has not asked any questions or sought to interfere in any way with either Marshall or myself while the crisis of the German counterattack has been going on."*[5]

Their faith in Eisenhower paid off, for in the Ardennes Hitler rolled snake eyes. Ike's men held on by their fingernails, but they blunted, then turned back the great armored onslaught. The north shoulder of the bulge held at Malmédy. Fierce resistance at Bastogne and St. Vith slowed Hitler's advance long enough for Patton's army to stab into the bulge from below. The day after Christmas, Patton's Fourth Armored Division shot its way into Bastogne, and Montgomery's British and American troops blunted the nose at the Meuse River, sixty miles from where the Germans started. By year's end it was clear to everyone that Hitler's gamble in the west had failed.

· · · ·

While Eisenhower juggled manpower, supply, and strategy problems, Field Marshal Montgomery lobbied for a shake-up in the Allied command structure.

Montgomery had begun complaining almost from the moment Eisenhower assumed personal command of Allied ground forces on September 1, and his cry was echoed in London by Field Marshal Brooke. Comparing the breathtakingly swift advance of August—when Monty had been ground force commander—to the slow grind of autumn, Monty told anyone who would listen that he, not Eisenhower, was the better man to run the ground war. Eisenhower, he said, should remain supreme commander, a "chairman of the board" coordinating land, air, sea, supply, and civil matters, but Montgomery should resume his place as the overlord of battle.[6]

Considering the matter to be Eisenhower's business, Marshall held his

412 ★ AMERICAN WARLORDS

tongue. But it was not easy to hold one's tongue with a persistent, tactless man like Montgomery. "I came pretty near to blowing off out of turn," Marshall said later. It was "very hard for me to restrain myself because I didn't think there was any logic in what he said but overwhelming egotism."

Yet Marshall did sound off to Ike. He cabled Eisenhower, "My feeling is this: Under no circumstances make any concessions of any kind whatsoever. You not only have our complete confidence but there would be a terrific resentment in this country following such an action. . . . You are doing a grand job and go on and give them hell."[7]

Montgomery's complaints to Brooke made their way to Churchill, who passed them along to the president. Roosevelt, like Marshall, refused to second-guess Eisenhower. "For the time being," Roosevelt counseled Churchill, "it seems to me the prosecution and outcome of the battles lie with our Field Commanders in whom I have every confidence."[8]

On the last day of the year, Stimson told Roosevelt of a clamor in London newspapers for Montgomery to replace Eisenhower as ground commander. He showed Roosevelt Marshall's "give them hell" telegram to Ike, and Roosevelt agreed they could not allow Eisenhower to be displaced by Montgomery or anyone else.

At Marshall's suggestion, Roosevelt added a word of support for Eisenhower to his "State of the Union" message to Congress. "The speed with which we have recovered from this savage attack is largely possible because we have one supreme commander in complete control of all the Allied armies in France," he told Congress. "General Eisenhower has faced this period of trial with admirable calm and resolution and with steadily increasing success. He has my complete confidence."[9]

★ ★ ★ ★

Other commanders held Roosevelt's confidence, and FDR intended to recognize them with more than just words. In early 1944, he decided to add another rung to the military ladder by giving Leahy, Marshall, and King a promotion.

The idea had lain dormant for some time. In 1942 Admiral King had suggested that the Joint Chiefs make a recommendation to the president standardizing the criteria for four-star ranks for men like Halsey and Eisenhower. To this King added, "We should also recognize the fact that there is a need to prepare for ranks higher than that of Admiral or General. As to such ranks, I suggest Arch-Admiral and Arch-General, rather than Admiral of the Fleet and Field Marshal."[10]

Nothing moved until January 1944, when the idea gained new life. This time, Roosevelt said he preferred "Chief Admiral" and "Chief General," while King suggested the perplexing titles "Captain Admiral" and "Colonel General." A nonplussed Secretary Knox had written King at the time, "Personally I don't care very much how the title is fixed so long as recognition is given to the pre-eminence of the command you [and Marshall] both exercise." But everyone who did care, other than King, thought the name "Captain Admiral" too confusing.[11]

According to the Washington grapevine, FDR's real motive behind the new stars was to promote his personal physician, Rear Admiral Ross McIntire—who, in addition to his jovial personality and deep-sea fishing prowess, had an exceptional gift for reassuring the press that his patient was in the best of health. To promote his cronies, FDR would have to bump up seven other admirals, which would require every senior admiral to move up by at least one more star. So Nimitz, Halsey, and Ingersoll would each become a five-star "Admiral of the Fleet," while Leahy and King would hold the six-star ranks of "Admiral of the Navy."[12]

The Army's supporters would naturally demand similar promotions, and it would be unthinkable for Congress to overlook *Time*'s Man of the Year. But Marshall frowned at the idea, since a promotion was the kind of favor some sharp-elbowed congressman would one day call in.

"I didn't want any promotion at all," Marshall explained later. "I didn't think I needed that rank and I didn't want to be beholden to Congress for any rank or anything of that kind. I wanted to be able to go there with my skirts clean and with no personal ambitions." Added to this was the odd ring of "Marshal Marshall," which had the same linguistic absurdity as, say, promoting a man named "Major" to the rank of major.*[13]

In early February Roosevelt told Stimson he intended to promote three rear admirals, including Dr. McIntire, to vice admiral. He would also approve fifth stars for King and Marshall, to keep parity with the British chiefs—Field Marshal Alan Brooke and Admiral of the Fleet Sir Andrew B. Cunningham.[14]

The impetus to pin more stars on the shoulders of America's war chiefs came to a screeching halt in April 1944, about the time the New York *Herald Tribune* ran an article under the headline "Marshall Asks That New Rank Bill Be Shelved." The article quoted a War Department source as saying General Marshall opposed the bill. It also claimed certain "friends of Admiral King were quick to point out that, like General Marshall, he too, was not in the least interested in 'gathering a new title.'"[15]

But King was interested in doing just that, for both organizational reasons and self-aggrandizement. When he read the story, he blew up and called in his

---

* Of course, Joseph Heller's Major Major Major, who bore a striking resemblance to Henry Fonda, was born to mediocrity, whereas Marshall was not.

director of public relations. He told the man to get over to the War Department and find the bastard who had talked to the press.

Marshall denied that anyone on his staff had contributed to the article, but he took the fall for killing the bill. Stimson told his diary, *"Poor Marshall said he was getting blamed by columnists as responsible for the holdup of Navy promotions and he told me that the President was hellbent on getting the five star promotions in large numbers for the Navy."*[16]

At the end of 1944, the idea was resurrected when Roosevelt again urged Congress to create a five-star rank. The Navy side was reasonably simple—the new rank would be called "Fleet Admiral"—but the Army side was a bit more complicated. "Field Marshal" was out, and Stimson and General McNarney looked for a name slightly less grandiose than Pershing's five-star rank, "General of the Armies." After kicking around various titles, they proposed "General of the Army," keeping Pershing's title unique while according Marshall preeminence among Pershing's successors.[17]

A bill to authorize eight five-star officers—four from the Army, four from the Navy—made its way through Congress in early December and was ready for the president's signature by mid-month. General Thomas Handy, Marshall's deputy chief, met with his opposite number at Main Navy, and together they worked out the line of seniority. The new ranks would keep the same pecking order they currently had, more or less. Admiral Leahy took first place, then came Marshall, then King, then MacArthur, Nimitz, Eisenhower, and Arnold.*[18]

Marshall, who had been away on an inspection trip during the latest fuss over promotions, had just arrived back in his Pentagon office when he and Hap were abruptly summoned to meet Stimson in his office. The two men entered to the welcoming smiles of nearly two dozen top generals holding champagne flutes. Warm congratulations were offered by the secretary of war, and the group toasted the Army's two newly minted Generals of the Army.[19]

Then, as always, everyone went back to work.

---

* The Navy reached an impasse over whom to nominate for its fourth slot. Forrestal favored Halsey, while to King it was a horse race between Halsey and Spruance. Ultimately Halsey won the fourth slot. After the war, Omar Bradley received his fifth star, becoming America's last five-star general. In 1976 Congress posthumously promoted Lieutenant General George Washington to "General of the Armies of the United States," a rank it decreed would be higher than any other Army rank, past or present. General Washington will always stand as America's highest-ranking soldier.

FIFTY

THE TSARINA'S BEDROOM

 ∽

"N O MORE LET US FALTER! FROM MALTA TO YALTA!"

In his whimsical prose Churchill wired Roosevelt in January 1945, as the western leaders prepared for their next meeting with Stalin. With decisions on the final drive into Germany, European borders, a world peace organization, and war with Japan coming at them fast, Roosevelt and Churchill agreed it was time for another meeting of the "Big Three."

Stalin said he would go no farther west than the Black Sea resort of Yalta, on the Crimean Peninsula. His role as Soviet military commander, he said, ruled out lengthy travel. He also said his health problems made any journey outside Russia's vast borders inadvisable; his doctors urged him not to travel too far.[1]

Roosevelt's own health had been an issue—a very public issue during the election—and he didn't relish the idea of traveling six thousand miles to the Black Sea. Most of the ports there had been destroyed by the Germans, and the health and sanitary conditions, he had heard, were abysmal. Churchill felt the same way; he groused to Harry Hopkins, "If we had spent ten years on research we would not have found a worse place in the world than Yalta."[2]

But Stalin was in no mood to bargain. His armies were almost to the Oder River, the last geographic barrier to Berlin. What Stalin wanted most, a security corridor on his western border, he had practically attained by force of arms. What he wanted after that—Allied recognition of his conquests—was of far less importance.

So the summit was set for Yalta in late January, shortly after Roosevelt's fourth inauguration. At Churchill's request, Americans and British would hold preliminary talks on Malta, Britain's island fortress in the eastern Mediterranean. Churchill chose the code name ARGONAUT for the meetings.[3]

*   *   *

On the second of February, the cruiser USS *Quincy* picked her way through the entrance of Malta's Grand Harbor to the cheers of what seemed like the island's entire population. Gliding serenely past a collage of ships and the limestone rubble of bombed-out buildings, the presidential flagship pulled into her berth as a beaming Winston Churchill stood at the siderail of a his own cruiser to bid the Americans welcome.[4]

Churchill later wrote, "I watched the scene from the deck of the *Orion*. As the American cruiser steamed slowly past us towards her berth alongside the quay wall I could see the figure of the President seated on the bridge, and we waved to each other. With the escort of Spitfires overhead, the salutes and the bands of the ships' companies in the harbor playing 'The Star-Spangled Banner,' it was a splendid scene."[5]

The Malta meetings of the Combined Chiefs had started three days earlier in the conference room of Montgomery House, a large, stately office building in a Valletta suburb named for the diminutive British general who had finalized his plans for the invasion of Sicily there. Marshall, King, and Major General Laurence Kuter, representing the ailing Hap Arnold, arrived on January 29 and took up the dwindling plate of military issues relating to the war against Germany and Japan.*[6]

By January 1945, most of the great decisions had been made. Execution of those decisions lay in the hands of Eisenhower, Jumbo Wilson, Lord Mountbatten, Admiral Nimitz, and General MacArthur. There was really only one unanswered question at the moment, and on that point Allied harmony broke down: Eisenhower's plan for the conquest of Germany.[7]

Eisenhower sent Lieutenant General "Beetle" Smith to Malta to explain his strategy. His plan was simple. Montgomery's Twenty-first Army Group in the north, Bradley's Twelfth Army Group in the center, and General Jacob Devers's Sixth Army Group in the south would close up against the Rhine together, destroying the enemy along the Siegfried Line and spreading Germany's thin forces to the breaking point. Once up to the Rhine, all three groups would make the

---

* Arnold was at Walter Reed recovering from his fourth heart attack of the war.

final push into Germany. Under Eisenhower's "broad front" plan, no army group would expose itself by dashing ahead of the others, a limitation the brash Montgomery found infuriating.[8]

Montgomery had argued, often and loudly, that his army group should be given the lion's share of ammunition, fuel, and air support. With enough resources, he assured the high command, he would drive into Berlin and plant the Union Jack on the Brandenburg Gate.

But Allied supply limitations and the enormous amount of fuel, food, and ammunition it took to move an army meant that to give Montgomery what he wanted, Eisenhower would have to shut down every other army group. To Eisenhower and his staff, pushing the Allied line on all fronts would prevent Hitler from concentrating his reserves against any single attack. In addition, the Allies could support only thirty-six divisions in Montgomery's area. With some eighty-five divisions available, he could afford to push Bradley through the center and Devers to the south.[9]

Even before first meeting at the Montgomery House, the Americans were in a foul humor. Brooke and Montgomery had been lobbying to displace Eisenhower as ground commander by inserting a new layer of command between him and his field generals. Marshall and King took this as a slap in the American face. Brooke's incessant pecking about American strategy had become tiresome and repetitive, and it did not help that after their long air journey to Malta, Marshall and King had endured a terrible night's rest in the Royal Artillery Barracks, an ancient billet so insufferably cold and damp that everyone slept fully clothed and wrapped in their topcoats.[10]

Shortly before the Combined Chiefs met to discuss Eisenhower's plans, Beetle Smith briefed Marshall, King, and Kuter on Ike's strategic reasoning and his problems with Montgomery. King, fully worked up, growled that the chiefs should sack Monty, and he insisted he would make that an item on their agenda. Marshall, barely less angry, tried to keep relations civil. "Please leave this to me. I will handle it," he said.[11]

At the meeting, Brooke, as expected, argued vehemently against Eisenhower's "broad front" strategy. The north, where Montgomery's group lay, was the decisive region, he said. It abutted the industrialized Ruhr and opened the most direct path to Berlin. In his rapid, grating manner, "Colonel Shrapnel" clicked off everything he found wrong with Eisenhower's approach. As he did, the Americans fumed.[12]

Marshall's patience finally snapped when Brooke claimed that the British chiefs were concerned that Eisenhower was being unfairly influenced by his

American associates Bradley and Patton. Marshall cleared the room and bellowed, "Well, Brooke, [the British chiefs] are not so nearly as much worried as the American Chiefs of Staff are about the immediate pressures and influence of Mr. Churchill on General Eisenhower. The president practically never sees Eisenhower, never writes to him—that is at my advice because he is an Allied commander—and we are deeply concerned by the pressures of the prime minister and of the fact of proximity of the British Chiefs of Staff, so I think your worries are on the wrong foot!"[13]

With that prologue, Marshall launched a full-throated attack on Montgomery, a supreme egotist whose attitude toward Eisenhower bordered on contempt of a superior officer. Montgomery had been pushing for months to supplant Eisenhower as Allied ground commander, and in his high-handed way, he had insisted on having all available gasoline, ammunition, and air to support his personal crusade to aggrandize himself.*[14]

Amidst heated words flying in both directions, Beetle did his best to clarify Eisenhower's plans. Ike did not intend to slight Montgomery, and he would allow Monty to charge over the Ruhr once everyone was, more or less, up against the Rhine. If something unexpected happened—a collapse in the Ruhr pocket, for instance—they would take advantage of the opportunities and unleash forces wherever they could be put to good use. Relieved to hear that Monty would not be held back while the Americans tore through central Germany, and stung by the venomous American response, Brooke, Portal and A.B.C. withdrew their objections to Eisenhower's plan and his retention as ground commander.[15]

Though the Combined Chiefs managed to settle the strategy questions, the meeting's sour taste lingered like a bad cigar. Two weeks later, Marshall was still fuming, and after the war Brooke described for his diary the bleak British viewpoint:

I did not approve of Ike's appreciation and plans, yet through force of circumstances I had to accept them. . . . In addition there was the fact that Marshall clearly understood nothing of strategy and could not even argue out the relative merits of various alternatives. Being unable to judge for himself he trusted and backed Ike, and felt it his duty to guard him from interference.[16]

---

* King was more circumspect than either Marshall or Eisenhower about Montgomery's prima donna attitude. "Of course [Montgomery] was a show-off (beret), like MacArthur (cap) or Patton (two pistols) and many other people (like Halsey or Ingram, etc.) and 'W.C.' and 'F.D.R.,'" he wrote years later.

. . .

After Roosevelt's flagship docked and protocol visits were completed, the American chiefs marched up *Quincy*'s gangplank to update the president, who sat placidly in a wicker chair near a slumbering antiaircraft gun.[17]

Roosevelt, King thought, looked terrible. His dark suit hung from his limbs and his tweed driving cap, almost comically wide now, balanced itself precariously over a thin face and body. Sea voyages usually buoyed him, but the sixty-three-year-old New Yorker looked like an aging groundskeeper on his last tour of the master's estate.[18]

Exchanging pleasantries, the group retired to Roosevelt's cabin to talk in private. FDR, almost somnolent, roused himself during the half-hour talk when Marshall brought up the fight over Eisenhower's strategy for central Germany. Perking up, Roosevelt called for a map of the region. Studying the map, he told his chiefs he had bicycled in that area as a boy. Momentarily energized by a story he repeated every time the discussion turned to Germany, he said he approved of Eisenhower's strategy, settling Allied strategy once and for all.[19]

The Anglo-American war chiefs had resolved other issues with the same give-and-take that had characterized the alliance since 1941. Churchill gave up his designs on northern Italy, and Roosevelt, seeing China for the dead end that it was, scaled back his support for Chiang.* Roosevelt agreed to allow General Sir Harold Alexander to join Eisenhower's headquarters team as deputy supreme commander, so long as Ike remained ground force commander. Since the move might be interpreted as disapproval of Eisenhower's handling of the Battle of the Bulge, Roosevelt insisted that they delay the announcement for several weeks, giving the Ardennes time to fade from the front pages.[20]

Next morning, the Allied high command piled into a fleet of C-54 transports for the 1,400-mile journey from Malta to the Black Sea coast. Accompanied by Harry Hopkins, his watchful daughter, Anna Boettiger, Admiral Leahy, and his cardiologist, Dr. Bruenn, FDR settled into a more luxurious version of the *Sacred Cow*, a dedicated presidential airliner boasting an office, a full-size bed, and elevators that allowed him to disembark without using tall, cumbersome wheelchair ramps.[21]

Sacred Cow carried Roosevelt over the Adriatic, across Greece and Mace-

---

* In October 1944, Chiang finally managed to rid himself of Stilwell. As Vinegar Joe was being shipped off for Washington—under Marshall's injunction to say nothing to the press—Chiang offered him China's highest military decoration. *"Told him to stick it up his arse,"* Stilwell told his diary.

donia, and over the Black Sea to Saki, a Soviet airfield near Sevastopol. It was Russian territory now, but as some spots had only recently been evacuated by the Germans, the pilots took no chances. They flew under cover of darkness, in radio silence, and the Navy stationed sea rescue vessels along the Adriatic and Black Sea portions of the flight, to pluck any unfortunates from the water.[22]

The air flotilla touched down in Saki and the entourage was welcomed by Commissar Molotov, Soviet Ambassador Andrei Gromyko, and a platoon of Red Army and Navy brass. An honor guard stood at attention, bayonets fixed, and behind them a Red Army band played passable renditions of "The Star-Spangled Banner," "God Save the King," and the minor-key Soviet anthem "Internationale."[23]

After inspecting the honor guard in an American jeep, Roosevelt transferred to a convertible Packard and was driven along the "Route Romanov" to Livadia Palace. The bumpy three-hour road trip took the Americans across the heart of the Crimean Peninsula. "The Russian drivers used only two speeds," King recalled. "Stop, and drive as fast as they could."

Winding along unnerving cliff roads, they passed surreal scenes of snow-capped mountains, burned-out German vehicles, and thousands of Red Army guards, men and women, stationed along the route. Rolling through the palace gates some sixteen hours after leaving Malta, the exhausted visitors bathed, ate a perfunctory meal, and fell into bed.[24]

Livadia Palace, Roosevelt's home for the next eight days, was a Romanov family mansion that had been converted into a seaside resort after the Bolsheviks revised Russia's organization chart. Built of white Inkerman granite in the Florentine style, the fifty-room palace commanded spectacular views of both sea and mountain. But for its more recent history, the palace would have made a picturesque setting for a gathering of world leaders.

Livadia had been occupied by the German Eleventh Army headquarters staff in 1941, and its Teutonic guests proved to be more than a little unruly. The palace had been stripped of its furniture and all ornamentation by the time the Red Army served its eviction notice in April 1944, and a few short weeks before the arrival of the Big Three, it remained a lice-ridden, filthy shell.[25]

Then Stalin's men went to work. With an urgency typical of Soviet wartime operations, workers had the palace deloused, scrubbed, furnished, and staffed in magnificent style. In less than three weeks, cars, dinnerware, linens, and furniture from hotels across Russia were shipped in on 1,500 railway cars and arranged by a regiment of bellhops, chambermaids, chauffeurs, cooks, and concierges drafted from Moscow's Metropol Hotel.[26]

Stalin's secret police, the NKVD, contributed to the palace renovations by installing hidden listening devices in the American quarters, and they transported a pool of bilingual women to the Crimea to transcribe recordings of each day's conversations. Those transcripts were reviewed daily by Commissar Lavrentii Beria, the NKVD head, and Stalin himself reviewed the important ones.[27]

During his stay, Franklin Roosevelt lived in reasonable comfort in a bedroom suite built for Tsar Nicholas II, just off the main ballroom where the main conferences would be held. His lieutenants were billeted in the tsarina's suite, on the second floor, while the British and Soviet deputations lived outside the palace, commuting to Livadia each day.

Up a flight of stairs, Marshall was assigned Tsarina Alexandra's main bedroom, while King was lodged in her boudoir. Marshall amused himself by making sure everyone knew King was sleeping in a room with a secret staircase reputedly built for Rasputin, the mad monk who held sway over the tsarina.

King took the ribbing in stride. "George," he asked, "what the hell would you do if Rasputin came through that window?"

"I'd call [Sergeant] Powder," said Marshall, referring to his burly enlisted orderly.[28]

No one found the bathroom arrangements amusing, however. The palace had been designed for the pious Nicholas without vodka toasts or heavy drinking in mind, and it was woefully short on lavatories. Only FDR had a private bathroom; everyone else had to stand in line. Each person carried at all times a card identifying the bathroom he was permitted to use. "Supplementary facilities," the Soviet information bulletin dryly informed them, "include outdoor latrines and as many wash basins, pitchers and buckets for bedrooms as are available." General Kuter later remarked, "Excepting only the war, the bathrooms were the most generally discussed subject at the Crimean Conference."[29]

On the afternoon of February 4, Stalin arrived at Livadia with Commissar Molotov and his military retinue. Surrounded by a phalanx of guards cradling submachine guns, he entered the palace and stopped by Roosevelt's study for a courtesy call.[30]

Believing, as always, that personal trust would be repaid in kind, FDR tried to set Stalin at ease by hinting at weakness and divisiveness within the Allied ranks. Turning over the same old card he played at Tehran, he tried to ingratiate himself with Stalin by complaining that the British were "peculiar people" who wanted to strengthen France artificially. He also paid a compliment to the Red Army by telling the dictator he had made a bet that the Russians would be in Berlin before the Americans captured Manila. Stalin mildly predicted that Roosevelt would lose his bet, as the Germans were fighting ferociously along the Oder River line.*[31]

After a few minutes, Roosevelt was wheeled into the palace ballroom. As stone-faced guards circled the upper balcony, burp guns at the ready, the "Big Three" took their seats at a large round table, flanked by foreign ministers, commanders, and interpreters. The first plenary session of the Yalta conference came to order.[32]

Germany was a comparatively easy issue. FDR observed that the European Advisory Commission had proposed a three-way partition of Germany between the Soviet Union in the east, the United States in the southwest, and Britain in the northwest. Churchill advocated an occupation zone for the French Provisional Government, to be carved out of Britain's territory. With some remonstration, Stalin agreed. Germany would be divided into four parts.[33]

Stalin's demands for reparations were as heavy as the Soviet heart. He wanted a two-year right to remove all factories, machine tools, rolling stock, and other means of production from Germany. Over ten years, Germany would be required to send its manufactured goods east. Steel, electrical, chemical, and military production would be eliminated. The cost of the reparations to Germany, he estimated, would be around ten billion dollars, and he was prepared to approve a similar scale of compensation for the west.[34]

A tired FDR said little, while Churchill argued that Stalin's demands were unrealistic. A stripped-down Germany would be unable to make manufactured goods or pay its debts. Much as he despised the Nazi leaders, said Churchill, "If you wished a horse to pull a wagon, you would at least have to give it fodder."

Stalin replied, "That was right, but care should be taken to see that the horse did not turn around and kick you." The Soviet government would take enough fodder to keep the horse thin and weak.[35]

At the second meeting Roosevelt announced that the Americans would not occupy Germany for long. "I can get the people and Congress to cooperate fully for peace but not to keep an army in Europe for a long time. Two years would be the limit," he said.[36]

It was no impromptu remark by a tired old man. America had a war to finish in the Pacific, and Roosevelt refused to maroon American soldiers in Central

* Stalin was right. The next day, MacArthur announced that his forces had entered Manila and were rapidly clearing the city. Fighting there continued for nearly another month, but the city was pacified before Berlin was breached.

Europe. America's fate would be linked to Europe diplomatically, economically, and through a world peace organization, but the United States had no interest in pinioning itself to a burned-out battlefield.

Churchill was horrified at Roosevelt's announcement. American might, he believed, was the West's best leverage against the Russian bear. Churchill had battled Bolshevism since 1917, and while he remained loyal to Stalin as an ally against Germany, he understood that gratuitous gestures and displays of weakness were a coin the Russians would never recognize, and the West could never redeem. Having lost more than fifty million citizens since 1914, they recognized strength and brushed aside sentiment.

Privately, Roosevelt's military leaders subscribed to Churchill's view. "The Russians are tough traders," King had told his confidants after the Tehran conference. "They do not become angry at the most direct talk or the toughest attitude on our part. If we are soft, they regard us as fools and they expect the other fellow to look out for himself—fully. If he doesn't, it's his own fault." Major General John Deane, head of the American military mission to Moscow since 1943, told Marshall the Russians "simply cannot understand giving without taking, and as a result even our giving is viewed with suspicion. Gratitude cannot be banked in the Soviet Union. Each transaction is complete in itself without regard to past favors." In Deane's view, "The party of the second part is either a shrewd trader to be admired or a sucker to be despised."[37]

FDR's hopes for a lasting peace transcended the "every man for himself" approach that had dominated geopolitics. He had won over antagonists like Joe Kennedy and Wendell Willkie, and the optimist in Franklin Roosevelt believed he could build a personal rapport with Stalin, a trust necessary to give him what he most desired—a world security institution.

Stalin, a tyrant without peer, had a soft spot in his granite heart for the capitalist leader. He later called Roosevelt "a great statesman, a clever, educated, far-sighted and liberal leader who prolonged the life of capitalism." He respected Roosevelt's broad vision and took a kind of rough pity on his inability to walk. After leaving Roosevelt's room one afternoon, he turned to Ambassador Gromyko and asked, "Why did nature have to punish him so? Is he worse than other people?"* No personal goodwill could change the decades-old struggle of two colliding economic and political systems, but Stalin was willing to give Roosevelt's experiment with a peace institution a try.[38]

Roosevelt had conceived of the "World Organization" as a gathering place for

---

* An astonished Gromyko wrote, "Despite his basic harshness of character, Stalin did just occasionally give way to positive human emotions."

all nations, a forum where grievances could be aired and solutions worked out. The world's military power would be vested in the "Big Four"—the United States, USSR, Britain and China—and charter membership would be held by the "United Nations," those countries that had signed the declaration of war against Germany. Others that maintained relations with the Axis, such as Ireland, Denmark, or Argentina, would be excluded from the first round but permitted to join later, and eventually all states would be granted membership to the general assembly. Stalin and Churchill agreed, and the Big Three scheduled a conference to inaugurate the world organization in San Francisco at the end of April.[39]

Of all the issues on the Crimean agenda, Poland proved the most intractable. It had been the opening battleground of the current war, and if the Big Three were not careful, it could become the first battleground of the next one.

It wasn't Poland's fault. The Polish people of the 1930s were industrious, intelligent, and democratic. They cut their military strength to levels that couldn't possibly threaten their neighbors, and did everything in their power to avoid giving offense that would encourage either Hitler or Stalin to invade their country. Which, of course, encouraged both Hitler and Stalin to invade their country.

In early January, Churchill had supported a power-sharing arrangement between the communist Polish resistance movement based in Lublin, in southeastern Poland, and the Polish government-in-exile that had fled to London in 1939. But to his shock, a few weeks before the Yalta summit the Red Army rushed into Warsaw, and the Soviet Union recognized the Lublin faction as the new provisional government of Poland.[40]

Churchill was obliged to champion the Polish cause. Poland had been Britain's *casus belli* in 1939, and England had given the exiles refuge. As a matter of British pride, Churchill could not let the London Poles be displaced by a communist puppet government. The prime minister argued vehemently that any government recognized by Britain must be installed on the basis of universal suffrage, secret ballot, and fair and free elections. Poland was, he told Stalin, a matter of Britain's national honor.

Stalin replied that to Russia, Poland was not just a matter of honor; it was also a matter of national security. During the war's final phases, the Red Army must keep all potential anticommunist partisans—including the London faction—from committing sabotage in its rear. After the war, he said, Poland would resume its place as a *cordon sanitaire* between Germany and the Soviet Union. Simply put, Stalin trusted the Lublin Poles and did not trust the Poles of London.[41]

Roosevelt saw Poland as a sideshow and said little. He favored free elections in Poland, as in other liberated countries, but while he paid lip service in deference to Polish-American voters, he would not fight hard or trade too much to preserve Poland's western borders or wring guarantees of fair elections from Stalin. To Roosevelt, the United Nations organization was far more important than the fate of any Eastern European country. Besides, with the Red Army occupying most of Poland, he and Churchill had few cards to play, and everyone there knew it.[42]

After several days of discussions among the leaders and their foreign ministers—Stettinius, Eden, and Molotov—the conferees agreed to free elections in which a broad sector of Polish leaders, including the London émigrés, would participate. They agreed to retain Poland's eastern border more or less at the boundary drawn after the First World War. For the western boundary, they discussed incorporating East Prussia into Poland and setting Poland's western border as far as the Oder and Neisse rivers. But they would study the matter and defer specifics to another time.[43]

The Polish formula drafted by the ministers rang with ambiguous precepts such as a "strong, free and independent Poland," "universal suffrage," a "secret ballot," and "free and unfettered elections." The western leaders also agreed to hand over thousands of Red Army prisoners liberated from German camps, together with a host of anticommunist partisans, White Army officers, and other "Soviet citizens" whom the NKVD would sort out and probably liquidate before they reached Russia.[44]

To Roosevelt, these concessions, affecting millions of lives, were the price of Soviet support, and he would swallow that price without complaint. Sitting at Roosevelt's side, Admiral Leahy took one look at the documents and whispered, "Mr. President, this is so elastic that the Russians can stretch it all the way from Yalta to Washington without ever technically breaking it."

Roosevelt turned to his chief of staff and whispered back, "I know, Bill, I know it. But it's the best I can do for Poland at this time."[45]

Roosevelt's only important military issue was Soviet entry into the war against Japan. Stalin's was what the USSR would get in return for its help.

To compensate the Soviet Union for opening up another war, Roosevelt offered Stalin the Kuriles and the southern half of Sakhalin Island, a Japanese possession since the Russo-Japanese War. Stalin said Roosevelt's concession was appreciated, but it would not be enough. The Soviet people understood why they had to go to war against Germany, but Japan had not declared war against them. Stalin claimed he and Molotov would have a difficult time convincing a war-weary nation to begin a new war. The price of his support, he said, was a warm-water port—somewhere along the Chinese coast—and use of the Manchurian railways running to the port of Darien on the Kwantung Peninsula, near Korea.

In private, the two men exchanged informal thoughts on the rest of eastern Asia. Roosevelt felt Korea should be placed in an international trusteeship, with no permanent troops stationed there. He also objected to giving Indochina back to the French—"the French had done nothing to improve the natives," he said— and Stalin predicted that the British similarly would be incapable of holding Burma.[46]

By week's end, Stalin agreed to enter the war against Japan within three months of Germany's capitulation. He would also support Chiang's Kuomintang faction and keep the Chinese communists from disrupting Chiang's war against Japan. In return, the Soviet Union would receive Sakhalin and the Kuriles, a naval base at Port Arthur, access to Darien, a lease of the Manchurian railway, and the continued independence of pro-Soviet Outer Mongolia.[47]

Marshall and King felt the price of Soviet participation against Japan was high, though at the time none of the chiefs expressed any regrets over Roosevelt's concessions. Marshall's staffers had foreseen Russian dominance over Central and Eastern Europe, though King believed Stalin was "too damn smart" to try to overrun Europe beyond eastern Poland.[48]

To Marshall, the big concession—Soviet entry into the war against Japan— was all that mattered; questions of Lend-Lease and postwar spheres of influence were political, not military concerns. Before the conference ended, Secretary Stettinius, standing on the Livadia Palace steps, said to him, "General, I assume you are very eager to get back to your desk."

To this Marshall replied, "Ed, for what we have got here, I would have stayed a month."[49]

Yet Leahy, who attended both military and political meetings, saw dark clouds in the offing. Mark Twain once quipped that the difference between a man and a starving dog is that a dog won't bite you after you feed it. Leahy agreed. He told his diary, *"One result of enforcing the peace terms accepted at this conference will be to make Russia the dominant power in Europe, which in itself carries a certainty of future international disagreements and prospects of another war."*[50]

As the discussions dragged on, most observers saw in Franklin Roosevelt a physically exhausted man whose mind was still sharp. "Chip" Bohlen, Roosevelt's interpreter, thought him up to the task of negotiating with the two most powerful men in the world, and Stalin's interpreter, Valentin Berezhkov, recalled, "Everybody who watched him said that in spite of his frail health, his mental potential was high."[51]

But the mind can only act through the body, and FDR's body had been crumbling for some time. After the session on Poland, his pulse became irregular and his color grew pale. Dr. Bruenn limited his workday, and his daughter, Anna, frightened at the deterioration of her father's heart, fought off all unnecessary visitors.[52]

The final agenda item was publicity, a matter dear to every politician's heart. On the afternoon of February 9, a battalion of photographers and cameramen assembled in the Livadia Palace courtyard to take a series of group portraits—the Big Three, the Big Three with diplomats, the Big Three with military, and so on—that would have their own page in history's enduring scrapbook.

The Crimean weather was blustery and overcast that day, and Roosevelt clutched his favorite navy boat cloak tight around his shoulders and drew heat from a cigarette. He was tired, and he looked it as his valet wheeled him to the center chair. He shifted himself into place as servant and wheelchair withdrew.

As the military commanders took their places behind the three leaders, a few photographers began snapping pictures as Roosevelt was in the process of adjusting his legs, suit, and cloak for a more natural look.

Without warning, FDR lit up in a rage. News cameramen had never been allowed to photograph him in his wheelchair, or being carried by Secret Service agents, or on crutches, or shifting from car or wheelchair to his seat. For nearly twenty-five years, he and his entourage had jealously guarded his public image, and he expected photographers to wait until they were told to begin. Blue eyes blazing, he barked out orders to stop all photographs.

Seeing this quiet, friendly old man burst into anger left a disturbing impression on some in the American contingent. "Rather embarrassing to see this great man who obviously was not mentally himself," commented one of King's staffers who watched him from the periphery. "Mentally upset, let's put it that way."[53]

But once he made himself ready for posterity, the peevish old man disappeared and the broad smile returned. He held a cigarette in his left hand and joked with Stalin and Churchill as cameramen wheeled their spools and photographers snapped, advanced film, and snapped again.

When the session was over, the group, less Roosevelt, moved on. Churchill, a black *shapka* covering his bald head, left with Stalin, joking together amidst a crowd of high-ranking brass. Roosevelt, momentarily left behind in his chair, waited patiently for his wheelchair and valet to arrive.

"The patio was deserted," remembered one naval staffer. "It was very sad, and you were embarrassed for him. You knew that he was embarrassed; he realized his condition. I think all the people who were close to him did."[54]

.   .   .

As the American warlords drifted back to Washington, they wondered whether the agreements on Poland, Soviet ports, or the United Nations organization would hold. Roosevelt could only hope that the Russian dictator was trustworthy and powerful enough to ensure that the foundations of peace laid at Yalta would carry long into the postwar years. He knew Stalin might not grant the liberated nations full democratic rights, and he foresaw resistance by Churchill to self-determination of Britain's imperial colonies.

But, as he had told Leahy, "It's the best I can do."

FIFTY-ONE

"O CAPTAIN"

࿇

W HEN MARSHALL RETURNED TO HIS CLEAN MAHOGANY DESK ON THE Pentagon's E Ring, he had not fully recovered from his three-week, 14,000-mile odyssey to Yalta and back. He was tired and felt much like Alice after a trip down the rabbit hole, but the Army's domestic problems would not wait for him to rest.

"Making war in a democracy is not a bed of roses," he wrote Eisenhower on March 6, after a week of fending off backseat drivers in the press and on Capitol Hill. "We are under attack of course for the inadequacy of our winter clothing and now for the charge that 75 percent of our material is inferior to that of the Germans. They grant that the jeep and the Garand rifle are all right but everything else is wrong." Senator Robert Taft lambasted the War Department for casting eighteen-year-olds into battle with too little training, while MacArthur complained of a shortage of fit combat troops in his theater.[1]

Eisenhower was sympathetic. "Misery loves company," he replied. "Sometimes when I get tired of trying to arrange the blankets smoothly over several prima donnas in the same bed I think that no one person in the world can have so many illogical problems. I read about your struggles concerning eighteen year old men in combat, and about the criticism of our equipment, and went right back to work with a grin."[2]

• • •

The final strategic debate on Germany bubbled up in a witches' brew of clashing visions. When General Bradley's First Army captured the Ludendorff railroad bridge over the Rhine River at Remagen on March 9, Eisenhower decided to push

429

Bradley's First and Third Armies through central Germany, where they would link up with the Red Army. Montgomery would protect Bradley's left flank, just as Bradley had guarded Montgomery's right in earlier campaigns. Eisenhower's plan left Germany east of the Elbe, including Berlin, to the tender mercies of the Red Army.[3]

As Ike saw it, Berlin was so bombed, shelled, surrounded, and gutted that it no longer held military significance. The city also lay deep within the postwar Soviet occupation zone. Berlin's capture might cost 100,000 casualties, and Eisenhower saw no reason to spend Allied lives for a piece of real estate he would have to hand over to Stalin when the shooting stopped. The original directive from the Combined Chiefs instructed him to destroy Germany's armed forces, not capture any particular city. That order left the question of Berlin in the supreme commander's hands, and the supreme commander believed the German capital was not a necessary target.[4]

Eisenhower's strategy also placed his old classmate, the quiet Omar Bradley, on the war's center stage. That rubbed Churchill and Montgomery the wrong way—Churchill, because the Allies were forsaking Berlin, and Montgomery, because the Allies were forsaking Montgomery. The main thrust, both argued, should be through Germany's northern plains, from Hamburg to the German capital.[5]

To the prime minister, the Soviets could not be permitted to take Berlin. He wrote to Roosevelt on March 30, "Will not their impression that they have been the overwhelming contributor to our common victory be unduly imprinted in their minds, and may this not lead them into a mood which will raise grave and formidable difficulties in the future? I therefore consider that from a political standpoint we should march as far into Germany as possible and that should Berlin be in our grasp, we should certainly take it."[6]

Eisenhower was not blind to the political significance of the German capital. Every man in his army, from private to general, had remarked at one time or other that the road home goes through Berlin. "I am the first to admit that a war is waged in pursuance of political aims," he wrote Marshall, and he promised to adjust his plans if the Combined Chiefs ordered him to take Berlin. But in absence of new and specific orders, he would bypass Berlin and send his men to destroy the enemy wherever the enemy could be found.[7]

He worried that he might find them in the south, near the Bavarian Alps, the cradle of Nazism. Rumors had been making the rounds at Army intelligence of a "National Redoubt," where hard-core Nazis would wage a guerrilla war until Hitler resurrected the Reich. Rooting them out of the mountainous region would take years, perhaps decades. Bradley, whose concerns of a redoubt coincided with

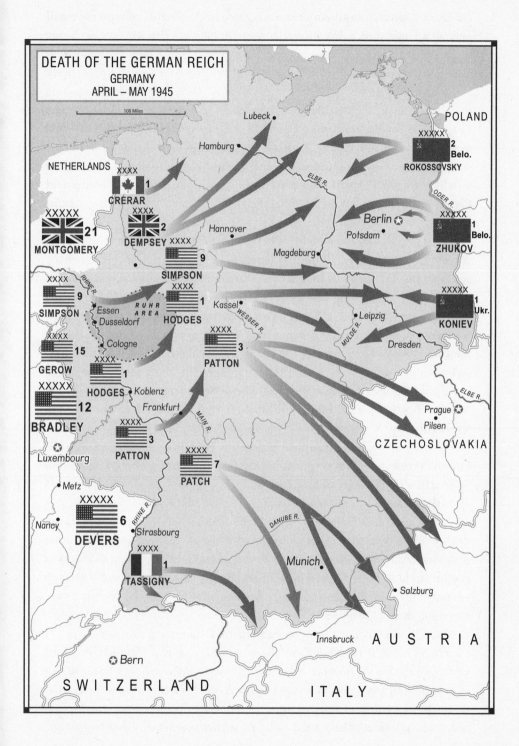

DEATH OF THE GERMAN REICH
GERMANY
APRIL – MAY 1945

100 Miles

POLAND

NETHERLANDS

Lubeck

Hamburg

ELBE R.

ODER R.

XXXX
1
CRÉRAR

XXXXX
21
MONTGOMERY

XXXX
2
DEMPSEY

Hannover

XXXX
9
SIMPSON

Magdeburg

XXXX
2 Belo.
ROKOSSOVSKY

Berlin

Potsdam

XXXXX
1 Belo.
ZHUKOV

XXXXX
9
SIMPSON

RHINE R.

RUHR
AREA

Essen
Dusseldorf

XXXX
1
HODGES

Cologne

XXXX
15
GEROW

XXXX
1
HODGES

Koblenz

Kassel

WESSER R.

Leipzig

MULDE R.

XXXXX
1 Ukr.
KONIEV

Dresden

XXXX
3
PATTON

ELBE R.

XXXXX
12
BRADLEY

Frankfurt

MAIN R.

XXXX
3
PATTON

XXXX
7
PATCH

Prague
Pilsen

CZECHOSLOVAKIA

Luxembourg

Metz

Nancy

XXXXX
6
DEVERS

RHINE R.

Strasbourg

DANUBE R.

Munich

Salzburg

XXXX
1
TASSIGNY

Innsbruck

AUSTRIA

Bern

SWITZERLAND

ITALY

his desire to give Hodges and Patton a big role in the final campaign, suggested making a major push toward Bavaria, Austria, and Czechoslovakia. Eisenhower agreed, and he ordered Bradley to send Hodges east, to the Elbe River, and Patton southeast toward Prague.[8]

Marshall again backed Eisenhower's strategy. While he personally doubted the existence of a "National Redoubt," he felt there was no reason to send Americans to die for Berlin. Unlike Churchill, who even then envisioned an "iron curtain" falling across Europe, Marshall's eye had already turned to Japan and the redeployment of divisions to the Far East. Central Germany was a logical region in which to bag the bulk of the German Army, and if that left Monty and Churchill out in the cold, he had no sympathy.[9]

* * *

While Marshall saw a strictly military problem—the task of finishing off Germany and arraying his forces against Japan—Roosevelt confronted an old political problem that was clawing its way out of the grave. Capitalists and communists had been enemies since the Bolsheviks took power in 1917. Red October begat a red scare that had permeated America from Washington to the heartland. In 1919, the United States had sent troops to aid White Russians against the Bolsheviks, and Washington had no diplomatic ties to the Union of Soviet Socialist Republics until 1933, when a first-term President Roosevelt appointed an ambassador to Moscow. It was not until Hitler turned east in 1941 that an alliance of necessity was born.

With every step toward Berlin, the necessity cementing that alliance receded. Dread of Hitler gave way to a surge of confidence, and the two sides began to wonder how far either could trust the other.

Minor mistakes could start brushfires. A cable Marshall sent to General Antonov advising of a German counterattack forming near Lodz, for instance, turned out to be a false alarm. A red-faced Marshall received a condescending reply from Antonov telling him that his report was probably German misinformation "to bluff both Anglo-American and Soviet Headquarters and divert the attention of the Soviet High Command from the area where the Germans were mounting their main offensive."[10]

Peace feelers in the Mediterranean stirred a simmering pot of distrust. In March, SS General Karl Wolff, an alleged representative of Field Marshal Kesselring, contacted the western Allies through Swiss intermediaries to discuss an armistice. Before any agreement could be explored, the Allies informed the Swiss that talks in Bern would be needed to verify Wolff's credentials as a peace emissary, which looked pretty thin in light of his Schutzstaffel affiliation.[11]

Stalin was not about to let peace break out in Italy—a peace freeing Germany to throw men at the Russian front. When FDR and Churchill informed him of the overtures, Molotov, replying for Stalin, demanded the presence of a Soviet at any discussions. Molotov's reply, wrote Stimson, *"bodes evil in the coming difficulties of the postwar scene."*[12]

To assuage Stalin's concerns, on April 1 FDR penned a letter intended to reassure the dictator. "No negotiations of surrender have been entered into," he wrote. "If there should be any negotiations, they will be conducted at Caserta with your representatives present throughout." The meeting in Bern, he emphasized, was simply to establish who was authorized to conduct negotiations on behalf of the Wehrmacht.[13]

"You affirm that so far no negotiations have been entered into. Apparently you are not fully informed," Stalin replied two days later. He said his sources "do not have any doubts that the negotiations have taken place and that they have ended in an agreement with the Germans, on the basis of which the German commander on the Western Front, Marshal Kesselring, has agreed to open the front and permit the Anglo-American troops to advance to the East, and the Anglo-Americans have promised in return to ease for the Germans the peace terms." This treaty, Stalin warned, "can in no way serve the cause of preservation or the strengthening of trust between our countries." Splashing gasoline on the fire, he added, "What we have at the moment is that the Germans on the Western Front have in fact ceased the war against Britain and America. At the same time they continue the war against Russia."[14]

Roosevelt passed Stalin's letter to Leahy, who forwarded it to Marshall to draft a reply. Marshall prepared an appropriately indignant letter that concluded, "It would be one of the great tragedies of history if at the very moment of the victory, now within our grasp, such distrust, such lack of faith should prejudice the entire undertaking after the colossal losses of life, materiel and treasure involved. Frankly, I cannot avoid a feeling of bitter resentment toward your informers, whoever they are, for such vile misrepresentations of my actions or those of my trusted subordinates."[15]

Marshall sent the draft to Stimson and Leahy. From Leahy it went to Roosevelt, who read it amid the loblolly pines of Warm Springs, Georgia. Roosevelt approved the letter and sent it to Stalin through his ambassador to Moscow, Averell Harriman.[16]

Stalin's reply was softer, but still cutting. He assured FDR that his informants were reliable, and to Marshall's embarrassment, he contrasted his information to General Marshall's letter to General Antonov warning of the phantom counterattack toward Lodz.[17]

More than anything else, Roosevelt wanted to keep the momentum building toward a United Nations charter meeting, and he was not about to let a squabble in Italy derail the consensus he had forged at Yalta. He prepared a conciliatory missive to Stalin, thanking him for a frank exchange of views and concluding, "There must not be mutual distrust, and minor misunderstandings of this character should not arise in the future." He sent the message to Harriman in Moscow, with instructions to pass it on to Stalin.

Like most Americans in Moscow, Harriman had grown frustrated by Soviet arrogance, to say nothing of the ingratitude toward the fortune America was spending to supply the Red Army. When he received the president's conciliatory cable, Harriman asked Roosevelt to edit it. He requested that the president eliminate the word "minor" before "misunderstandings," so as not to downplay the seriousness of Stalin's charges. He remarked to Roosevelt, "I must confess that the misunderstanding appeared to me to be of a major character."

Harriman sent his message to the Map Room, which forwarded it to Roosevelt at Warm Springs in the early-morning hours of April 12. Admiral Leahy, in Washington, knew Roosevelt wanted to keep bad blood from poisoning the United Nations conference, and drafted a suggested reply from the president to Harriman that concluded, "I do not wish to delete the word 'minor' as it is my desire to consider the Bern misunderstanding a minor incident."[18]

Roosevelt's approval of Leahy's message, downplaying the disagreement among friends, arrived at the White House at 1:06 p.m.

* * *

As March gave way to April, a battered Franklin Roosevelt parsed out his fading energies on the postwar peace. He had spent seven years managing the clinic of Dr. New Deal, and another five in the office of Dr. Win-the-War. Having served longer than any of his predecessors, he was willing to push the old body one more time, to get the United Nations off the ground before hanging up his spurs.

The political doctor took immense satisfaction in the clinical signs of his patient's robust health: unemployment was nil, inflation had not mushroomed, returning veterans would enjoy unprecedented educational and housing benefits, and America had displaced Europe as the world's manufacturing center. Negro soldiers fought alongside white soldiers in a few places, many factories had opened their doors to women and blacks, and the fundamental reforms of the New Deal had survived the draft, the war, and rationing. And the crown jewel of his foreign ambitions, the United Nations conference, would open in San Francisco on April 25.

Twelve years at the helm had taken a ghastly toll, for Franklin Roosevelt had emerged from nearly three and a half years of war gaunt, tired, and thin. He had lost twenty-five pounds in recent months, and his clothes sagged on his wiry frame. In meetings, his eyes would glaze and his jaw would droop. He lost interest in driving his Ford or mixing drinks during Children's Hour, his evening cocktail time. After his doctors recorded his blood pressure at 260 over 150, they banned him from swimming, which removed him another step from the vibrant, active yachtsman who had nothing to fear but fear itself.

His mind still carried a bright flame that blazed with intensity from time to time, but the flame now flickered. In late March Bill Hassett, Roosevelt's secretary, told Dr. Bruenn, "He is slipping away from us, and no earthly powers can keep him here."[19]

But FDR had no intention of slipping away, and he knew of places that could restore his body. Perhaps, he thought, a short vacation to Warm Springs would do him some good.

"Come right in," Roosevelt said, nodding to the dark-haired woman cradling a palette, paintbrushes, and pad.

Lucy Mercer Rutherford and Elizabeth Shoumatoff stepped inside Roosevelt's rustic cottage amid the pines of Warm Springs. The two women— Roosevelt's former lover and her artist friend—waited with FDR in the small living room as he finished his paperwork. His head was bent, and he sat in a leather-backed chair over a paper-strewn card table, his feet casually propped on a short wicker stool.

Shoumatoff set up her easel and watched the president intently as he tended his paper garden: a reply to Ambassador Harriman, edits to his United Nations speech, a bill to extend the Commodity Credit Corporation, appointments of small-time postmasters. He signed his name on official documents with a heavy stroke of a fountain pen that was air-dried, never blotted. Seeing the artist waiting, he quipped with an apologetic grin, "I have to stay here until my laundry dries."[20]

His secretary, Bill Hassett, removed the laundry and spread the papers on nearby tables to dry. Hassett would check the signatures and ensure that all documents were properly dated: April 12, 1945.

In his gray suit and crimson tie, Roosevelt looked surprisingly well to Shoumatoff. His color was good—a feature she drew out in her painting—and he was looking forward to a hot bowl of Brunswick stew at an afternoon barbecue to be thrown for him by the mayor of Warm Springs.

But Shoumatoff had an odd feeling. She offered to come back later, as the president was obviously busy, but Roosevelt wouldn't hear of it. He'd had a slight headache that morning and a stiff neck, but otherwise he felt fine.[21]

She began painting, asking him to turn his head this way and that to capture different angles of the long, famous face. In her portrait, the business suit would be cloaked with a Navy cape, the kind he had worn at Yalta, so she concentrated her brushstrokes on his face. The room grew quiet as everyone drifted off to lunch except Shoumatoff, cousin Daisy Suckley, who crocheted on the sofa, Laura "Polly" Delano, Lucy, and Fala.[22]

As Shoumatoff worked through her outline—the high cheekbones, broad shoulders, thick fingers holding a roll of papers—she chatted with Roosevelt, hoping her eye would catch some interesting aspect of his face. About one o'clock, he glanced at his watch and remarked, "We've got fifteen minutes more."[23]

She managed to complete a rough outline of the suit and a more detailed rendition of the blue eyes, the dark eyebrows, the somber, almost haunting expression. A picture of a figure staring up from a hazy dream, a man determined to walk to the end of some predestined line.[24]

Roosevelt returned to his work while Shoumatoff's brush flicked over the paper. Their cook, Irineo Esperancilla, known to the family as Joe the Filipino, set the table for lunch.

About 1:15 Roosevelt's head craned forward and his hands fumbled in his lap. Daisy walked over from the sofa and stooped below his face. When she asked him if he had dropped his cigarette, he put his left hand to the back of his skull and said, "I have a terrific pain in the back of my head."[25]

Daisy asked him to lean back in his chair, so he would not topple forward, but he couldn't move. The room sprang to life as Lucy and Polly tilted the chair back and yelled for help. Daisy found a Secret Service agent and sent him to find Dr. Bruenn.

Joe the Filipino and Arthur Prettyman, Roosevelt's valet, picked up the president and carried him to his bedroom. They laid him on the bed and covered him with blankets. As they loosened his tie and fanned his face, his eyes rolled from side to side, flashing around the room with no sign of recognition.[26]

Fifteen minutes later, Dr. Bruenn arrived to find the president cold and unresponsive. As the Roosevelt cousins withdrew to the cottage living room, Lucy Rutherford and Elizabeth Shoumatoff quietly gathered their things and left.[27]

Over the next hour, Franklin Roosevelt's body endured a slow collapse. Bruenn surmised that he had suffered a cerebral hemorrhage, and knew he could do nothing more. Roosevelt's heart raced, his blood pressure skyrocketed, his pupils dilated unevenly, and he lost bladder control. As he lay in his rough cottage bed, his chest rose in deep, steady gasps—gasps that grew more shallow with each push of his straining lungs.[28]

．　　．　　．

One of young Franklin's favorite pastimes was to pilot a small schooner up the coast to Campobello Island. He loved standing behind the wheel for hours at a time, inhaling the musky salt air, feeling the tilt of the deck under his feet as he tacked among the waves.

After a long day aligning the boat's bow with the winds, he would squint through the setting sun and pick out Campobello's wooden pier. With a quick step, he would steady the rudder and hop from one line to the next, letting wind spill from the sails as the sheets played through his big, strong hands. As the schooner eased toward the rocky shore, he would clamber to the rail, ready to cast the last line over the beckoning moorings.

The cable thrown, the boat tied to the dock, Frank Roosevelt could allow himself a moment of blissful, satisfied peace. He had stood his turn at the wheel through deep waters and brought his vessel home.

At last, in the comfort of his old, familiar harbor, he could rest those tired legs.

PART THREE

Swords, Plowshares, and Atoms

1945

∽

To entrust the winning of the war and the framing of the peace
into the hands of any man with a limited outlook and without the
experience needed for such a job would be the sheerest folly.

—HARRY TRUMAN, FEBRUARY 22, 1944

FIFTY-TWO

─────────────

TRUMAN

ᔥ

T HE PLANE FROM WASHINGTON FLEW HIGH OVER THE HUDSON RIVER AS
it threaded its way through upstate New York. Passing thick woods, hills,
boulders, and towns that were ancient in Washington Irving's time, the Air Force
transport settled down on the runway at Stewart Field near West Point. Henry
Stimson, George Marshall, and Ernie King clambered into a convoy of automobiles and rode along the Hudson's west bank, through Cornwall, through New-
burgh, across the river to Poughkeepsie, and finally to the tree-lined hamlet of
Hyde Park. They were going to the final resting place of their wartime commander-
in-chief.[1]

The death of a president, like his life, is not entirely his own. When a president dies in office, his passing becomes a joint venture in which the public, a
stakeholder, plays a dual role. It pours out its collective heart to the memory of a
fallen leader, then casts him aside to get on with the nation's business.

In this operation, Marshall played the role of executive officer: Transporting
Mrs. Roosevelt to Warm Springs. Safeguarding the president's body as it returned
by train to Washington. The procession from Union Station. The flag-draped
casket on a caisson. The memorial service in the East Room. Banks of flowers,
foreign dignitaries, press, and condolences from around the world. The last ride of
Franklin Roosevelt, to his flower garden on Springwood's placid grounds.[2]

The ritual of committing Roosevelt's mortal coil was presided over by the old
rector of St. James' Episcopal Church, the Roosevelt family chapel. At the burial
site, a cordon of young West Point cadets braced at attention. Roosevelt's casket,
hastily purchased in Atlanta, was lowered into the ground and three precise volley

cracks split the air. Beyond the hedges surrounding the garden, a band struck up a farewell rendition of "Hail to the Chief."[3]

Then he was gone.

"My dear general," Eleanor wrote Marshall the night of Roosevelt's funeral, *"I want to tell you tonight how deeply I appreciate your kindness & thoughtfulness in all the arrangements made. My husband would have been grateful & I know it was all as he would have wished it. He always spoke of his trust in you & his affection for you."*[4]

The personal loss of the man they followed for more than twelve years left the American warlords with a lingering sadness, though not a profound one. Death had been stalking Roosevelt for some time, and most of his war leaders, civilian and military, commented at one time or another on his physical decline.

Yet they could not shake—for the moment—the sense that the country had lost a friend. At Woodley a melancholy Henry Stimson wrote, *"For all his idiosyncrasies our Chief was a very kindly and friendly man and his humor and pleasantry had always been the life of Cabinet meetings. I think every one of us felt keenly the loss of a real personal friend."* Leahy told his diary FDR's death was *"a personal bereavement to me in the loss of a devoted friend whom I have known and admired for thirty-six years."*[5]

But the war would not tarry for any man's death, and in the White House Cabinet Room on April 12, a short, bespectacled man who thought he would be playing poker that night was holding a Bible and repeating the oath of office to Chief Justice Harlan Stone. Marshall, Stimson, King, and Leahy prepared to explain to an unplanned president what they had been doing to win the war.[6]

Minutes after learning of Roosevelt's death, an anxiety-ridden Harry Truman paid his respects to Eleanor in her second-floor sitting room. The First Lady, though obviously shaken, was, in her moment of grief, a tower of dignity and strength. With the uplifting spirit that had given heart to thousands of wounded soldiers and millions of America's poor, she stood when Truman entered and put her hand on his shoulder.

"Is there anything I can do for you?" Truman asked her.

"Is there anything *we* can do for *you*?" she answered. "You are the one in trouble now."[7]

Harry Truman was, in nearly every way, a different man from the glib glad-hander who had ruled the White House since 1933. Short, quick, plainspoken, he made no small talk, avoided no question. At ease among close friends, he lacked the instant familiarity that Roosevelt offered freely to one and all.

Unlike Roosevelt, he believed the buck stopped in the Oval Office. As he later told Anthony Eden, "I am here to make decisions, and whether they prove right or wrong, I am going to make them."[8]

He had known the job of president might be thrust upon him prematurely. In August, before the election, FDR had his new running mate over to the White House for lunch and a photo session on the South Lawn. When he got back to his Senate office, a rattled Harry Truman told a friend that when Roosevelt "tried to pour cream into his tea, more went into the saucer than into the cup." He added, "It doesn't seem to be a mental lapse of any kind, but physically he's going to pieces."[9]

The "juggler," as FDR had called himself, refused to see the pieces crumbling, and he was, by his own design, the only man who knew what the right and left hands of his administration were doing. For twelve years he had fragmented information, ignored jurisdictional lines, and overlapped responsibilities. Secretive while being genial, concealing everything when being garrulous, Roosevelt rendered no concessions to his political or physical mortality by taking his vice president—any vice president—into his confidence.

A year earlier, when deriding the qualifications of Roosevelt's Republican opponents, Truman told the newspapers, "To entrust the winning of the war and the framing of the peace into the hands of any man with a limited outlook and without the experience needed for such a job would be the sheerest folly." Yet FDR never showed Truman any of his secret cables with Stalin and Churchill. Truman had never seen the Map Room or even been told of its existence. He knew nothing of Allied strategy or the atomic bomb. Stimson knew the new president would be *"laboring under the terrific handicap of coming into such an office where the threads of information were so multitudinous that only long previous familiarity could allow him to control them."*[10]

So did Truman. After ad-libbing "So help me God" in the Cabinet Room on the night of April 12, he asked Stimson to bring his war chiefs to the Oval Office the next morning.[11]

Roosevelt's footsteps echoed across the stage, but most of the actors sitting in the Oval Office on the morning of April 13 had changed since the war's first dreadful weeks. Stimson was still there, but in the secretary of the navy's chair sat Jim Forrestal. With Marshall was Admiral King, who had taken the place of the exiled Betty Stark. Leahy had been ambassador to Vichy when Pearl Harbor was attacked, and Arnold, on inspection in Europe that day, was a full member of the Joint Chiefs of Staff, a name that did not exist in 1941.[12]

And for the first time since the war began, the president's chair was occupied by a man other than Franklin Delano Roosevelt.

Leahy explained the function of the Joint Chiefs to the new president. At Stimson's suggestion, Marshall and King gave short summaries of the state of the war. Marshall said Eisenhower would send Patton's army into Czechoslovakia. Bradley was driving toward the Elbe River for his rendezvous with the Red Army, and Montgomery was moving on Lübeck on the Baltic coast.

King described the Pacific. Air bases were operating on northern Luzon and cutting off Japan's oil, food, and metal. Iwo Jima's capture three weeks earlier had put American fighter escorts within range of Japan, and the Tenth Army was making steady progress on Okinawa. Total casualties for all services to date were nearly 800,000.

Truman told the group he was impressed by the high command's effectiveness. "If the South had had a staff organization like that," he said, "the Confederates would have won the Civil War." He asked a few questions, then thanked the men for their help. He said he was satisfied with the war's progress and wanted no major changes in its prosecution. The meeting was brief and efficient, and the war chiefs left the overwhelmed president to face his next meeting.[13]

On their way back to the Pentagon, Stimson and Marshall quietly compared impressions of the new commander-in-chief. Stimson was cautious but optimistic. Marshall was more guarded. "We will not know what he is really like until the pressure really begins to be felt," he said.

King, for his part, felt good about the new man. Truman was a straight shooter, like himself, and apparently didn't give a damn whose feathers he ruffled when he was right. King liked that. Three months into Truman's tenure he would tell Lord Moran, "Watch the president. This is all new to him, but he can take it. He is a more typical American than Roosevelt, and he will do a good job, not only for the United States but for the whole world."[14]

Unlike FDR, Truman was an Army man. He had been a battery commander at the Meuse-Argonne in the First World War. He kept his commission in the Army Reserve, and between the wars, he commanded the reservist 381st Artillery Regiment. Shortly after Pearl Harbor, he spoke with Marshall about resigning from the Senate to serve in the Army—a path Marshall advised against—and as chairman of the Truman Committee, he kept a watchful eye on military affairs.[15]

That Army influence was felt as Truman put his stamp on the executive office. Naval scenes in the presidential study were replaced with prints of airplanes; an Andrew Jackson bust took the place of a Dutch ship model. Knickknacks on FDR's desk—pigs, donkeys, and other odd mementos—were swept away, replaced by a model cannon, a clock, and pen sets. From a glance around the study, Navy men like Forrestal and King could see that their greatest ally, the meddling, whimsical late commander-in-chief, had vanished.

Truman also turned his broom on the old guard. Harold Ickes, Frances Perkins, and Francis Biddle were soon gone. Henry Morgenthau resigned shortly after Truman refused to bring him to Germany to meet with the Big Three.* Others soon left. The country was the same, the policies were the same, but Harry Truman's tone would be very different from the one Roosevelt's New Dealers had grown up with.[16]

While Roosevelt's monumental ego had been at home among the Churchills and MacArthurs of the world, the humble Harry Truman didn't cotton to showmen in uniform. He valued the Marshalls and Eisenhowers, men who put duty above status, far more than the *"Mr. Prima Donna, Brass Hat, Five Star Mac-Arthur,"* as he told his diary. *"Don't see how a country can produce such men as Robert E. Lee, John J. Pershing, Eisenhower, and Bradley and at the same time produce Custers, Pattons, and MacArthurs."*[17]

But since the Pattons and MacArthurs were fighting the war's battles for the moment, Harry Truman would leave that burden to Henry Stimson and George Marshall.

On April 23, Truman called Stettinius, Stimson, Forrestal, Harriman, and his military chiefs into the Oval Office. It was Poland, he told them. The Russians were calling for recognition of the Lublin faction at the United Nations conference, and United Nations recognition would legitimize the pro-Soviet party in the eyes of the world. It was a blatant violation of the Yalta agreement, which required Lublin's communists to include exiles and other non-communist groups.[18]

Truman was furious at Soviet temerity. Leahy had shown him some blunt, almost nasty cables from Molotov and Stalin, and he was ready to return those sentiments in Kansas City street language. The Americans, he thundered, would proceed to San Francisco exactly as planned, and would not recognize the Lublin delegation. "If the Russians did not wish to join us they could go to hell," he said.[19]

Four years earlier—two days after Hitler invaded Russia—Senator Truman had told reporters, "If we see that Germany is winning we ought to help Russia and if Russia is winning we ought to help Germany and that way let them kill as many as possible." President Truman was constrained to deal with Stalin in a way that Senator Truman was not, but now the two big pillars of the U.S.-Soviet alliance—a common enemy and Roosevelt's personal diplomacy—had tumbled to the ground. Truman was a different man, and he was not going to let the red dictator step one inch beyond the line Roosevelt had drawn.[20]

---

* Truman told one White House staffer that Morgenthau "didn't know shit from apple butter."

The diplomat in Stimson saw the United States heading down a dangerous path. He explained to Truman that while the Soviets had been frustrating to deal with in small matters, on large ones, such as military offensives, they had kept every promise and were often better than their word. Without knowing the importance the Soviets attached to Lublin faction, the men sitting in the president's office could not foresee the consequences of a hard line on Poland. Besides, Stimson added, outside the United States and Great Britain, virtually no country had any real understanding of what free and fair elections really meant.[21]

A sour-looking Forrestal told Truman the Soviets wanted all of Eastern Europe, and Stalin obviously did not think the United States would respond with force. "If the Russians were to be rigid in their attitude," he said, "we had better have a showdown with them now than later." Stettinius, Harriman, and Leahy came down on Forrestal's side.[22]

Outnumbered by the hawks, a worried Stimson saw the administration pushing a confrontation with a great power while they had a war against Japan to finish. *"Then to my relief,"* he wrote, *"a brave man and a wise man spoke and said that he, like me, was troubled."*[23]

Marshall, who had sat quietly through the discussion, said he would like to mention some military considerations. For a time, Soviet participation in the war against Japan would be useful. He remarked, "The Russians had it within their power to delay their entry into the Far Eastern war until we had done all the dirty work." A break with the Soviets now, he said, would have serious military consequences as the Americans prepared to invade Japan and had to think of a million Japanese soldiers in China and Manchuria.[24]

As before, the meeting ended with Truman thanking his men for their views. Siding with Stettinius, Leahy, and Forrestal, he concluded that in this early stage, America should hold fast "to our understanding of the Crimean agreements." The emerging diplomatic war, like the dying World War, would begin with Poland.[25]

• • •

As Truman was meeting with his war chiefs, SS Reichsführer Heinrich Himmler was informing Sweden's Count Bernadotte that Adolf Hitler had suffered a cerebral hemorrhage and had only a few days to live. Himmler was correct in a manner of sorts, though the cause of Hitler's cerebral hemorrhage was a piece of lead spinning through his dark and troubled brain.* On the second of May, Stalin

---

* Der Führer's resignation followed by two days the death of his collaborator, Benito Mussolini—of similar causes, though unlike Hitler's, Il Duce's pills were not self-administered.

announced the capture of Berlin, and General Alexander in Italy notified London that German troops there, some 600,000 in all, were being directed to surrender.[26]

The collapse of lines east, south, and west impelled Hitler's successor, Grossadmiral Karl Dönitz, to put out peace feelers to the west. Terrified of a Russian occupation, the German high command hoped to leave an open door for fellow Germans to escape a revenge-fueled spree of pillage, rape, and murder by Stalin's conquerors. Dönitz's emissaries begged Eisenhower for a truce in the west that would let them fight a rearguard action in the east.

Following to the letter the "unconditional surrender" formula laid down by Roosevelt, Eisenhower refused to accept a separate peace. The Third Reich's will to resist crumbled, and General Alfred Jodl, on behalf of the German high command, signed the instrument of surrender. On the morning of May 7, Eisenhower cabled Marshall:

THE MISSION OF THE ALLIED FORCE WAS FULFILLED AT 0241, LOCAL TIME, MAY 7TH, 1945.

FIFTY-THREE

DOWNFALL

∽

"ANOTHER HARD JOB STILL LIES AHEAD IN THE PACIFIC," ANNOUNCED Ernie King in a press statement the day Germany surrendered.[1]

In late 1944 MacArthur had garnered the headlines, wading ashore at Leyte and announcing to the world, in a rare use of the first person, "I have returned."* But over the last twelve months, it was the Navy's string of victories in the Philippine Sea, the Leyte Gulf, and the East China Sea that closed the ring around Japan. The Navy had severed Japan's supply lines. The Navy had cut off more than a million Japanese soldiers in China from their homes. And Japan lay within the Pacific Ocean Areas. *Navy* territory.

The capture of the main Mariana Islands—Saipan, Tinian, and Guam— gave Hap Arnold's Twentieth Air Force a platform from which to smash what little Japanese industry had not been starved by lack of raw materials. In March, Major General Curtis LeMay, Arnold's Pacific bomber commander, ordered a change in tactics from precision bombings to mass wave attacks. The bombers, he decreed, would go in with no machine-gun ammunition and at low altitudes, enabling them to squeeze more weight into their bomb bays. They would carry mostly incendiary bombs, not explosives, and would torch entire sections of cities, rather than specific plants or shipyards. Japan's factories—along with the workers who labored in the factories, and the homes of those laborers, and the families living in those homes—would be turned to charcoal.[2]

LeMay's flaming scythe began swinging on the night of March 9–10. Three

---

* MacArthur frequently referred to himself as "MacArthur."

hundred B-29s carrying 2,000 tons of incendiaries flew low and level over Tokyo and set the city ablaze. Wooden houses went up like kindling, superheated winds sucked oxygen out of living beings, and a firestorm—wildfire hot enough to generate storm-force winds—ripped through the metropolis.

In the inferno, 83,000 people were killed, more than 40,000 wounded, and another million left homeless. Crews in the last bomber waves could smell burning flesh at 5,000 feet, and as the silver dragons flew back to the Marianas, tail gunners could make out the city's glow from 150 miles away. Radio Tokyo described the capital as "a sea of flames."[3]

The scene repeated itself over eleven nights. Nagoya. Osaka. Kobe. Six of Japan's largest industrial centers, with a combined population of eight million people, were lit up in LeMay's firebombing campaign. Between a quarter and two-thirds of each city was burned or flattened, and in the carnage a quarter million or more died. The destruction was so complete that LeMay's staff estimated that by October 1, 1945, there would be no strategic targets left. Japan was on the verge of annihilation.[4]

Marshall was not so sure. The air campaign over Germany had been sustained much longer than LeMay's effort. Firestorms had turned Hamburg, Dresden, and swaths of Berlin into briquette pits, and for the war's last year bombers had flailed Germany from west and south. Yet the Third Reich refused to die until it was overrun on the ground.

Japan was even more militant than Germany. Its young samurai killed or were killed; the *bushido* code allowed no other fate. Long after fighting had ended on Mindanao, Luzon, and Saipan, isolated soldiers would pop out of the forest for a final *banzai* charge. Japanese civilians—women and children—threw themselves off cliffs and into the sea. Japan's battleships sallied on one-way missions, and thousands of kamikaze attacks in flying rockets and converted trainers demonstrated that Nippon's pool of men willing to die for their emperor had not yet run dry.

"There wasn't any question that the toughest individual fighter we ran into in this war was the Jap," said Marshall's planning chief years later. "Here were small groups of them on those islands out there who were eating roots and there were even reports they were resorting to cannibalism . . . but every time we cleaned up one of [those islands] we got a bunch of people killed."[5]

On Monday afternoon, June 18, Harry Truman sat at his desk and listened carefully as Leahy, Stimson, Marshall, and King spoke of death. Death the nation could expect to suffer in Operation DOWNFALL, the two-stage invasion of Japan.

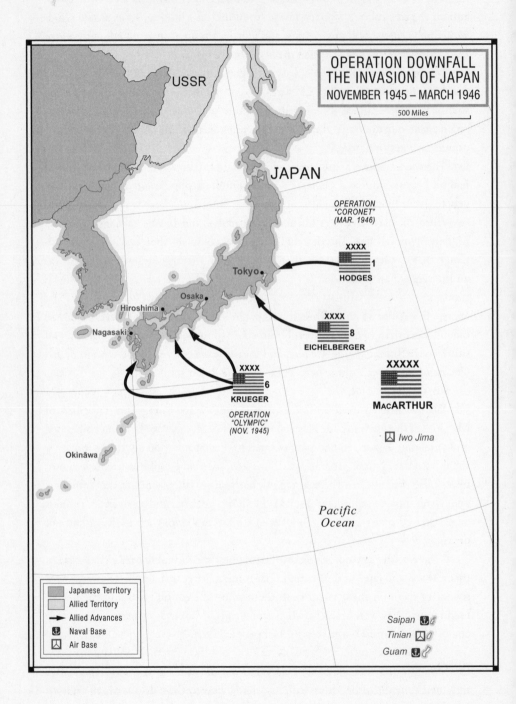

OPERATION DOWNFALL
THE INVASION OF JAPAN
NOVEMBER 1945 – MARCH 1946

500 Miles

USSR

JAPAN

OPERATION
"CORONET"
(MAR. 1946)

Tokyo

XXXX
1
HODGES

Osaka

XXXX
8
EICHELBERGER

Hiroshima

XXXXX
MacARTHUR

Nagasaki

XXXX
6
KRUEGER

Iwo Jima

OPERATION
"OLYMPIC"
(NOV. 1945)

Okinawa

Pacific
Ocean

Japanese Territory
Allied Territory
Allied Advances
Naval Base
Air Base

Saipan
Tinian
Guam

The Joint Chiefs and MacArthur had planned Operation OLYMPIC, the invasion of Kyushu, to commence in November. About four months later, they would launch another invasion, code-named Operation CORONET, against Japan's main island of Honshu. MacArthur would command the first invasion with an army of 750,000 soldiers. For the second, he would lead just over a million.

The question weighing on Truman's mind was how many of those American soldiers would be buried in Japanese soil before the war ended.

Earlier that month, former president Herbert Hoover had sent Truman a memorandum suggesting that an invasion of Japan would cost one million American lives. Truman forwarded the memo to Stimson for analysis, since the Joint Chiefs had been working with a number closer to half a million *total* casualties—killed, wounded, and missing.[6]

All of these numbers were little more than guesswork. At the June 18 meeting Marshall told Truman there was no reliable projection of U.S. killed and wounded. Considering only Kyushu, the smaller of the two islands, Marshall gave Truman a rather vague ratio of American and Japanese losses, from which he concluded that during the first thirty days on the island American casualties would not exceed the proportionate price the nation paid for Luzon. That was about 31,000 casualties. Of course, those casualties were only a down payment for the first thirty days. In the end, no one could say whether the entire population of Japan would rise up against the invaders in a national hara-kiri.[7]

Leahy approached the question of casualties from a different angle. On Okinawa, he pointed out, U.S. forces suffered a 35 percent casualty rate. He asked Marshall the size of the force he planned to use for OLYMPIC, and the general replied that 766,000 troops would be used. No one got out a pencil and paper to do the math, and King pointed out that Kyushu offered more attack options than Okinawa did. But it was easy to see that if the Okinawa ratio held, total casualties on Kyushu might top 268,000.[8]

And Kyushu, of course, was only Step One. Another 1.2 million enemy soldiers awaited on Honshu, and that meant more Americans would die.[9]

Japanese military casualties would depend on how many soldiers the Emperor fielded. The math itself was easy: take the total number of Japanese soldiers—350,000 on Kyushu—and subtract a tiny fraction of those left alive but too badly injured to commit suicide. That would give you the number of Japanese combat deaths. Civilian deaths could not be predicted, but as in all wars, they would be far more numerous. Many would be self-inflicted; at Saipan around a thousand old men and mothers with children, terrified of the American murderers and rapists they had heard so much about, leaped to their deaths from "Suicide Cliff" and other promontories.[10]

452 ★ AMERICAN WARLORDS

Truman's war chiefs were divided on the wisdom of invasion. Hap Arnold's planners argued that conventional air bombardment would bring Japan to her knees. By the time of the invasion, he predicted, there would be virtually no industrial centers left; large portions of Japan's major cities would simply cease to exist.*[11]

Admiral Leahy believed a naval blockade, combined with air bombardment, would force the island nation to surrender on reasonable terms. Nippon no longer had access to oil or steel. She had little in the way of food, and the Joint Chiefs were considering a plan to drop salt over rice paddies to spoil crops as they were ripening. A blockade might take more time and more money, but it would cost fewer American lives.[12]

King agreed with Leahy. Letting fortified islands starve had worked magnificently for Rabaul and Truk. Blockading Honshu would simply be a "die on the vine" strategy writ large. King later wrote of his private objections to invasion: "I have said many times we (U.S.) didn't have to get in such a fix if we could merely wait for the effective Naval Blockade to starve the Japanese into submission for lack of oil, rice and other essentials. The Army, however, with its complete lack of understanding of 'Sea Power' insisted on a direct invasion and occupational conquest of Japan proper—which I still contend was wrong!"[13]

Marshall would not countenance a war lasting into late 1946. As in Europe, he looked for the fastest and surest way to end the fighting. A week before the June 18 meeting, in a speech to the Maryland Historical Society, he spoke words close to his heart: "In a war, every week of duration adds tremendously, not only to the costs, but in casualties measured in lives and mutilation. . . . The full impact of the war comes more to me, I think, in some respects than it does to anyone in the country. The daily casualty lists are mine. They arrive in a constant stream, a swelling stream, and I can't get away from them."[14]

As with Nazi Germany, Marshall believed Japan's will to resist would be broken quickest by invasion, not by blockade or bombardment. Fanatical resistance on Okinawa and Saipan, Japan's legendary contempt for surrender, and its apparent indifference to LeMay's firebombing in March convinced him that a blockade would lengthen the war, producing more casualties than would be saved. Airmen shot down and naval victims of kamikaze attacks, to say nothing of Russian and Japanese casualties in Manchuria, would pile up as Japan held out for months, perhaps years.[15]

---

* The following month, at Potsdam, General Arnold bet Pug Ismay two dollars that the war with Japan would last into 1946. "Odd that he didn't figure in Hiroshima and Nagasaki," Ismay mused years later.

Marshall later described his mind-set the day he urged invasion on Truman: "We had to assume that a force of 2.5 million Japanese would fight to the death, fight as they did on all those islands we attacked. We figured that in their homeland they would fight even harder. We felt this despite what generals with cigars in their mouths had to say about bombing the Japanese into submission. We killed 100,000 Japanese in one raid in one night, but it didn't mean a thing insofar as actually beating the Japanese."[16]

However mixed their inner feelings, at the White House meeting the Joint Chiefs appeared united. General Arnold's representative, Lieutenant General Ira Eaker, fell in line with Marshall, pointing out that air casualties would run about 30 percent per month if bombers attacked an enemy able to focus all its domestic resources on antiaircraft defense. King voiced his general agreement with Marshall, and emphasized the importance of taking Kyushu as a springboard for air and naval operations. Leahy, who had told the Joint Chiefs he would go along with the invasion only if they persuaded President Truman, questioned Marshall's estimates, but never openly disagreed with Marshall's premise that an invasion was a prerequisite to Japan's surrender.[17]

Stimson, recovering from a migraine headache that morning, said he favored invasion. But before taking that last step, he suggested that the Allies find a way to reach out to what he called "the submerged class" in Japan. He believed a silent majority opposed the war; this underclass would fight tenaciously if their islands were invaded by white soldiers, but they also might force Japan to give up before an invasion became necessary.[18]

After listening to his war chiefs, Truman summarized the consensus. He said he understood the Joint Chiefs were unanimous in recommending an invasion of Kyushu in November as the "best solution under the circumstances." The chiefs said they agreed.[19]

With nothing left to discuss, the meeting began to break up. As the participants stood up to leave, Truman noticed Jack McCloy standing silently off to one side. Realizing that the undersecretary had not voiced any opinion, he asked McCloy what he thought.

McCloy said they should look for a political solution that did not require an invasion. Fixing his eyes on Truman, he referred obliquely to the "new weapon."

His reference to a "new weapon" brought shocked looks from around the room. "It was just like mentioning Skull and Bones at Yale," he said later. "You shouldn't have said that out loud, yet everyone knew it was there."[20]

McCloy had brought the atomic bomb into the question of Japan's fate, and Harry Truman told his advisers to sit down again. They had one more item to discuss.

FIFTY-FOUR

"COME AND SEE"

H ENRY STIMSON'S SEVENTY-SEVEN-YEAR-OLD MIND KEPT RETURNING TO THE S-1 project. He and Mabel never had any children, so he never had to contemplate the life his grandchildren would inherit. But other men had children, and grandchildren, and Stimson's thoughts swirled around the mystery of what a nuclear bomb meant to a world that would soon outlive him.[1]

The idea that mankind could tap into a fantastic source of power reached into Stimson's aged chest and electrified his mind. The project humbly known as S-1, whose crinkled file pages lined a section of his office safe, would change man's relationship to nature and nations' relationships to their neighbors.

Picturing an atomic arsenal as a diplomatic tool as much as a military one, at the end of 1944 he had asked FDR to reveal the secret to the secretary of state. In his last meeting with Roosevelt, Stimson also suggested that new controls over the weapon should be proposed "before the first projectile is used."[2]

At the close of Truman's first cabinet meeting, Henry Stimson referred obliquely to a powerful weapon under development. But in the commotion of power's transition—and the jostling among cabinet members who had not been cleared for the secret—he could say little else.

After giving Truman several days to adjust to his new office, Stimson felt that "S-1" could wait no longer. On April 24, he wrote the new president:

> I think it is very important that I should have a talk with you as soon as
> possible on a highly secret matter. I mentioned it to you shortly after you
> took office but have not urged it on account of the pressure you have been

under. It, however, has a bearing on our present foreign relations and has such an important effect upon all my thinking in this field that I think you ought to know about it without much further delay.[3]

Truman summoned his war secretary to the White House the next day.

Acquainting Truman with details of the bomb, Stimson said its destructive power made it "the most terrible weapon ever known in human history." In time, he said, its secrets would undoubtedly spread to other nations. Statesmen would be called upon to develop "thorough-going rights of inspection and internal controls" sufficient to prevent the weapon from being used irresponsibly. With these controls, he said, a new relationship between mankind and the atom offered an opportunity "to bring the world into a pattern in which the peace of the world and our civilization can be saved."[4]

The Missourian was transfixed by what he heard. Henry Stimson, the old man of the cabinet, was talking about nuclear proliferation, international inspections and controls, and peace through nuclear deterrence. Other men focused on the tactical effects of the bomb; only Stimson, a man born shortly after the Civil War's end, seemed to grasp its long-term implications.

The meeting left a deep impression on Truman. In his memoirs, Truman would write, "I listened with absorbed interest, for Stimson was a man of great wisdom and foresight. . . . Byrnes had already told me that the weapon might be so powerful as to be potentially capable of wiping out entire cities and killing people on an unprecedented scale. . . . Stimson, on the other hand, seemed at least as much concerned with the role of the atomic bomb in the shaping of history as in its capacity to shorten this war."[5]

Long before he met with Truman, Stimson concluded that the weapon should be used if Japan did not surrender first. Stimson's task was to end the war, and in October 1942 he had ordered General Leslie Groves to produce a working bomb "at the earliest possible date so as to bring the war to a conclusion." If a single day of fighting could be saved, he told Groves, "save that day."[6]

In May 1945 he asked Marshall whether an invasion of Japan could be postponed long enough to try ending the war with atomic weapons. Marshall mulled that one over. General Groves was on track to have a gun-method weapon ready by August 1. The gun version, nicknamed "Little Boy," was essentially a long tube that fired a wedge of uranium-235 into a larger piece of the same material. When the pie became whole, the mass of uranium would become unstable and explode with the force of 20,000 tons of TNT. Or, perhaps, with the force of 2,000 tons of TNT. Or perhaps nothing at all. No one really knew, though the scientists thought it would produce a big reaction. And it could be dropped without a preliminary test.

A second bomb, nicknamed "Fat Man," used conventional explosives to squeeze plutonium from a big, loose ball into a small, compact one. This implosion method would require testing before it could be used in combat, but Marshall told Stimson he thought it would be possible to have both bombs ready before "the locking of arms came and much bloodshed."[7]

On May 29, Stimson, Marshall, and McCloy discussed ways to warn Japan and which targets to choose. Marshall suggested hitting a purely military target, such as a naval installation, with the first bomb. He also believed some kind of warning should be given to Japan's citizens, designating a number of manufacturing areas that the United States intended to destroy with a short time to evacuate. "We must offset by such warning," he said, "the opprobrium which might follow from an ill-considered employment of such force."[8]

Planning military operations around an unprecedented weapon was not easy, given the veil of secrecy concealing it even from those charged with its use. Admiral King, for instance, could not tell anyone on his staff—not even his intelligence chiefs—of the bomb's existence. In early summer he called in Captain William Smedberg, his intelligence head, and told him, "Smedberg, now this is very, very secret, what I'm going to say to you. I want you to go back and I want you and your staff to work and in the next two or three days I want you to tell me when you think the Japanese will surrender if the most awful thing you can imagine happens to them in, say, the next two or three months."[9]

A baffled Smedberg returned to his office with no idea of what King was talking about. "The most awful thing you can imagine," he supposed, was a big earthquake. Japan had a long history with them, and he knew the Allies had discussed packing a line of freighters with explosives and detonating them along a fault line. Perhaps that was what King meant.[10]

. . .

Truman approved Stimson's suggestion to create an advisory committee on the political and policy effects of the bomb. Stimson's "Interim Committee" comprised himself, an assistant secretary of state, a Navy undersecretary, and Jimmy Byrnes, the incoming secretary of state. Stimson also invited four eminent scientists: J. Robert Oppenheimer, head of the Los Alamos complex, and Nobel laureates Enrico Fermi of Columbia, Arthur Compton of Chicago, and Ernest Lawrence of the University of California. Sitting with them as invited guests were Generals Marshall and Groves.

Stimson's committee met periodically throughout May and held its climactic discussion on the last day of the month. In his diary that night, an excited Stimson wrote, *"I told them that [the Army] did not regard it as a new weapon merely but as a revolutionary change in the relations of man to the universe; that the project might even mean the doom of civilization or it might mean the perfection of civilization; that it might be a Frankenstein which would eat us up or it might be a project by which the peace of the world would be helped in becoming secure."*[11]

Deferring to the generals on military matters, the scientists joined in a discussion of the bomb's effects. The gaunt Oppenheimer described his vision of the hellish blast. "The visual effect would be tremendous," he promised. "It would be accompanied by a brilliant luminescence which would rise to a height of 10,000 to 20,000 feet." The radiation spewing from the blast "would be dangerous to life for a radius of at least two-thirds of a mile."[12]

Stimson said he preferred to use the bomb with no advance warning, in an area sufficiently populated as to "make a profound psychological impression on as many of the inhabitants as possible." On this point there was little dissent. Oppenheimer went further, suggesting several simultaneous strikes, though Groves argued that each bombing would provide valuable information that could make subsequent attacks more effective.[13]

As Stimson recalled, the committee carefully considered alternatives such as the bomb's use on an uninhabited area, or an advance warning. A specific warning, they feared, might prompt the Japanese Army to move American prisoners to the target. Moreover, some in the administration did not believe the bomb would work. Admiral Leahy, for one, called it "a professor's pipe dream."* It might stiffen Japanese resistance if the United States announced the use of a fantastic new weapon and the fantastic new weapon turned out to be a dud.[14]

On June 1 the Interim Committee submitted its recommendations to Truman. It proposed using the bomb against Japan as soon as possible, without warning, and against a target that would demonstrate its "devastating strength." With some discussion and hand-wringing, the committee reaffirmed its recommendation "that the weapon be used against Japan at the earliest opportunity . . . on a dual target, namely, a military installation or war plant surrounded by or adjacent to homes or other buildings most susceptible to damage."[15]

Thus, on June 18, when Truman called his warlords back into conference, he wanted to know what options the atomic bomb gave him. Jack McCloy had men-

---

* After the war, Leahy declared his opposition to the "barbarous" weapon, though the only objection voiced to Truman at the time was that the bomb might not work.

tioned the prospect of a political solution—terms of surrender that would end the war without further bloodshed—but after a brief discussion of surrender formulas, politics were laid aside and the bomb's use became the group's focus. The American high command reached no final decision, but the implied consensus was that the bomb would be used unless something dramatic happened.[16]

In discussions over the next month, no responsible person in either the United States or Great Britain seriously considered invasion or indefinite blockade preferable to using the atomic bomb. As Churchill wrote afterward, "The decision whether or not to use the atomic bomb to compel the surrender of Japan was never an issue. There was unanimous, automatic, unquestioned agreement around our table; nor did I ever hear the slightest suggestion that we should do otherwise."[17]

The question remained *where* to drop the bomb. According to Marshall, the basic idea was to drop one or two bombs on Japan's mainland for shock value. If an invasion were still necessary, third and fourth bombs would be dropped behind the front lines to knock out supporting forces. Exactly where the weapon would be dropped was a question for the bombing experts.[18]

General Groves and his staff searched reconnaissance reports for suitable targets. The first criterion held that the target must be a military or industrial location not already pounded to rubble by LeMay's Superforts. That eliminated Tokyo and many other large cities. Other criteria included the presence of surrounding hills to maximize the effect of the bomb blast, and a well-known location, so that news of the explosion would spread to the Japanese people regardless of government propaganda. That narrowed down the list to Kyoto, Hiroshima, and Niigata.[19]

Stimson found carpet bombing abhorrent, and he had been repelled when the Army Air Forces turned Dresden into a crematory. He understood why the strategy was more useful in Japan—Nippon's war industries were scattered throughout residential areas—but he was still troubled enough to write Truman about indiscriminate destruction. *"I told him I was anxious about this feature of the war for two reasons,"* he told his diary. *"First, because I did not want to have the United States get the reputation of outdoing Hitler in atrocities; and second, I was a little fearful that before we could get ready [for the atomic bomb] the Air Force might have Japan so thoroughly bombed out that the new weapon would not have a fair background to show its strength. . . . [The President] laughed and said he understood."*[20]

As he scanned the list of bomb targets, one name caught Stimson's eye. He and Mabel had visited the city of Kyoto when he was governor-general of the Philippines. He had strolled its stone avenues and marveled at its Tokugawa castles and Shinto shrines. To General Groves, Kyoto was a city with a million square feet of factory floors, but to Stimson it was a place of beauty, humanity, religion, and culture—a treasure to the world as much as to the Japanese people. He ordered Arnold and Groves to strike Kyoto from their target list.[21]

On July 21, Stimson received another cable sent at Groves's request. The message read, "Your local military advisors engaged in preparation definitely favor your pet city and would like to feel free to use it as first choice."[22]

It was typical bullheaded Groves. Annoyed at the general's persistence, Stimson went to Truman. Catching him between meetings, he told Truman if the Americans destroyed a cultural center like Kyoto, the world would never forgive them. He wanted Truman's blessing to spare the city.[23]

Truman respected the old Republican, and he agreed to veto Kyoto. In triumph, Stimson cabled Groves back, "Give name of place or alternate places, always excluding the particular place against which I have decided. My decision has been confirmed by the highest authority."[24]

Kyoto was safe. The new list of targets read: Hiroshima, Kokura, Niigata, and Nagasaki. One of them would take Kyoto's place as the sacrificial lamb.[25]

·   ·   ·   ·

FDR rarely brought his cabinet ministers to conferences with Allied leaders, for he liked to think that he was his own cabinet. When preparing for TERMINAL, the last Big Three conference of the war, Truman had not asked Stimson to attend, on account of his age. But when Stimson asked to come anyway, Truman, seeking all the help he could get, cordially invited Stimson to join him.[26]

Truman arrived at Potsdam, a suburb of Berlin, on July 15. He was lodged in a three-story stucco mansion at Number 2 Kaiserstrasse. The mansion was lately owned by the head of a German movie company, "who is now with a labor battalion somewhere in Russia," as the president's official log surmised. To ensure that the presidential staff caused no friction with their hosts, Truman called his people together and told them the Russians had gone to great trouble to fix up their accommodations. He wanted everything there when they arrived to be sitting there when they left. There might be hard talk at the negotiation tables, but there would be no souvenir hunting in the presidential quarters.[27]

As at Yalta, political decisions crowded the agenda, and Stimson, Marshall and King, the military specialists, remained on the periphery of the big discussions. On the periphery, that is, until 7:30 p.m. on July 16, when a message from Washington arrived for Stimson by special courier. It read:

OPERATED ON THIS MORNING. DIAGNOSIS NOT COMPLETE BUT
RESULTS SEEM SATISFACTORY AND ALREADY EXCEED
EXPECTATIONS. . . . DR. GROVES PLEASED. HE RETURNS
TOMORROW. I WILL KEEP YOU POSTED.[28]

The message, decrypted yet cryptic, told Stimson that the implosion-type plutonium bomb had been successfully tested in the New Mexico desert. Beaming at the news, Stimson rushed over to the "Little White House," Number 2 Kaiserstrasse, to tell the president.[29]

Five days later, a more descriptive report arrived. The test bomb, Groves reported, had exceeded all expectations. An explosion the equivalent of fifteen to twenty thousand tons of TNT carved a crater into the desert nearly a quarter of a mile wide and 130 feet deep. The fireball could be seen in El Paso, Albuquerque, Santa Fe, and points nearly 200 miles away; one witness, newspaper reports claimed, was a blind woman.*[30]

Stimson reviewed the report with Marshall, then presented it to Truman and Byrnes. Both were ecstatic at the result. *"The President was tremendously pepped up by it and spoke to me of it again and again when I saw him,"* wrote Stimson. *"He said it gave him an entirely new feeling of confidence."*[31]

He would need that confidence to get through talks with Stalin. The United States and the Soviet Union had come a long way from the dark days of 1942, when they fought shoulder to shoulder—metaphorically, as the Russians reminded them—against the German peril. To Truman, Russia was becoming the world's emerging problem. Stalin's ministers demanded ports in China. They wanted influence in Austria, Romania, Bulgaria, Yugoslavia, the Dardanelles, and of course Poland. They wanted reparations, they wanted bases in Manchuria and Turkey, and they wanted a trusteeship over Korea—potentially even sole control over that peninsula.[32]

Truman was not about to let Stalin face him down. With the Red Army, Stalin was holding a straight, but when the Los Alamos reports arrived, Truman held a royal flush. Between meetings, he told Stalin his army had developed a new, more powerful weapon than any other built to date. Then he waited for the dictator's reaction.

---

* A news service article, ascribing the explosion to an accident at an Army arsenal, was planted to let authorized persons at Potsdam know the test had been a success.

It was not what he had expected. The news didn't seem to register with Stalin. His expression never changed, and he continued the conversation in a lighthearted vein, merely replying that he hoped America would make "good use of it against the Japanese." Stalin did not appear greatly interested in what Truman told him, and some wondered whether he even grasped the significance of Truman's revelation.*[33]

Truman was a cagey poker player, and he had come to believe that most Soviet demands were sheer bluff. But he still needed the Russians to help him end the war with Japan. Or did he?

He asked Stimson to double-check with Marshall, and Stimson passed the question on: Did the United States need the Red Army to invade Manchuria in order to defeat Japan?

Marshall hedged. If the Americans invaded Japan, they probably needed the Red Army to pin down the fifty divisions of Japan's Kwantung Army in Manchuria. Even if Japan succumbed to an atomic bomb without invasion, the Russians could still march into Manchuria and take whatever they wanted. But the Joint Chiefs were shifting their views from the days of Tehran and Yalta, and were less inclined to beg for Stalin's help.

Stimson inferred from Marshall's answer that America could probably go it alone, now that the arsenal of democracy included nuclear weapons. To Stimson, it was not a clear answer, but in war most answers are not. He accepted that answer and informed Truman.[34]

·   ·   ·

For Henry Stimson, there was one last battle to fight. As with Germany, this one placed the secretary of war in the odd position of opposing the secretary of state by favoring diplomacy over force.

Since the spring, Stimson had considered whether the war could be ended by giving Japan one concession: the retention of her emperor under a modified constitutional arrangement. It would mean altering the "unconditional surrender" formula laid down by FDR at Casablanca two years earlier, but Stimson thought it a reasonable concession. As he told Truman, "Japan is not a nation composed wholly of mad fanatics of an entirely different mentality from ours. On the contrary, she has within the past century shown herself to possess ex-

---

* He had. Stalin's spies had kept him informed of the Manhattan Project's progress since 1943, and he had learned of the scheduled test date in New Mexico earlier in the month.

tremely intelligent people. . . . I think the Japanese nation has the mental intelligence and versatile capacity in such a crisis to recognize the folly of a fight to the finish and to accept an offer of what will amount to an unconditional surrender."[35]

Stimson spoke to Truman about terms permitting the Japanese to keep their emperor if the nation were purged of its militaristic influences. Truman listened but gave Stimson no answer.[36]

It was a harder question than it looked. In June, Truman had warned Japan of massive destruction if she refused to surrender, and Secretary Byrnes believed the principle of unconditional surrender had become too deeply ingrained with the American public to let it go now. Like Truman, Byrnes had been a respected senator before joining Roosevelt's wartime team, and the new president put a lot of stock in the secretary's political opinion. Besides, a June Gallup poll showed that a third of Americans wanted Hirohito to hang as a war criminal, and only 7 percent believed he should keep his throne.[37]

On July 12, messages between Tokyo and the Japanese embassy in Moscow, decrypted by MAGIC men in Washington, told a story of a nation on the brink of extinction, desperately seeking Soviet help in mediating an end to the war, yet unwilling to agree to unconditional surrender. Truman told his diary, *"Believe Japs will fold up before Russia comes in. I am sure they will when Manhattan appears over their homeland."*[38]

At Stimson's urging, Truman planned to issue an ultimatum to Japan, signed by himself, Churchill, and Chiang.* The message would demand Japan's unconditional surrender, or it would face "prompt and utter destruction." The draft statement declared that justice would be meted out to war criminals, and the leaders who had drawn Japan into war would be removed. As for the Emperor, Stimson included in his draft a proviso that the new government "may include a constitutional monarch under the present dynasty if it be shown to the complete satisfaction of the world that such a government never again will aspire to aggression."[39]

The draft Truman approved was revised by Secretary Byrnes. It was close to Stimson's version, but it eliminated any reference to the Emperor's continued employment. That omission worried Stimson, because the document also called for the removal of Japan's militarists. The Japanese were bound to think one of those intended militarists was their divine leader.

---

* The USSR would not declare war on Japan until August 9, so it would not be a party to the declaration.

The next day Stimson met with Truman at Number 2 Kaiserstrasse. He stressed the importance of preserving Shōwa dynasty to induce Japan to agree to what was otherwise an unconditional surrender. He told his diary, *"I had felt that the insertion of that in the formal warning was important and might be just the thing that would make or mar their acceptance."*[40]

But by the time Stimson pleaded his case to Truman, the president had approved the declaration without the "emperor provision" and sent it to Chiang for his signature. Circulation of the draft made any softening language impossible, so Stimson asked Truman to assure the Japanese through other, indirect channels that the Shōwa period would continue. *"He said he had that in mind, and that he would take care of it,"* Stimson wrote.[41]

With nothing left to do in Potsdam, Henry Stimson boarded a plane to Munich. There he saw his old friend General Patton, then flew home to Highhold for a long-overdue rest.[42]

The Potsdam Declaration, broadcast over the radio and dropped in leaflet form over Japan, fell on deaf ears and blind eyes. Prime Minister Suzuki Kantarō pointedly refused to respond, and while Japan made behind-the-scenes efforts to broker a truce through Stalin, Harry Truman saw no reason to think Japan would surrender anytime soon.

He would leave the next move in the hands of his airmen.

The first bomb, "Little Boy," had been transported to Tinian aboard the cruiser *Indianapolis* on July 26. Final assembly was completed on the island, down to the dark gray paint job, and rehearsal drops were completed by July 31. As Little Boy sat in his bomb pit, dark and surly, his cheery-looking brother, "Fat Man," was painted a festive yellow. His bright nose was stenciled with the legend "JANCFU"—GI slang for "Joint Army-Navy Combined Fuck-Up"—and his side and fins bore the penciled signatures of technicians who had put the hellish device in weapon form. Fat Man was wheeled to his pit and lowered, there to await a big bomber to roll over the pit and open its bay doors.

Marshall ordered his acting chief of staff, General Handy, to draft an order directing the use of one or more atomic bombs as soon after August 3 as weather would permit. The order arrived in Potsdam on July 24 and was sent back to Washington by Marshall six hours later with the approval of Stimson and Truman. It listed four targets: Hiroshima, Kokura, Niigata, and Nagasaki.[43]

The flight of the *Enola Gay*, the B-29 assigned to usher the world into the nuclear age, was scheduled for the morning of August 4, Tokyo time. But the skies over Japan, critical to measuring the bomb's effect, were overcast, so *Enola Gay*'s flight was delayed until the next day, and then the next.[44]

. . .

Exhausted by his military and diplomatic whirlwind, a weary Henry Stimson relaxed at Highhold with Mabel. The president was at sea, returning from Potsdam. Marshall and King were in Washington. And high above the Pacific, death rode the wings of a silver bird.

FIFTY-FIVE

"THIS IS A PEACE WARNING"

〰

O N SUNDAY, AUGUST 5, HARRY TRUMAN WAS CROSSING THE ATLANTIC aboard King's old flagship, the cruiser *Augusta*. As the ship plowed gracefully through the Atlantic's swells early in the noon watch, teletype machines in the ship's communications center clacked out a "Top Secret" message for the president.

The message, when decoded, went to Captain Frank H. Graham, the Map Room officer on duty. Graham handed the message to the president, who was eating lunch with the crew in the ship's mess hall.

It read:

BIG BOMB DROPPED ON HIROSHIMA AUGUST 5 AT 7:15 P.M. WASHINGTON TIME. FIRST REPORTS INDICATE COMPLETE SUCCESS WHICH WAS EVEN MORE CONSPICUOUS THAN EARLIER TEST.*

Truman's face brightened. Like a man whose horse has just won by two lengths, he looked at Graham and beamed with joy. "This is the greatest thing in history!" he said, pumping the bewildered captain's hand. Then he stood up and called for the crew's attention.

---

* The bomb fell at 8:15 a.m. on August 6, 1945, Hiroshima time.

"Please keep your seats and listen for a moment. I have an announcement to make," he said in his crisp Missouri accent. "We have just dropped a new bomb on Japan which has more power than twenty thousand tons of TNT. It has been an overwhelming success!"

The mess hall erupted into cheers. Truman, exuberant as any midshipman, left to spread the gospel of America's new power.[1]

. . .

On Wednesday the eighth, Naotake Satō, Japan's ambassador to the Union of Soviet Socialist Republics, was ushered into the office of Foreign Minister Molotov. The purpose of Satō's visit was to enlist Soviet assistance in mediating the war against the Allies. To Satō's shock, Molotov brusquely informed him that as of the following day, the Soviet Union would consider itself at war with the Empire of Japan.

On the far side of the globe, Stimson met in Washington with Truman and Byrnes to refine the surrender terms. He argued, once again, that the war would end sooner if the terms were not so punitive as to stiffen the resolve of Tokyo hardliners. "When you punish your dog you don't keep souring on him all day after the punishment is over," he said.[2]

But Japan had not surrendered, and the punishment was not over. Over two days following the Hiroshima bombing, 550 of *Enola Gay*'s sisters dropped thousands of tons of conventional bombs on the beleaguered island of Honshu. Then, on the morning of August 9, three silver emissaries flew over the port city of Nagasaki, a city of a quarter million souls and the setting of Puccini's tragic opera *Madame Butterfly*. The trailing bomber, named *Bockscar*, opened its bay doors and dropped what, from the ground, appeared to be a small dot.

Forty-three seconds later, another yellow-orange ball lit up the morning sky, and seventy thousand more lives were blotted out.[3]

. . .

On Friday morning, August 10, Stimson was about to leave Washington for a short vacation when Colonel McCarthy, Marshall's aide, called to say that Radio Tokyo had announced that the Japanese government accepted the Potsdam surrender terms. The capitulation, it said, was subject to the very point Stimson had raised in Potsdam and Washington: the Emperor must retain his title as sovereign of Japan.

Stimson rushed to the White House, where he and Leahy haggled with Truman and Byrnes over a formula acceptable to the United States, the Allies, and Japan. Stimson and Leahy urged Truman to accept the condition and call it "unconditional surrender" anyway. Leahy thought the Emperor's tenure a ridiculous point to stand on if the war could be ended immediately. Agreeing with Leahy, Stimson pointed out that the United States would need the Emperor's help to maintain order in occupied Japan. Besides, he said, there were millions of Hirohito's soldiers spread out across Asia. Having the Emperor issue surrender orders, drafted by the Allies, would prevent "a score of bloody Iwo Jimas and Okinawas all over China and the New Netherlands."

Byrnes, as usual, worried about American public reaction to the Emperor's retention. Stimson grumbled to his diary, *"There has been a good deal of uninformed agitation against the Emperor in this country, mostly by people who know no more about Japan than has been given them by Gilbert and Sullivan's 'Mikado,' and I found today that curiously enough it had gotten deeply embedded in the minds of influential people in the State Department."*[4]

By lunchtime, Stimson and Byrnes had worked out a compromise. The Emperor would retain his title, subject to orders of Supreme Commander of the Allied Powers General Douglas MacArthur. Byrnes asked Stimson to draft the instruments of surrender, and Stimson, McCloy, and Marshall began working on a document that would, in time, make its way to the quarterdeck of USS *Missouri*.

• • •

While Admiral King had gone along with the plan to invade Kyushu, he had never been persuaded of the wisdom of invading Honshu. The bomb, however, headed off another fight with the Army, and when he learned of the Tokyo announcement, he cabled Nimitz, "THIS IS A PEACE WARNING. TOKYO HAS INDICATED IN ULTRA CHANNELS THAT JAPAN WISHES TO BRING ABOUT PEACE IMMEDIATELY."

On Tuesday morning, August 14, Radio Tokyo announced that Japan accepted the Allied terms of surrender. The war in the Pacific was over. King's final wartime message to Nimitz read: "SUSPEND ALL OFFENSIVE HOSTILE ACTION. REMAIN ALERT."[5]

Hostile action against Japan was suspended, but there was one last battle the admiral would fight. As arrangements were being made for the surrender ceremony in Tokyo Bay, to take place aboard USS *Missouri*, King learned that a planning committee chaired by the State Department intended to recommend that General MacArthur represent the United States. The Navy had won the Pacific war, but MacArthur's signature would rest on the instrument of surrender for all posterity to see.

King wasn't about to let MacArthur elbow in for all the credit. He ordered the pugnacious Savvy Cooke to ensure that the Navy was not shoved off its own battleship while the Empire of Japan surrendered to MacArthur.

Savvy ordered Rear Admiral Robert Dennison, a member of the Joint War Plans Committee, to set the surrender ceremony committee straight. Dennison protested that he had no connection to the committee; he wasn't a member and had no authority to represent the Navy in such matters.

"That doesn't make any difference," Savvy told the admiral. "Just go down there and tell them that you're a member, and then get in there and get this thing changed."

Dennison obediently showed up at the next meeting, where the chaos of peace had tossed the surrender protocol into a tepid pot of indifference. "I did carry out my instructions and I stepped on a lot of toes doing it," recalled Dennison afterward. "I had the backing of the people in the Navy, and the other people could not have cared less. Nobody really knew who the hell I was, and they probably didn't care."[6]

They didn't care, so they gave way to King's man. The committee recommended that at the surrender ceremony Admiral Nimitz would represent the United States of America, while General MacArthur would represent the supreme commander for the Allied powers. That is to say, MacArthur would represent himself.

America's war ended, as it began, on a battleship. On September 2, 1945, the enemy came dressed in worn black top hats and pressed coats with long tails. They came not as conquerors, but as supplicants.

After the actors took their places on the stage, and performed their roles for the final scene of mankind's great tragedy, MacArthur, as usual, had the last soliloquy. "Let us pray that peace be now restored to the world and that God will preserve it always," he declared in his benediction.

Facing his new subjects, the ministers of Japan, he announced, "These proceedings are closed."[7]

EPILOGUE

℁

ROOSEVELT AND HIS WARLORDS HAD DONE SOMETHING THOUGHT IMPOS-
sible in a democracy. Four men fundamentally different in personality, background, and training had weathered storms of defeat, recrimination, professional
bias, personality difference, and political division. They came together to defeat
two of history's most murderous empires and changed the face of their world.
And they did it without suspending the freewheeling, adversarial, often stupe-
fying system that has characterized American democracy since the founding of
the Republic.

The genius in Roosevelt's approach—a uniquely American method of war
leadership—was finding a group of headstrong fighters, meddling when he needed
to meddle, and letting them do their jobs when he didn't. That these men could
"get along," as King and Marshall put it in 1942, was both a testament to their
patriotism and a sign of how dark were the storms clouds that hung over the
nation from May 1940 to November 1942. That they could preserve their artificial unity through the victories of Midway, Normandy, Saipan, and beyond was
a near miracle.

.   .   .

When the guns fell silent, the time came for the American warlords to exit the
stage and make way for younger men who would lead America through a new and
uncertain era. Henry Stimson remained in Washington until his seventy-eighth
birthday, September 21, 1945, nineteen days after V-J Day. He held a perfunctory
press conference, said farewell to his Pentagon staff, then rode to the White

House, where a smiling President Truman presented the lifelong Republican with the Distinguished Service Medal. Stimson's last act as secretary of war was to lead a cabinet discussion about future controls of nuclear weapons.

Stimson and Mabel took a car to National Airport for their final return to Highhold. There the old soldier was nearly brought to tears by a double-line formation of every general and senior staffer in Washington. Nineteen howitzers boomed as he and Mabel took the long walk past saluting officers to the plane's stairway. An Army band played "Happy Birthday" for Stimson, followed by "Auld Lang Syne."

Fighting "an emotional and coronary breakdown," Stimson shook hands with George Marshall, his partner in so many struggles on the winding road to victory. Then he followed Mabel up the steps of the waiting plane, waved good-bye, and was gone.[1]

As the silver DC-3 set him down at his beloved Highhold, a new and vibrant America was awakening. Fifteen million Americans had left their homes for jobs in industrial centers that supplied Stimson's Army. Twelve million more had been inducted into the Army and Air Forces. America's gross national product had more than doubled, and the G.I. Bill had thrown open the doors of the middle class to men and women who served under Stimson, Marshall, and King.[2]

America, at the height of its global dominance, sauntered into a world of opportunities and dangers without Henry Stimson's guiding hand. The establishment Republican who left his mark serving a liberal Democrat died at Highhold five years later with Mabel at his bedside.[3]

He left the "youngsters" to figure out what it all meant.

. . . .

With Truman's appointment of General Eisenhower as Army Chief of Staff, George Marshall went into retirement on November 26, 1945. His retirement lasted less than a day. As he carried his bags into Dodona Manor, home of his beloved gardens, President Truman phoned to ask him to go to China as his special envoy. Unable to resist the call to service, Marshall agreed, and the gardens would wait another six years.

After China, Marshall served as Truman's secretary of state, then as chairman of the American Red Cross, then as secretary of defense. He became the only professional soldier to be awarded the Nobel Peace Prize, which he earned for his role in promoting the phenomenally successful European Recovery Program—an economic lightning bolt that he refused to call by its popular name: the Marshall Plan.[4]

As the years went by, Marshall was fondly remembered by statesmen and generals as the giant of their era. He and Katherine were regular guests at the Eisenhower White House, and in his free hours he enjoyed shooting, riding, and, of course, tending his flower beds. The fears of the past—defense of the Americas, the Army's budget, fights with the Navy, the catastrophes of Pearl Harbor, the Philippines, Kasserine Pass, and the Bulge—faded into distant memories. The strategic plans he now made were the defense of his marigolds from marauding blackbirds, and staying out of Walter Reed Hospital.[5]

With growing frequency, Marshall's reunions with old friends were held beside flag-draped coffins: Hap Arnold died in 1950, Jonathan Wainwright in 1953, Brehon Somervell in 1955. Occasionally these reunions would take place in Walter Reed's presidential suite.

In January 1959, a medical corpsman stationed outside Marshall's bedroom heard choking gasps. Bursting into the room, he found Marshall writhing in the throes of a stroke. Bedridden at home, the old general suffered a second stroke a month later.

Army doctors moved him from Dodona Manor to Walter Reed, where Truman and Eisenhower, now political enemies, separately visited him from time to time. On his last visit, President Eisenhower brought with him an eighty-five-year-old Winston Churchill, who stood in the doorway of Marshall's suite but did not enter. Tears welled in Churchill's eyes as he watched the faded general taking life's last, slow march.

As the months wore on, Marshall gradually withdrew from the world he helped shape. His speech, sight, and hearing faded, and eventually he lapsed into a coma, suspended between this life and the next. The end came on October 16, 1959.

In keeping with Marshall's request, there was no state funeral. President Eisenhower ordered the nation's flags to fly at half-mast, and the general's body lay for twenty-four hours at the National Cathedral, guarded by a squad of cadets from his alma mater, Virginia Military Institute. His funeral service was conducted at the small post chapel at Fort Myer.

George Catlett Marshall was committed to the earth at Arlington National Cemetery on October 20, not far from the resting place of his wartime friend and collaborator Sir John Dill. Eisenhower and Truman sat side by side, sharing memories of a man of no politics, a general who bridged a gulf that divided Democrat and Republican. The burial ceremony ended with a straightforward coda for a straightforward soldier: nineteen guns, a rifle volley, and "Taps."[6]

*   *   *

While Ernie King loved history, there was one story from ancient times that may have escaped his notice. As a boy, the Greek admiral Themistocles was said to have been taken by his father to a deserted beach, where his father showed him the carcasses of old war galleys lying sun-baked, prostrate, and neglected. That, his father told him, is how a democracy treats its leaders when they no longer have use for them.

King had once objected to a wartime pay raise for soldiers, sailors, and officers. When the shooting stopped, he said, a grateful nation would distribute just rewards to the men who had brought them safely through the fire. When asked if he would write a book about the war, King replied that while he would do it, the book would have only two words: "We won."[7]

The admiral who shaved with a blowtorch had given no thought to life after the war. Like Patton, Grant, Sherman, and other men who stare transfixed into the bonfires of Mars, King settled into the realization that on the day Japan's emissaries signed the surrender documents, he had accomplished his life's work.

"King was a lost soul when the war was over," said one friend. "He had served his purpose. He had done what he had set out to do. He had won his part of the war."[8]

There would be a massive demobilization as the Navy returned its men to civilian life. The Pearl Harbor inquiry would become public, Congress would slash the Navy's budget, and old salts like himself would be put out to pasture, to make way for younger admirals.

With Forrestal as navy secretary, King knew retirement would follow quickly. He had gotten along with Knox only because the Chicago newsman knew nothing about the Navy, admitted it, and stayed out of King's way. Forrestal would not. During the war, King had cursed Forrestal out in the halls of the Navy Department, and had browbeaten him into staying out of naval operations. "I didn't like him, and he didn't like me," King said.[9]

After the war, King heard that Forrestal had told a senator, "King had the brains, all right, but I hated his guts."

King replied, "I hated his guts, too."[10]

As King's tenure as COMINCH-CNO drew to a close, he requested a relief date of January 15, 1946, to give the incoming chief of naval operations, Admiral Nimitz, a well-deserved rest before taking on his new assignment. At a minimum, King wanted to hang on to his job until December 17, the fourth anniversary of his promotion to COMINCH.

Forrestal was too sour to keep King at Main Navy a day longer than necessary. "Mr. Forrestal got 'mad' about the matter and ordered the change of command moved up to December 15th," King remarked with some bitterness. On that day, King's flag secretary recalled, the old admiral's hands trembled as he accepted the orders for his relief.[11]

Five-star officers do not retire; they remain on active duty for as long as they draw breath. But unlike Marshall, King's government service had drawn to an irrevocable close. Retirement in all but name would bring banquets, honorary degrees, medals, reminiscences, and, to his sadness, the ghosts of Pearl Harbor. Newspaper headlines in November 1948 read: "Admiral King Would Clear Stark, Kimmel: Withdraws Condemnation of Two Officers Made in Report on Pearl Harbor."

King had concluded that Admirals Stark and Kimmel were no more to blame for the disaster than Generals Marshall and Short. "I'll never forgive the Army for not taking at least part of the blame for Pearl Harbor," the flinty admiral told his biographer, Commander Walter Muir Whitehall. "That was why I didn't like Stimson."[12]

King's oaken hull began to split in 1947, when he suffered a stroke. His mind remained alert, but his iron-plated timbers began to creak and sag. He moved into a suite at Bethesda Naval Hospital for full-time care, and at one point he shared a floor with the acutely depressed James Forrestal, who ended his life by jumping from his sixteenth-floor window in 1949.[13]

King spent the next seven summers at the naval hospital in Portsmouth, New Hampshire. He slipped his moorings and sailed over the bar on June 25, 1956, at the age of seventy-eight. He was buried at Annapolis, home of the United States Naval Academy.[14]

The only hymn sung at his funeral was a Navy anthem, an old favorite of Franklin Roosevelt's: "Eternal Father, Strong to Save."

SELECTED ALLIED CODE NAMES

Unlike modern military operations, whose names are chosen for their public relations value, operations in World War II were christened on the governing principle that the name should give no hint of the objective. To this, Winston Churchill added a second requirement: operations should not be given boastful or frivolous monikers. As he told Pug Ismay, "Intelligent thought will already supply an unlimited number of well-sounding names that do not suggest the character of the operation or disparage it in any way and do not enable some widow or mother to say that her son was killed in an operation called 'BUNNYHUG' or 'BALLYHOO.'"

There were exceptions, for men did lose their lives in Operations TOENAILS (New Georgia) and SLAPSTICK (Taranto).

ANAKIM	Allied invasion of Burma near Rangoon (never attempted)
ANVIL	Allied invasion of southern France (later DRAGOON), August 1944
ARCADIA	Allied conference, Washington, December 1941–January 1942
ARGONAUT	Allied-Soviet conferences, Malta-Yalta, January–February 1945
AVALANCHE	Allied landings at Salerno, Italy, September 1943
BAYTOWN	Allied landings at Calabria, Italy, September 1943
BOLERO	Allied force buildup in U.K., 1942–1944
BUCCANEER	Allied attack on Burma's Andaman Islands (never attempted)
CARTWHEEL	U.S. drive against Rabaul, in New Britain (formerly ELKTON)
COBRA	U.S. attack to break out of Normandy, July 1944

CORONET	MacArthur's plan to invade Honshu, Japan (never attempted)
DEXTERITY	MacArthur's landing on New Britain, December 1943
DRAGOON	Allied invasion of southern France, August 1944 (formerly ANVIL)
ELKTON	MacArthur's Rabaul campaign (never attempted)
EUREKA	Allied-Soviet conference, Tehran, November–December 1943
FORAGER	U.S. Marine-Army invasion of the Mariana Islands, June 1944
GRANITE	Nimitz's campaign plan for the Central Pacific, 1944
HUSKY	Allied invasion of Sicily, July 1943
JUPITER	Churchill's plan to invade Norway (never attempted)
MAGIC	Decryption of Japanese diplomatic messages
MATTERHORN	Strategic bombing of Japan, June 1944–August 1945
NEPTUNE	Allied landings in Normandy, France, June 1944 (part of OVERLORD)
OCTAGON	Allied conference, Quebec, September 1944
OLYMPIC	MacArthur's plan to assault Kyushu, Japan (never attempted)
OVERLORD	Allied invasion of France, June 1944
POINTBLANK	U.S.-British bomber offensive against Germany, 1943–1944
QUADRANT	Allied conference, Quebec, August 1943
RAINBOW	U.S. prewar strategic plans
RECKLESS	MacArthur's landing at Hollandia, New Guinea, April 1944
RENO	MacArthur's campaign for the Southwest Pacific, 1944–1945
ROUNDUP	Allied cross-Channel invasion (later OVERLORD)
SEXTANT	Allied conference, Cairo, November–December 1943
SHINGLE	Allied landing near Anzio, Italy, January 1944
SLEDGEHAMMER	Allied contingency plan to invade France, 1942 (never attempted)
SYMBOL	Allied conference, Casablanca, January 1943
TERMINAL	Allied-Soviet conference, Potsdam, July–August 1945

TORCH	Allied invasion of French North Africa, November 1942
TRIDENT	Allied conference, Washington, May 1943
ULTRA	U.S. decryption of encoded foreign messages
WATCHTOWER	U.S. assault against Tulagi and Guadalcanal, August 1942

ACKNOWLEDGMENTS

I am indebted to a diverse battalion of technical experts, archivists, historians, writers, editors, and war veterans who have done their level best to keep me on a steady course as this book crawled from concept to jumble of granular facts to completed story. I owe a great deal to the editorial geniuses of Hal Elrod, Allegra Jordan (*The End of Innocence*), and Penguin's Brent Howard, as well as the incisive Richard Story, Jim Hornfischer (*Neptune's Inferno*), Kirk Patterson, Sally Jordan, Andy Hill, Warren Kimball (*The Juggler*), the Churchill Centre's Richard Langworth, Kate Jordan, and Jennifer Elrod, all of whom provided critiques, questions, insight, or encouragement. I am also grateful to Paul Barron and Jeffrey Kozak of the George C. Marshall Research Library; and to the staffs of the Franklin D. Roosevelt Presidential Library, the Naval War College, the Library of Congress, the Navy Historical Foundation, the Eisenhower Library, the Harry S. Truman Presidential Library, the U.S. Army Military History Institute, and Murray State University's Forrest C. Pogue Library. I must also express my appreciation to the legion of history bloggers and forum participants from armchairgeneral.com and other sites who have sparred over, critiqued, or dealt death blows to many hypotheses that went into or were left out of this book.

SELECTED BIBLIOGRAPHY

᠁

PRIMARY SOURCES

★ *Archival Materials* ★

Amherst College, Amherst, Massachusetts
John J. McCloy Diary

Dwight D. Eisenhower Presidential Library, Abilene, Kansas
Ruth Briggs
Harold Bull
Gilbert Cook
Norman Cota
Dwight D. Eisenhower (Pre-Presidential Papers)
Dwight D. Eisenhower (Presidential Papers)
Dwight D. Eisenhower (Post-Presidential Papers)
Alfred Gruenther
Floyd Parks
Walter Bedell Smith (Personal Papers)
Barbara Wyden

Forrest C. Pogue Library, Murray State University, Murray, Kentucky
Forrest Pogue

Franklin Delano Roosevelt Presidential Library, Hyde Park, New York
Atomic Bomb File
Adolf Berle
Wilson Brown
Howard Bruenn
James Byrnes
Albert Wayne Coy

George M. Elsey
Federal Bureau of Investigation
George Fox
Franklin Delano Roosevelt Foundation
Anna Roosevelt Halsted
Herbert Hoover
Harry Hopkins
Anthony Johnstone
Pare Lorentz
Map Room Files
John McCrea
A. J. McGehee
Ross McIntire
William McReynolds
Lowell Mellett
Henry Morgenthau Jr.
Robert Ogg
Charles Palmer
Fabienne Pellerin
William Rigdon
Elliott Roosevelt
FDR—Official File
FDR—President's Personal File
FDR—President's Secretary File
James Roosevelt
Fred Schneider
Elizabeth Shoumatoff
Lela Mae Stiles
Grace Tully
Sumner Welles

George C. Marshall Foundation Library, Lexington, Virginia
Harvey A. DeWeerd
George Elsey
Thomas Handy
George C. Marshall
Frank McCarthy
Reminiscences File
Paul Robinett

William Sexton
Reginald Winn

Library of Congress, Washington, D.C.
Frank Maxwell Andrews
Henry Arnold
Felix Frankfurter
William Halsey
Everett Hughes
Harold Ickes
Ernest King
Frank Knox
William Leahy
Robert Patterson
George Patton
Carl Spaatz

National Archives and Records Administration, College Park, Maryland

U.S. Army Military History Institute, Carlisle Barracks, Pennsylvania
Clay and Joan Blair
Omar Bradley
Robert Coffin
Hobart Gay
Chester Hansen
Courtney Hodges
John Lucas
Office of the Center of Military History

U.S. Naval War College Library, Newport, Rhode Island
Ernest King

Yale University Library, New Haven, Connecticut
Henry Stimson Diaries
Henry Stimson Papers

★ *Oral Histories and Interviews* ★

Adams, Claude (GCML)
Arnold, Eleanor (PL)

Baruch, Bernard (PL)

Bolte, Charles (PL)

Bonesteel, Charles (PL)

Bradley, Omar (PL)

Brady, Dorothy (FDRL)

Brooke, Alan (PL)

Bryden, William (GCML)

Buchanan, Kenneth (GCML)

Bull, Harold (PL)

Bundy, Harvey (PL)

Byrnes, James (PL)

Carter, Marshall (PL)

Christiansen, James (GCML)

Clark, Mark (PL)

Clay, Lucius (PL)

Collins, Joseph (PL)

Coningham, Arthur (PL)

Cooke, Charles (PL)

Cunningham, Andrew (PL)

De Gaulle, Charles (PL)

Deane, John (PL)

Dennison, Robert (HSTL)

DuBois Jr., Josiah (HSTL)

Eden, Anthony (PL)

Eisenhower, Dwight (PL)

Elsey, George (HSTL)

Farley, James (FDRL)

Frankfurter, Felix (PL)

Franklin, John (FDRL)

Gerow, Leonard (PL)

Groves, Leslie (PL)

Halsted, Diana Hopkins (FDRL)

Halsted, James (FDRL)

Handy, Thomas (PL)

Harriman, W. Averell (HSTL)

Hassett, William (FDRL, PL)

Hastie, William (HSTL)

Heffner, William (GCML)

Hull, John (PL)

Ismay, Hastings (PL)

Johnson, Louis (PL)

Johnson, Thomas (GCML)

King, Ernest (PL)

Kittridge, T. B. (GCML)

Kuter, L. S. (GCML)

Leahy, William (FDRL, PL)

Lovett, Robert (HSTL, PL)

MacArthur, Douglas (PL)

McCarthy, Frank (PL)

McCloy, John (PL)

McNarney, Joseph (PL)

Marshall, George (GCML, PL)

Marshall, Katherine (PL)

Martyn, John (GCML)

Miles, Sherman (PL)

Montgomery, Bernard (PL)

Morgan, Frederick (PL)

Mountbatten, Louis (PL)

Nimitz, Chester (PL)

Ogg, Robert (FDRL)

Pasco, H. Merrill (GCML)

Perkins, Frances (COHP)

Portal, Charles (PL)

Powder, James (GCML)

Rayburn, Sam (PL)

Rigdon, William (HTSL)

Roosevelt, Curtis (FDRL)

Roosevelt, Eleanor (FDRL, PL)

Roosevelt, Franklin D., Jr. (FDRL)

Roosevelt, James (FDRL)

Rosenman, Samuel (FDRL)

Royce, Charles (GCML)

Sexton, William (GCML)

Singer, John J., Mrs. (GCML)

Smith, Truman (PL)

Smith, Walter Bedell (PL)

Smith, William (PL)

Stark, Harold (FDRL, PL)

Stayer, M. C. (GCML)

Suckley, Margaret (FDRL)

Thomas, Cora (GCML)

Truman, Harry (PL)

Wallace, Henry (FDRL)

Watson, Edwin, Mrs. (GCML)

Wedemeyer, A. C. (GCML)

★ *Other Sources* ★

Halifax, Edward F.L., Earl of, "Lord Halifax Diary," University of York, at https://
 dlib.york.ac.uk/yodl/app/collection/detail?id=york%3a814833&ref=browse.

Low, Francis S., "A Personal Narrative of Association with Fleet Admiral Ernest J.
 King, U.S. Navy," 1961, U.S. Naval War College, Newport, Rhode Island.

"The World War II Diaries of Joseph W. Stilwell, 1941–1945," Hoover Institution, at
 www.hoover.org/library-and-archives/collections/east-asia/featured-collections/
 joseph-stilwell.

U.S. Joint Strategic Survey, *Interrogations of Japanese Officials*, OpNav P-03-100,
 accessed at www.ibiblio.org/hyperwar/AAF/USSBS/IJO/index.html#pageVI.

★ *Books* ★

Arnold, Henry H., *Global Mission* (New York: Harper Bros., 1949).

Baldwin, Hanson, *Battles Lost and Won* (New York: Harper & Row, 1966; Avon,
 1968 ed.).

Bradley, Omar N., *A Soldier's Story* (New York: Henry Holt, 1951; New York:
 Modern Library, 1991 ed.).

Bradley, Omar N., and Clay Blair, *A General's Life* (New York: Simon & Schuster,
 1983).

Brooke, Alan, *War Diaries 1939–1945*, Alex Danchev and Daniel Todman, eds.
 (London: Wedenfield & Nicholson, 1957; London: Butler & Tanner, 2001 ed.).

Butcher, Harry C., *My Three Years with Eisenhower* (New York: Simon & Schuster,
 1946).

Byrnes, James F., *All in One Lifetime* (New York: Harper & Bros., 1958).

———, *Speaking Frankly* (New York: Harper & Bros., 1947).

Capra, Frank, *The Name Above the Title* (New York: Macmillan, 1971; Da Capo,
 1997 ed.).

Churchill, Winston S., *The Grand Alliance: The Second World War* (New York:
 Houghton Mifflin, 1948; New York: RosettaBooks, 2010 ed.).

———, *Closing the Ring: The Second World War* (New York: Houghton Mifflin, 1951;
 New York: RosettaBooks, 2010 ed.).

———, *Triumph and Tragedy: The Second World War* (New York: Houghton Mifflin, 1953; New York: RosettaBooks, 2010 ed.)

———, *The Churchill War Papers*, Martin Gilbert, ed. (3 vols., New York: Norton, 1993–2001).

Churchill, Winston S. and Franklin Delano Roosevelt, *Churchill and Roosevelt: The Complete Correspondence*, Warren Kimball, ed. (3 vols., Princeton, NJ: Princeton University Press, 1984).

Clark, Mark W., *Calculated Risk* (New York: Harper & Bros., 1950).

Compton, Arthur, *Atomic Quest* (Oxford: Oxford University Press, 1956).

Cunningham, Andrew B., *The Cunningham Papers* (2 vols., Aldershot: Ashgate, 2006).

———, *A Sailor's Odyssey* (London: Dutton, 1951).

Dykes, Vivian, *Establishing the Anglo-American Alliance*, Alex Danchev., ed. (London: Brassey's, 1990).

Eisenhower, David, *Eisenhower at War 1943–1945* (New York: Random House, 1986; New York: Wings, 1991 ed.)

Eisenhower, Dwight D., *Crusade in Europe* (Garden City: Doubleday, 1949).

———, *Papers of Dwight David Eisenhower*, Alfred Chandler, Jr., ed. (5 vols., Baltimore: Johns Hopkins University Press, 1970).

———, *Eisenhower Diaries*, Robert Ferrell, ed. (New York: Norton, 1981).

Forrestal, James, *The Forrestal Diaries*, Walter Millis, ed. (New York: Viking, 1951).

Galbraith, John Kenneth, *A Life in Our Times* (New York: Houghton Mifflin, 1981).

Gromyko, Andrei, *Memoirs* (New York: Doubleday, 1990).

Halsey, William F., *Admiral Halsey's Story* (New York: McGraw-Hill, 1947).

Hansell, Haywood S., Jr., *The Strategic Air War Against Germany and Japan: A Memoir* (Washington: U.S. Government, 1986).

Harriman, Averell, and Elie Abel, *Special Envoy to Churchill and Stalin* (New York: Random House, 1975).

Hassett, William D., *Off the Record with F.D.R.* (New Brunswick: Rutgers University Press, 1958).

Hull, Cordell, *Memoirs of Cordell Hull* (2 vols., New York: Macmillan, 1948).

Ickes, Harold L., *The Secret Diary of Harold L. Ickes: The Lowering Clouds* (New York: Simon & Schuster, 1955).

Ismay, Hastings Lionel, *Memoirs of General Lord Ismay* (New York: Viking, 1960).

Ito, Nobutaka, ed., *Japan's Decision for War* (Stanford: Stanford University Press, 1967).

Jackson, Robert H., *That Man* (New York: Oxford University Press, 2003; New York: Oxford University Press, 2004 ed.).

Janeway, Eliott, *The Struggle for Survival* (New Haven: Yale University Press, 1951).

King, Ernest J., *Official Reports to the Secretary of the Navy* (Washington: United States Navy Department, 1946).

King, Ernest J., and Walter Muir Whitehill, *Fleet Admiral King, A Naval Record* (New York: W. W. Norton, 1952).

Leahy, William D., *I Was There* (London: Victor Gollancz, Ltd., 1950).

MacArthur, Douglas A., *Reports of General MacArthur: Japanese Operations in the Southwest Pacific Area* (2 vols.: Tokyo: U.S. Government, 1950; Washington: Center of Military History, 1994 ed.).

———, *Reminiscences* (New York: McGraw-Hill, 1964; Annapolis: Naval Institute Press, 2001 ed.).

McCloy, John J., *The Challenge to American Foreign Policy* (Cambridge: Harvard University Press, 1953).

Marshall, George C., *Selected Speeches and Statements of General of the Army George C. Marshall*, H.A. DeWeerd, ed. (Washington: Infantry Journal, 1945; Da Capo, 1973 ed.).

———, *Papers of George Catlett Marshall*, Larry Bland and Mark A. Stoler, eds. (6 vols., Baltimore: Johns Hopkins, 1981–2013).

Marshall, Katherine Tupper, *Together: Annals of an Army Wife* (New York: Tupper & Love, 1946; Chicago: Peoples Book Club, 1948 ed.).

Montgomery, Bernard Law, *Memoirs* (London: World Publishing, 1958; Signet, 1959 ed.).

Moran, Lord, *Churchill: Taken from the Diaries of Lord Moran: The Struggle for Survival, 1940–1945* (Cambridge: Riverside, 1966; Norman S. Berg, 1976 ed.).

Morgenthau, Henry, Jr., *From the Morgenthau Diaries: Years of Urgency*, John Blum, ed. (Boston: Houghton Mifflin, 1965).

———, *From the Morgenthau Diaries: Years of War*, John Blum, ed. (Boston: Houghton Mifflin 1967).

Murphy, Robert, *Diplomat Among Warriors* (New York: Doubleday, 1964).

Perkins, Frances, *The Roosevelt I Knew* (New York: Viking, 1946; New York, Penguin, 2011 ed.).

Rigdon, William M., with James Derieux, *White House Sailor* (Garden City, NY: Doubleday, 1962).

Reilly, Michael F., *Reilly of the White House* (New York: Simon and Schuster, 1947).

Reynolds, Quentin, *Only the Stars Are Neutral* (New York: Random House, 1942).

Richardson, K. D., ed., *Reflections of Pearl Harbor* (West Point, CT: Praeger, 2005).

Roosevelt, Eleanor, *This I Remember* (New York: Harper & Bros., 1949; Praeger, 1975 ed.).

———, *The Autobiography of Eleanor Roosevelt* (New York: Harper & Bros., 1961; Da Capo, 1992 ed.).

Roosevelt, Eleanor and Anna Boettiger, *Mother and Daughter*, Bernard Asbell, ed. (New York: Coward, McCann & Geoghegan, 1982).

Roosevelt, Elliott, *As He Saw It* (New York: Duell, Sloan & Pearce, 1946).

Roosevelt, Franklin D., *F.D.R.: His Personal Letters*, Elliott Roosevelt, ed. (2 vols., New York: Duell, Sloan & Pearce, 1950).

———, *Public Papers and Addresses of Franklin D. Roosevelt*, Samuel Rosenmann, ed. (13 vols., New York: Random House 1938, Macmillan 1941, Harper 1950).

Franklin D. and Winston S. Churchill, *Roosevelt and Churchill: Their Secret Wartime Correspondence*, Francis L. Loewenheim, Harold D. Langley, and Manfred Jonas, eds. (New York: Dutton, 1975; New York: Da Capo, 1990 ed.).

Roosevelt, Franklin D. and Josef Stalin, *Dear Mr. Stalin: The Complete Correspondence Between Franklin D. Roosevelt and Joseph V. Stalin*, Susan Butler, ed. (New Haven: Yale University Press, 2005).

Roosevelt, James, and Bill Libby, *My Parents: A Differing View* (Chicago: Playboy Press, 1976).

Roosevelt, James, and Sydney Shallett, *Affectionately, F.D.R.* (New York: Harcourt, Brace, 1959).

Rosenman, Samuel, *Working with Roosevelt* (New York: Harper & Bros., 1952).

Sherwood, Robert E., *Roosevelt and Hopkins* (New York: Harper & Bros., 1948).

Smith, A. Merriman, *Thank You, Mr. President: A White House Notebook* (New York: Harper & Bros., 1946).

Stettinius, Edward R., Jr., *Lend-Lease: Weapons for Victory* (New York: Macmillan, 1944).

Stilwell, Joseph W., *The Stilwell Papers*, Theodore H. White, ed. (New York: Sloane Associates, 1948; Da Capo, 1991 ed.).

Stimson, Henry L., *On Active Service in Peace and War* (New York: Harper & Bros., 1947).

Suckley, Margaret, *Closest Companion*, Geoffrey C. Ward, ed. (New York: Simon & Schuster, 1995).

Tedder, Arthur, *With Prejudice* (Boston: Little, Brown, 1966).

Truman, Harry S., *Year of Decisions* (Garden City, NY: Doubleday, 1955).

Tully, Grace, *F.D.R.: My Boss* (New York: Charles Scribner's Sons, 1949; Chicago: Peoples Book Club, 1949 ed.).

U.S. Atomic Energy Commission, *In the Matter of J. Robert Oppenheimer, Transcript of Hearing Before Personnel Security Board* (Washington: Government Printing Office, 1954).

U.S. Congress, Joint Committee on the Investigation of the Pearl Harbor Attack, *Hearings Before the Joint Committee on the Investigation of the Pearl Harbor Attack* (39 vols., Washington: Government Printing Office, 1946).

U.S. Department of State, *Peace and War: United States Foreign Policy, 1931–1941* (Washington: U.S. Government, 1943).

————, *Foreign Relations of the United States: The Conferences at Malta and Yalta 1945* (Washington: Government Printing Office, 1955).

————, *Foreign Relations of the United States: The Potsdam Conference* (2 vols., Washington: Government Printing Office, 1960).

————, *Foreign Relations of the United States: The Conferences at Cairo and Tehran, 1943* (Washington: Government Printing Office, 1961).

————, *Foreign Relations of the United States: The Conferences at Washington, 1941– 1942, and Casablanca, 1943* (Washington: Government Printing Office, 1968).

————, *Foreign Relations of the United States: The Conferences at Washington, 1943, and Quebec, 1943* (Washington: Government Printing Office, 1970).

————, *Foreign Relations of the United States: The Conference at Quebec, 1944* (Washington: Government Printing Office, 1972).

U.S. Joint Chiefs of Staff, Joint History Office, *World War II Inter-Allied Conferences* (Washington: U.S. Government, 2003).

U.S. Strategic Bombing Survey, *Interrogations of Japanese Officials* (Washington: U.S. Government, 1946).

USSR Ministry of Foreign Affairs, *Correspondence Between Stalin, Roosevelt, Truman, Churchill, and Atlee During World War II* (New York: Capricorn, 1958; Minerva, 2001 ed.).

Vandenberg, Arthur H., Jr. and Joe Alex Morris, *The Private Papers of Senator Vandenberg* (New York: Houghton Mifflin, 1952).

Ward, Geoffrey, ed., *Closest Companion* (New York: Houghton Mifflin, 1995; New York: Simon & Schuster Paperbacks, 2005 ed.).

Wedemeyer, Albert C., *Wedemeyer Reports!* (New York: Henry Holt, 1959).

★ *Periodicals* ★

Morton, Louis, "The Decision to Use the Atomic Bomb," *Foreign Affairs* (January 1957).

Smith, Alice Kimball, "Behind the Decision to Use the Atomic Bomb," *Bulletin of the Atomic Scientists* (October 1958).

Stimson, Henry L., "The Decision to Use the Bomb," *Harper's* (February 1947).

★ *Photographic Collections* ★

Forrest C. Pogue Library
Franklin Delano Roosevelt Presidential Library
George C. Marshall Research Library
Library of Congress
National Archives and Records Administration
Naval War College

SECONDARY SOURCES

★ *Books* ★

Allen, Paul, *Katyn: Stalin's Massacre* (Bloomfield: Northern Illinois University Press, 2010).

Ambrose, Stephen E., *Eisenhower and Berlin 1945* (New York: W. W. Norton, 2000 ed.).

Andrew, Christopher and Vasili Mitrokhin, *The Sword and the Shield* (New York: Basic Books, 1999).

Arakaki, Leatrice R. and John R. Kuborn, *7 December 1941* (Hickam AFB, Hawaii: Pacific Air Forces Office of History, 1991).

Atkinson, Rick, *An Army at Dawn* (New York: Henry Holt, 2002).

———, *The Day of Battle* (New York: Henry Holt, 2007).

———, *Guns at Last Light* (New York: Henry Holt, 2013).

Baptiste, Fitzroy André, *War, Cooperation & Conflict* (Westport: Greenwood Press, 1988).

Barnhart, Michael A., *Japan Prepares for Total War* (Ithaca: Cornell University Press, 1987).

Beria, Sergo, *My Father*, Brian Pearce, trans. (London: Duckworth, 2001).

Beschloss, Michael, *The Conquerors* (New York: Simon & Schuster, 2002).

Bird, Kai, *Chairman of the Board* (New York: Simon & Schuster, 1992).

Blair, Clay, *Hitler's U-Boat War: The Hunters, 1939–1942* (New York: Random House, 1996).

Blumenson, Martin, *Mediterranean Theater of Operations: Salerno to Cassino* (Washington: U.S. Government, 1967).

Borneman, Walter R., *The Admirals* (New York: Little Brown, 2012).

Brown, Anthony Cave, *Bodyguard of Lies* (Guilford: Lyons Press, 2007 ed.).

Bryant, Arthur, *Turn of the Tide* (New York: Doubleday, 1957).

Buell, Thomas B., *Master of Sea Power* (Boston: Little, Brown, 1980).

Cannon, M. Hamlin, *The War in the Pacific: Leyte: Return to the Philippines* (Washington: Center of Military History, 1993 ed.).

Cantril, Hadley and Mildred Strunk, *Public Opinion, 1935–1946* (Princeton: Princeton University Press, 1951).

Cienciela, Anna M., Wojciech Materski, and N. S. Lebedeva, *Katyn* (New Haven: Yale University Press, 2008).

Cline, Ray S., *Washington Command Post: The Operations Division* (Washington: Center of Military History, 2003 ed.).

Coffey, Thomas M., *Hap* (New York: Viking, 1982).

Congressional Quarterly, Inc., ed., *Presidential Elections 1789–1996* (Washington: Congressional Quarterly, Inc., 1997).

Conn, Stetson, Rose C. Engleman, and Byron Fairchild, *The Western Hemisphere: Guarding the United States and its Outposts* (Washington: Center of Military History, 2000 ed.).

Conn, Stetson, and Byron Fairchild, *The Western Hemisphere: The Framework of Hemisphere Defense* (Washington: Center of Military History, 1989 ed.).

Costigliola, Frank, *Roosevelt's Lost Alliances* (Princeton: Princeton University Press, 2012).

Craven, W. F., and J. L. Cate, eds., *The Army Air Forces in World War II* (7 vols., Washington: U.S. Government, 1947–1958).

Crosswell, D. K. R., *Beetle* (Lexington: University of Kentucky Press, 2010).

Crowl, Philip A., *War in the Pacific: Campaign in the Marianas* (Washington: U.S. Government, 1959).

Crowl, Philip A. and Edmund G. Love, *The War in the Pacific: Seizure of the Gilberts and Marshalls* (Washington: U.S. Government, 1955).

Dalleck, Robert, *Franklin D. Roosevelt and American Foreign Policy* (New York: Oxford University Press, 1995 ed.).

D'Este, Carlo, *Eisenhower* (New York: Henry Holt, 2002).

DeFelice, Jim, *Omar Bradley* (Washington: Regenery, 2011).

Dobbs, Michael, *Six Days in 1945* (New York: Alfred A. Knopf, 2012).

Dyer, George C., *The Amphibians Came to Conquer* (Washington: U.S. Government, 1969).

Eisenhower, John S. D., *Allies* (New York: Doubleday 1982; Da Capo, 2000 ed.).

———, *General Ike* (New York: Free Press, 2003).

Fairchild, Byron and Jonathan Grossman, *Army and Industrial Manpower* (Washington: Center of Military History, 1988 ed.).

Fisher, Ernest F., *Mediterranean Theater of Operations: Monte Cassino to the Alps* (Washington: Center of Military History, 1989 ed.).

Fullilove, Michael, *Rendezvous with Destiny* (New York: Penguin Press, 2013).

Furer, Julius Augustus, *Administration of the Navy Department in World War II* (Washington: U.S. Government, 1959).

Garand, George W., and Truman R. Strobridge, *History of U.S. Marine Corps Operations in World War II: Vol. IV: Western Pacific Operations* (Washington: U.S. Government, 1971).

Garland, Albert N. and Howard M. Smyth, *The Mediterranean Theater of Operations: Sicily and the Surrender of Italy* (Washington: Center of Military History, 1993 ed.).

Giangreco, D. M., *Hell to Pay* (Annapolis: Naval Institute Press, 2009).

Gilbert, Martin, *Winston S. Churchill: The Road to Victory, 1941–1945* (New York: Houghton Mifflin, 1986).

Goldberg, Harold J., *D-Day in the Pacific: The Battle of Saipan* (Bloomington: Indiana University Press, 2007).

Goodwin, Doris Kearns, *No Ordinary Time* (New York: Simon & Schuster, 1994).

Greenfield, Kent Roberts, Robert R. Palmer, and Bell I. Wiley, *The Army Ground Forces: The Organization of Ground Combat Troops* (Washington: Center of Military History, 1987).

Hamilton, Nigel, *Mantle of Command* (New York: Houghton Mifflin Harcourt, 2014).

Hancock, W. K. and M. M. Gowing, *British War Economy* (London: HMSO, 1949).

Hardesty, Von, *Air Force One* (Northword Press, 2003).

Harrison, Gordon A., *European Theater of Operations: Cross-Channel Attack* (Washington: U.S. Government, 1950).

Hewlett, Richard G. and Oscar E. Anderson, Jr., *The New World* (University Park: Pennsylvania State University Press, 1962).

Hirshson, Stanley P., *General Patton* (HarperPerennial, 2003 ed.).

Hodgson, Godfrey, *The Colonel* (New York: Knopf, 1990).

Hoffman, Carl W., *Saipan* (Washington: Government Printing Office, 1950).

Hopkins, William B., *The Pacific War* (Minneapolis: Zenith, 2008).

Hough, Frank O., Verle E. Ludwig and Henry I. Shaw, *History of U.S. Marine Corps Operations in World War II: Vol. I: Pearl Harbor to Guadalcanal* (Washington: U.S. Government, 1958).

Hunt, Frazier, *The Untold Story of Douglas MacArthur* (New York: Signet Books, 1964).

James, D. Clayton, *A Time For Giants* (New York: Franklin Watts, 1987).

Jones, Vincent C., *MANHATTAN: The Army and the Atomic Bomb* (Washington: Center of Military History, 1985).

Jordan, Jonathan W., *Brothers Rivals Victors* (New York: NAL/Caliber, 2011).

Jungk, Robert, *Brighter Than a Thousand Suns* (New York: Houghton Mifflin, 1958).

Kimball, Warren F., *The Juggler* (Princeton: Princeton University Press, 1991).

Kirkpatrick, Charles E., *An Unknown Future and a Doubtful Present* (Washington: Center for Military History, 1992).

Korda, Michael, *Ike* (New York: Harper Perennial, 2008 ed.).

Kort, Michael, *The Columbia Guide to Hiroshima and the Bomb* (New York: Columbia University Press, 2007).

Lacey, James, *Keep from All Thoughtful Men* (Annapolis: Naval Institute Press, 2011).

Larrabee, Eric, *Commander-in-Chief* (New York: Harper & Row, 1987).

Lerwill, Leonard L., *The Personnel Replacement System in the United States Army* (Washington: U.S. Government, 1954).

Lomazo, Steven, and Eric Fettman, *FDR's Deadly Secret* (New York: PublicAffairs, 2009).

MacGregor, Morris J., *Integration of the Armed Force, 1940–1965* (Washington: Center of Military History, 1985).

McLaughlin, John J., *General Albert C. Wedemeyer* (Philadelphia: Casemate, 2012).

Manchester, William and Paul Reid, *The Last Lion: Winston Spencer Churchill: Defender of the Realm, 1940–1965* (New York: Little, Brown, 2012).

Meacham, Jon, *Franklin and Winston* (New York: Random House, 2003).

Miller, Edward S., *Bankrupting the Enemy* (Annapolis: Naval Institute Press, 2007).

Miller, John, Jr., *War in the Pacific: Guadalcanal: The First Offensive* (Washington: U.S. Government, 1948).

———, Jr., *War in the Pacific: CARTWHEEL: The Reduction of Rabaul* (Washington: Center of Military History, 1968 ed.).

Moe, Richard, *Roosevelt's Second Act* (New York: Oxford University Press, 2013).

Morgan, Ted, *F.D.R.* (New York: Touchstone, 1986 ed.).

Morison, Elting E., *Turmoil and Tradition* (New York: History Book Club, 2003 ed.).

Morison, Samuel Eliott, *History of the United States Naval Operations in World War II* (15 vols., Boston: Little, Brown, 1947–1962).

Morton, Louis, *War in the Pacific: The Fall of the Philippines* (Washington: U.S. Government, 1952).

Nasal, David, *The Patriarch* (New York: Penguin, 2012).

Newfeld, Michael J., and Michael Berenbaum, eds., *The Bombing of Auschwitz* (New York: St. Martin's, 2000).

Nofi, Albert A., *To Train the Fleet* (Washington: Government Printing Office, 2010).

O'Brien, Francis A., *Battling for Saipan* (New York: Random House Digital, 2003).

Olson, Lynne, *Those Angry Days* (New York: Random House, 2013).

Parker, Frederick D., *A Priceless Advantage* (Washington: Center for Cryptologic History, National Security Agency, 1993).

Parrish, Thomas, *Roosevelt and Marshall* (New York: William Morrow, 1989).

Parshall, Jonathan and Anthony Tully, *Shattered Sword* (Washington: Potomac, 2005).

Perry, Mark, *Partners in Command* (New York: Penguin, 2007).

Persico, Joseph, *Roosevelt's Centurions* (New York: Random House, 2013).

Petty, Bruce M., *Saipan: Oral Histories of the Pacific War* (Jefferson, NC: McFarland & Co., 2002).

Pogue, Forrest C., *George C. Marshall: Education of a Soldier, 1880–1939* (New York: Viking, 1963).

———, *George C. Marshall: Ordeal and Hope, 1939–1942* (New York: Viking, 1966).

———, *George C. Marshall: Organizer of Victory, 1943–1945* (New York: Viking, 1973).

Prange, Gordon W., et al., *At Dawn We Slept* (New York: Penguin, 1991 ed.).

Potter, E. B., *Nimitz* (Annapolis: Naval Institute Press, 1976).

Ritchie, Donald A., *Reporting from Washington* (New York: Oxford University Press, 2005).

Roberts, Andrew, *Masters and Commanders* (New York: HarperCollins, 2009).

———, *Storm of War* (New York: HarperCollins, 2011).

Romanus, Charles F. and Riley Sunderland, *China-Burma-India Theater: Stilwell's Mission to China* (Washington: U.S. Government, 1952).

———, *China-Burma-India Theater: Time Runs Out in CBI* (Washington: U.S. Government, 1958).

Roscoe, Theodore, *United States Destroyer Operations in World War II* (Annapolis, MD: Naval Institute Press, 1953).

Shaw, Henry I., Jr., Bernard C. Nalty, and Edwin T. Turnbladh, *History of the U.S. Marine Corps Operations in World War II: Central Pacific Drive* (Washington: U.S. Government, 1966).

Shirer, William L., *The Rise and Fall of the Third Reich* (New York: Fawcett Crest, 1992 ed.).

Simpson, B. Mitchell III, *Admiral Harold R. Stark* (Columbia: University of South Carolina Press, 1989).

Smith, Jean Edward, *FDR* (New York: Random House Trade, 2008).

Smith, Robert Ross, *The War in the Pacific: Triumph in the Philippines* (Washington: Office of the Chief of Military History, 1991 ed.).

———, *The War in the Pacific: The Approach to the Philippines* (Washington: Office of the Chief of Military History, 1996 ed.).

Stevenson, Charles A., *Warriors and Politicians* (Oxford: Taylor & Francis, 2006).

Stoler, Mark A., *Allies and Adversaries* (Chapel Hill: University of North Carolina Press, 2000 ed.).

———, *Allies in War* (Bloomsbury Academic, 2007 ed.).

Symonds, Craig L., *The Battle of Midway* (Oxford: Oxford University Press, 2011).

Taaffe, Stephen R., *Marshall and His Generals* (Lawrence: University Press of Kansas, 2011).

Thompson, Harry and Lidas Mayo, *The Ordnance Department: Procurement and Supply* (Washington: Center of Military History, 1960).

Tobin, James, *The Man He Became* (New York: Simon & Schuster, 2013).

Toland, John, *The Rising Sun* (New York: Modern Library, 2003).

Toll, Ian, *Pacific Crucible* (New York: W. W. Norton, 2012).

Tomblin, Barbara Brooks, *With Utmost Spirit* (Lexington: University Press of Kentucky, 2004).

Toughill, Thomas, *A World to Gain* (Forest Row: Clairview, 2004).

Unger, Debi, Irwin Unger, and Stanley Hirshson, *George C. Marshall* (New York: Harper, 2014).

Vogel, Steve, *The Pentagon: A History* (New York: Random House, 2007).

Watson, Mark, *Chief of Staff: Prewar Plans and Preparations* (Washington: Government Printing Office, 1950).

Weigley, Russell F., *Eisenhower's Lieutenants* (Bloomington: Indiana University Press, 1981).

Weintraub, Stanley, *15 Stars* (New York: Free Press, 2007).

———, *Final Victory* (Philadelphia: Da Capo, 2012).

Wohlstetter, Roberta, *Pearl Harbor: Warning and Decision* (Stanford: Stanford University Press, 1962).

★ *Other Sources* ★

Berlin, Jon Kent, "'A Supreme Act of Faith'" (University of Virginia master's thesis, 1990).

Coatney, Louis Robert, "The Katyn Massacre" (Western Illinois University master's thesis, 1993).

Davis, Vernon, "The History of the Joint Chiefs of Staff in World War II: Organizational Development: Origin of the Joint and Combined Chiefs of Staff" (Historical Division, Joint Secretariat, Joint Chiefs of Staff, 1972).

Fischer, Benjamin B., "The Katyn Controversy," *Studies in Intelligence* 43:4 (2000).

Hanyok, Robert J., "Catching the Fox Unaware," *Naval War College Review* (Autumn 2008).

Kern, Gary, "How 'Uncle Joe' Bugged FDR," *Studies in Intelligence* 47:1 (2003).

Kimball, Warren F., "Whodunnit? Listening in on Roosevelt and Churchill," *Finest Hour* (Summer 2006).

Lunghi, Hugh, "Troubled Triumvirate," *Finest Hour* (Summer 2007).

Smith, Abbott, "Battle of the Atlantic" (chapter VI of draft JCS history, "The War Against Germany," NARA RG 218, Manuscripts, Transcripts and Drafts, box 6).

U.S. Navy, Bureau of Naval Personnel Historical Section, "The Negro in the Navy," first draft, n.d.

Williams, Clarence Hughes III, "We Have . . . Kept the Negroes' Goodwill and Sent Them Away" (Texas A&M University master's thesis, 2008).

ENDNOTES

★   *Endnote Abbreviations*   ★

AP	Henry H. Arnold Papers, Library of Congress
BCS	British Chiefs of Staff
CCS	Combined Chiefs of Staff
COHP	Columbia Oral History Project
DAM	Douglas A. MacArthur
DDE	Dwight D. Eisenhower
EJK	Ernest J. King
EL	Dwight D. Eisenhower Presidential Library
EP	*Eisenhower Papers*
ER	Eleanor Roosevelt
FDR	Franklin Delano Roosevelt
FDRL	Franklin D. Roosevelt Presidential Library
FDR-PL	FDR, *Personal Letters*
FDR-PP	FDR, *Public Papers and Addresses*
FRUS	*Foreign Relations of the United States*
GCM	George C. Marshall
GCML	George C. Marshall Library
HHA	Henry H. Arnold
HLH	Harry L. Hopkins
HLS	Henry L. Stimson
HST	Harry S. Truman
HSTL	Harry S. Truman Presidential Library
JCS	Joint Chiefs of Staff
KP	Buell-Whitehill Collection of Ernest J. King Papers, NWC
LC	Library of Congress
MP	*The Papers of George Catlett Marshall*

MP-GCML	George C. Marshall Papers, GCML
NARA	National Archives and Records Administration II
NHC	Naval Historical Center
NHC-OL	NHC Online Archives
NWC	Naval War College
NYT	*The New York Times*
PHH	Joint Congressional Committee, Pearl Harbor Hearings
PL	Forrest Pogue Collection, Murray State University
PP-LC	Patton Papers, LC
PSF	President's Secretary's File, FDRL
SEP	*The Saturday Evening Post*
SP	Henry L. Stimson Papers, Yale University
USAMHI	U.S. Army Military History Institute
WP	*The Washington Post*
WSC	Winston S. Churchill

Dates without corresponding descriptions represent diary entries.

PROLOGUE ★ "THERE MUST BE SOME MISTAKE"

1 Pharmacist's Mate 2d Class Lee Soucy, interview, n.d., NHC-OL ("crazy Marines"); "Spectacular Air Drama Plays at the Waikiki," Honolulu *Star-Bulletin*, 12/7/41; "Navy Is Superior to Any, Says Knox," NYT, 12/7/41; CINCPAC to Secretary of the Navy, "Report of Japanese Raid on Pearl Harbor," 2/15/42, NARA/NHC-OL; Arakaki and Kuborn 62–67.

2 James Powder, interview, 10/19/59, GCML; GCM, testimony, 12/7/45, 12/13/45, PHH 3:1108–10, 3:1509; GCM, interview, 3/29/54, GCML; GCM to Madge Brown, 12/5/41, MP 2:694–95; Toland 201. There are many discrepancies regarding the timing of Marshall's movements that morning. This account is based on the testimony of Marshall and the few associates who were with him that day.

3 GCM, testimony, 12/7/45, 12/13/45, PHH 3:1108–10, 3:1509; Katherine Marshall 98–99; "Army Pearl Harbor Investigation," PHH 3:1448; Pogue 2:229, 3:57.

4 Japanese Government to United States Government, 12/7/41, *Department of State Bulletin*, vol. 3, no. 129, 12/13/41; Sherman Miles to GCM, 12/15/41, PHH 2:929; Miles, testimony, 11/29/45, PHH 2:930, 938.

5 GCM, testimony, 12/7/45, PHH 3:1108–10; Sherman Miles to GCM, 12/15/41, PHH 14:1410; J. R. Deane to W. B. Smith, 6/8/42, PHH 14:1411; Sherman Miles, "Sunday Morning, December 7, 1941," 12/15/41, PHH 2:929; Leonard T.

Gerow, "Memorandum for the Record," 12/15/41, PHH 14:1409; W. B. Smith, "Memorandum for the Record," 12/15/41, PHH 14:1410; Toland 191, 202; Prange 506; Morison 3:98–127.

6　CINCPAC to Secretary of the Navy, 2/15/42, NARA/NHC-OL; Richardson 2–19, quoting Robert E. Thomas, Oswald S. Tanczos, Dave Smith, and Tom A. Beasley, Howard Asa Price, Jr., interviews; Toland, *Rising Sun*, 212–14.

7　HLH, memorandum, 12/7/41, in Sherwood 430–31; FDR, Jr., interview 1/11/79, FDRL; Toland, *Rising Sun*, 223 ("mistake"); Smith, FDR, 508.

8　Commanding Officer, USS *Utah*, to CINCPAC, 12/15/41, NHC-OL. *Utah* had been commissioned as a battleship, but was a target ship by the time of Pearl Harbor, and not moored on Battleship Row.

9　CINCPAC to Secretary of the Navy, 2/15/42, NARA/NHC-OL; Commanding Officer, USS *Oklahoma*, to CINCPAC, 12/18/41, NHC-OL; "Report by the Secretary of the Navy to the President," 12/14/41, FDRL (PSF 59).

10　Commanding Officer, USS *Tennessee*, to CINCPAC, 12/11/41, NHC-OL; Commanding Officer, USS *West Virginia*, to CINCPAC, 12/11/41, NHC-OL; Commander, Battleships, Battle Force to CINCPAC, 12/19/41, NHC-OL.

11　"Report by the Secretary of the Navy to the President," 12/14/41, FDRL (PSF 59); Prange 506; Toland, *Rising Sun*, 216–19.

12　HLS, 12/6–7/41; Pogue 2:222. *Note: all HLS diary entries are found in Stimson Papers, Yale.*

13　Foreign Office to Nomura, 11/15/41, PHH 12:137; HLS, 12/6–7/41.

14　HLS, 12/7/41 ("Have you"); Felix Frankfurter, interview, 2/20/58, PL; Morison 529.

15　Toland, *Rising Sun*, 219.

16　EJK, interview, 5/6/46, KP (box 2); George Russell, interview, 12/11/74, KP (box 11); Francis S. Low, "A Personal Narrative of an Association with Fleet Admiral Ernest J. King, U.S. Navy," 20, unpublished MS, 1961, KP (box 10); Buell, excerpts from "CINCLANT History," 1:257, KP (box 5); D. L. Madeira to Thomas Buell, 8/9/74, KP (box 2) ("Hell").

17　EJK, portrait, NARA; EJK, interview, 8/27/50, KP (box 8); EJK, interview, 5/6/46, KP (box 2); Neil Dietrich, interview, 12/10/74, KP (box 11); Whitehill 349; Buell 135.

18　CINCPAC to Walter B. Howe, 12/27/41, NARA/NHC-OL; CINCPAC to CINCUS, "Damages Sustained by Ships as a Result of the Japanese Raid," 12/7/41, NARA/NHC-OL.

19 Memorandum, 12/7/41, FDRL (Official File 4675: World War II, 1941, box 1).

20 ER, *This I Remember*, 233; Goodwin 290, citing James Roosevelt, interview.

21 Usher's Log, 12/7/41, FDRL; FDR and Morgenthau, transcript, 12/7/41, FDRL (Morgenthau Papers, box 515); Tully 254–55; Ickes 3:661; Claude R. Wickard, 12/7/41, FDRL (Claude R. Wickard Papers; Department of Agriculture Files; Cabinet Meetings, 1941–1942, box 13).

22 HLH, 12/7/41, in Sherwood 432–33; Francis Biddle, 12/7/41, FDRL (Biddle Papers, Cabinet Notes 1941, box 1); Smith 536; Tully 256 ("Sit down"); FDR, Draft No. 1, 12/7/41, FDRL ("world history"). "Infamy" was added in the next draft.

23 Claude R. Wickard, 12/7/41, FDRL (Wickard Papers, Cabinet Meetings, box 13); Francis Biddle, 12/7/41, FDRL (Biddle Papers, Cabinet Notes 1941, box 1); Ickes 3:662; HLS, 12/7/41 ("1861"); HLH, 12/7/41, in Sherwood 432–33; Perkins 124; Frances Perkins, interview, part viii, 63, COHP ("berth"); Tully 257.

ONE ★ "NEW POWERS OF DESTRUCTION"

1 Ickes, 5/12/40, LC (Ickes Papers); Sherwood 500.

2 Samuel Rosenman, interview, 4/9/59, 160–64, FDRL; FDR, inscription in personal copy of *Mein Kampf*, 1933, FDRL.

3 "Roosevelt's Valet Joins Navy," NYT, 2/22/43; Tully 111; "Irvin M'Duffie, 63, Roosevelt's Valet," NYT, 1/31/46.

4 Tully 76–77; Sherwood 206; Leahy 123; "24-Hour Day at the White House," NYT, 5/26/40; Goodwin 17, citing *New Yorker*, 6/16/34; Charles Hurd, "As His Third and Hardest Term Begins," NYT, 1/19/41.

5 Ickes, 5/19/40, LC (Ickes Papers).

6 Ickes, 5/12/40, 5/26/40, 3:175, 190; Goodwin 31, quoting Morgenthau, memorandum, 5/10/40, Morgenthau Papers, FDRL.

7 FDR, Jr., interview, 1/11/79, FDRL; "Hopkins Disbursed Billions on Relief," NYT, 8/25/40.

8 ER, interview, 7/13/54, FDRL (Oral Histories, box 1); Goodwin 31 ("ill-fed").

9 Sherwood 203–04, 212 ("extraordinary").

10 W. A. Jansen to HLH, 5/10/40, FDRL (Hopkins Papers, box 302); Holt to Carroll Wilson, 5/9/40, FDRL (Hopkins Papers, box 302); "Turning Point," *Time*, 5/20/40; Ickes, 5/12/40, LC (Ickes Papers); "H.F." to HLH, 5/15/40, FDRL

(Hopkins Papers, box 302); Ickes, 5/12/40, 3:173–75; Grosvenor M. Jones to James W. Young, 5/9/40, FDRL (Hopkins Papers, box 302); Carroll L. Wilson to HLH, 5/16/40, FDRL (Hopkins Papers, box 302).

11 Ickes, 5/12/40, LC (Ickes Papers) ("drunk").

12 "President Solemn," NYT, 5/17/40; Goodwin 41, citing WP 5/17/40; "Asks $1,820,000,000 for Defense," *Corsicana Daily Sun*, 5/16/40; ER, interview, 9/3/53, FDRL (Oral Histories, box 1); Ickes, 5/19/40, LC (Ickes Papers); Samuel Rosenman, interview, 4/9/59, 183, FDRL; Tobin 327–28 n.69.

13 FDR, address, 5/16/40, FDR-PP, 1940, 199–202 ("ominous").

14 FDR, address, 5/16/40, FDR-PP, 1940, 202; Smith, "Battle of the Atlantic," 12; Heinrichs 10.

15 James Cate and E. Kathleen Williams, "The Air Corps Prepares for War, 1939–41," in Craven and Cate 1:107; Goodwin 41.

16 FDR, address, 5/16/40, FDR-PP, 1940, 203–04 ("There are some").

17 GCM, "Biennial Report of the Chief of Staff of the United States Army, July 1, 1939, to June 30, 1941, to the Secretary of War," 3 ("third-rate"); GCM, speech to American Historical Association, 12/28/39, *Vital Speeches*, 6:268–70; Smith, *Economic Mobilization*, 24–36, 119–126.

18 DDE, interview, 6/28/62, PL ("helluva"); GCM, "Biennial Report of the Chief of Staff of the United States Army, July 1, 1939, to June 30, 1941," 3; Pogue 2:7.

19 W. B. Smith, interview, 7/29/58, PL; Leonard T. Gerow, interview, 2/24/58, PL; Thomas Handy, interview, 3/23/59, PL; Parrish 357, quoting Mona Nason, interview, 6/26/86; M. C. Stayer, interview, 1/20/60, GCML; MP 1:684 n.2.

20 Mrs. J. J. Singer, interview, 2/7/60, GCML.

21 Parrish 49–51, quoting Frank H. Partridge to Edgar F. Puryear Jr., 9/10/62; Cray 103.

22 Helen Bailey, "The Office of the Chief of Staff," 4/11/01, GCML (World War II–Korean War Memories Project); Parrish 55, 93–94; Taaffe 3.

23 Felix Frankfurter, interview, 2/20/58, PL; GCM, interview, 7/9/47, GCML; Bernard Baruch, interview, 4/15/57, 3/14/61, PL.

24 Usher's Log, 4/23/39, FDRL; Parrish 137, quoting James T. Williams, Jr., GCML ("When I disapprove"); Parrish 97–98 ("unpleasant"); GCM, interview, 7/9/47, GCML; "Marshall Named as Chief of Staff," NYT, 4/27/39.

TWO ★ THREE MINUTES

1 Thomas Handy, interview, 7/9/70, PL; Louis Johnson, interview, 10/28/57, PL; Parrish 121-22.

2 Pogue 1:271; Bernard Baruch, interview, 10/15/57, 3/14/61, PL.

3 Bernard Baruch, interview, 3/14/61, PL; GCM, testimony, 2/23/40, MP 2:163–64 ("blazes"); Pogue 2:17; Goodwin 25 ("After the war").

4 Hassett 1/8/42; Beschloss 46–50 ("only Jew"); FDR to GCM, 5/24/40, FDR-PL 2:1030.

5 Usher's Log, 5/13–14/40, FDRL; Blum, *Years of War*, 2:140–41 ("filed your protest"); GCM, interview, 11/15/56, GCML ("three minutes"); Parrish 133–34.

6 Truman Smith, interview, 10/15/59, PL; John Hull, interview, 5/5/70, PL; GCM, interview, 11/21/56, GCML; HLS, 11/16/42; Vogel 284; Thomas Handy, interview, 3/23/59, PL; Pogue 3:59, citing Marjorie Payne Roberts Lunger, interview, 6/20/69, GCML; McCarthy to GCM, 6/25/42, GCML (McCarthy Papers, box 17); Pogue 3:60, 65–66; Crosswell 207–09; Parrish 171.

7 Helen Bailey, "The Office of the Chief of Staff," 4/11/01, GCML (World War II–Korean War Memories Project); Matthew Ridgway, "My Recollections of General of the Army George C. Marshall," 10/3/80, GCML (Reminiscences File, box 1); Truman Smith, interview, 10/15/59, PL; John Hull, interview, 5/5/70, PL; GCM, interview, 11/21/56, GCML; HLS, 11/16/42; Vogel 284; Thomas Handy, interview, 3/23/59, PL; Pogue 3:59, citing Marjorie Payne Roberts Lunger, interview, 6/20/69, GCML; Pogue 3:60, 65–66; Crosswell 207–09.

8 James Powder, interview, 10/19/59, GCML; GCM to Morrison C. Stayer, 1/15/39, MP 1:682–84; Pogue 3:60–61; Larrabee 102–05; Crosswell 207–09, 223–24.

9 Helen Bailey, "The Office of the Chief of Staff," 4/11/01, GCML (World War II–Korean War Memories Project); Felix Frankfurter, interview, 2/20/58, PL ("solve").

10 Albert Wedemeyer, interview, 2/1/58, PL; Parrish 368, quoting Mona Nason, interview, 6/26/86 ("It's difficult"); Parrish 358, citing Mona Nason, interview, 6/26/86 ("Wedemeyer").

11 GCM to HLS, 5/16/41, MP 2:508–10; FDR-PL 2:1303; Richard G. Davis, "Hap: Henry H. Arnold, Military Aviator" (Washington: U.S. Government, 1997); HHA, *Global Mission*, 186.

12 "Five-Alarm Rib," SEP, 12/9/44; "Prankster, Actor Vince Barnett Dies," Pittsburgh *Post-Gazette*, 8/10/77.

13 Pogue 3:327–28; Parrish 345.

14 Parrish 136 ("In the first place").

15 Cray 401 ("most brilliant"); Parrish 354–55; Pogue 3:218–19.

16 DDE, interview, 6/28/62, PL ("outburst").

17 DDE, interview, 6/28/62, PL; Crosswell 209–10; Parrish 138–39.

18 James Powder, interview, 10/19/59, GCML; Frank McCarthy, interview, 6/56, PL; "Marshall's Big Sergeant," *St. Petersburg Times*, 10/18/59; Pogue, speech to U.S. Air Force Association, Harmon Memorial Lecture #10, 1968 ("No one").

19 "General Marshall's Greatest Satisfaction As Told to a Friend," GCML (Reminiscences File, box 1); James Powder, interview, 10/19/59, GCML; Thomas Handy, interview notes, 8/21/56, PL; William M. Kerrigan to Frank Cash, Jr., 9/22/83, GCML (Reminiscences File, box 1); Felix Frankfurter, interview, 2/20/58, PL; Truman Smith, interview, 10/15/59, PL; W. B. Smith, interview, 7/29/58, PL; Mark Clark, interview, 11/17/59, PL; Bernard Baruch, interview, 10/15/57, PL ("tired").

20 Pogue 2:302 ("pants"); GCM, interview, 2/11/57, GCML ("fearful").

21 "Promotion Issue Stirs Army, Navy," NYT, 7/10/39 ("The President").

22 Bundy, *Active Service*, 238; Pogue 2:22–23; HLS, 11/25/40; Morison 488–89; Stoler 16.

23 GCM, interview, 11/13/56, GCML ("cigarette-holder"); Pogue 1:324 ("dinner table").

24 W. B. Smith, interview, 7/29/58, PL ("disposal").

25 GCM, interview, 2/11/57, GCML ("voluble").

26 FDR to Asa Singleton, 11/22/39, MP 2:107–08 ("bored"); DDE, interview, 6/28/62, PL.

27 FDR to Hopkins, 11/4/42, MP 3:423–24.

THREE ★ "THE HAND THAT HELD THE DAGGER"

1 WSC to FDR, 5/15/40.

2 GCM, interview, 1/15/57, GCML; Pogue 2:50 ("shortage"); Pogue 2:48–50; Watson, *Chief of Staff*, 305–08.

3 Ickes, 6/5/40, 3:200 ("guess wrong"); Watson, *Chief of Staff*, 303, quoting GCM, memorandum, 2/21/40 ("adverse"); Matloff, *1941–1943*, 17 ("lamp post"); Pogue 2:52, quoting WBS to GCM, 6/11/40.

4   Larabee 644, quoting Gunther, *Roosevelt in Retrospect*, 326 ("he decides"); FDR, fireside chat, 12/29/40.

5   WSC to FDR, 5/15/40; FDR to WSC, 5/16/40; Rosenman 192; "Last Call," *Time*, 8/5/40.

6   FDR to WSC, 5/16/40; Rosenman 190; Watson, *Chief of Staff*, 310.

7   Morgenthau 6/5/40, in Blum, *Years of Urgency*, 155 ("extra push"); Pogue 2:51.

8   Chief of Ordnance to GCM, 5/22/40, Watson, *Chief of Staff*, 309; Pogue 2:51 (citing Hall, North American Supply, 134–35); Kent 39–41; GCM, interview, 12/7/56, GCML ("duplicity"); MP 2:262; Louis Johnson, interview, 10/28/57, PL.

9   U.S. Army Corps of Engineers, Fact Sheet, "FUDS: Former Raritan Arsenal, Edison, New Jersey," www.nan.usace.army.mil/Portals/37/docs/civilworks/projects/nj/fuds/Raritan/raritan0410.pdf; "Nazi Ring Closing in on Paris," *New York Times*, 6/13/40; Stettinius, *Lend-Lease*, 27.

10 Watson, *Chief of Staff*, 304; Morison 477; Goodwin 23; Pogue 2:20.

11 FDR to Harry Woodring, 6/25/40, FDR-PL 2:1042–44; Stettinius, *Lend-Lease*, 28; Goodwin 66

12 "The Tenth of June," *Time*, 6/17/40; HLS, 12/29/40; William Safire, "What's in a Phrase: 'Stab in the Back,'" *New York Times Magazine*, 5/21/89.

13 Rosenman 347; HLS, 12/29/40 ("conscience").

14 FDR, address, 6/10/40, FDR-PP, 1940, 263–64 ("dagger," "all roads"); "The Tenth of June," *Time*, 6/17/40.

15 "Walsh Navy Bill Passed by Senate," NYT, 6/21/40 ("dangerous"); Walsh, *Congressional Record*, 76 Cong., 3rd Sess, 6/21/40, 8784 ("American boy"); Charles Lindbergh, address, 5/19/40, *Vital Speeches* 6:484–85 ("hysteria").

16 Ickes, 5/19/40, LC (Ickes Papers, reel 2).

17 HLS to FDR, 5/21/40, PSF box 106, FDRL ("Goebbels"); "Turning Point," *Time*, 5/20/40 ("fold up"); Sherwood 165.

18 "24-Hour Day at the White House," NYT, 5/26/40.

19 Felix Frankfurter, interview, 2/20/58, PL ("baby").

20 Ickes, 5/19/40, LC (Ickes Papers, reel 2); Goodwin 23, quoting ER, interview, FDRL ("angry"); Pogue 2:20; Morison 477–78.

FOUR ★ "FEWER AND BETTER ROOSEVELTS"

1 HLS, 2/16/41; Bundy, *Active Service,* 94–99; Hodgson 8–9.

2 Morison 12–13.

3 Bundy, *Active Service*, 235–37, 249; Morison, *Turmoil*, 385–87, 472; Hodgson 218–19; Hassett 4/5/43.

4 HLS, 5/8/40 ("half-baked"); Hodgson 210–11, citing HLS, 11/9/32, 11/16/32.

5 Felix Frankfurter, interview, 2/20/58, PL; Ickes, 5/19/40, LC (Ickes Papers, reel 2); FDR to Harry Woodring, 6/19/40, FDR-PL 2:1041; Hodgson 61, 221–23; Morison 480–82; Watson, *Chief of Staff*, 190–95.

6 Annie Knox, "Pattern of Life of Frank Knox from 1898–1944," n.d., LC (Knox Papers, box 1); "U.S. at War," *Time*, 9/7/42, LC (Knox Papers, box 4); Frances Perkins, interview, part viii, 64, COHP; James Forrestal to Knox, 1/20/44, LC (Knox Papers, box 4); "Text Book of the Republican Party, 1936," Republican National Committee, 22.

7 Knox to FDR, 12/15/37, FDRL (PPF, box 4083) ("drunk"); "Secretary Knox," *Life*, 3/10/41; Sherwood 163; "Turning Point: Moving Away from Isolationism," *Time*, 5/20/40; "Bulletin," *The Daily Times-News* (Burlington, NC), 5/17/40; Smith, *FDR*, 365 ("fewer").

8 Ickes, 7/5/40, LC (Ickes Papers, reel 2); Annie Knox, "Pattern of Life of Frank Knox, 1898–1944," n.d., LC (Knox Papers, box 1); "Roosevelt Signs Two-Ocean Navy Bill," NYT, 7/21/40; EJK, Report to the Secretary of the Navy, 3/27/44, 12–13; Reilly 87; Morison 1:27; Morison 482, quoting Grenville Clark, interview, 8/17/54, and Felix Frankfurter, interview, 9/29/54; HLS, 6/25/40; Bundy, *Active Service*, 323–24; Larrabee 108.

9 H. R. Stark, interview, 3/13/59, PL; Leahy, interview, 5/24/48, 4, KP (box 9); Knox to FDR, 1/4/43, KP (box 9).

10 HLS, 6/25/40; Bundy, *Active Service*, 323–24; Morison, *Turmoil*, 481; Hodgson 214–15; Smith, *FDR*, 450.

11 "Stimson and Knox Disowned by Party," NYT, 6/21/40 ("treachery"); "Scores Hamilton on Knox," NYT, 6/26/40; "Two Appointments," *Time*, 7/1/40 ("If there"); Ickes, 7/5/40, LC (Ickes Papers, reel 2); Hodgson 224.

12 "Turning Point: Moving Away from Isolationism," *Time*, 5/20/40; *PM*, 10/22/41, 10/24/41, University of California San Diego (Dr. Seuss Collection, MSS 230); Casey 28, citing *Fortune*, 7/21/40.

13 HLS, 7/9/40; Hodgson 224.

14   GCM to HLS, 1/22/27, MP 1:322; HLS to GCM, 1/21/28, MP 1:322; GCM
      to HLS, 1/21/29, MP 1:322; GCM, interview, 4/11/57, GCML; Greg Franke,
      "Henry Stimson and George Marshall: An Enduring Friendship," GCML;
      "Henry L. Stimson Dies at 83 in His Home on Long Island," NYT, 10/21/50;
      Hodgson 231.

15   GCM to Katherine Marshall, 6/28/40, MP 2:252 n.1 ("They are"); GCM to
      Frank McCoy, 6/26/40, MP 2:252.

16   Morison, *Turmoil*, 508 n. 7, citing HLS, 8/14/41; Hodgson 231.

17   HLS, 7/29/40; Hodgson 241; "Henry L. Stimson Dies at 83 in His Home on
      Long Island," NYT, 10/21/50; HLS, 7/29/40; Bundy, *Active Service*, xxii.

18   Smith, "Battle of the Atlantic," 12; Roscoe 25.

19   Keegan, *Atlas*, 35.

20   WSC to FDR, 7/31/40 ("Mr. President").

21   "Last Call," *Time*, 8/5/40; Rosenman 204; Goodwin 138–39 ("survival").

22   Stark to FDR, 9/28/38, KP (box 9); FDR, memorandum, 8/2/40, FDR-PL
      2:1050–51; HLS, 8/3/40; "In the Open," *Time*, 8/12/40; Simpson 52–53, quoting
      Stark to Knox, 8/17/40; Jackson 97; Ickes, 7/5/40, LC (Ickes Papers, reel 2);
      Naval Expansion Act, 6/28/40, in Baptiste 53; FDR to Sumner Welles, 6/1/40,
      FDR-PL 2:1036; FDR to Frank Knox, 7/22/40, FDR-PL 2: 1048–49; H. R. Stark,
      interview, 3/13/59, PL; Stevenson 105 ("not essential"); Simpson 52.

23   Jackson 87, 97; FDR, memorandum, 8/2/40, FDR-PL 2:1050–51; HLS, 8/2/40;
      FDR, memorandum, 8/2/40, FDR-PL 2:1050–51; HLS, 8/3/40; "In the Open,"
      Time, 8/12/40; Simpson 52–53, quoting Stark to Knox, 8/17/40.

24   "No Legal Bar Seen to Transfer of Destroyers," NYT, 8/11/40; HLS, 8/12/40;
      Jackson 95.

25   HLS, 8/13–14/40.

26   FDR, press release, 8/31/35, FDRL (Berle Papers, box 58); HLS, 8/21/40; Simpson
      52–54.

27   Jackson 74, quoting Robert Jackson, interview, COHP ("The President").

28   Stark, interview, 5/26/48, KP (box 9) ("I'll be breaking").

29   FDR, memorandum, 8/13/40, FDR-PL 2:1052; HLS, 8/17/40; "Roosevelt to See
      MacKenzie King," NYT, 8/17/40; FDR to David Walsh, 8/22/40, FDR-PL
      2:1056–57; Jackson 91; "US-Canada Ties Welded," NYT, 8/18/40; Goodwin
      142–43.

30 HLS, 8/19/40 ("funny").

31 FDR, press conference, 9/3/40, FDR-PP 1940, 376 ("Louisiana Purchase"), 379 ("all over").

FIVE ★ THE NEW DEAL WAR

1   Sherwood 159, quoting Harold Smith ("total war"); Hassett, 2/3/42; Tully 17.

2   FDR, press conference, 8/7/42, FDR-PP, 1943, 326 ("home front"); Janeway 71 ("New Deal war").

3   Hodgson 227, citing Kennedy, *The Rise and Fall of the Great Powers*, 355 & n.595; Hodgson 224, citing Hillman, "Comparative Strength of the Great Powers," in Arnold Toynbee (ed.), *The World in March 1939*, 443.

4   FDR, address, 10/31/36, FDR-PP, 1936, 568–59 ("hatred").

5   "Arming America," NYT, 6/2/40; FDR, press conference, 5/28/40, FDR-PP, 1940, 245–48; Ickes, 6/2/40, LC (Ickes Papers, reel 2).

6   Sherwood 161–62; "Members of the President's National Defense Commission," Washington *Evening Star*, 5/29/40; Rosenman 224; Goodwin 55–56; FDR, press conference, 5/28/40, FDR-PP, 1940, 242 ("cosmetics").

7   FDR to Grenville Clark, 5/18/41, FDR-PL 2:1026.

8   Rosenman 225; Goodwin 139; FDR to William White, 12/14/39, FDR-PL 2:967 ("to get"); Murphy 69 ("Mothers").

9   HLS, 8/1/40; Ickes, 7/5/40, LC (Ickes Papers, reel 2); Stark, interview, 5/26/48, KP (box 9) ("You could").

10 Rosenman, *Working with Roosevelt*, 167 ("It is a terrible").

11 GCM, interview, 1/22/57 ("crude").

12 Earl Rickard, "Marshall Builds the Defense Army," *WWII History* (September 2012) 50; Pogue 2:58–59, citing Senate Comm. on Mil. Aff., *Compulsory Military Training and Service*, 7/12/40.

13 FDR, press conference, FDR-PP, 1940, 320 ("We figured").

14 "First Lady Expands Her Views on Draft," NYT, 9/17/40; "First Lady Ready for Sons' Drafting," NYT, 9/17/40.

15 Ickes, 7/4–5/42, LC (Ickes Papers, reel 2); HLS, 7/19/40; Harold Ickes to Grace Tully, 7/14/50, FDRL (FDR Memorial Foundation, Ickes folder); FDR, Jr., interview, 1/11/79, FDRL, 4.

16   FDR, address, 7/19/40, FDR-PP, 1940, 296–97 ("lying awake").

17   "The Crowd at Ellwood," *Time*, 8/26/40 ("cannot ask"): "Text of Willkie Speech," NYT, 8/18/40.

18   Goodwin 146.

19   HLS, 8/22/40, 9/9/40; "Exchange Blows in House," NYT, 9/6/40; "The Bitter End," *Time*, 9/16/40.

20   FDR, proclamation, 9/16/40, FDR-PP, 1940, 429–31; Photograph, 9/16/40, GCML; Pogue 2:62.

21   "The Draft: How It Works," *Time*, 9/23/40; Pogue 2:63.

22   GCM, radio address, 9/16/40, MP 2:311–12 ("for the first time").

SIX ★ "ONE-FIFTY-EIGHT"

1   HLS, 9/9/40 ("agonizing"); Pogue 2:63.

2   Pogue 2:64, quoting HLS, 9/27/40 ("chest").

3   HHA, *Global Mission*, 184–86 ("Guam"); "Plan Exports Stir Capitol Tempest," NYT, 3/13/40.

4   Watson, *Chief of Staff*, 306 ("even-Steven"); HLS, 11/8/40, 9/23/41.

5   Pogue 2:64, citing HLS, 9/27/40 ("chart"); Pogue 2:67, citing COS Conference, 12/2/40.

6   FDR, handwritten memo, 11/12/40, in HLS, 11/12/40; HLS, 11/12/40 ("only peg"), 11/24/40; Marshall, interview, 11/15/56, GCML; Pogue 2:66.

7   HLS, 11/13/40; GCM, interview, 1/15/57, GCML ("ashamed"); HLS, 11/15/40.

8   HLS, 8/5/41, 11/12/41 ("topsy-turvy"), 12/18/40 ("vagrant beam"); Morison 508–09.

9   HLS, 9/17/41 ("boom"), 12/16/40 ("marvelous").

10   "Army: The Problem," *Time*, 10/23/40; Bureau of Naval Personnel, *Negro in the Navy*, 1–2; Lee 42, 47, 69; Goodwin 165, citing Jean Byers, "A Study of the Negro in Military Service," War Department Study, June 1947, 67; MacGregor, ch. 3 ("bellhops").

11   Stenographer's Log, 9/27/40, FDRL.

12   FDR, conversation, 9/27/40, at http://whitehousetapes.net/clips/1940_0927_randolph/.

13   Frances Perkins, interview, part viii, 64, COHP; "Fragments on Back-Room Politics and Civil Rights," in R. J. C. Butow, "The Story Behind the Tapes," *American Heritage* (February/March 1982), 24.

14   GCM to Henry Cabot Lodge, Jr., 9/27/40, MP 2:336–37.

15   HLS, 9/27/40, 1/24/42 ("crime"); Bundy, *Active Service*, 461–64; Morison 555.

16   Frank McCarthy to Edwin Watson, 5/6/43, GCML (McCarthy Papers, box 27).

17   Lee 76 ("The policy").

18   Goodwin 171, citing Press Release, 10/9/40, and NAACP release, 10/11/40 ("stab"), FDRL; "Army: The Problem," *Time*, 10/29/40 ("Jim Crow").

19   Goodwin 171, quoting White House statement, FDRL.

20   "Army: Blunder & Precedent," *Time*, 11/4/40; HLS, 10/25/40; 10/28/40.

21   HLS, 10/23/40 ("There is").

22   HLS, 10/22/40; Lee 79–80.

23   HLS, 3/5/41, 10/25/50, SP ("I had"); William Hastie, interview, 1/5/72, HSTL, 23–24.

24   "Promises Broken," NYT, 10/25/40 ("count").

25   "Mr. Lewis for Willkie," NYT, 10/26/40; Sherwood 190, 192–93, 198; Goodwin 182–84, citing Matthew and Hannah Josephson, *Sidney Hillman: Statesman of American Labor,* 488 ("scared").

26   "Peace in Strength," NYT, 10/30/40.

27   FDR, address, 10/29/40, FDR-PP, 1940, 510–11 ("forced"); "Peace in Strength," NYT, 10/30/40; Newsreel, 10/29/40 ("one-fifty-eight").

28   Rosenman 240 ("Any"); HLS, 10/29/40.

29   FDR, speech, 10/31/40, FDR-PP, 1940, 517 ("I have said").

30   Sherwood 191; Rosenman 242 ("attacked").

31   Smith, *Thank You*, 76; "Roosevelt Looks to 'Difficult Days,'" NYT, 11/6/40 ("demeanor").

32   Reilly 66.

33   Sherwood 200; "Roosevelt Looks to 'Difficult Days,'" NYT, 11/6/40; Congressional Quarterly, *Presidential Elections*, 113.

34   "Roosevelt Looks to 'Difficult Days,'" NYT, 11/6/40; Sherwood 200 ("Safe"); Goodwin 189.

35 Congressional Quarterly, *Presidential Elections*, 113; Sherwood 200; "Roosevelt Rests in Hyde Park," NYT, 11/6/44.

SEVEN ★ THE PARABLE OF THE GARDEN HOSE

1 "Treasurer to the Democracies," NYT, 6/22/41.

2 HLS, 12/10/40; FDR to Cordell Hull, 1/11/41, FDR-PL 2:1103–04; Hancock and Gowing 232.

3 Pogue 2:68–69.

4 Goodwin 190.

5 "Log of the President's Inspection Cruise Through the West Indies, 3–14 December, 1940," FDRL (Tully Papers, box 7); "President and His Dog, Falla, Sail on 15-Day Rest Cruise in the Caribbean," *Life*, 12/16/40; Goodwin 191.

6 "Log of the President's Inspection Cruise Through the West Indies, 3–14 December, 1940," FDRL (Tully Papers, box 7); Sherwood 223.

7 WSC to FDR, 12/7/40; "President and His Dog, Falla, Sail on 15-Day Rest Cruise in the Caribbean," *Life*, 12/16/40; Sherwood 222–23.

8 WSC to FDR, 12/7/40, FDRL.

9 Sherwood 224.

10 FDR, press conference, 12/17/40, FDR-PP, 1940, 607 ("Let me").

11 Sherwood 224 ("convincing"); Goodwin 193 ("clairvoyant"); Sherwood 224.

12 FDR to Russell Leffingwell, 3/16/42, FDR-PL 2:1298; FDR to Mary Norton, 3/24/42, FDR-PL 2:1300; Casey 35, citing Hooper Ratings Chart, FDRL, and McIntyre to Hassett, 3/16/42, FDRL (Office File, box 857); FDR, fireside chat, 12/29/40 ("arsenal of democracy").

13 HLS, 11/25/40 ("short of war"), 12/16/40, 12/29/40, 12/31/44.

14 Stoler 4–5; Watson 103.

15 Stark to Knox, 11/12/40, FDRL; Pogue 2:126; Morison 1:42–44; Morton, *Fall of the Philippines*, 61; Watson 122–24; Stoler 29–33.

16 Leighton and Coakley, *1941–1943*, 43, 77.

17 HLS, 1/17/41; "Rotten, Dastardly," NYT, 1/15/41 ("plow"); "President Angered by Sen. Wheeler," Ludington (MI) *Daily News*, 1/15/41; Goodwin 210.

18 "Kennedy Says Democracy All Done," *Boston Globe*, 11/10/40; Goodwin 211.

19   White House Usher's Log, 1/16/41, FDRL; Nasaw 512–13; Goodwin 211.

20   FDR to WSC, 1/19/41; "Mr. Willkie Lands," *Time*, 2/3/41; "Mr. Willkie's Tes-
      timony," NYT, 2/12/41 ("He was").

21   FDR to HLS, 2/25/41, FDR-PL 2:1128; "Roosevelt Signs Aid Measure," NYT,
      3/12/41; "Final Swift Step," NYT, 3/12/41; WSC, speech, 4/17/45, "Churchill Eu-
      logizes Roosevelt," NYT, 4/18/45 ("unselfish"); Pogue 2:70.

22   Watson, *Chief of Staff*, 122–23; Pogue 2:127, citing Joint Planning Board to Joint
      Board, 12/21/40.

23   HLS, 3/14/41; Smith, *Army and Economic Mobilization*, 441; Goodwin 217,
      citing *Fortune*, May 1941, 58, 62.

24   HLS, 4/15/41; Lee 43; Millet 5; Pogue 2:108.

25   HLS, 4/15/41.

26   HST, interview, 11/14/60, PL; Pogue 2:109, quoting GCM to Deputy COS,
      4/18/41 ("patriotic").

27   HST, interview, 11/14/60, PL; Robert Dennison, interview, 82, KP (box 10);
      Charlotte Pihl, interview, 3/16/74, 2, KP (box 10); Pogue 2:108.

EIGHT ★ INCHING INTO WAR

1   "U-Boats Attack Eight Ships in Convoy," NYT, 12/3/40; Blair 1:210–11; Hein-
     richs 30; Keegan, *Atlas*, 35; FDR, address, 5/27/41, FDR-PP, 1940, 181–84; Smith,
     "Battle of the Atlantic," 73; WSC, radio address, 4/27/41, in Gilbert, *War Papers*,
     3:553 ("Battle").

2   Frank Knox to EJK, 1/9/41, KP (box 2).

3   Charles Wellborn, interview, 2/6/74, 141, KP (box 10); Betsy Matter, interview,
     12/9/74, 3, KP (box 10); George Russell, interview, 12/11/74, KP (box 11); Morison,
     Battle of the Atlantic, 51; Buell 107.

4   EJK, interview, 7/5/50, KP (box 7); "King's Way to Tokyo," *Saturday Evening
     Post*, 12/9/44, KP (box 9); Whitehill 21–23; Buell 11–12, 43–44, 56, 77, 100.

5   Whitehill, "Memorandum of Conversation," 5/29/46, KP (box 6); D. W. Knox,
     interview, 5/31/46, KP (box 13); Paul Pihl, interview, 3/9/74, 4, KP (box 10); Buell
     22, 30–31, 37, 62–63, 88–90, 210.

6   Reginald Winn, memoirs, 92, GCML (Small MS Collections, box 3); Buell 74,
     quoting George van Deurs, interview ("liquor"), 78 ("trust").

7   Buell 78; Charlotte Pihl, interview, 3/9/74, 7–10, KP (box 10); James Russell to
     John Mason, Jr., 9/28/76, KP (box 7).

8 Robert S. Quackenbush Jr. to Thomas Buell, 11/19/74, KP (box 8) ("There are"); Malcolm Schoeffel to Thomas Buell, 9/9/74, KP (box 2); Howard Orem, interview, 12/12/74, KP (box 11); Reginald Winn, memoirs, 75, GCML (Small MS Collections, box 3); Buell 81.

9 EJK, interview, 7/31/49, KP (box 8); Neil Dietrich, interview, 12/10/74, KP (box 11); Nofi 231.

10 Chester Nimitz to EJK, 1/11/41, KP (box 2); EJK, 6/13/46, KP (box 9); Charlotte Pihl, interview, 3/16/74, 7, KP (box 10); Buell 114.

11 EJK, 6/13/46, KP (box 9); Low 14; Buell 76 ("ass").

12 Alan R. McFarland to Thomas Buell, 9/3/74, KP (box 2); Buell 117.

13 David C. S. Kline to Thomas Buell, 8/6/74, KP (box 2) ("ecru"); Charles Moses to Thomas Buell, 8/74, KP (box 2).

14 EJK, interview, 7/29–31/50, KP (box 7); Frank Knox, 5/13/41, LC (Knox Papers, box 1); "Navy Establishes 3 Separate Fleets," NYT, 1/9/41; Buell 120; Stoler 27–28.

15 Chester Nimitz to EJK, 1/11/41, KP (box 5); EJK to Frank Knox, 1/17/42, KP (box 5); Buell 120.

16 Buell 120 ("maintain").

17 HLS, 4/8/41; GCM to Leonard Gerow, 1/17/41, MP 2:391–92; Goodwin 233–34.

18 Gallup poll, 4/23/41, at http://ibiblio.org/pha/Gallup/Gallup%201941.htm.

19 FDR, announcement, 4/10/41, FDR-PP, 1941, 96–97; Stenographer's Log, 4/10/41, FDRL; HLS, 4/10/41; FDR to Cordell Hull, 4/18/41, FDR-PL 2:1142–43.

20 Stenographer's Log, 4/15/41, FDRL; HLS, 4/15/41; Fairchild and Conn 105–06; EJK, "Operations Plan 3-41," 4/18/41, in Smith, "Battle of the Atlantic," 23; Pogue 2:129; FDR, press conference, 4/25/41, FDR-PP, 1941, 133; Sherwood 295.

21 HLS, 4/22/41, 5/22/41 ("accidental shot").

22 U.S. Army Japan, *Japanese Monograph No. 45: History of Imperial General Headquarters, Army Section* (Tokyo 1959, rev. ed.), 15, accessed at www.ibiblio.org/hyperwar/Japan/Monos/JM-45/index.html; Keegan, *Atlas*, 21; HLS, 2/10/41 ("Japs").

23 Stenographer's Log, 4/24/41, FDRL; HLS, 4/24/41 ("impregnable"); Thomas Handy, interview notes, 3/23/59, PL.

24 HLS, 4/24/41.

25 "Most Reassuring," *Time*, 7/21/41; HLS, 4/24/41 ("clearly hostile").

26 HLS, 5/13/41.

27 Buell 125 ("pointed").

28 EJK, interview, 8/29/49, KP (box 7) ("FDR's").

29 Gallup poll, 5/21/41, at http://ibiblio.org/pha/Gallup/Gallup%201941.htm; "National Affairs: Patrols and Convoys," *Time*, 5/12/41.

30 Knox, 5/13/41, LC (Knox Papers, box 1); HLS, 5/27/41.

31 Sherwood 279–80; HLS to FDR, 5/25/41, in HLS, 5/25/41; Morison 519; Rosenman, *Working*, 282–84 ("ammunition").

32 "Text of the President's Address Depicting Emergency Confronting the Nation," NYT, 5/28/41 (photo); Rosenman, *Working*, 284; ER, "My Day," 5/29/41; "A New Mark in Radio," NYT, 6/1/41; Reilly 76–77.

33 FDR, speech, 5/27/41, FDR-PP, 1941, 181–84.

34 Usher's Log, 5/27/41, FDRL; ER, "My Day," 5/29/41; Rosenman, *Working*, 287; Sherwood 298.

NINE ★ BEGGARS BANQUET

1 Sherwood 303; Ickes, 6/22/41, LC (Ickes Papers, reel 1) ("attitude").

2 WSC, *Grand Alliance*, 331–33 ("hell," "Nazidom").

3 FDR to HLS, 5/28/41, 8/30/41, FDR-PL 2:1162, 1201–02; FDR to Wayne Coy, 8/2/41, FDR-PL 2:1195 ("burr"); HLS, 7/21/41; Executive Order 8428, 9/8/39; Office of War Information, *U.S. Government Manual 1945* (Washington: U.S. Government, 1945) 60; Pogue 2:71–73.

4 Blum, *Years of Urgency*, 275–76 ("dislocates").

5 HLS, 6/3/41; Leonard T. Gerow, interview, 2/24/58, PL; E. Kathleen Williams, Louis E. Asher, "The Air Corps Prepares for War, 1939–41," in Craven and Cate 1:115; Jackson 105; Pogue 2:72; Stoler 51; Kimball 25–29.

6 FDR to Leahy, 6/26/41, FDR-PL 2:1177 ("diversion"); FDR to Edwin Watson, 7/25/41, FDR-PL 2:1189; HLS to FDR, 6/23/41, in HLS, 6/23/41; HLS, 7/31/41; Stoler 52–54.

7 HLS, 7/31/41, 8/1/41; GCM to HLS, 8/29/41, in Watson 329 ("Our entire"); Morgenthau, 8/1/41, in Blum, *Years of Urgency*, 2:264 ("miserable"); Ickes, 8/1/41, in Ickes, 3:592 ("worst").

8 HLS, 8/1/41 ("hoity-toity"), 8/4/41 ("at his worst"), SP.

9   FDR, recording, 10/8/40, FDRL ("God").

10   HLS, 7/19/40, 8/1/40; FDR to ER, 11/13/40, FDR-PL 2:1077; "Embargo Put on
     Oil," NYT, 7/26/40; Morison 520 n.4, citing HLS to FDR, 6/23/41, and HLS
     6/30/41; Toland 86; Heinrichs 7, 10; Yergin 291.

11   FDR to Ickes, 7/1/41, in Ickes, 7/5/41, LC (Ickes Papers, reel 4) ("important");
     Ickes, 7/27/41, LC (Ickes Papers, reel 4) ("noose"); Bundy, *Active Service*, 383–85;
     Stoler 63.

TEN ★ LAST STAND OF THE OLD GUARD

1   Watson 16 (chart); 216–17; Pogue 2:146.

2   HLS, 7/11/41; "General Marshall's Testimony," NYT, 7/16/41; GCM, interview,
     1/15/57, GCML; Pogue 2:147.

3   HLS, 6/21/41, 7/9–11/41.

4   Gallup poll, 8/6/41, at http://ibiblio.org/pha/Gallup/Gallup%201941.htm.

5   GCM, interview, 1/22/57 ("hatred," "few"); Pogue 2:149–53, quoting Wad-
     sworth to Marshall, 8/25/41.

6   Elizabeth Anne Oldmixon, *Uncompromising Positions* (Georgetown University
     Press, 2005), 156 ("When you"); HLS, 8/6/41, 7/10–21/41.

7   HLS, 8/6/41 ("wild rumor").

8   EJK, "synthetic diary," 1946, KP (box 2); C. B. Lanman, "Memorandum for
     Fleet Admiral King," 4/22/48, KP (box 2); HHA, "Notes of Roosevelt-Churchill
     Conference," 8/3/41, LC (Arnold Papers, box 2).

9   "Log of the President's Cruise Aboard the USS Potomac and USS Augusta,
     3–16 August, 1941," FDRL; HHA, "Notes of Roosevelt-Churchill Conference,"
     8/4/41, LC (Arnold Papers, box 2); EJK, interview, 6/29/49, KP (box 2); C. B.
     Lanman, "Memorandum for Fleet Admiral King," 4/22/48, KP (box 2); Reilly
     118; Buell 128.

10   HHA, "Notes of Roosevelt-Churchill Conference," 8/7–8/41, LC (Arnold
     Papers, box 2); "Log of the President's Cruise Aboard the USS Potomac and
     USS Augusta, 3–16 August, 1941," FDRL ("ugly fish"); C. B. Lanman, "Memo-
     randum for Fleet Admiral King," 4/22/48, KP (box 2); Buell 128; Reilly 65.

11   Pogue 2:142; "Statement of Admiral Stark," n.d., LC (Arnold Papers, box 2);
     Parrish 186–87.

12   HHA, "Notes of Roosevelt-Churchill Conference," 8/9/41, LC (Arnold Papers,

box 2); EJK, "Synthetic Diary," 1946, KP (box 2) (8/9/41 entry); Whitehill 332–36; Buell 130; Pogue 2:142; HHA, *Global Mission*, 253–54.

13   Reilly 120; Goodwin 265 ("at last").

14   "Log of the President's Cruise Aboard the USS Potomac and USS Augusta, 3–16 August, 1941," FDRL, 9; EJK, "Synthetic Diary," 1946, KP (box 2) (8/10/41 entry); J. R. Beardsall to Military Staff, 8/9/41, LC (Arnold Papers, box 2); C. B. Lanman, "Memorandum for Fleet Admiral King," 4/22/48, KP (box 2); Roberts 53; Buell 128; Morison 1:70 n.19.

15   Goodwin 17, 267; Reilly 120, 198.

16   "Log of the President's Cruise Aboard the USS Potomac and USS Augusta, 3–16 August, 1941," FDRL; Sherwood 353; Buell 130.

17   WSC, *Grand Alliance*, 384 ("great hour"); Smith, *FDR*, 501.

18   Reid and Manchester 19 ("pillar"); Elliott Roosevelt, *As He Saw It*, 33 ("with God's help"); Smith, *FDR*, 501.

19   Sumner Welles, "Memorandum of Conversation," 8/10/41, PHH, 14:1269; HHA, "High Lights of Churchill's Speech," 8/9/41, LC (Arnold Papers, box 2); Pogue 2:143; Hough, Ludwig, and Shaw 1:35–37.

20   HHA, "Notes of Roosevelt-Churchill Conference," 8/14/41, LC (Arnold Papers, box 2) ("pants"); "Statement of Admiral Stark," n.d., LC (Arnold Papers, box 2).

21   Watson 405; Buell 131.

22   Sumner Welles, "Memorandum of Conversation," 8/11/41, PHH, 14:1275–90.

23   FDR and WSC, signed statement, 8/14/41, FDRL (Berle Papers, box 54); WSC, *Grand Alliance*, 394 ("far-reaching").

24   "The Congress: State of Mind," *Time*, 8/25/41; Pogue 2:152–54.

25   *Congressional Record*, 8/12/41, 87 (part 7), H7074–75; "The Congress: State of Mind," *Time*, 8/25/41.

26   *Congressional Record*, 8/12/41, 87 (part 7), H7074–75; "The Congress: State of Mind," *Time*, 8/25/41; HLS, 8/13/41; Goodwin 267–68.

ELEVEN ★ YEAR OF THE SNAKE

1   Ickes, 5/25/41, LC (Ickes Papers, reel 3); HLS, 7/21/41; Morison 1:84–85; Ickes 3:643.

2   FDR, fireside chat, 9/11/41, FDR-PP, 1941, 384–85, 389–90 ("rattlesnake").

3 Morison, *Battle of the Atlantic*, 79–81; Buell 132; Heinrichs 166–68.

4 "Gallup Poll, 1941," at http://ibiblio.org/pha/Gallup/Gallup%201941.htm; Goodwin 278.

5 *Ottawa Citizen*, 11/1/41; Morison, *Battle of the Atlantic*, 92–94; *Dictionary of American Fighting Ships*, "*Reuben James*," NHC; "Hull Warns Law Shackles Defense," NYT, 10/14/41; "Sea Fight Related," NYT, 10/19/41; U.S. Atlantic Fleet, "Revised Escort-of-Convoy Instructions," 11/17/41, KP (box 2).

6 EJK to L. F. V. Drake, 10/24/41, KP (box 2) ("Kearny").

7 EJK to Joseph J. Thorndike, extract, 11/21/41, KP (box 2); Buell 135, citing *Life*, 11/24/41.

8 Sherman Miles to Marshall, 10/17/41, PHH, 14:1359–60; Pogue 2:177.

9 Henry Morgenthau, notes of conversation with T. V. Soong, 11/27/41, FDRL (Morgenthau Papers, box 515).

10 Reynolds 170 ("hate us").

11 FDR to Ickes, 6/23/41, FDR-PL 2:1173–74; FDR to Ickes, 6/18/41, and Ickes to FDR, 6/20/41, in Ickes, 6/28/41, LC (Ickes Papers, reel 4); Ickes 558–60; WSC to FDR, 5/15/40 ("dog"); Heinrichs 132; Pogue 2:177; Larrabee 49, 63–64; Jackson 116–17.

12 FDR to Ickes, 7/1/41, in Ickes, 7/5/41, LC (Ickes Papers, reel 4) ("drag-down").

13 "Marshall and Stark Present at Talk at White House," NYT, 10/17/41; Sherman Miles to GCM, 7/17/41, PHH 14:1342–43; Toland 85; Heinrichs 134–35.

14 FDR, Executive Order 8832, 7/26/41; FDR to WSC, 7/26/41, FDR-PL 2:1189; Morgenthau, notes of conversation with FDR, 11/25/41, FDRL (Morgenthau Papers, box 515); Ickes, 6/28/41, LC (Ickes Papers, reel 4); Toland 85–86; Heinrichs 141; Barnhart 229; Dalleck 274.

15 Joint Board, "Joint Board Estimate of Over-All Production Requirements," 9/11/41, and appendices, reprinted at http://www.alternatewars.com/WW2/VictoryPlan/VictoryPlan.htm; Watson, *Chief of Staff*, 330–37, 343–46; Kirkpatrick, *Unknown Future*, 81–100; Matloff, *1941–1943*, 59.

16 HLS, 9/25/41, 12/4/41; Ickes 3:659; Pogue 2:160–61, quoting *Chicago Tribune*, 12/4/41, and *Washington Times-Herald*, 12/4/41 ("July"); Thomas Handy, interview notes, 8/21/56, PL; Stoler 57.

17 HLS, 12/5/41 ("infernal"); Ickes 3:659–60.

18 HLS, 12/4/41; FBI Report, 12/5/41, and HLS to FDR, 3/2/42, FDRL (PSF, box 82); Thomas Handy, interview notes, 8/21/56, PL; Albert Wedemeyer, interview, 2/1/58, PL.

19 Sherwood 493 ("The Army").

20 Katherine Marshall 124–26; GCM to Mrs. J. J. Winn, 9/4/41, MP 2:597–98.

TWELVE ★ *KIDO BUTAI*

1 HLS, 10/16/41; "Marshall and Stark Present at Talk at White House," NYT, 10/17/41; "Paraphrase of Code Radiogram Received at the War Department," 10/20/41, FDRL (Hopkins Papers, box 180); Smith, *FDR*, 520–22, and Hull 2:1024–25.

2 FDR to WSC, 10/16/41, FDR-PL 2:1223–24; Ickes 3:629; HLS, 10/16/41 ("radically"); Toland 132 ("hara-kiri").

3 GCM to Leonard Gerow, 1/17/41, MP 2:391–92; GCM to George Grunert, 2/8/41, MP 2:414–16; Morton, *Fall*, 62–64.

4 GCM, *Biennial Report, 1941–1943*, 6; Morton, *Fall*, 12–18; Pogue 2:178, quoting GCM to Stark, 11/29/40.

5 GCM, interview, 1/15/57, GCML; GCM and Stark to FDR, 11/5/41, PHH 14:1061; Pogue 2:174–77, 183, 186–89; Morton, *Fall*, 18; Stoler 58–59.

6 GCM and Stark to FDR, 11/5/41, PHH 14:1061–62 ("weaken"); Pogue 2:159.

7 Joint Board, minutes, 11/3/41, PHH 14:1064 ("It appeared"); Pogue 2:159.

8 Foreign Office to Nomura, 11/4/41, PHH, 12:91–95 ("bargain"); Toland 134–35; Keiichiro Komatsu, "Misunderstanding and Mistranslation in the Origins of the Pacific War 1941–1945," presentation to Japan Society, 7/5/01.

9 Nomura to Foreign Office, 11/14/41, PHH, 12:127–28 ("pacification").

10 Foreign Office to Nomura, 11/22/41, PHH 12:165 ("things"); FDR to WSC, 11/24/41, FDR-PL 2:1245.

11 FDR, handwritten note, n.d., PHH 14:1109; Hull to FDR, 11/26/41, PHH 14:1176; FDR to WSC, 11/24/41, FDR-PL 2:1245–446; Nomura to Foreign Office, 11/17–18/41, PHH 12:146–48; Foreign Office to Hong Kong, 11/14/41, PHH 12:126; Henry Morgenthau, notes of conversation with FDR, 11/22/41, 11/25/41, FDRL (Morgenthau Papers, box 515); Henry Morgenthau, notes of conversation with T. V. Soong, 11/27/41, FDRL (Morgenthau Papers, box 515); HLS, 11/6/41; Adolf Berle, 12/1/41, FDRL (Berle Papers, box 213); Toland 94, 137–39; Hull, memorandum, 11/25/41, PHH 14:1167; FDR, press conference, 11/18/41, FDR-PP, 1941, 502; Perkins, interview, Part VIII, 30, COHP ("If Cordell Hull"); Ickes 3:650; Tully 174.

12 Office of Naval Intelligence, Intelligence Report, 11/26–27/41, PHH 15:1886, 1889; Sherman Miles to Marshall, 11/26/41, PHH 14:1366; HLS, 11/25/41.

13 HLS, 11/25/41 ("jumped," "maneuver").

14 Hull to Nomura, 11/26/41, *Department of State Bulletin*, 12/13/41.

15 GCM and Stark to FDR, 11/27/41, PHH, 14:1083 ("Japan may"); WSC to FDR, 11/26/41, PHH, 14:1300; FDR, press conference, 11/28/41, FDR-PP, 1941, 501; HLS, 11/27/41.

16 Adolf Berle, 11/27–28/41, 12/1/41, FDRL (Berle Papers, box 213); HLS, 11/27/41 ("washed"); Toland 173.

17 HLS, 12/1/41; Morison 529; Pogue 2:201.

18 Office of Naval Intelligence, "Memorandum: Japanese Fleet Locations," 12/1/41, PHH 15: 1896; "Memorandum for the President," 12/5/41, PHH 20:4116; Hanyok 114–18.

19 GCM to Short et al., 11/27/41, PHH 11:5424; HLS, 11/27/41 ("negotiations").

20 Stenographer's Diary, 11/28/41, FDRL; HLS, 11/28/41.

21 HLS, 11/28/41 ("the opinion").

22 U.S. Department of State, "Outline of Proposed Basis for Agreement Between United States and Japan," 11/25/41, PHH 14:1155; Hull, "Proposed Message from the President to the Emperor of Japan," 11/29/41, PHH, 14:1224; Ickes 3:654–55; Hull 2:188–89; Tully 248–49; Toland 144–45.

23 Hull 2:1089–90; Ickes 3:655; Joseph Grew to Hull and Welles, 12/1/41, PHH, 14:1302; "Hull Relays Data," NYT, 11/30/41 ("exploitation"); "U.S. Principles Rejected," NYT, 12/1/41; Toland 180–81.

24 "President Is Grim," NYT, 12/1/41; "Battle Stations," *Time*, 12/8/41; Tully 251 ("last time").

25 Foreign Office to Nomura, 11/15/41, PHH 12:137; Stark to CINCPAC, 12/3/41, PHH, 14:1407.

26 Frances Perkins, interview, part viii, 50, COHP ("expect"); Sherman Miles, testimony, 11/29/45, PHH 2:790; GCM to FDR, 2/2/44, MP 4:293–94; Wohlstetter 180; Toland 188.

27 FDR to Hirohito, 12/6/41, FDRL (PSF, box 43) ("dynamite"); Tully 253.

28 Sherman Miles, testimony, 11/29/45, PHH 2:941; GCM, testimony, 12/7/45, 12/13/45, PHH 3:1108–10, 3:1509; W. B. Smith, interview, 7/29/58, 12, 14 PL ("safe").

29 Leonard T. Gerow, "Memorandum for the Record," 12/15/41, PHH 14:1409; W. B. Smith, "Memorandum for the Record," 12/15/41, PHH 14:1410; Sherman Miles to GCM, 12/15/41, PHH 14:1410; Pogue 2:227; Sherman Miles, testimony, 11/29/45, PHH 2:931–35.

30 Sherwood 434; W. B. Smith, interview, 7/29/58, PL; Leonard T. Gerow, interview, 2/24/58, PL; Toland 202.

THIRTEEN ★ KICKING OVER ANTHILLS

1 FDR, address, 12/8/41, FDR-PP, 1941, 514–16 ("December 7").

2 "Unity in Congress," NYT, 12/9/41; Ickes 3:665–66.

3 "What the People Said," *Time*, 12/15/41 ("every American"); Olson 443.

4 FDR to Stimson and Knox, 12/15/41, FDR-PL 2:1259; Smith, *Thank You*, 117; Reilly 4–5, 26–29; Tully 258–59; Goodwin 298, quoting Lorena Hickok, manuscript, FDRL ("Fala").

5 Claude R. Wickard, 12/7/41, FDRL (Wickard Papers, Cabinet Meetings, box 13); HLS, 12/7/41; Sherwood 430; Shirer 1169, fn.

6 FDR, fireside chat, 12/9/41, FDR-PP, 1941, 522 ("reinforce"); Shirer 1173–77 ("considers").

7 Leahy, interview, 5/24/48, FDRL (Oral History Collection, box 1); EJK, interview, 5/6/46, KP (box 2); R. S. Edwards, interview, 1/30/46, KP (box 7); Knox, flight log, 12/7/41, LC (Knox Papers, box 1); EJK, interview, 5/6/46, KP (box 2); EJK, interview, 4/28–30/48, KP (box 2); FDR to EJK, 12/20/41, KP (box 5); Knox to EJK, 12/20/41, KP (box 5).

8 EJK, 6/13/46, KP (box 9); EJK, interview, 5/6/46, 4/28–30/48, KP (box 2).

9 EJK, interview, 5/6/46, KP (box 2); Buell 139.

10 EJK, interview, 12/11/46, KP (box 7) ("sink us"); Low 24; Neil Dietrich, interview, 12/10/74, KP (box 11); Buell 139.

11 EJK to Harold Stark, 11/5/41, KP (box 5); J. R. Topper to Thomas Buell, 9/27/74, KP (box 2); Harry Sanders, "Assorted Notes re FAdm King," n.d., KP (box 2); EJK, interview, 7/4/50, KP (box 7); Neil Dietrich, interview, 12/10/74, KP (box 11); Ernest K. Lindley to EJK, 8/20/42, 11/4/42, KP (box 5).

12 EJK, interview, 8/26/50, KP (box 7) ("Marshall would"); Pogue 2:149–50; Pogue 3:505, citing Frederick L. Allen, "Marshall, Arnold, King: Three Snapshots," *Harper's Magazine* (Feb. 1945), 287.

13 Neil Dietrich, interview, 12/10/74, KP (box 11); George Russell to Thomas Buell, 11/18/74, KP (box 2) ("I have"); George Russell, interview, 12/11/74, KP (box 11).

14 Neil Dietrich, interview, 12/10/74, KP (box 11).

15 EJK, interview, 8/24/49, KP (box 4); EJK, interview, 8/27/50, KP (box 6); Neil Dietrich, interview, 12/10/74, KP (box 11).

16 EJK, interview, 12/11/46, KP (box 8); Executive Order No. 8984, 12/18/41, in Buell, Appendix III.

17 Buell 140 ("appreciate").

18 Ruthven E. Libby, interview, 2/8/70, 5, KP (box 13) ("decrepit"); Ruthven E. Libby, interview, 6/7/70, 222, KP (box 13); Whitehill 353; Buell 138, quoting Harry Sanders, "King of the Oceans," *U.S. Naval Institute Proceedings* (August 1974) ("anthill").

19 J. R. Topper to Thomas Buell, 9/27/74, KP (box 2); George Russell to Thomas Buell, 11/18/74, KP (box 2) ("His desk"); George Russell, interview, 12/11/74, KP (box 11) ("miserable housekeeper").

20 Forrest Davis, "King's Way to Tokyo," *Saturday Evening Post*, 12/3/44, 103, KP (box 9) ("If I").

21 A. R. McCann, interview, 9/74, KP (box 10); Howard Orem, interview, 12/12/74, KP (box 11); G. A. McLean to Thomas Buell, 8/9/74, KP (box 2); Eunice H. Rice, 7/2/76, KP (box 7); George C. Dyer, interview, 8/1/45, KP (box 13); C. M. Keyes, interview, 8/8–10/74, KP (box 2); Buell 230. "Memorandum for Captain S. E. Morison," 1/25/47, KP (box 5) ("Washington mentality").

22 C. M. Keyes, interview, 8/8–10/74, KP (box 2); Robert B. Pirie, interview, 147, KP (box 10) ("He didn't").

23 Low 19; R. D. Shepherd to Thomas Buell, 8/9/74, KP (box 2); Forrest Davis, "King's Way to Tokyo," *Saturday Evening Post*, 12/3/44, 103, KP (box 9); Howard Orem, interview, 12/12/74, KP (box 11); Neil Dietrich, interview, 12/10/74, KP (box 11); Malcolm Schoeffel to Thomas Buell, 9/27/74, KP (box 2) ("?"); Buell 160, 211.

24 J. R. Topper to Thomas Buell, 9/27/74, KP (box 2); Omar T. Pfeiffer, interview, 233, KP (box 10); Buell 215–16.

25 William Smedberg to Thomas Buell, 2/10/75, KP (box 2) ("lone wolf"); J. R. Topper to Thomas Buell, 9/27/74, KP (box 2); Kenneth Knowles to Thomas Buell, 9/22/74, KP (box 2) ("armor-plated steel").

26 William Smedberg to Thomas Buell, 2/10/75, KP (box 2) ("within a few minutes"); William R. Smedberg III, interview, 6/9/76, 43, KP (box 11).

27 C. M. Keyes, interview, 8/8–10/74, KP (box 2).

28 John Hyland to Thomas Buell, 5/5/76, KP (box 9); John Hyland, interview, 3/8/76, 126, KP (box 13); Buell, notes on Cadillac, n.d., KP (box 7) (citing EJK to CNO, 4/13/46, and EJK to H. R. Stark, 12/22/41).

29 EJK to Fleet Commanders, 12/30/41, KP (box 5); "Memorandum for Captain

S. E. Morison," 1/25/47, KP (box 5); EJK to C. F. Grisham, 1/2/42, KP (box 5); "Adm. King's Flag is Hoisted on Yacht 'Made in Germany,'" n.d., KP (box 7).

30　EJK to C. F. Grisham, 1/2/42, KP (box 5); EJK to Fleet Commanders, 12/30/41, KP (box 5); "Memorandum for Captain S. E. Morison," 1/25/47, KP (box 5); "US at War," *Time*, 9/7/42; EJK to Staff, 1/9/42, KP (box 5); G. B. Turnbull to EJK, 5/27/42, KP (box 5); EJK to C. F. Grisham, 1/2/42, KP (box 5); Eunice H. Rice to Buell, 7/2/76, KP (box 7); Forrest Davis, "King's Way to Tokyo," *Saturday Evening Post*, 12/3/44, 103, KP (box 9); Neil Dietrich, interview, 12/10/74, KP (box 11); Charles Griffins to Buell, 12/19/74, KP (box 2); Howard Orem, interview, 12/12/74, KP (box 11); George Dyer, interview, 8/1/45, 465–66, KP (box 10); George Russell to Buell, 11/18/74, KP (box 2); Buell 143–44, 379, 392, 397; A. R. Taylor to Thomas Buell, 9/25/74, KP (box 2).

FOURTEEN ★ "DO YOUR BEST TO SAVE THEM"

1　Richard L. Watson, "Pearl Harbor and Clark Field," in Craven and Cate 1:212–13; Sherman Miles to Edwin Watson, 12/12/41, FDRL (Map Room Papers, box 38); Keegan, *Atlas*, 57.

2　DDE, *Crusade*, 21–22 ("your best").

3　HLS, 12/12/41, 12/14/41; Pogue 2:242–46 ("out-shipped").

4　Pogue 2:240, 246, citing MacArthur to GCM, 2/4/42.

5　DDE, 1/19/42, EL (DDE Diaries, box 1) ("big a baby"); DDE, 2/3/42, DDE, Prewar Diaries 46 ("nerve").

6　Cray 119.

7　DDE, 1/8/42, DDE, Prewar Diaries 42; MP 3:51–52, quoting DDE, memorandum, 1/9/42 ("You may").

8　Morton, *Fall*, 399; Pogue 2:242–43.

FIFTEEN ★ "O.K. F.D.R."

1　Goodwin 302, quoting Mrs. Charles Hamlin, "Memories," FDRL, and "Old River Friend," *New Republic*, 4/15/45 ("common cause").

2　GCM, memorandum, 12/23/41, FRUS, *Arcadia*, 69–72; CCS, minutes, 12/24/41.

3　HLS, 12/24/41; Sherwood 442; WSC, *Grand Alliance*, 663 ("maps").

4　Goodwin 311, quoting James Roosevelt, interview ("wonderful").

5 WSC, *Grand Alliance*, 662 ("outstanding"); Goodwin 302–03.

6 Reilly 125 ("awe"); Hassett 5/27/43 ("fish").

7 Goodwin 302, quoting Elliott Roosevelt, interview.

8 Betty Hight to family, 8/20/42, FDRL (Betty Hight Papers); Goodwin 311, quoting Justine Polier, interview.

9 HLS, 12/24/41.

10 HLS, 12/25/41 ("resignation").

11 HLS, 1/16/42, 12/25/41 ("This incident").

12 Reginald Winn, memoirs, 2, 48, GCML (Small MS Collections, box 3); ARCADIA Minutes, 12/24/41, *World War II Inter-Allied Conferences*, 1; GCM, interview, 10/29/56 ("best mind"); EJK, interview, 8/28/49, KP (box 7); Davis, *History of the Joint Chiefs*, 139; Lawrence Kuter, interview, 11/10/60, in Pogue 2:271; Robinett, 12/24/41, in Pogue 2:271.

13 Roberts 71, quoting Thomas T. Handy, interview, 1974 ("babes").

14 Paul Robinett, 12/24/41, GCML (Robinett Papers, box 16); Paul Robinett, "Saving National Face," GCML (Robinett Papers, box 16).

15 HLS, "Memorandum of Decisions at White House, Sunday, December 21, 1941," in HLS, 12/21/41; CCS, Minutes, 12/24/41; Paul Robinett, 12/24/41, GCML (Robinett Papers, box 16); Davis 145; Roberts 73; Buell 146.

16 WSC, strategy memoranda, 12/16–20/41, FRUS, *First Washington Conference*, 21–31; BCS, "American-British Strategy," in CCS, minutes, 12/24/41, Annex 1; BCS, "WW-1," FDRL (PSF Safe File, box 1); BCS, General Strategy Review, 7/31/41, in Watson, *Chief of Staff*, 403.

17 HHA, notes, 12/22/41, in FRUS, *Arcadia*, 66; Matloff and Snell, 113 n.59.

18 HLS to FDR, 12/20/41, FRUS, *Arcadia*, 44; Joint Board, No. 325, 12/21/41, FRUS, *Arcadia*, 50–51; HLS, 1/3/42.

19 GCM, notes, 12/23/41, FRUS, *Arcadia*, 72 ("considered").

20 Katherine Marshall 100–102; GCM, interview, 2/20/57; Parrish 221.

21 CCS, minutes, 12/25/41 ("I am convinced").

22 CCS, minutes, 12/25/41 ("highest authority"); Robinett, 12/25/41, in Pogue 2:276–77; Davis 152.

23 CCS, minutes, 12/26/41, FRUS, *Arcadia*, 102–04; DDE, "Notes Taken at Jt. Conference of Chiefs of Staff, on afternoon, 25 Dec. 41," EP 1:25–26; HLS, 12/27/41; Davis, *History of the Joint Chiefs*, 153, 156.

24 EJK to H. R. Stark, 1/14/41, KP (box 2); EJK to C. S. Freeman, 1/5/42, KP (box 7) ("I have"); EJK to J. G. Scrugham, 2/11/42, KP (box 7).

25 HLS, 12/28/41; Pogue 2:278; Buell 148, quoting Robinett, 12/27/41 ("When King"); BCS, "Procedure for Assumption of Command by General Wavell," 1/10/42, *World War II Inter-Allied Conferences*, 135.

26 HLS, 12/28/41 ("bay steers"); Beaverbrook to HLH, n.d., FDRL (Hopkins Papers, box 136) ("You should").

27 Pogue, interview notes, 10/5/56 and 11/21/56; Roberts 80.

28 Pogue, interview notes, 10/5/56 ("Frobisher").

29 Moran, *Struggle for Survival*, 20, 22 ("Marshall remains").

30 Davis, *History of the Joint Chiefs*, 154–55; Pogue, interview notes, 10/5/56 and 11/21/56.

31 Lord Privy Seal to WSC, 12/29/41, FDRL (Hopkins Papers, box 136); CCS, minutes, 12/29/41, *World War II Inter-Allied Conferences*, 66–67; Sherwood 467; Davis 156–57.

32 HLS, "Memorandum of Decisions at White House, Sunday, December 21, 1941," in HLS, 12/21/41; CCS, minutes, 12/29/41; Matloff, *Strategic Planning*, 125; Buell 151

33 CCS to FDR, 12/29/41, Sherwood 467 ("existing machinery"); CCS, minutes, 1/13/42; BCS, "Post-ARCADIA Collaboration" (WW-8), 1/8/42; CCS, "Post-ARCADIA Collaboration" (ABC-4/CS-4), 1/14/42; EJK to Thomas Dobyns, 9/18/50, KP (box 8).

34 CCS, minutes, 12/29/41; HLH to Stark, 12/30/41, in *World War II Inter-Allied Conferences*, 80–81; Moran 20–21; Sherwood 468.

35 Sherwood 469; CCS, "Post-ARCADIA Collaboration" (ABC-4/CS4), 1/14/42; Leahy 130; Davis 163; Larrabee 17–18; Moran 23; Roberts 95 and Crosswell 246, quoting Thomas T. Handy, interview, USAMHI.

36 HLS, 1/27/42, 5/12/43; Davis 180–81; Pogue 2:283; Roberts 60.

37 GCM to FDR, 3/16/43, MP 3:589–90; Pogue 2:283; GCM, interview, 2/14/57, GCML ("I tried").

38 Executive Order 9082, 2/28/42, FDR-PP, 1942, 140–41; Davis 235–37; Morison 543–44.

39 Leahy to FDR, 6/16/43, FDRL (PSF, box 6); CCS, minutes, 1/23/42; EJK to Harriet Wedeen, 1/8/43, KP (box 6); Davis 226–32, 252.

40   Leahy to Harold D. Smith, 7/16/43, KP (box 10); FDR to Leahy, 7/16/43, KP
      (box 10); Leahy 126 ("chit"); Stoler, *Allies*, 108.

41   Bryant 234, quoting Dill to Brooke, 1/3/42 ("George Washington").

42   CCS, minutes, 1/23/42; HLS to FDR, 3/20/42, in Davis, *History of the Joint
      Chiefs*, 191 n.22.

43   Davis, *History of the Joint Chiefs*, 256 ("O.K."); Buell 164.

SIXTEEN ★ "THERE ARE TIMES WHEN MEN HAVE TO DIE"

1   Neil Dietrich, interview, 12/10/74, KP (box 11); EJK, "Note," n.d., KP (box 7);
     EJK, interview, 1/5/46, KP (box 7); EJK to Knox, 2/20/42, KP (box 5).

2   Glenn Perry to E. P. Bartnett, 11/30/42, KP (box 11); Buell 176 ("No fighter").

3   Hayes 51, 88, citing War Department minutes, 2/16/42.

4   Coordinator of Information, "The War This Week," 3/5–12/42, FDRL (Hopkins
     Papers, box 330); FDR to WSC, 1/18/42, Sherwood 510, Matloff 165–66; FDR to
     WSC, 2/18/42, FDRL (Hopkins Papers, box 136); Hayes 88–89, citing FDR
     to WSC, 2/18/42 (draft initialed by HLS and GCM), and War Department
     minutes, 2/16/42; Davis, *History of the Joint Chiefs*, 200–03; CCS, minutes,
     2/17/42, 2/23/42, 3/3/42; Hayes 90–91, citing Admiralty to BAD Washington,
     2/23/42; WSC to FDR, 2/21/42, FDRL (Hopkins Papers, box 136); FDR to
     WSC, 3/9/42.

5   DAM, *Japanese Operations*, 106–10.

6   "U.S. Speeds Men, Arms to Far East," NYT, 1/23/42; GCM to FDR, 3/3/42,
     FDRL (PSF Safe File, box 3); FDR to WSC, 3/7/42; HLS, 1/23/42.

7   Frances Perkins, interview, part VIII, 47, COHP; Stark to FDR, 11/12/40,
     FDRL (PSF File); Watson 413–15; Simpson 65, 75.

8   Manuel Quezon to FDR, 2/8/42, accessed at http://www.gov.ph/1942/02/08/
     telegram-of-president-quezon-to-president-roosevelt-february-8-1942/.

9   DAM to GCM, 2/8/42, in HLS, 2/13/42 ("The temper"); Morton, *Fall*, 354.

10   HLS, 2/8–9/42; Hayes 92.

11   Morison 550, quoting George Roberts, interview, 9/14/55 ("There are times");
     HLS, 2/9/42; Bundy, *Active Service*, 404.

12   HLS, 2/9/42; FDR to Quezon, 2/9/42, in HLS, 2/13/42; Pogue 3:70, 316.

13   GCM, interview, 11/14/56; Pogue 2:23, 247–48 ("great man").

14 FDR to Quezon, 2/9/42, in HLS, 2/13/42 ("flag"); FDR to DAM, 2/9/42, in HLS, 2/13/42 ("mandatory"); Morton, *Fall*, 354; Bundy, *Active Service*, 400–04.

15 DAM to Adjutant General, 2/11/42, in HLS, 2/13/42 ("share the fate").

16 Pogue 2:249.

17 HLS, 2/12/42; *Congressional Record*, 77th Cong., 2d. Sess., vol. 88, part 1, 1249, and part 2, 1266, in Pogue 2:249; "The People: Bring Home MacArthur!" *Time*, 2/23/42 ("the air").

18 Stenographer's Diary, Usher's Log, 2/22/42, FDRL; Hayes 93–94, quoting GCM to DAM, 2/22/42; Sherwood 509; Buell 171.

19 Pogue 2:252–53.

20 *The Adelaide Advertiser* (South Australia), 3/21/42 ("I shall return"); Pogue 2:251.

SEVENTEEN ★ *"INTER ARMA SILENT LEGES"*

1 FDR, address, 1/11/42, FDR-PP, 1942, 37; HLS, 1/12/42.

2 Sherwood 472 ("numbers"), 474 ("really try"); Thompson and May 234; Stoler 67.

3 HLH, 1/14/42, in Sherwood 475; HLS, 1/16/42; Goodwin 315.

4 HLS, 1/28/42.

5 GCM to FDR, 2/18/42, FDRL (PSF, box 3) ("war effort"); Roberts 286.

6 John McCrea, interview, 3/19/73, FDRL (McCrea Papers).

7 John McCrea, interview, 3/19/73, FDRL (McCrea Papers); George M. Elsey, "Remarks at the 'Castle' of the Smithsonian Institution," 10/28/04, FDRL (Elsey Papers, box 1); George Elsey to Riley Sunderland, 11/14/78, FDRL (Elsey Papers, box 1); George Elsey, memorandum, 1/19/55, FDRL (Elsey Papers, box 1); Tully 262; Parrish 236.

8 George Elsey, interview, 2/10/64, HSTL ("The walls").

9 George Elsey, interview, 2/10/64, HSTL; Rigdon 7–8; Leahy 122.

10 John McCrea, interview, 3/19/73, 22, FDRL (John McCrea Papers); "Types and Sources of Information for White House Map Room," 12/43, FDRL (Elsey Papers, box 1); George M. Elsey, "Remarks at the 'Castle' of the Smithsonian Institution," 10/28/04, FDRL (Elsey Papers, box 1); Wilson Brown to Vardaman, 5/3/45, FDRL (Elsey Papers, box 1); Leahy 122; GCM to Sumner Welles, 3/5/42, MP 3:123–24.

11 "Roosevelt to Warn U.S. of Danger," NYT, 2/21/42; Rosenman 330 ("I'm going").

12 "President Speaks," NYT, 2/24/42.

13 FDR, fireside chat, 2/23/42, FDR-PP, 1942, 105–08, 115.

14 Williams, "We Have Kept," 88–109; Goodwin 328, citing Dennis Nelson, "Negro in the Navy," n.d., FDRL (Office file, box 93).

15 HLS, 1/17/42 ("The Navy").

16 HLS, 1/17/42 ("spoiled child").

17 FDR to Knox, 1/9/42, FDRL (PSF, box 7) ("I think"); General Board to Knox, 2/3/42, FDRL (PSF, box 7) ("Men on board"); Bureau of Naval Personnel, *Negro in the Navy*, 4–8.

18 Goodwin 329–33, quoting FDR to Fraternal Council of Negro Churches, 3/16/42, FDRL (Office File, box 93); FDR, Executive Order 8802, 6/25/41, FDRL.

19 Marriner S. Eccles, *Beckoning Frontiers* (New York: Knopf, 1951), 336 ("feather").

20 FDR to Knox, 2/9/42, FDRL (PSF, box 7) ("opinion"); Bureau of Naval Personnel, *Negro in the Navy*, 4–8.

21 "Negro Hero Identified," NYT, 3/13/42.

22 Knox to FDR, 3/27/42, FDRL (PSF, box 7) ("boldly"); General Board to Knox, 3/20/42, FDRL (PSF, box 7); Bureau of Naval Personnel, *Negro in the Navy*, 6–7.

23 U.S. Government, 1940 Census, Table C-7; Conn, *Guarding*, 115.

24 Conn, *Guarding*, 117–18, quoting John DeWitt and Allen Guillion, telephone conversation, 12/26/41 and Mark Clark to JAG, 1/24/42; HLS, 2/3/42; Morison 547; Pogue 3:140–41.

25 HLS to Attorney General, 1/25/42, in Conn, *Guarding*, 121; HLS, 7/16/40, 3/31/41, 1/30/42, 2/3/42; "Stimson to Explain Naval Treaty to Senators Today," *Chicago Tribune*, 5/12/30; Thomas Handy, interview notes, 8/21/56, PL.

26 "Scare on the Coast," *Time*, 2/16/42 ("All along"); Walter Lippmann, "The Fifth Column on the Coast," *Los Angeles Times*, 2/15/42 ("It is a fact"); Pogue 3:144.

27 HLS, 2/26/42, 7/7/42; Morison 549, quoting HLS to FDR, 7/7/42, FDRL; John DeWitt and Maj. Bendetsen, 1/29/42, in Conn, *Guarding*, 122.

28 "U.S. Uproots Jap Aliens," *Life*, 3/9/42 ("Officials"); Goodwin 323, quoting *New Republic*, 6/15/42, 822–23 ("Japs").

29 HLS, 2/3/42, 2/10–11/42; "Army Gets Power to Move Citizens and Aliens Inland," NYT, 2/21/42; "Map Tighter Rule over West Coast Aliens," NYT,

2/20/42; "Coast Japanese Split on Ouster," NYT, 2/21/42; John DeWitt, *Final Report: Japanese Evacuation from the West Coast, 1942* (Washington: Government Printing Office, 1943) 9–10; Morison 547–48; John DeWitt and John McCloy, telephone conversation, 2/3/42, in Conn, *Guarding*, 125–26.

30 Allen Guillion to John McCloy, 2/5/42, Conn, *Guarding*, 128–29; HLS, 2/3/42, 2/10/42 ("underestimating," "difficult proposition").

31 HLS, 2/11/42; McCloy and Bendetsen, telephone transcript, 2/11/42, Conn, *Guarding*, 132 ("reasonable").

32 HLS, 4/23/42 ("*Inter arma*"), 5/22/42, 9/15/42; Bundy, *Active Service*, 406.

33 Executive Order 9066, 2/19/42; HLS 2/18–20/42; Conn, *Guarding*, 136–37.

34 HLS, 5/17/44, 5/26/44; Conn, *Guarding*, 137–41; GCM and EJK to FDR, 7/15/42, FDRL (PSF, box 4).

EIGHTEEN ★ ROLLING IN THE DEEP

1 Cora Thomas, interview, 3/10/61, GCML; HLS, 3/21/42.

2 HHA, "Notes of Roosevelt-Churchill Conference," 8/4/41, LC (Arnold Papers, reel 2); GCM, interview, 10/5/56, GCML ("sore"); DDE, 3/14/42, DDE, Prewar Diaries, 51 ("That's").

3 Clark, interview, 11/17/59, PL ("Clark").

4 Stilwell, 1/20/42 ("high-powered"); DDE, 3/10/42, DDE, Prewar Diaries, 50 ("shoot King"); Roberts 97, quoting Paul Caraway, interview, 1971 ("time of day").

5 Bundy, *Active Service*, 281, 504 ("Neptune").

6 EJK, interview, 8/26/50, KP (box 7) ("Sometimes"); EJK to T. B. Kittredge, 2/1/50, KP (box 8) ("General Marshall"); EJK, interview, 7/4/50 ("very agreeable"), 7/30–31/49, KP (box 7).

7 EJK, interview, 7/4/50, KP (box 8) ("didn't know," "damn thing") 7/30–31/49, 8/28/49, KP (box 7) ("I am sure").

8 GCM, interview, 2/14/57 ("can't afford").

9 EJK to CNO and JCS, 2/18/42, KP (box 4); Stoler 68, quoting EJK to Knox, 2/8/42 ("hold"); Matloff 154.

10 GCM to EJK, 2/24/42, KP (box 4).

11 EJK to GCM, 3/2/42, KP (box 4).

12 Pogue 2:378, citing DDE to GCM, 4/21/42.

13   EJK to FDR, 3/5/42, KP (box 4).

14   FDR to Churchill, 3/7/42; Hayes 783 n.14.

15   DDE to GCM, 2/28/42, EP 1:149–55; GCM to Brett, 3/10/42, FDRL (PSF Safe File, box 3); BCS to JCS, 3/7/42, Hayes 97; GCM to JCS, 3/9/42, Matloff 169–70; Buell 173–74.

16   EJK to FDR, 4/5/42, Matloff 169 n.82; Buell 174.

17   Hayes 99; Buell 175.

18   JCS, minutes, 3/16/42, in Matloff 169; EJK to David Lawrence, 9/28/42, KP (box 5); Hayes 99; Buell 175, 178, 181.

19   Morison 1:128–30; MP 3:119–20 n.1; Charles M. Sternhell and Alan M. Thorndike, "ASW in World War II," Operations Evaluation Group Report No. 51 (Washington: U.S. Government, 1946), 25–26; Keegan, *Atlas*, 77 (table); WSC to FDR, 7/14/42; FDR to WSC, 3/18/42; Sherwood 600; Buell 271–72.

20   Arthur B. Ferguson, "The AAF in the Battle of the Atlantic," Craven and Cate 1:522; Morison 1:134, 154–56; Charles M. Sternhell and Alan M. Thorndike, "ASW in World War II," Operations Evaluation Group Report No. 51 (Washington: U.S. Government, 1946), 26; Frank Knox to George Sample, 8/11/42, LC (Knox Papers, box 4).

21   Morison 1:200–01.

22   EJK to HHA, 2/20/42, KP (box 4); HHA to EJK, 2/25/42, KP (box 4); Morison 1:237; HHA to EJK, 3/16/42, KP (box 4); Ferguson, in Craven and Cate 1:539–40, citing HHA to EJK, 3/9/42.

23   FDR to HLS and Knox, 3/16/42, FDRL-PL 2:1297 ("The point"); HLS, 3/16–17/42, 4/21/42 ("didn't know"); Morison 569; Ferguson, in Craven and Cate 1:541–42; Morison 1:242–43.

24   EJK, "Notes Taken From Flight Log No. 3," 8/13/43 entry, KP (box 4) ("able statesman"); EJK, interview, 8/26/50, KP (box 4) ("I've been"); EJK, interview, 11/27/50, KP (box 8) ("Stimson was").

25   EJK, "Notes from Air Log—August—December 1945," n.d., KP (box 4); EJK, interview, 11/27/50, KP (box 8); EJK, interview, 7/4/50, KP (box 4) ("The Army").

26   EJK to HHA, 3/18/42, KP (box 4) ("All of us," "I think").

27   HLS, 7/7/42; Ferguson, in Craven and Cate 1:547; GCM to EJK, 6/19/42, KP (box 4) ("losses").

28   EJK to GCM, 6/21/42, Morison 1:309–10 ("invulnerable").

NINETEEN ★ SHARKS AND LIONS

1 Commanding Officer, USS *Enterprise*, AAR, 4/28/42; James H. Doolittle, AAR, 6/5/42; Halsey to CINCPAC, 4/24/42.

2 F. S. Low to F. H. Schneider, n.d., KP (box 13); Cornelius Bull, "The Fifth Seminar," 6/6/43, KP (box 6); EJK, interview, 7/4/50, KP (box 4); Low 25; EJK to Quentin Reynolds, 3/19/52, KP (box 13); EJK to Grace Hayes, 11/20/51, KP (box 13); Ruthven E. Libby, interview, 6/7/70, 237, KP (box 13); Glenn Perry to E. P. Bartnett, 6/8/43, KP (box 11).

3 "American Planes Bomb Tokyo," Bismarck *Tribune*, 4/18/42; "Tokyo Factories Hit in Raid," NYT, 4/20/42; HLS, 4/18/42 ("a very good").

4 FDR, press conference, 4/21/42, FDR-PP, 1942, 214 ("Shangri-La"); Hassett, 4/20/42.

5 DDE, 1/22/42, EP 1:66 ("We've"); HLS, 10/7/41; Pogue 2:304.

6 Stenographer's Diary, 3/25/42, FDRL; FDR to WSC, 7/8/42, FDRL (Hopkins Papers, box 136); Morison 581, 582 n.26; DDE, *Crusade*, 42–48; Cline 152–53; Matloff and Snell 177–87, Appendix A; Pogue 2:305–06; HLS, 3/25/42; Roberts 143; Bundy, *Active Service*, 419; Marshall to FDR, "Basis for Preparation of Attached Outline Plan for Invasion of Western Europe," n.d., FDRL.

7 HLS, 3/25/42 ("debauch").

8 HLS, 3/5/42 ("orthodox").

9 Morison 581.

10 HLS to FDR, 3/27/42, in HLS, 3/27/42; Morison 582.

11 GCM to Leonard Gerow, 1/17/41, MP 2:391–92 ("we must"); Stoler, *Allies*, 75.

12 HLS to FDR, 3/27/42; HLS 4/1/42; FDR to WSC, 4/3/42; GCM, interview, 1/14/57, GCML; Pogue 2:306–08; GCM to WSC, 4/28/42, MP 3:175–76; Sherwood, 522–28; Roberts 129, 141.

13 W. B. Smith, interview, 7/29/58, PL.

14 Brooke, 4/15/42 ("a good general").

15 GCM, interview, 2/11/49, GCML; GCM, interview, 10/5/56, GCML ("Somme"); WSC, *Closing the Ring*, 514; Ismay 250; Croswell 259, quoting War Cabinet, "Comments on General Marshall's Memorandum," 4/13/42, GCML; Roberts 213–14.

16 Sherwood 523, quoting HLH, meeting notes, 4/8/42; GCM, interview, 10/5/56, GCML.

17   WSC, *Hinge of Fate*, 319 ("march ahead").

18   GCM to HLS, 4/15/42, MP 3:162–63; GCM to Joseph McNarney, 4/12/42, 4/13/42, MP 3:159–61; GCM to FDR, 4/18/42, MP 3:164–65; McNarney to FDR, 4/12/42, FDRL (PSF, box 83); WSC to FDR, 4/12/42, 4/17/42; Hastings Ismay, interview, 12/17/46, PL; Pogue 2:318, Sherwood 534–38; HLS 4/20/42; Crosswell 259; Roberts 153–57.; Parrish 270.

19   Ismay 249–50 ("unfortunate"); Hastings Ismay, interview, 10/18/60, PL ("swept"); Brooke, interview, 1/28/47, PL; Stoler, *Allies*, 77.

20   Symonds 160–61; Thomas Buell, "King/Nimitz Conferences," n.d., KP (box 8) (April 25–27, 1942); G. L. Russell, interview, 5/6/48, KP (box 8); Buell 180.

21   EJK, interview, 7/31/49, KP (box 8) ("fixer"); Charlotte Pihl, interview, 3/16/74, 18, KP (box 10) ("Damn"); Buell 342.

22   EJK to Nimitz, 4/24/42, and Nimitz, "Running Summary," 4/24/42, in Nimitz Papers, NHHC (box 1:409, 411), and "Minutes of Conversation Between CominCh and CinCPac, Saturday, April 25, 1942," EJK Papers, NHHC (Series 11, box 10), Symonds 162; Nimitz, Gray Book, 1:409, 416.

23   FDR to Mackenzie King, 5/18/42, FDR-PL 2:1320; Commander, TF 17.2.2 to CINCPAC, 5/17/42, NHCO; Masatake Okumiya and H. Sekino, interrogation, 10/17/45, *Interrogations of Japanese Officials*, 29–31; Hayes 128; Toland 322.

24   "Preliminary Report: USS Lexington Loss in Action," 5/8/42, USN (online); Commanding Officer, USS *Lexington* to CINCPAC, 5/15/42, USN (online).

25   Forrest Davis, "King's Way to Tokyo," *Saturday Evening Post*, 12/3/44, 101, KP (box 9); Buell 182–83.

26   EJK to Nimitz, 5/14/42, Nimitz, Gray Book, 1:463 ("loss of").

27   EJK to Nimitz, 5/15/42 and 5/17/42 and 5/17/42 entry, Nimitz, Gray Book, 1:463–64, 468, 483, 489–90 ("attritional tactics"); Glenn Perry to E. P. Bartnett, 11/7/42, KP (box 11); Cornelius Bull, "The First Seminar," 11/6/42, 4, KP (box 6); Hayes 128; Buell 183.

28   Nimitz to EJK, 5/14/42, Nimitz, Gray Book, 1:465–66, 530; Nimitz to EJK, 5/16/42, Nimitz, Gray Book, 1:471; Nimitz, Gray Book, 1:482; Buell 184; CINCPAC, "Estimate of the Situation Attack on Hawaiian and Alaskan Bases," 5/26/42, Nimitz, Gray Book, 1:506; Joseph Rochefort, interview, 10/5/69, Symonds 195.

29   Cornelius Bull, "The First Seminar," 11/6/42, 3, KP (box 6); Office of Naval Intelligence, "The Japanese Story of the Battle of Midway," *ONI Review* (May 1947).

30 Symonds 195, quoting Nimitz to EJK, 5/29/42 ("expected guests"); Cornelius Bull, "The First Seminar," 11/6/42, 4, KP (box 6).

TWENTY ★ "LIGHTS OF PERVERTED SCIENCE"

1 Morison 565 ("[Some of my]").

2 HLS, 9/23/42; Henry de Wolf Smyth, *Atomic Energy for Military Purposes* (Princeton, 1947), § 5.23; GCM, interview, 2/11/57, GCML; Leslie Groves, interview, 5/7/70, PL; FDR to Vannevar Bush, 6/15/40, FDRL (Hopkins Papers, box 132); Jungk 113–14; Jones, *Manhattan*, 31.

3 Jones, *Manhattan*, 37, quoting Vannevar Bush to FDR, 3/9/42; Hewlett and Anderson 46, 51.

4 GCM, interview, 2/11/57, GCML ("first money"); Pogue 4:12–13.

5 McCullough, *Truman*, 365, quoting HLS-HST, telephone conversation, 6/17/43.

6 GCM to FDR, 11/23/42, MP 3:449–50.

7 Capra 327.

8 Capra 327–28; Reilly 59.

TWENTY-ONE ★ MIDWAY'S GLOW

1 *Washington Times-Herald*, 6/7/42; Dina Goren, "Communication Intelligence and Freedom of the Press," *Journal of Contemporary History* (Oct. 1981); Weintraub, *Final Victory*, 155.

2 GCM to EJK, 6/7/42, MP 3:227–29; Cornelius Bull, "The First Seminar," 11/6/42, 5, KP (box 6) ("The only"); "Report of a Press Conference Held at 1700 June 7, 1942, in Admiral King's Office," n.d., KP (box 8) ("save face"); Buell 185.

3 "Report of a Press Conference Held at 1700 June 7, 1942, in Admiral King's Office," n.d., KP (box 8) ("compromises").

4 Arthur H. McCollum, interview, 9/71, 474, KP (box 10) ("traitor," "Hayman").

5 Buell 186–87.

6 Thomas B. Allen, "Midway: The Story That Never Ends," *Proceedings* (June 2007); Buell 187 ("please"); FDR to Winant, 10/6/42, FDRL (Map Room Papers, box 12) ("lies"); FDR to WSC, 10/6/42.

7   Cornelius Bull, "The First Seminar," 11/6/42, 4–5, KP (box 6); Glenn Perry to Edmond. P. Bartnett, 2/19/44, KP (box 11); EJK to Thomas Dobyns, 9/18/50, KP (box 8); EJK to Hanson Baldwin, 6/12/50, GCML (Baldwin Papers, box 8).

8   DAM to GCM, 6/8/42, in MP 3:234–35 n.1.

9   GCM to DAM, 6/10/42 and DAM to GCM, 6/11/42, MP 3:234–35 n.2 ("delicacy").

10   EJK to GCM, 6/11/42, KP (box 7) ("amphibious character").

11   Matloff, *Strategy and Command*, 295, citing GCM to EJK, 6/12/42; Buell 199.

12   Pogue 2:380, citing EJK to GCM, 6/25–26/42, and Handy to GCM, 6/24/42, and GCM to EJK, 6/26/42, and EJK to Nimitz, 6/27/42; Buell 200.

13   EJK to GCM, 6/25/42, MP 3:252–54 n.1; EJK, interview, 8/18/49, KP (box 4); EJK to T. B. Kittredge, 2/1/50, KP (box 8); Pogue 2:380, citing EJK to Nimitz, 6/27/42; Haynes 763–64 n.48.

14   GCM to DAM, 6/1/42, in Morton, *Strategy and Command*, 294; GCM to EJK, 6/26/42, MP 3:252–54.

15   EJK to GCM, 6/26/42, MP 3:252–54 n.3.

16   DAM to GCM, 6/28/42, MP 3:255–56 ("Navy retains"); EJK, interview, 7/4/50, KP (box 4).

17   DAM to GCM, 6/29/42, MP 3:255–56; GCM to EJK, 6/29/42, MP 3:254; DAM to GCM, 7/1/42 in Pogue 2:381; Hayes 146, quoting EJK to GCM, 7/2/42.

18   DAM to GCM, 7/1/42, MP 3:256 n.2; Hayes 146–47, citing GCM to EJK, 7/1/42; Buell 201; JCS, 7/2/42, Pogue 2:382; EJK to Merritt A. Edson, 9/29/49, KP (box 4); Morton, *Strategy and Command*, 302.

19   GCM to DAM, 7/3/42, MP 3:256 ("every conceivable").

20   Ghormley and DAM to COMINCH and COS, 7/8/42, in Morton, *Strategy and Command*, 305–06 ("doubts"); Buell 203.

21   Buell 203–04; Pogue 2:382, citing EJK to GCM, 7/10/42; Hayes 148.

22   Glenn Perry to E. P. Bartnett, 11/7/42, KP (box 11); Pogue 2:382; EJK to Merritt A. Edson, 9/29/49, KP (box 4); Miller, *Guadalcanal*, 32–33.

TWENTY-TWO ★ "THE BURNED CHILD DREADS FIRE"

1   WSC to FDR, 5/28/42 ("gymnast").

2   FDR to JCS, HLH, HLS and Knox, 5/6/42, EL (PSF, box 83) ("killing").

3   FDR to GCM and EJK, 5/6/42, FDRL (PSF, box 83) ("disturbed").

4   ER, *Autobiography*, 235 ("Mr. Molotov"); Samuel H. Cross, meeting notes, 5/30/42, in Sherwood 561; Alonzo Fields, 5/28/42, HSTL (Alonzo Fields Papers); Lela Mae Stiles to "Mother," 6/15/42, FDRL (Lela Mae Stiles Papers, box 16) ("Brown"); Hassett 6/1/42.

5   FDR to WSC, 3/18/42 ("guts").

6   Usher's log, 5/30/42, FDRL; Samuel H. Cross, meeting notes, 5/30/42, in Sherwood 563.

7   GCM, interview, 2/11/49, GCML ("Murmansk?"); Samuel H. Cross, meeting notes, 5/30/42, in Sherwood 564.

8   HLH, memorandum, 5/31/42, FDRL (Hopkins Papers, box 194).

9   HLH, memorandum, 5/31/42, FDRL (Hopkins Papers, box 194); Molotov to FDR, 6/12/42, FDRL (PSF, box 49); FDR to WSC, 5/31/42, FDRL (Hopkins Papers, box 194); Sherwood 569 ("anxious"); Samuel H. Cross, meeting notes, 6/1/42, in Sherwood 574–75.

10   FDR to JCS, HLH, HLS and Knox, 5/6/42, EL (PSF, box 83); Hopkins, notes, in Sherwood 577.

11   Stenographer's Diary, 6/9/42, FDRL; Mountbatten to FDR, 6/15/42, FDRL (Hopkins Papers, box 194); Sherwood 582–83; HHA, 6/2/42, LC (Arnold Papers, box 2); Pogue 2:327; Roberts 181, 202, quoting Mountbatten, interview ("nightmare").

12   GCM, interview, 10/5/56, GCML ("stand pat"); HLS, 6/20/42.

13   Usher's Log, 6/17/42, FDRL; MP 3:242–46; Matloff and Snell, 235 n.6; Pogue 2:329; HLS, 6/17/42 ("King wobbled").

14   HLS, 6/20/42 ("can't help").

15   WSC, *Hinge*, 383–84.

16   WSC to FDR, 6/20/42; Sherwood 589; Pogue 2:328–30.

17   FDR to GCM and EJK, 6/20/42, in Sherwood 588; McCrea to GCM and EJK, 7/20/42, in Pogue 332.

18   Usher's Log, 6/21/42, FDRL; Hastings Ismay, interview, 10/18/60, PL; HLS, 6/21/42; Brooke, 6/20/42.

19   John McCrea, interview, 3/19/73, 32, FDRL (John McCrea Papers).

20   Ismay 254 ("wince"); John McCrea, interview, 3/19/73, 32, FDRL (John McCrea Papers); WSC, *Hinge of Fate*, 344 ("defeat").

21 Ismay 255; WSC, *Hinge of Fate*, 344 ("chivalry"); Brooke, in Danchev, ed., 269 ("sympathy"); Ismay, interview, 10/18/60, PL.

22 Ismay, interview, 10/18/60, PL ("Mr. President").

23 Ismay 255; HLS, 6/22/42 ("friend"); Pogue 2:332.

24 CCS, memorandum, 6/19/42, and HLS to FDR, 6/21/42, FRUS, *Washington-Quebec*, 426–27, 459–60 ("spit"); HLS, 6/21/42; Sherwood 592.

25 Stenographer's Log, 6/22/42, FDRL; Ismay, memorandum, 6/21/42, in FRUS, *Washington*, 434–35; HLS, 6/22/42; M. C. Stayer, interview, 1/20/61, GCML; WSC, *Hinge of Fate*, 344; Matloff and Snell 198–200; Pogue 2:331–34, citing WBS to GCM, 7/21/42, forwarding draft notes by Ismay; R. E. Libby to EJK, 2/12/46, KP (box 4); Roberts 201–02.

26 GCM and Alan Brooke to FDR, 6/22/42, FDRL (Map Room Files, box 165); HLS, 6/28/42; Roberts 201–02.

27 WSC to FDR, 7/7/42 ("No responsible").

28 HLS, 7/10/42 ("weakening").

29 GCM, interview, 9/28/56, 11/13/56, GCML ("entertained").

30 DDE to GCM, 3/25/42, EP 1:207 ("backs").

31 JCS, minutes, 7/10/42, in Matloff and Snell 268 ("concentrating").

32 Brooke, 4/15/42 ("drain"); JCS, minutes, 7/10/42, in Matloff and Snell 268; Stoler, *Allies*, 79.

33 GCM and EJK to FDR, 7/10/42, MP 3:269–70 ("Our view"); HLS, 7/10–12/42; R. E. Libby to EJK, 2/12/46, KP (box 4); EJK comments to draft memo, 7/10/42, in Hayes 786 n.53.

34 GCM to FDR, 7/10/42, MP 3:271–72 ("to force"); GCM to EJK, 7/15/42, MP 3:276–77; EJK, comments to chapter IV, Joint Chiefs of Staff History of the War Against Japan, n.d., KP (box 8); GCM, interview, 10/5/56, GCML; *but see* Stoler, *Allies*, 79–80.

35 Matloff and Snell 270.

36 John R. Deane to EJK, 7/12/42, MP 3:272–73; HLS, 7/12/42; Stoler, *Allies*, 84.

37 GCM, EJK and HHA to FDR, 7/12/42, MP 3:272–73; HLS, 7/12/42.

38 FDR to GCM, undated draft, FDRL (McCrea Papers) ("disapproved"); FDR to GCM and EJK, telephone message, n.d., FDRL (Map Room Files, box 165); Stoler, *Allies*, 85.

39 Usher's Log, 7/15/42, GCML; HLS, 7/15/42 ("dishes"); Stoler, *Allies*, 88–89.

40 HLS, 7/15/42 ("rough day").

41 HLS, 7/15/42; Roberts 234–40.

42 FDR to GCM, undated draft, FDRL (McCrea Papers); FDR to Hopkins, GCM and EJK, 7/16/42, in Sherwood 603; FDR to GCM, EJK and Hopkins, 7/18/42, FDRL (PSF Safe File, box 3); Hopkins, notes, 7/15/42, in Sherwood 602–03 ("do not believe"); Hayes 152, quoting FDR to GCM, 7/14/42.

43 FDR, edits to GCM and EJK to FDR, 7/15/42, FDRL (PSF, box 4) ("highest").

44 Sherwood 608–09.

45 Cornelius Bull, 9/16/43, KP (box 6) ("The burned child"); Brooke, 7/21/42; Roberts 252 ("defensive"); Pogue 2:345.

46 Sherwood 606.

47 FDR to Hopkins, GCM, and EJK, n.d. and 7/24/42, FDRL (PSF, box 4); Sherwood 611; Cray 333; Pogue 2:345.

48 Pogue 2:346–47, citing JCS to GCM, 7/24/42, in EJK and GCM to FDR, 7/28/42; MP 3:277–78; Crosswell 274.

49 GCM, interview, 9/28/56, GCML ("remarkable").

50 Crosswell 275, quoting Vivian Dykes, 7/25/42 ("attack," "Dakar").

51 GCM, interview, 9/28/56, GCML ("blew hell"); GCM to BCS, 7/24/42, MP 3:278–79; Stoler, *Allies*, 89.

52 Pogue 2:347; HLS, 7/25/42; Sherwood 611; Crosswell 277, quoting Vivian Dykes, 7/30/42; Stoler, *Allies*, 89–90.

53 Pogue 2:348, citing CCS, 7/25/42 meeting minutes; Matloff, *Strategic Planning*, 281.

54 DDE, *Crusade*, 71–72; Butcher 32; Pogue 2:348.

55 DDE, 10/16/62, PL; Pogue 2:408, quoting GCM to DDE, 9/28/42.

56 "64% Wish Military to Direct the War," NYT, 8/7/42; HLS, 8/7/42.

57 HLS, 8/7/42 ("happy faculty").

58 HLS to FDR, 8/10/42, in HLS, 8/10/42.

59 HLS, 8/10/42 ("manly").

60 HLS, 8/10/42.

TWENTY-THREE ★ CHAIRMAN OF THE BOARD

1   EJK, interview, 7/4/50, KP (box 4).

2   EJK, interviews, 12/11/46 and 2/12/46, KP (box 8); Davis 237–38.

3   Leahy 128–29 ("high opinion"); FDR to Carl Vinson, 3/26/42, KP (box 9); FDR to WSC, 3/9/42; Harold Stark to Frank Knox, 5/2/42, LC (Knox Papers, box 4); EJK, interview, 4/28–30/42, KP (box 2); Whitehill 356–57; Buell 161; Roberts 88.

4   Executive Order 9096, 3/12/42, FDR-PP, 1942, 157–59; Buell 161, 500–02 (Appendix IV); EJK, interview, 8/27/50, KP (box 6); EJK, 8/27/49, KP (box 8).

5   EJK, interview, 7/30/50, KP (box 8) ("didn't have time"); Buell 236.

6   EJK, interview, 7/30/50, KP (box 8) ("Why should I"); EJK, interview, 8/26/50, KP (box 7); Lyle Wilson to Robert Sherrod, 11/6/64, KP (box 11); Buell 218.

7   John McCrea, "Secretary Knox and Ad. King Collide, Spring 1942," 5/16/77, KP (box 8) ("Admiral King . . . Good day."); COMINCH office photo, 3048 Navy Department, n.d., KP (box 13).

8   EJK, interview, 7/4/50, KP (box 8); Harold R. Stark to FDR, 3/17/39, KP (box 9); Harold R. Stark, interview, 5/26/48, KP (box 9); FDR to Knox, 12/23/40, FDR-PL 2:1088–89; Buell 226.

9   FDR to Knox, 8/31/42, KP (box 9) ("Star Spangled Banner"); Buell 228.

10 EJK to Knox, 5/15/42, FDRL (PSF Papers, box 62) ("advisable").

11 EJK, interview, 11/29/50, KP (box 7); FDR to Knox, 5/22/42, FDRL (PSF Papers, box 62) ("I wish"); FDR to Knox, 6/13/41, 6/23/41, 10/25/41, FDR-PL 2:169, 1171–72, 1226.

12 FDR to Stark, 2/17/42, FDR-PL 2:1286.

13 EJK, interview, 7/30/50, 11/29/50, KP (box 7) ("craziest"); Buell 292.

14 FDR to Lister Hill, 3/17/42, FDR-PL 2:1299; FDR, press conference, 3/17/42, FDR-PP, 1942, 165–68; HLH, 1/24/42, Sherwood 491–92; "Text of Willkie's Lincoln Day Talk Before Boston Club," NYT, 2/13/42; HLS, 3/20/42; Davis 246, 252; Larrabee 20.

15 EJK, interview, 7/4/50, KP (box 4) ("He used").

16 EJK, interview, 7/4/50, KP (box 4); Buell 340 ("I'm going").

17 EJK to M. A. Edson, 8/24/49, KP (box 4); Leahy 119, 282; EJK to J. G. Scrugham, 2/11/42, KP (box 7); EJK, interview, 8/29/48, KP (box 7) ("quite sure"); Davis 259; Forrestal, "Statement Before Senate Naval Affairs Committee," 5/1/46, KP

(box 7); EJK, "Statement Before Senate Naval Affairs Committee," 5/7/46, KP (box 7); Buell 165, 167.

18　GCM to HLH, 11/4/42, MP 3:423–24; Davis 106, quoting GCM to James Byrnes, 7/10/43 ("After Cabinet"); GCM, interview, 11/13/56; Davis 256–58; Stoler, *Allies*, 108; Buell 167.

19　GCM, interview, 11/13/56, GCML ("Superman"); GCM, interview notes, 10/5/56, PL; HLS, 2/24/42.

20　Leahy, interview, 5/24/48, FDRL (Oral History Collection, box 1); GCM, interview, 11/13/56, GCML; HLS, 2/25/42, 3/20/42; Thomas Handy, interview notes, 8/21/56, PL; Albert K. Murray, interview, 9/29/80, 122, 12/1/80, 80, KP (box 13) ("four degrees"); Davis 259; Stoler, *Allies*, 104–05; Roberts 315.

21　EJK to Thomas Dobyne, 6/22/51, KP (box 4); EJK, interviews, 7/31/49 ("fixer"), 7/4/50, 8/27/50, and 11/28/50, KP (box 8); Charlotte Pihl, interview, 3/16/74, KP (box 10).

22　Stenographer's Diary, 7/7/42, FDRL; FDR to Leahy, 7/20/42, Leahy 117, 119–20.

23　FDR, press conference, 7/21/42, FDR-PP, 1942, 301–02 ("foggiest," "order to duty"); Davis 260.

24　John McCloy, interview, 10/15/57, PL; John J. Hyland to Thomas Buell, 5/5/76, KP (box 9); Robert L. Dennison, interview, 69–70, 72, KP (box 10); Buell 251.

25　Charles J. Moore, interview, 762, 772, KP (box 11); Robert L. Dennison, interview, 69–70, KP (box 10) ("Well George," "kept falling").

26　GCM to FDR, 9/1/43, MP 4:113.

TWENTY-FOUR ★ ALONG THE WATCHTOWER

1　H. L. Goodwin, compilation of marine songs, 1943, LC (Folklore Archive); see also William Bruce Johnson, *The Pacific Campaign in World War II: Pearl Harbor to Guadalcanal* (Routledge, 2005), 242. Buell 204–05; Pogue 2:382–83.

2　Morton, *Strategy and Command*, 318; Pogue 2:383.

3　Morton, *Strategy and Command*, 322, citing EJK to GCM, 8/1/42, and GCM to Thomas Handy, n.d., attached to Handy to GCM, 8/5/42; GCM to BCS, 7/24/42, MP 3:278–79; Pogue 2:383.

4　Morton, *Strategy and Command*, 326, citing GCM to DDE, 7/20/42.

5　Pogue 2:383, citing EJK to GCM, 8/1/42, and GCM to Thomas Handy, n.d., attached to Handy to GCM, 8/5/42; Morton, *Strategy and Command*, 322; Hayes 170.

6   M. Matsuyama, interrogation, n.d., *Interrogations of Japanese Officials*, 255; David
     E. Quantock, "Disaster at Savo Island, 1942," 4/9/02, U.S. Army War College
     paper, www.ibiblio.org/hyperwar/USN/rep/Savo/Quantock/index.html.

7   William R. Smedberg III, interview, 5/8/75, 77, KP (box 13); Forrest Davis,
     "King's Way to Tokyo," *Saturday Evening Post*, 12/3/44, 101, KP (box 9).

8   George Russell, interview, 12/11/74, KP (box 11); Alan McFarland to Thomas
     Buell, 9/3/74, KP (box 2).

9   George Russell, interview, 12/11/74, KP (box 11) ("Admiral").

10   Buell 205–06.

11   "Guadalcanal Risk Explained by King," NYT, 10/22/45; George Russell, in-
     terview, 12/11/74, KP (box 11); George Russell to Thomas Buell, 11/18/74, KP
     (box 2) ("can't thank").

12   Glenn Perry to E. P. Bartnett, 11/7/42, KP (box 11) ("expected").

13   "Guadalcanal Risk Explained by King," NYT, 10/22/45; Buell 206.

14   Buell 206.

15   FDR to EJK, 8/12/42, KP (box 4) ("blowtorch"); Usher's Log, 8/11/42, FDRL
     (Mrs. Harry Hopkins); EJK, interview, 4/27/48, KP (box 7).

16   Cornelius Bull, "The First Seminar," 11/6/42, 9, KP (box 6).

17   Hanson Baldwin, interview, (interview 4), 359, KP (box 10) ("I object!" "hell of
     a way"); Charles J. Moore, interview, 781, KP (box 11).

18   Morton, *Strategy and Command*, 323, 326–27, citing EJK to GCM, 8/8/42,
     8/13/42; Matloff, *Strategic Planning*, 302; Pogue 2:384; Stoler, *Allies*, 90.

19   GCM, interview, 11/21/56, GCML; Morton, *Strategy and Command*, 325; Pogue
     2:384, citing EJK to GCM, 8/20/42 and Arnold to JCS, 8/21/42; Pogue 2:387,
     quoting McNarney to EJK, 9/3/42, and Arnold to EJK, 9/3/42; Stoler, *Allies*, 92.

20   Morton, *Strategy and Command*, 324–36, citing GCM to DAM, 8/10/42; Pogue
     2:384, citing DAM to GCM, 8/12/42, and GCM to DAM, 8/13/42; Matloff,
     Strategic Planning, 303–04, quoting Leahy to GCM, 8/22/42.

21   Pogue 2:386, citing DAM to GCM, 8/28/42, 8/30/42, and 9/6/42; Morton,
     Strategy and Command, 339–42.

22   GCM to DAM, 8/31/42, MP 3:330; DAM to GCM, 8/30/42, MP 3:330 n.1;
     Morton, *Strategy and Command*, 328, 331; HHA, 9/18/42, LC (Arnold Papers,
     box 2) ("Listen"); Pogue 2:386, citing GCM to FDR, 9/2/42, and GCM to
     DAM, 8/31/42.

23 HLS, 9/2/42 ("discouraged," "lets up"); HHA, 9/16/42, LC (Arnold Papers, box 2) ("subterfuge").

24 HHA, 9/16/42, LC (Arnold Papers, box 2); Morton, *Strategy and Command*, 334–35, citing Joint Planning Staff, memorandum, 8/28/42, and JCS, minutes, 9/15/42; Hayes 186, citing JCS, minutes, 9/15/42.

25 Pogue 2:392, quoting EJK to GCM and Leahy, 10/3/42; DAM to GCM, 10/17/42.

TWENTY-FIVE ★ GIRDLES, BEER, AND COFFEE

1 "The Pot Boils," *Time,* 10/26/42 ("chance," "mad").

2 HLS, 10/29–30/42 ("Fifth Column"); Pogue 2:394–95, citing GCM to FDR, 8/12/42; Henry I. Shaw, Jr., *First Offensive: The Marine Campaign for Guadalcanal* (Washington: Marine Corps Historical Center, 1992).

3 GCM to FDR, 10/18/42, MP 3:404–05; FDR to Queen Wilhelmina, 10/17/42, FDR-PL 2:1355; FDR to WSC, 10/19/42; Morton, *Strategy and Command*, 358, citing John McCrea to GCM, 10/14/42; FDR to JCS, 10/24/42, in Sherwood 624–25 ("every possible weapon"); Hassett 129; Stoler, *Allies*, 95–96.

4 Hassett, 11/1/42 ("withheld").

5 GCM, interview, 10/5/56, GCML ("Please"); GCM, interview, 7/25/49, GCML; DDE, *Crusade*, 195; Pogue 2:402.

6 Leahy 141; Murphy, 102 ("No one") 115–18; DDE to GCM, 10/17/42, EP 1:622–23.

7 Murphy 115–18.

8 HLS, 11/4/42; "Occupation Duties Taught to Officers," NYT, 4/12/43; Pogue 3:456.

9 Pogue 3:456.

10 GCM, interview, 2/14/57; "Says Rule by Army Will End on Peace," NYT, 12/30/42.

11 HLS, 8/27/42 ("cherubs"); Pogue 3:458–59, quoting John Hilldring, interview, 3/30/59 ("priesthood").

12 HLS, 11/4/42 ("Klein").

13 HLS, 11/6/42 ("difficulties").

14 GCM, 2/14/57, GCML ("bitter," "shocked"); Pogue 3:456, citing Julius Klein to Pogue, 2/14/70, GCML.

15 Leighton, *1940–1943*, 201–02, 602; HLS, 10/23/42.

16 Laura Bellew Hannon, Ph.D. dissertation, "The Stylish Battle: World War II and Clothing Design Restrictions in Los Angeles" (December 2012) 2–3; "The Nation," NYT, 1/23/44.

17 HLS, 11/25/42; FDR, fireside chat, 6/12/42, FDR-PP, 1942, 270–71.

18 John Kenneth Galbraith, "What I've Learned," *Esquire* (January 2002); Galbraith, *Life in Our Times*, 155.

19 "WPB Bars Rubber for Corset Field," NYT, 2/26/42; "Return is Hinted of Elastic Yarns," NYT, 7/17/43; "Fears of Milady for Girdles Fade," NYT, 7/10/42; "New Corsets End Fears of Women," NYT, 9/30/42; "Elastic in Girdles Ordered Reduced," NYT, 4/23/42; Goodwin 357, quoting Raymond Clapper, *Watching the World*, 155, and Rubber Survey Committee, Report, 9/10/42, 6, FDRL (OF 150).

20 FDR to Bernard Baruch, 7/29/42, FDR-PL 2:1334; Frank Kluckhohn, "Rubber Situation Still Confused," NYT, 6/13/42; "Campaign at Hand for Scrap Rubber," NYT, 6/7/42; FDR, radio address, 6/12/42, FDR-PP 1942, 270–71; Goodwin 357.

21 FDR, radio address, 6/12/42, FDR-PP 1942, 270–71 ("about rubber").

22 Goodwin 357–58.

23 HLS, 10/31/42; EJK to FDR, 10/23/42, KP (box 5) ("at your service").

24 EJK to FDR, 10/23/42 and FDR to EJK, KP (box 13) ("So what"); Whitehill 412.

25 Glenn Perry to Thomas Buell, 6/25/75, KP (box 11); Glenn Perry to E. P. Bartnett, 11/7/42, KP (box 11); HLS, 10/30/42.

26 EJK, testimony, *Senate Hearings: Navy Appropriation Bill for 1944*, 2 (Washington: U.S. Government, 1943) ("pussyfoot"); Buell 213.

27 HLS, 10/30/42, 12/4/43.

28 Glenn Perry to Thomas Buell, 6/25/75, KP (box 11); Robert Sherrod to Thomas Buell, 4/8/75, KP (box 11); Glenn Perry, "King's Press Circle," n.d., KP (box 11).

29 Glenn Perry to Thomas Buell, 6/25/75, KP (box 11) ("hate newspapermen"); Phelps Adams to Thomas Buell, 6/10/75, KP (box 11); Arthur B. Krock, "In the Nation," NYT, 6/28/56; Robert Sherrod to Thomas Buell, 7/11/75, KP (box 11); Perry, *Dear Bart*, 83.

30 HLS, 10/31/42; Glenn Perry to Thomas Buell, 6/25/75, KP (box 11).

31 Cornelius Bull, "The First Seminar," 11/6/42, KP (box 6); Glenn Perry, "King's Press Circle," n.d., KP (box 11); Phelps Adams to Lloyd Graybar, n.d., KP (box 11) ("expletives").

32   Phelps Adams to Lloyd Graybar, n.d., KP (box 11); Buell 244.

33   "War Songs," *Time*, 10/26/42; "OPA To Regulate Our Coffee Habit," NYT, 11/1/42; "Outstanding Events and Main Trends of the Year in Review," NYT, 12/27/42; Rosenman, *Working*, 361.

34   Goodwin 385, quoting *U.S. News*, 11/13/42.

35   Rosenman, *Working*, 349–50; Reilly 67.

TWENTY-SIX ★ THE DEVIL'S BRIDGE

1   Tully 264; Rosenman, *Working*, 363.

2   Tully 264 ("Thank God.").

3   JCS, minutes, 11/9/43, KP (box 9); Brown, *Bodyguard*, 529 ("to Gib").

4   Pogue 2:420; Jordan, *Brothers*, 100–01.

5   Blum 148–49 ("most ruthless"); Jordan 97 ("sleeping"); HLS, 11/16/42 ("starry"); WSC to FDR, 11/15–17/42.

6   FDR to WSC, 11/16/42, FDRL (Hopkins Papers, box 136) ("You may").

7   Pogue 2:420, quoting GCM to DDE, 11/20/42 ("Do not worry"); GCM to Davis, 12/12/42, MP 3:480–81; HLS, 11/17/42, 12/22/43 ("harassed"); DDE, interview, 10/15/62, PL.

8   HLS, 11/16/42; GCM, interview, 2/15/57, GCML; Parrish 332–33 ("incredibly stupid").

9   Rosenman, *Working*, 363.

10   FDR, "Statement on the Temporary Political Arrangements in North and West Africa," 11/17/42, FDR-PP, 1942, 479–80 ("expedient"); Parrish 333.

11   FDR, press conference, 11/17/42, FDR-PP, 1942, 479 ("children"); Stalin to WSC, 11/28/42, in WSC to FDR, 12/2/42, FRUS, *Casablanca*, 493.

12   HLS, 9/17/42; Lang to HLH, 12/31/42, FDRL (Hopkins Papers, box 194); Thompson and Mayo 238–40, 255.

13   Lang to HLH, 12/31/42, FDRL (Hopkins Papers, box 194); EJK, statement, "Personnel Situation in the Navy," n.d., KP (box 7); Greenfield et al. 209–11; Pogue 2:425; Roberts 297.

14   JCS to FDR, 1/8/43, W. M. Jeffers to FDR, 12/1/42, FDRL (PSF, box 6); HLS, 1/6/43; "Timetable for the Draft," *Time*, 2/1/43; Goodwin 395; Blum, *V for Victory*, 123.

15 WSC to FDR, 10/31/42, FDRL (Hopkins Papers, box 136); JCS, minutes, 11/9/42, 12/10/42, KP (box 9).

16 Sherwood 665.

TWENTY-SEVEN ★ "HOLLYWOOD AND THE BIBLE"

1 "Dirty Gertie Promises Cleanup for Unintentional Army Lyricist," *Milwaukee Journal*, 8/25/43; John W. Alexander, "Debunking Gertie from Bizerte," SEP, 12/11/43; Bull, "The First Seminar," 11/6/42, KP (box 6); Glenn Perry to E. P. Bartnett, 11/30/42, KP (box 11).

2 Bull, "The Third Seminar," 2/19/43, KP (box 6); Glenn Perry to E. P. Bartnett, 2/22/42, KP (box 11); Bull, "The Second Seminar," 11/20/42, KP (box 6); EJK to Hanson Baldwin, 6/12/50, GCML (Baldwin Papers, box 2); Romanus and Sunderland 82.

3 Glenn Perry to E. P. Bartnett, 11/30/42, KP (box 11) ("underbelly"); Cornelius Bull, "The Fourth Seminar," 4/5/43, KP (box 6); Glenn Perry to E. P. Bartnett, 2/23/43, KP (box 11).

4 Buell 248.

5 Bull, 12/18/43, KP (box 6) ("The Germans"); Cornelius Bull, 11/30/42, KP (box 6); Buell 249.

6 Glenn Perry to E. P. Bartnett, 2/22/42, KP (box 11) (Russian front); Cornelius Bull, "The Third Seminar," 2/19/43, KP (box 6) ("last analysis"); Glenn Perry to E. P. Bartnett, 2/23/43, KP (box 11) ("nine-tenths").

7 GSP, "Description of the Visit of the Commanding General and Staff to General Nogues and the Sultan of Morocco," 11/16/42, PP-LC (box 10) ("Hollywood"); Reilly 151.

8 "Log of the Trip of the President to the Casablanca Conference, 9–31 January, 1943," FRUS, *Casablanca*, 522; FDR to WSC, 1/2/43, FDRL (Map Room Papers, box 3); HHA, 1/13/43, LC; Hardesty 37.

9 Reilly 149–52; "Roosevelt-Churchill Conversation," 1/16/43, FRUS, *Casablanca*, 579; "Log of the Trip of the President to the Casablanca Conference, 9–31 January, 1943," FRUS, *Casablanca*, 522; Pogue 3:17; WSC to FDR, 1/2/43, FRUS, *Casablanca*, 503 ("Frankland").

10 FRUS, *Casablanca*, 491–500; WSC to FDR, 12/21/42, FRUS, *Casablanca*, 501; EJK, "Notes Taken From Flight Log No. 3," 8/22/50, KP (box 4) ("Every").

11 Hassett, 2/7/43 ("prima donna").

12   "Minutes of a Meeting at the White House," 1/7/43, KP (box 9); Pogue 3:16.

13   "Minutes of a Meeting at the White House," 1/7/43, KP (box 9); "Joint Chiefs of Staff Minutes of a Meeting at the White House," 1/7/43, FRUS, *Casablanca*, 510; Pogue 3:16.

14   HLS, 1/7/42 ("Marshall said"); Stoler, *Allies*, 102.

15   EJK, "My Notes Assisted from My 'Air-Log' of *1945*," n.d., 1/29/45 entry, KP (box 4);Pogue 3:17.

16   EJK, interview, 8/26/50, KP (box 7) ("cost you"); EJK, "Notes Based on Flight Log Book No. 3," n.d., KP (box 4) ("When we"); Buell 253.

17   Frank McCarthy, interview, 10/17/57, 1/18/61, PL; Reginald Winn, memoirs, 4, GCML (Small MS Collections, box 3); Pogue 3:17–18, citing McCarthy, memorandum, 1/18/61.

18   Pogue 3:20.

19   EJK, interview, 7/29/50, KP (box 7); EJK, interview, 7/30–31/49, KP (box 7); CCS, minutes, 1/14/43; Brooke, 1/14/43.

20   Pogue 3:19, quoting Ian Jacob, 1/14/43 ("shoot his line"); Roberts 317.

21   WSC to FDR, 12/30/42, FRUS, *Casablanca*, 501; Wedemeyer to Handy, 1/22/43, in Matloff 107 ("locusts"); Pogue 3:21.

22   EJK, interview, 7/4/50, KP (box 7); Parrish 386, citing Charles Donnelly, ms, USAMHI; Buell 254.

23   Pogue 3:27, quoting Jacob, 1/14/43 ("testator"); Roberts 303.

24   CCS, minutes, 1/16/43.

25   Charles Portal, 2/7/47, PL ("tied"); Brooke, interview, 1/28/47, PL; Buell 255; EJK, "Comments of Fleet Admiral E.J. King to Chapter II of J.C.S. History," n.d., KP (box 7); Pogue 3:7, quoting C.E. Lambe, interview, 2/26/47, MHI ("rock"); Brooke, 4/15/42 ("clever").

26   CCS, minutes, 1/15/43.

27   CCS, minutes, 1/15/43.

28   CCS, minutes, 1/18/43; EJK, interview, 7/7/47, PL; Brooke, 1/14–19/43; Charles Portal, interview, 2/7/47, PL; Pogue 3:30–31, 84, quoting GCM to McNair, 7/28/43; JCS, minutes, FDR-JCS meeting, 1/16/43, FRUS, *Casablanca*, 596; Matloff 23; Stoler, *Allies*, 97.

29   Roberts 337, quoting Jacob ("lovely").

30 EJK, interview, 7/3/50, KP (box 7); Buell 257; Roberts 77.

31 Buell 256, quoting Ian Jacob, 1/14/43 ("protective").

32 Brooke, 1/14/43 ("nicely lit"); GCM, interview, 11/13/56, GCML; HHA, 1/14/43, LC.

33 Charles Portal, 2/7/47, PL ("tied"); Brooke, interview, 1/28/47, PL; Buell 255; EJK, "Comments of Fleet Admiral E. J. King to Chapter II of J.C.S. History," n.d., KP (box 7); Pogue 3:7, quoting C. E. Lambe, interview, 2/26/47, MHI ("rock").

34 Roosevelt, *As He Saw It*, 82; CCS, minutes, 1/14/43; EJK, interview, 7/7/47, PL; EJK to Thomas C. Dobyns, 1/12/51, KP (box 7); EJK, interview, 11/27/50, KP (box 6); Glenn Perry to E. P. Bartnett, 4/5/43, KP (box 11); Buell 252–55; EJK to Hanson Baldwin, 6/12/50, KP (box 4).

35 CCS, minutes, 1/14/43; Pogue 3:23; EJK, "Mr. Roosevelt Versus Chiang Kai-shek," n.d., KP (box 4); Buell 258; Mark D. Sherry, *China Defensive 1942–1945* (CMH Pub. 72-38), 4; Romanus and Sutherland, *Time Runs Out*, 49.

36 Chiang Kai-shek to WSC, 12/28/42, in WSC to FDR, 1/7/43, FRUS, *Casablanca*, 518; FDR to Chiang, 1/9/43, FRUS, *Casablanca*, 516; Pogue 3:24, quoting McNarney to FDR, 1/9/43; CCS, minutes, 1/17/43.

37 Cornelius Bull, "Second Seminar," 11/30/42, KP, 14 (box 6); CCS, minutes, 1/14/43; Ruthven E. Libby, interview, 6/7/70, 227, KP (box 13); Buell 257.

38 CCS, 1/18/43 ("If and when"); Buell 256.

39 CCS, minutes, 1/18/43.

40 DDE to Thomas Handy, 1/28/43, EP 2:928 ("One of").

41 EJK, interview, 7/3/50, KP (box 7); John Hull, interview, 6/25/68, PL; Thomas Handy, interview, 3/23/59, PL; Thomas Handy, interview notes, 8/21/56, PL; Buell 257; Roberts 77.

42 CCS, minutes, 1/18/43; Brooke, 1/18/43 notes; GCM to FDR, 2/20/43, MP 3:557–58.

43 Brooke, 6/26/42 notes ("Winston"); Stoler, *Allies*, 121 ("Some Americans").

44 CCS, minutes, 1/17–18/43; CCS 155/1, 1/19/43; "Roosevelt-Churchill Dinner Meeting," 1/17/43, FRUS, *Casablanca*, 612; Matloff 36.

45 CCS, minutes, 1/18/43 and 1/19/43; CCS, "Conduct of the War in 1943," 155/1, 1/19/43.

46 CCS, minutes, 1/18/43.

47 Roosevelt, *As He Saw It*, 82–83.

48 Usher's log, 1/21/42, FDRL; Roosevelt, *As He Saw It*, 107 ("rarin'").

49 CCS, minutes, 1/15/43.

50 Brooke, 1/20/43 notes ("We were carrying").

51 Sherwood 689 ("he would not").

52 DDE to GCM, 1/17/43, EP 2:908; Jordan 115–16.

53 John McCrea, "Roosevelt-Giraud Conversation," 1/17/43, FRUS, *Casablanca*, 610 ("very splendid"); Robert Murphy to Cordell Hull, 1/13/43, FRUS, *Casablanca*, 518.

54 Anthony Eden to WSC, 1/17/43, FRUS, *Casablanca*, 814–15.

55 FDR to Cordell Hull, 1/18/43, FRUS, *Casablanca*, 816.

56 FDR, Jr., interview 1/11/79, 13 FDRL ("payday").

57 Reilly 157–58; Eden to WSC, 1/20/43, FRUS, *Casablanca*, 819; Hassett 152–53; FRUS, *Casablanca*, 821 n.2.

58 Reilly 159–60 ("cameras"); FDR, press conference, 2/12/43, FDR-PP, 1943, 84; "Roosevelt-Churchill Meeting with De Gaulle and Giraud," 1/24/43, FRUS, *Casablanca*, 725; Hopkins, notes, 1/24/43, FRUS, *Casablanca*, 839–40.

59 HLS, 9/25/41; Pogue 3:34.

60 "Joint Chiefs of Staff Minutes of a Meeting at the White House," 1/7/42, FRUS, *Casablanca*, 505; FRUS, *Casablanca*, 506; "Minutes of a Meeting at the White House," 1/7/45, KP (box 9); EJK, "Notes Based on Flight Log Book No. 3," 6/15/50, KP (box 4) ("As the war").

61 FDR, press conference, 1/24/43, FRUS, *Casablanca*, 725 n.1, 727.

62 Harriman and Abel 188–89 ("Grant"); Kimball 232 n.52.

63 Jackson 108; Pogue 3:33–34; GCM, interview, 2/11/57, GCML; Parrish 337.

TWENTY-EIGHT ★ "A WAR OF PERSONALITIES"

1 HLS, 1/30/43; Wedemeyer to Thomas T. Handy, 1/22/43, in Matloff 106 ("shirts"); Stoler, *Allies*, 109–10.

2 DAM to GCM, 10/17/42, MP 3:402–03; GCM, interview, 11/21/56; Pogue 2:393.

3 Glenn Perry to E. P. Bartnett, 4/5/43, KP (box 11) ("little use").

4 Leahy 183–84; Pogue 3:626 n.23, quoting HLS, 11/22/44 ("will not").

5   GCM, interview, 11/21/56, GCML ("vicious war"); Pogue 3:168.

6   HLS, 3/11/43; Thomas Handy, interview 7/9/70, PL; Matloff, *Strategy and Command*, 390–91; Charles J. Moore, interview, 757, KP (box 11).

7   Wedemeyer to GCM, 3/16/43, MP 3:604–07; Plan ELKTON, 2/28/43, in Morton, *Strategy and Command*, App. I; Pogue 3:171, citing minutes, Pacific Military Conference, 3/12/43, 10:30 a.m. session; Morton, *Strategy and Command*, 390–93; Thomas Handy, interview 7/9/70, PL.

8   Charles J. Moore, interview, 765, KP (box 11) ("really shocked"); Buell 283.

9   Pogue 3:173, citing minutes, JCS, 3/19/43, and Matloff, *Strategic Planning, 1943–44*, 95; Morton, *Strategy and Command*, 394–96.

10   Buell 283 ("Whether"); Matloff 95, citing JCS, minutes, 3/19/43, and George Kenney, *General Kenney Reports*, 215–16.

11   Morton, *Strategy and Command*, 390, citing EJK to GCM, 2/18/43, and GCM to EJK, 2/19/43.

12   EJK to JCS, 3/27/43, in Matloff 96; GCM to JCS, 3/26/43, in Morton, *Strategy and Command*, 397; JCS, minutes, 3/28/43; Pogue 3:175; *Fleet*, 430.

13   Leahy 183–84; Pogue 3:176.

14   JCS to DAM, Nimitz, and Halsey, 3/28/43, in Morton 641; Miller, *Cartwheel*, 15; JCS 238/5, 4/28/43, in Morton, *Strategy and Command*, App. K; Cooke and Wedemeyer to JCS, 3/16/43, in Morton, *Strategy and Command*, 393, 395, 398.

15   Morton, *Strategy and Command*, 399.

16   GCM to DDE, 2/16/43, MP 3:553–54 ("disturbed").

17   GCM to Alexander D. Surles, 5/8/43, MP 3:686 ("You can tell"); Pogue 3:191.

18   Glenn Perry to E. P. Bartnett, 2/22/43, 4/5/43, KP (box 11); Cornelius Bull, "The Fourth Seminar," 4/5/43, KP (box 6); Morison 1:405, 410–11.

19   HLS, 12/27/42 ("conservatism"), 1/25/43.

20   HLS, 3/5/43 ("Knox"), 3/24/43.

21   Memorandum of Operational Changes," n.d., in HLS, 3/10/43; HLS, 3/25/43; HLS to FDR, 3/26/43, in HLS, 3/26/43.

22   EJK to Knox, 6/23/41, KP (box 2).

23   Low 35 ("one solution").

24   HLS to Knox, n.d., in HLS, 3/31/43; HLS, 4/1/42 ("think").

25   HLS, 4/5/43, SP; Morison 574, citing Knox to HLS, 4/5/43, and EJK to Knox, 4/5/43; L. A. Thackrey to EJK, 2/4/48, KP (box 4).

26   HLS, 4/8/43, 4/17/43, 4/28/43, 5/6/43, 5/14/43, 5/26/43, 5/28/43, and 6/14/43; EJK to L. F. V. Drake, 10/24/41, KP (box 2); Cornelius Bull, 7/24/43, KP (box 6). Morison 574–76; GCM to EJK, 2/25/42, KP (box 4); C. E. Weakley to Secretary of the Navy, 1/30/48, KP (box 4).

27   Whitehill 463, quoting EJK to JCS, 5/1/43 ("control"); A. R. McCann, interview, 9/74, KP (box 10); HLS, 5/3/43.

28   HLS, 4/30/43; Low 33–35; EJK to GCM, 6/3/43, KP (box 4); EJK, interview, 7/30/50, KP (box 7); EJK, "Notes About the First Draft of Two Brief Sub-Sections . . . ," n.d., KP (box 7); Whitehill 463; Morrison 1:244-45.

29   Bundy, *Active Service*, 516 ("almost perfect").

30   EJK to GCM, 6/14/43, KP (box 4); Low 30–33; L. A. Thackrey to EJK, 2/4/48, KP (box 4), citing EJK to GCM, 6/5/43, and GCM to EJK, 6/15/43; GCM to EJK, 6/15/43, KP (box 4).

31   EJK to GCM, 6/19/43, KP (box 4); HLS to McNarney, 6/25/43, in HLS, 6/25/43; L. A. Thackrey to EJK, 2/4/48, KP (box 4).

32   HLS to McNarney, 6/25/43, in HLS, 6/25/43; GCM to EJK, 6/28/43, KP (box 4) ("If the matter"); L. A. Thackrey to EJK, 2/4/48, KP (box 4); HLS, 6/30/43.

33   HLS, 7/5/43; GCM to EJK, 7/3/43, 7/9/43, KP (box 4); EJK to GCM, 7/12/43, KP (box 4); L. A. Thackrey to EJK, 2/4/48, KP (box 4); EJK, interview, 8/26/50 ("Boy"); EJK, "Random Notes," n.d., KP (box 7); Morison 1:245–46.

34   EJK to Chief of Naval Personnel, 6/27/42, KP (box 9) ("stop-gap"); EJK, memorandum, 4/1/43, KP (box 9); Buell 357, citing EJK to E. C. Hammer, 4/10/43; Betsy Matter, interview, 12/9/74, 6, KP (box 10).

35   Buell 357.

36   William R. Smedberg III, interview, 6/9/76, 22, KP (box 11); Buell 357.

37   Howard Orem, interview, 12/12/74, KP (box 11); Alan R. McFarland to Thomas Buell, 9/3/74, KP (box 2) ("bus drivers"); Paul Pihl, interview, 3/9/74, 13, KP (box 10) ("None of us"); Eunice H. Rice, 7/2/76, KP (box 7).

38   Thomas Webb to Thomas Buell, 11/9/74, KP (box 2); William Smedberg to Thomas Buell, 2/10/75, KP (box 2); Buell 357 ("clarifying").

39   Buell 284 ("strenuous").

40 Charlotte Pihl, interview, 3/9/74, 11, KP (box 10) ("The farm").

41 Charlotte Pihl, interview, 3/9/74, KP (box 10) ("Ernest"); John A. Moreno, interview, 8/19/76, KP (box 10) ("boudoir athlete"); William Smedberg to Thomas Buell, 2/10/75, KP (box 2); Robert Sherrod to Thomas Buell, 4/8/75, KP (box 11).

42 Kirkpatrick to EJK, 5/24/44, KP (box 7); EJK to Betsy Matter, 10/18/42, 2/7/45 KP (box 7); Betsy Matter, interview, 12/9/74, 4, 8 KP (box 10).

43 Betsy Matter, interview, 12/9/74, KP (box 10) ("There would be times," "Just before").

44 Betsy Matter, interview, 12/9/74, KP (box 10) ("When King").

45 Toland 440–41; Prange 11.

46 Davis 299–300; Toland 441–42.

47 FDR to Widow Yamamoto, 5/24/43, FDRL (Grace Tully Papers, box 11).

TWENTY-NINE ★ BLIND SPOTS

1 GCM to McCloy, 2/4/43, MP 3:528–30.

2 Goodwin 411, quoting Frankfurter, *Diaries*, 168 ("had better").

3 HLS, 2/3/43.

4 FDR to WSC, 6/17/43 ("fed up"); Pogue 3:237, quoting FDR to DDE, 6/17/43.

5 Pogue 3:239, citing Wilson Brown to JCS, 7/3/43.

6 Leahy 188; CCS, minutes, 5/12/43; Cornelius Bull, "The Fourth Seminar," 4/5/43, KP (box 6) ("In China"); EJK, interview, 3/28/46, KP (box 7); Glenn Perry to E. P. Bartnett, 6/8/43, KP (box 11).

7 Stilwell, *Stilwell Papers*, 321 ("hates").

8 "Chiang," *Time*, 3/1/43 ("tears"); Leahy 184–85.

9 HLS, 2/23/43 ("beguiling"), 5/4/43; ER to Anna Boettiger, 2/21/43, in Asbell 156 ("afraid").

10 HLS, 12/15/42; Glenn Perry to E. P. Bartnett, 6/8/43, KP (box 11).

11 Glenn Perry to E. P. Bartnett, 6/8/43, KP (box 11); Glenn Perry to Keats Speed, 7/26/43, KP 9 (box 11); Leahy 188.

12 HLS to FDR, 5/3/43, in HLS 5/3/43; HLS, 12/15/42, 5/1/43.

13 FDR to GCM, 10/3/42, FDR-PL 2:1350; FDR to GCM, 3/8/43, MP 3:584–86 ("My first").

14 Leahy 189; Glenn Perry to E. P. Bartnett, 6/8/43, KP (box 11).

THIRTY ★ STICKPINS

1 FDR, press conference, 2/12/43, FDR-PP, 1943, 82 ("stick").

2 HLS, 4/28/43; WSC to FDR, 4/29/43, and FDR to WSC, 5/2/43.

3 Usher's Log, 5/3/43, FDRL; HLS, 5/3/43; Pogue 3:195.

4 Usher's Log, 5/9/43, FDRL; Matloff 125, citing Leahy 157–58; HLS, 5/10/43; JCS, "Recommended Line of Action at Coming Conference," 5/8/43, FDRL ("United States," "if the British").

5 HLS, 5/3/43, 5/10/43; JCS, "Recommended Line of Action at Coming Conference," 5/8/43, FDRL; Joint Staff Planners, "Conduct of the War in 1943–1944," 5/8/43, FDRL; Joint Strategic Survey Committee, "Current British Policy and Strategy in Relation to the U.S.," 5/8/43, FDRL; Leahy 157–58 ("pin down").

6 HLS, 5/10/43 ("counterpoise").

7 GCM to HHA, 5/13/43, MP 3:691–92 ("trivial").

8 Coffey 306–09.

9 Usher's Log, 5/11/43, FDRL; Hassett 6/18/42; Matloff 126.

10 Usher's Log, 5/12/43, FDRL; CCS, minutes, 5/12/43 ("doom," "fatal").

11 CCS, minutes, 5/12/43 ("Much"); Leahy 189.

12 BCS, "Conduct of the War in 1943–44," CCS, minutes, 5/13/43, Annex "B" ("unthinkable").

13 CCS, minutes, 5/13/43 ("deeply regret," "prolongation").

14 CCS, minutes, 5/13/43 ("1946").

15 Parrish 385, citing Charles Donnelly, ms, USAMHI.

16 Reginald Winn, memoirs, 34, GCML (Small MS Collections, box 3).

17 Frank McCarthy, "Room & Orderly Assignments," GCML (McCarthy Papers, box 27); CCS, minutes, 5/15/43; Roberts 364; GCM to HHA, 5/14/43, MP 3:693–94; Gerald Bath, "Report on British High Command's Visit to Williamsburg, Virginia, May 15–16, 1943," GCML (McCarthy Papers, box 27);

"Williamsburg, Virginia Guide Map," n.d., GCML (McCarthy Papers, box 27); Roberts 364; Frank McCarthy, interview, 9/28/58, PL.

18 Gerald Bath, "Report on British High Command's Visit to Williamsburg, Virginia, May 15–16, 1943," GCML (McCarthy Papers, box 27); Frank McCarthy, notes, n.d., GCML (McCarthy Papers, box 27); Hastings Ismay, interview, 10/18/60, PL; Frank McCarthy, interview, 9/29/58, PL; *Matthew* 6:19–34.

19 BCS staff, "Defeat of the Axis Powers in Europe," 5/17/43; JCS staff, "Defeat of the Axis Powers in Europe," 5/17/43; CCS, minutes, 5/19/43, 5/21/43; HLS, 5/19/43; Matloff 133; Parrish 350.

20 CCS, minutes, 5/12/43, 5/14/43; Leahy 186.

21 HLS, 5/20/43; CCS, minutes, 5/20/43.

22 CCS, minutes, 5/20/43, 3:30 p.m. closed session; HLS, 5/20/43.

23 CCS, minutes, 5/21/43; Glenn Perry to E. P. Bartnett, 4/5/43, KP (box 11).

24 CCS, 5/21/43; Forrest Davis, "King's Way to Tokyo," *Saturday Evening Post*, 12/3/44, 103, KP (box 9); EJK, interview, 8/29/49, KP (box 8).

25 EJK to Thomas Dobyns, 9/18/50, KP (box 8); CCS, minutes, 5/21/43; Glenn Perry to E. P. Bartnett, 4/5/43, KP (box 11).

26 CCS, "Operations in the Pacific and Far East in 1943–44," CCS 239/1, 5/21/32; CCS, minutes, 5/21/43; Buell 317.

27 CCS, minutes, 5/21/43, 5/24/43.

28 Brooke 5/24/43; Roberts 471 ("thousand pities").

29 CCS, minutes, 5/25/43; CCS 242/6, 5/25/43; CCS 244/1, 5/24/43; Leahy 193.

30 CCS 242/6, 5/25/43; Leahy 193.

31 Leahy, 5/24/43, LC (Leahy Papers, reel 4) ("permanent"); HLS, 5/27/43 ("spoiled"); Brooke, 5/24/43 ("angels").

32 Leahy 195.

33 Goodwin 439.

THIRTY-ONE ★ THE FIRST CASUALTY

1 "Good for Goebbels," *Time*, 4/26/43; U.S. Department of Defense, "Facts and Documents Concerning Polish Prisoners of War Captured by the USSR During the 1939 Campaign," 2/46, NARA (RG 319), Coatney 11; Allen i; Cienciela 215.

2   Stalin to FDR, 4/21/43, FDRL (Map Room box 8) ("certain pro-fascist"); Sumner Welles to FDR, draft telegram, 4/24/43, NARA-OL (RG 59); Brown to Hull, 4/19/43, NARA-OL.

3   "Soviet-Polish Rift Laid to Nazi Guile," NYT, 5/2/43; "Russo-Polish Quarrel Involves Other Allies," NYT, 5/2/43; "A Lesson in Maneuver," *Time*, 5/10/43; Coatney 15, citing Chicago *Tribune*, 4/28/43 and *Public Opinion Quarterly* (Summer 1943), 334.

4   "Davis Warns Nation of Axis Propaganda," NYT, 5/5/43 ("very fishy"); "Cairo to MILID-AGWAR," 5/30/43, NARA-OL (RG 319); OWI Special Guidance, 4/25/43, NARA; Henry Szymanski to George Strong, 5/29/43, NARA; FDR to Stalin, 4/36/43, FDRL (Map Room box 8) ("stupid").

5   O'Malley to Anthony Eden, 5/31/43, FDRL; WSC to FDR, 4/25/43; "Eden Seeks to Heal Breach," NYT, 5/5/43; Coatney 10–15.

6   O'Malley to Anthony Eden, 5/31/43, FDRL ("We have").

7   WSC to FDR, 8/13/43, FDRL (PSF box 49); WSC, *Hinge of Fate*, 759 ("if they are dead"); Cienciela 193–94.

8   "Report Condemns Lewis Strike View," NYT, 4/3/43; "Coal Parlays to Resume Today," NYT, 4/5/43; "Mr. Lewis's Defiance," NYT, 5/1/43; "Stoppage of Coal Hits War Output," NYT, 5/1/43.

9   "Strike Three," *Time*, 6/28/43; HLS, 2/26/43, 6/9/43.

10 FDR to HLS, 6/7/41, FDR-PL 2:1167–68; HLS, 11/7/41 ("We hate"), 6/19–22/43; "Army Will Start Seizures in Coal Zones," NYT, 5/1/43; "Expect New Test of Lewis's Power," NYT, 5/6/43.

11 HLS, 6/9/43, 6/25/43; FDR, message, 6/25/43, FDR-PP, 1943, 268–69 ("99").

12 "Strike Three," *Time*, 6/28/43; HLS, 6/25–28/43 ("The President").

13 HLS, "Memorandum re Race Riots in Detroit," 6/21/43; HLS, 6/21/43, 6/25/43; "Domestic Messes Hurt Our Standing in the World," NYT, 6/27/43 ("Our President"); "Single Food Chief Urged by Hoover," NYT, 6/4/43; "The Nation," NYT, 6/27/43.

14 HLS, 5/28/43 ("man of courage"); Ickes 3:610.

THIRTY-TWO ★ LANDINGS, LUZON, AND *LADY LEX*

1   HLS, 7/1/43; GCM, interview, 2/11/57, GCML.

2   Roberts 385, citing CCS, minutes, 7/16/43.

3   HLS, 7/1/43.

4 EJK to Nimitz, 5/27/43, Nimitz, Gray Book; Buell 336–37, 383.

5 Pogue 3:251–53, quoting EJK to GCM, 6/11/43, and EJK to GCM, 6/14/43.

6 Pogue 3:251–52, quoting JCS, memorandum, 6/14/43 ("tantamount"), and JCS, minutes, 6/15/43.

7 Pogue 3:252–53, quoting GCM to DAM, 6/14/43, and DAM to GCM, 6/20/43 ("I am").

8 Cornelius Bull, "The Fourth Seminar," 4/5/43, KP (box 6); Glenn Perry to E. P. Bartnett, 4/5/43, KP (box 11); Buell 338–39; JCS, minutes, 6/29/43, in Pogue 3:253 ("great service").

9 EJK, interview, 8/29/49, KP (box 8); HLS, 7/7/43; EJK, "Notes Based on Flight Book No. 3," 8/33/50, KP (box 4) ("The reason"); EJK to Hanson Baldwin, 6/12/50, GCML (Baldwin Papers, box 2).

10 Pogue 3:253, citing EJK to GCM, 7/22/43, and GCM to EJK, 7/29/43, and GCM, memo, 8/4/43; Matloff, *1943–1944*, 192; EJK, interview, 8/27/50, KP (box 8); Shaw et al. 26; Buell 339.

11 EJK, interview, 8/27/50, KP (box 8); Matloff, *1943–1944*, 255.

THIRTY-THREE ★ "A VITAL DIFFERENCE OF FAITH"

1 HLS to FDR, 8/4/43, in HLS, 8/4/43; HLS, 7/12/43.

2 HLS to FDR, 8/4/43, in HLS, 8/4/43 ("foul blow"); GCM, interview, 2/11/49, GCML.

3 HLS, 7/12/43 ("pledge"); HLS to FDR, 8/4/43, in HLS, 8/4/43 ("unless").

4 GCM, interview, 2/11/49, GCML.

5 HLS, 7/12/43, 7/19/43.

6 HLS to FDR, 8/4/43, in HLS, 8/4/43; HLS, 7/22/43 ("Roundhammer").

7 HLS, 7/22/43 ("eye," "start").

8 HLS, 7/23/42 ("fatigued"), 8/9/43.

9 HLS to FDR, 8/10/43, in HLS, 8/10/43 ("We cannot" "vital difference").

10 HLS to FDR, 8/10/43, in HLS, 8/10/43.

11 "Minutes of a Meeting at the White House," 8/10/43, KP (box 9); HLS, 8/10/43 ("whole hog" "astonished").

12 JCS, minutes, 8/10/43, FRUS, *Washington-Quebec*, 501 ("preponderance");
 HLS, 8/10/43.

13 HLS, 11/1/43.

THIRTY-FOUR ★ PLAINS OF ABRAHAM

1 William Rigdon, "Log of the President's Visit to Canada," FDRL (Hopkins
 Papers, box 329).

2 WSC to FDR, 7/21/43, FRUS, *Quebec 1943*, 394; Wilson Brown, draft of *Four
 Presidents*, 161, FDRL (Brown Papers); V. D. Long to EJK, 2/10/46, KP (box 4).

3 William Rigdon, "Log of the President's Visit to Canada," 3, FDRL (Hopkins
 Papers, box 329); Leahy 209; FRUS, *Quebec 1943*, v–ix; Reilly 68, 124.

4 WSC, *Grand Alliance*, 596–97 ("American mind").

5 EJK to Ruthven Libby, 8/31/43, KP (box 5) ("show-downs"); Leahy, 8/16/43, LC
 (Leahy Papers, box 2) ("undiplomatic").

6 Reginald Winn, reminiscences, 92, GCML (Small MS Collections, box 3)
 ("Come on"); Leahy 208–09; CCS, minutes, 8/15/43; Pogue 3:245, quoting JCS,
 minutes, 8/16/43; Roberts 393; Buell 373–74; Betsy Matter, interview, 12/9/74, 4,
 KP (box 10).

7 Brooke 8/15–16/43; Brooke, interview, 5/5/61, PL.

8 Brooke, 8/15–16/43.

9 CCS 319/5, "Final Report," 8/24/43; Buell 375.

10 EJK, "22 January 1943," KP (box 4); CCS, minutes, 8/19/43; EJK, "Notes Taken
 From Flight Log No. 3," 8/13/43 entry, KP (box 4).

11 Brooke, 8/19/43; EJK, interview, 7/31/49, KP (box 9) ("damn fool"); EJK,
 "Notes Taken From Flight Log No. 3," 8/13/43 entry, KP (box 4) ("one of those");
 Buell 375.

12 CCS 315/2, "Habakkuk," 8/29/43; CCS, minutes, 8/19/43; EJK, "Notes Taken
 From Flight Log No. 3," 8/13/43 entry, KP (box 4); JCS, "Minutes of Meeting,"
 11/15/43, KP (box 9) ("better leave"); EJK, interview, 7/31/49, KP (box 9); Leahy
 212–13; G. A. McLean to Thomas Buell, 8/9/74, KP (box 2); Buell 375.

13 WSC to HLH, 2/27/43, FDRL (Hopkins Papers, box 132).

14 WSC to HLH, 2/16/43, 2/27/43, 4/1/43, FDRL (Hopkins Papers, box 132); WSC
 to FDR, 2/27/43 ("breach").

15   FDR and WSC, "Agreement Relating to Atomic Energy," 8/19/43, FRUS, *Quebec-Washington 1943*, 1117.

16   Pogue 3:249; Wedemeyer, *Wedemeyer Reports!*, 245.

17   CCS 319/5, "Final Report," 8/24/43; CCS, minutes, 8/23/42; Morton, *Pacific Strategy*, 596.

18   CCS 319/5, "Final Report," 8/24/43; Brooke, 8/23/43; Leahy 210–11; Roberts 407; Buell 377.

THIRTY-FIVE ★ THE INDISPENSABLE MAN

1   HLS, 9/6/43; EJK, interview, 8/25/50, KP (box 6); EJK, interview, 7/4/50, KP (box 7).

2   Ickes, 6/21/43, Vogel 251; Thomas Handy, interview 7/9/70, PL; GCM, interview, 11/13/56, GCML; Sherwood 760–61; Pogue 3:265, citing Somervell to GCM, 9/12/43.

3   Vogel 252 ("ruthless").

4   Pogue 3:264–65, citing *Army and Navy Journal*, 9/4/43, Somervell to GCM, 9/12/43; Sherwood 759–61.

5   EJK, interview, 7/7/47, PL; EJK, "Remarks About 'Quadrant' Which I've Forgotten to Write Down," n.d., KP (box 4) ("no time"); EJK, interview, 7/4/50, KP (box 7); Glenn Perry to Edmond P. Bartnett, 12/20/43, KP (box 11); Buell 388.

6   Glenn Perry to Thomas Buell, 6/25/75, KP (box 11) ("Heaven and Earth"); "Marshall to England?" *Time*, 9/27/43.

7   "Marshall Rumor Brought to House," NYT, 9/20/43 ("American rights"); Sherwood 759–60, quoting *Army and Navy Journal*, 9/18/43 ("powerful"); Pogue 3:268, citing Newark *Star-Ledger*, 9/19/43; Washington *Times-Herald*, 9/19/43 ("technical"); HLS, 9/20/43, 9/28–29/43 ("upstairs"); "The Rumored Transfer of General Marshall," NYT, 9/20/43; Frank McCarthy, interview, 9/29/58, PL; Glenn Perry to Thomas Buell, 6/25/75, KP (box 11); Pogue 3:267; Leahy 226–27; Buell 388–89; Pogue 3:270–71, citing *Congressional Record*, 9/27/43, Detroit *Free Press*, 9/23/43, New York *Herald Tribune*, 9/22–23/43, New York *Sun*, 10/1/43, and Washington *Times-Herald*, 9/26/43 and 10/19/43; Pershing to FDR, 9/16/43, in Katherine Marshall, 156–57.

8   HLS, 9/28/43 ("We are asking"), 9/29/43 ("reputation").

9   HLS, 9/7/43 ("We both").

10   Katherine Marshall 165–66; "Two-Year Report," *Time*, 9/13/43; HLS, 9/15/43.

11 HLH to FDR, 9/16/42, 10/4/42, FDRL (Hopkins Papers, box 332); HLS, 9/15/43.

12 Sherwood 761 ("Dear Harry").

13 Katherine Marshall 175; Truman Smith, interview, 10/5/59, PL; Pogue 3:275–77, citing Thomas P. Handy to GCM, 10/5/43, GCM to Jacob Devers, 9/24/43; Sherwood 761.

14 FDR to WSC, 10/3/43 ("The newspapers"); WSC to FDR, 10/31/43; Leahy 226–28.

15 Leahy, 10/28–29/43, LC (Leahy Papers, box 2) ("stalling").

16 FDR to Pershing, 9/20/43, GCML ("He is").

THIRTY-SIX ★ "DIRTY BASEBALL"

1 GCM, interview, 7/25/49, GCML ("desire to preserve").

2 FDR, press conference, 7/23/43, FDR-PP, 1943, 309 ("I don't believe").

3 FDR to WSC, 7/25/43 ("my thought"), 7/30/43; HLS, 1/22/45; "Our Terms Stand," NYT, 7/26/43; FDR, press conference, 7/27/43, FDR-PP, 1943, 323; "President Rebukes OWI Broadcast," NYT, 7/28/43; Pogue, *Supreme Command*, 340–43; Persico, *Centurions*, 301, quoting FDR to DDE, FDRL (Map Room Papers, box 12).

4 Harold Callender, "Terms for Italy Unchanged," NYT, 7/28/43; Leahy 204, 214; Crosswell 503–04.

5 HLS, 9/8/43; Leahy 215.

6 HLS, 9/8/43.

7 Garland and Smith 544–52.

8 "Italy Will Declare War on Germany," NYT, 10/13/43; Leahy 216–17.

9 WSC to DDE, 10/7/43; HLS, 10/12/43; WSC to Wilson, 9/9/43, in WSC, *Closing the Ring*, 205; Matloff 254 ("dare"); Leahy 217; WSC to FDR, 10/7/43.

10 Leahy 217; GCM, interview, 11/13/56, GCML; Stoler, *Allies*, 121, 172.

11 FDR to WSC, 10/8/43; Pogue 3:639 n.47.

12 HLS, 10/12/43 ("childish").

13 HLS, 10/25–28/43; Pogue 3:294, quoting Deane to JCS, 10/24/43, 10/28/43.

14 HLS, 10/28/43, 11/2/43; WSC to FDR, 10/23/43 ("hypothetical"); Pogue 3:294.

15 HLS, 10/28/43 ("Jerusalem!").

16 HLS, 10/29/43 ("Balkans"), 10/31/43 ("dirty baseball").

17 FDR to WSC, 11/10/43, FDRL (Hopkins Papers, box 332); Pogue 3:273–77; Pogue, *Supreme Command*, 23–33; Roberts 428.

18 HLS, 4/8/43; Leahy 180–81; Pogue 3:176–78.

19 GCM, interview, 2/18/53, GCML ("lap").

20 HLS, 4/8/43; "Something About a Soldier," *Time*, 5/17/43; Pogue 3:177; Hunt 282 ("win-the-war"); Pogue 3:281, citing Washington *Post*, 9/22/43, and Washington *Times-Herald*, 9/22/43; Hunt 282 ("Roosevelt"); Pogue 3:283, quoting Vandenberg, 9/30/43 ("These people").

21 "Will You Love Me in November?" *Time*, 5/31/43; Cantril 630–33, 666; "M'Arthur Strong in Test by Gallup," NYT, 9/19/43; "Trial Heat Goes to the Democrats," NYT, 5/17/43; "Dewey Leads Poll for '44 Nomination," NYT, 9/18/43.

THIRTY-SEVEN ★ VINEGAR JOE AND PEANUT'S WIFE

1 Romanus 374, citing Chennault to FDR, 9/5/43; Pogue 3:284.

2 HLS, 10/18/43; GCM, interview, 10/20/56, 10/29/56, GCML ("duck soup"); Albert Wedemeyer, interview, 2/1/58, PL; Parrish 443–44; Pogue 3:285.

3 GCM to Stilwell, 10/19/43, MP 4:137; HLS, 10/18–19/43.

4 HLS, 10/19–20/43 and 11/1/43; Stilwell 232–35; Pogue 3:286, citing Somervell to GCM, 10/24/43; Romanus 376–78; GCM, interview, 2/11/49.

THIRTY-EIGHT ★ A RUSSIAN UNCLE

1 JCS, "Sextant Information Bulletin No. 8," 11/9/43, FDRL (Hopkins Papers, box 331); JCS, "Sextant Information Bulletin No. 4," 11/7/43, FDRL (Hopkins Papers, box 331); EJK, "To Cairo To Tehran and Back to Washington, D.C.," n.d., KP (box 4).

2 FRUS, *1941*, 4:752; FRUS, *Cairo-Tehran*, 3; Moran 325 ("good democrat").

3 Stalin to FDR, 9/8/43, FRUS, *Cairo*, 23–24; Stalin to WSC and FDR, 9/12/43, FRUS, *Cairo*, 25; Hull to FDR, 10/21/43, FRUS, *Cairo*, 34–35; WSC to FDR, 10/26/43; Stoler, *Allies*, 166.

4 JCS, "Sextant Information Bulletin No. 2," 11/5/43, FDRL (Hopkins Papers,

box 331); JCS, "Sextant Information Bulletin No. 4," 11/7/43, FDRL (Hopkins Papers, box 331); Brooke, 11/20/43.

5   Brooke, 10/8/43.

6   Brooke, 10/27/43, 11/1/43.

7   GCM to EJK, 11/4/43, MP 4:176–78 ("We have").

8   B. H. Bieri to EJK, 11/7/43, MP 4:176–78 n.3 ("privacy").

9   EJK, "To Cairo to Tehran and Back to Washington, D.C.," n.d., KP (box 4); Robert J. Coffee, "Recollections of Dr. Robert J. Coffey Aboard U.S.S. *Iowa* with Presidential Party," 5/77, KP (box 9); Whitehill 500.

10   Wilson Brown, draft of *Four Presidents*, 130, FDRL (Brown Papers); "Log of the President's Trip," FRUS, *Cairo*, 274, 277; Leahy 230–31.

11   Goodwin 409, quoting ER, interview, FDRL (Graff Papers); FDR to ER, 11/18/43, FDR-PL 2:1469; Leahy 232–33 ("Mr. President,").

12   EJK, "To Cairo To Tehran and Back to Washington, D.C.," n.d., KP (box 4); Leahy 231; Robert J. Coffee, "Recollections of Dr. Robert J. Coffey Aboard U.S.S. *Iowa* with Presidential Party," 5/77, KP (box 9); Whitehill 500.

13   Reilly 59.

14   "Birds-Eye View of History," (Orangeburg, SC) *Times and Democrat*, 11/30/03 (interview with Herbert Lindahl); Robert J. Coffee, "Recollections of Dr. Robert J. Coffey Aboard U.S.S. *Iowa* with Presidential Party," 5/77, KP (box 9); "Log of the President's Trip," FRUS, *Cairo*, 275 ("Welcome").

15   HHA, 11/14/43, LC (Arnold Papers, reel 2); HLH, notes, n.d., in Sherwood 768; Robert J. Coffee, "Recollections of Dr. Robert J. Coffey Aboard U.S.S. *Iowa* with Presidential Party," 5/77, KP (box 9) ("Torpedo defense!"); V. D. Long to EJK, 4/28/46, KP (box 7).

16   HLS, notes, n.d., in Sherwood 768; Leahy 232; Whitehill 501.

17   "Log of the President's Trip," FRUS, *Cairo*, 278; Leahy 232; V. D. Long to EJK, 4/28/46, KP (box 7) ("hot and straight"); Reilly 167.

18   Robert J. Coffee, "Recollections of Dr. Robert J. Coffey Aboard U.S.S. *Iowa* with Presidential Party," 5/77, KP (box 9); "Log of the President's Trip," FRUS, *Cairo*, 279–80; EJK, "To Cairo To Tehran and Back to Washington, D.C.," n.d., KP (box 4) ("It seemed"); "Bird's-Eye View of History," *Times and Democrat*, 11/30/03; Whitehill 501; HLH, notes, n.d., in Sherwood 768; Arnold 455.

19   Kit Bonner, "The Ill-Fated USS *William D. Porter*," *Retired Officer Magazine* (March 1994) ("Don't shoot!"); Parrish 380.

20 "Log of the President's Trip," FRUS, *Cairo*, 285; Robert J. Coffee, "Recollections of Dr. Robert J. Coffey Aboard U.S.S. *Iowa* with Presidential Party," 5/77, KP (box 9) ("The crew").

21 Frank McCarthy, interview, 9/29/58, PL; Reilly 161; FDR, Trip Log, 11/20/43, FDRL; EJK, "Memorandum for the Record," 4/2/51, KP (box 7); EJK, interview, 7/3/50, KP (box 7); V. D. Long to EJK, 4/28/46, KP (box 7); "Log of the President's Trip," FRUS, *Cairo*, 286.

22 "Log of the President's Trip," FRUS, *Cairo*, 287; FDR and JCS, meeting minutes, 11/15/43, FRUS, *Cairo*, 195; Leahy 235; DDE, *Crusade*, 196; Whitehill 504; EJK, "To Cairo To Tehran and Back to Washington, D.C.," n.d., KP (box 4) ("The time has come"); Butcher, 11/23/43, EL (Pre-Pres Papers, box 167) ("consoled").

23 EJK, "To Cairo To Tehran and Back to Washington, D.C.," n.d., KP (box 4); DDE, *Crusade*, 196 ("embarrassed"); Butcher 446 ("General Marshall").

24 D'Este, *Eisenhower*, 464; Cray 11 ("obey").

25 "Log of the President's Trip," FRUS, *Cairo*, 287–89; Parrish 382–83; Pogue 3:303.

26 Roberts 432.

27 Sherwood 770 ("Ike").

28 "The President's Log at Cairo," FRUS, *Cairo*, 293; V. D. Long to EJK, 4/28/46, KP (box 7); EJK, "To Cairo To Tehran and Back to Washington, D.C.," n.d., KP (box 4); Leahy 236; Reilly 164.

29 "Roosevelt-Chiang Meeting," FRUS, *Cairo*, 350; CCS, minutes, 11/24/43; Chinese Government, Chinese Summary Record of FDR-Chiang Meeting, 11/23/43, FRUS, *Cairo*, 322–25; Leahy 236; Pogue 3:304; EJK, "To Cairo To Tehran and Back to Washington, D.C.," n.d., KP (box 4) ("F.D.R."); Goodwin 474; Parrish 389.

30 GCM, interview, 10/29/56 ("sold"); WSC, *Closing the Ring*, 328–29.

31 CCS, minutes, 11/23–26/43; Brooke, 11/23/43; Ismay, interview, 12/17/46, PL.

32 Parrish 387, quoting JCS, minutes, 11/19/43, FDRL.

33 CCS, minutes, 11/23–26/43; Stilwell, 11/23/43 ("bastard"); Roberts 437 ("insults," "father"); HHA, 11/23/43, LC (Arnold Papers, reel 3); Whitehill 511; Leahy 237.

34 Brooke, 11/18/43 ("[Churchill]"); Buell 408.

35 GCM, interview, 11/13/56 ("muskets," "God forbid," "horrified").

36   Frank McCarthy, interview, 9/2/58, PL; EJK, "To Cairo To Tehran and Back to Washington, D.C.," n.d., KP (box 4); "The President's Log at Tehran," FRUS, *Cairo*, 459–60; Pogue 3:309.

37   Frank McCarthy, interview, 9/29/58, PL; Kimball, "Whodunnit?" 20–21; Danny Mander, interview, *Finest Hour* (Spring 2008); Reilly 172, 175–80, 182.

38   Charles Bohlen, "Roosevelt-Stalin Meeting," 11/28/43, FRUS, *Cairo*, 483 ("I am glad"); Reilly 179; Leahy 240.

39   Pogue 3:313, quoting GCM, interview, 11/15/56 ("rough SOB"); Brooke, 8/14/42 ("doom").

40   Leahy, 11/29/43, LC (Leahy Papers, reel 2); Brooke, 11/28/43 ("brain").

41   Bohlen, "Roosevelt-Stalin Meeting," 11/28/43, FRUS, *Cairo*, 483.

42   HHA, 11/28/43, LC (Arnold Papers, Reel 2); Sherwood 778; Bohlen, "First Plenary Meeting," 11/28/43, FRUS, *Cairo*, 487; HLS, 11/19/43; Leahy 240; "The President's Log at Tehran," FRUS, *Cairo*, 463–65; EJK, "To Cairo To Tehran and Back to Washington, D.C.," n.d., KP (box 4); WSC, *Closing the Ring*, 347; Pogue 3:309; Buell 409.

43   WSC, *Closing the Ring*, 347–48; Sherwood 778; CCS, "Minutes of First Plenary Meeting," 11/28/43, FRUS, *Cairo*, 497.

44   CCS, "Minutes of First Plenary Meeting," 11/28/43, FRUS, *Cairo*, 497.

45   Bohlen, "First Plenary Meeting," 11/28/43, FRUS, *Cairo*, 489; Sherwood 778–81.

46   Bohlen, "First Plenary Meeting," 11/28/43, FRUS, *Cairo*, 489–90.

47   Bohlen, "First Plenary Meeting," 11/28/43, FRUS, *Cairo*, 489–90.

48   Bohlen, "First Plenary Meeting," 11/28/43, FRUS, *Cairo*, 502–03.

49   EJK, "Stalin vs. Doodles," 5/14/48, KP (box 7); EJK, interview, 8/13/49, KP (box 7); EJK, interview, 7/31/49, KP (box 9); EJK, "To Cairo To Tehran and Back to Washington, D.C.," n.d., KP (box 4).

50   Bohlen, "First Plenary Meeting," 11/28/43, FRUS, *Cairo-Tehran*, 505–08.

51   HLS, 12/5/43 ("Lord"); Stoler, *Allies*, 169, quoting McCloy to HLS, 12/2/43; Pogue 3:311.

52   HHA, 11/28/43, LC (Arnold Papers, reel 2); Leahy 240; "The President's Log at Tehran," FRUS, *Cairo*, 463–65; EJK, "To Cairo To Tehran and Back to Washington, D.C.," n.d., KP (box 4); Sherwood 778; Pogue 3:309–10.

53   Glenn Perry to Edmond. P. Bartnett, 2/19/44, KP (box 11) ("pitchforked"); GCM, interview, 11/15/56 ("hose"); Pogue 3:310–11.

54   FDR, Stalin and WSC, "Military Conclusions of the Tehran Conference,"
      12/1/43, FRUS, *Cairo*, 652; EJK, "Notes Assisted from My 'Air-Log' of *1945*,"
      n.d., 2/13/45 entry, KP (box 4) ("eating"); FRUS, *Cairo*, 524–25, 533–52; Pogue
      3:311–12.

55   Perkins 80–81 ("discouraged").

56   Bohlen, memorandum, 12/43, FRUS, *Cairo*, 836–38; Perkins 83–84; Lunghi
      17–18; Stoler, *Allies*, 123–24.

57   Perkins 81–82 ("tease"); Dobbs 39–40.

58   Beria 92–93; Lunghi 17–18; CCS, "Memorandum for Information No. 165,"
      12/1/43; Leahy 247; Pogue 3:314; Stoler, *Allies*, 167–69; Roberts 445.

59   Cray 434 ("There I sat").

60   Pogue 3:312 ("Who will"); JCS, minutes, 11/29/43, FRUS, *Cairo*, 537, 541 ("Then
      nothing"); Leahy 246 ("Bolshevik"); Bohlen, "Second Plenary Meeting,"
      11/29/43, FRUS, *Cairo*, 535.

61   Sherwood 791; Bohlen, "Roosevelt-Churchill-Stalin Luncheon Meeting,"
      11/30/43, FRUS, *Cairo*, 565; CCS minutes, 11/30/43; GCM, interview, 11/15/56,
      GCML.

62   CCS, minutes, 12/3/43; CCS M-165, "Military Conclusions of the 'Eureka'
      Conference," 12/2/43; Buell 413.

63   CCS, minutes, 11/30/43; CCS 428 (Revised), "Relation of Available Resources
      to Agreed Operations," 12/15/43; GCM, interview, 2/11/49, GCML.

64   CCS, minutes, 11/29/43 ("rubber boat"); GCM, interview, 2/11/49, GCML.

65   CCS, minutes, 12/5/43; JCS, minutes, 11/28/43, KP (box 9); Buell 413; EJK,
      "Notes About 'Quadrant' Which I've Forgotten to Write Down," n.d., KP
      (box 4) ("I knew").

66   CCS, minutes, 12/4/43.

67   Leahy, 12/5/43, LC (Leahy Papers, reel 1); CCS, minutes, 12/4/43; FDR to
      Chiang, 12/5/43, in Sherwood 801; Craven and Cate 5:25–26.

68   Stilwell, *Stilwell Papers*, 251 ("mule").

69   Stilwell, 12/6/43, Hoover Library ("God-awful").

70   James Roosevelt, *My Parents*, 167 ("best politician"); Pogue 3:320.

71   Rosenman, interview, 4/9/59, FDRL; Ingersoll 48 ("stand up").

72   GCM to Sherwood, in Sherwood 801 ("He need"), 803; GCM, interview, 11/15/56, GCML; Leahy 227; Pogue 3:321.

73   Pogue 3:321.

74   GCM, interview, 11/15/56, GCML.

75   GCM, interview, 11/15/56, GCML ("I was determined," "I just repeated").

76   GCM to Sherwood, Sherwood 803 ("sleep").

77   Morison 595, citing HLS, 12/16–18/43; FDR to Stalin, 12/6/43, FRUS, *Cairo*, 819 ("Eisenhower"); GCM to DDE, 12/6/43, MP 4:197 ("memento").

THIRTY-NINE ★ RENO AND GRANITE

1   On a flat plane, the shortest distance between two points is a straight line (hypotenuse). On a curved surface, like the Earth, the shortest distance often appears as a curve on a map.

2   Crowl 13; Cornelius Bull, "The Second Seminar," 11/30/42, 13, KP (box 6); EJK to Thomas Dobyns, 9/18/50, KP (box 8); Hayes 545; Buell 418.

3   CCS 417, "Over-all Plan for the Defeat of Japan," 12/2/43; MP 4:276–80, quoting "Campaign Plan GRANITE," 1/13/44; Hays 545; Whitehill 532; Buell 417, citing EJK-Nimitz, minutes, 1/3–4/44; Matloff, *Strategic Planning*, 455.

4   Bull, 12/18/43, KP (box 6) ("wallop").

5   "United States," NYT, 12/2/43 (quoting Pacific Fleet communique #21); Shaw et al., 102 & Appendix H; Crowl and Love 156.

6   "Grim Tarawa Defense a Surprise," NYT, 12/3/43 ("It has"); "Navy's Tarawa Job Defended by Knox," NYT, 12/4/43.

7   Bull, 2/18/44, KP (box 6) ("Nobody").

8   Frank McCarthy, itinerary, 12/15/43, GCML (McCarthy Papers, box 8); Hansell 149–50; DAM, *Reminiscences*, 183 ("long and frank"); MP 4:199–200.

9   Hansell 149 ("various maps"); MP 4:276–80; DAM, *Reports of General MacArthur*, 1:168–69; Pogue 3:439–40.

10   Hansell 150; DAM, *Reminiscences*, 184; GCM to DAM, 12/23/43, 1/24/44, MP 4:201, 245–46.

11   HLS, "Message Sent to the Secretary of War by General MacArthur Through General Osborn," in HLS, 2/17/44 ("pride").

12 CCS 417/2, "Overall Plan for Defeat of Japan," 12/2/43; Morton, *Pacific Strategy*, 598, citing JSSC 535/8, 11/8/43.

13 Hayes 546–48; Craven and Cate 4:552; MP 4:276–80, citing Thomas Handy to GCM, 2/7/44; Matloff, *Strategic Planning*, 455.

14 Leahy 264; Buell 420; Hayes 549, quoting DAM to GCM, 2/2/44 ("All available"); Crowl 16, quoting JCS, minutes, 2/8/44; Hayes 546–48.

15 Hayes 549, quoting EJK to Nimitz, 2/8/44 ("I have read" "The idea"); Crowl 19.

16 MP 4:276–80, quoting EJK to GCM, 2/8/44 ("optimistic"); Leahy 264; Hayes 550; Crowl 17; Buell 240.

17 Crowl 17 and Hayes 554, citing DAM to JCS, 3/5/44.

18 MP 4:329–31 n.1, quoting Nimitz to GCM, 2/24/44; Hayes 564, citing GCM to DAM, 3/20/44; Leahy, "Oral Report by Rear Admiral Shafroth," in Leahy, 3/11/44, LC (Leahy Papers, reel 1); Hayes 564.

19 MP 4:329–31 n.1, quoting DAM to GCM, 2/27/44; Pogue 3:441–43; Hayes 564.

20 GCM to DAM, 3/9/44, MP 4:329 ("Your professional integrity"); Hayes 564–65.

21 Leahy, "Oral Report by Rear Admiral Shafroth," in Leahy, 3/11/44, LC (Leahy Papers, reel 2) ("suffering from"); Nimitz to EJK, 4/2/44, KP (box 8); Pogue 3:441, quoting DAM to GCM, 2/27/44; Halsey, *Halsey's Story*, 188–90 ("lumped"); Hopkins 217–18.

22 Dill to Allen Brooke, 1/4/44, in Bryant, *Triumph in the West*, 105–06 ("his war").

23 Stenographer's Log, 3/11/44, FDRL; Leahy, 3/11/44, LC (Leahy Papers, reel 2).

24 GCM to DAM and Nimitz, 3/12/44, MP 4:336–38; Hayes 559–60, 577–79 quoting JCS to DAM and Nimitz, 3/12/44; Crowl 17–18; MP 4:323, citing JCS 713; Smith, *Approach*, 208.

25 Nimitz to EJK, 4/2/44, KP (box 12) ("blew up").

26 EJK, interview, 12/11/48, KP (box 7); Leahy 220; Isidore Lubin, "Gains and Losses—Monthly Merchant Shipping Available to United Nations," 10/14/43, FDRL (Hopkins Papers, box 219); Buell 222, quoting EJK, speech, 10/25/43.

27 EJK, interview, 8/29/49, KP (box 4) ("Forrestal believed"); EJK, interview, 8/26/50, KP (box 7) ("hated").

28 EJK, interview, 8/27/50, KP (box 8); Buell 222.

29 Vinson to Knox, 2/11/44, KP (box 8); Buell 228–29.

30 Knox to FDR, 2/11/44, KP (box 8) ("When I").

FORTY ★ "CONSIDERABLE SOB STUFF"

1 GCM to DDE, 12/28/43, MP 4:215-16 ("You will"); W. B. Smith, interview, 5/9/47, PL; DDE to GCM, 1/22/44, EP 3:1671–72; DDE to CCS, 1/23/44, EP 3:1673–74; Harrison, 164–70.

2 Butcher 275 ("coffin"); Leighton 2:10–11.

3 DDE to CCS, 1/23/44, EP 3:1674.

4 Bull, 2/18/44, KP (box 6) ("dry-shod").

5 Leighton 2:11–12, 17–18, citing JCS, minutes, 9/15/43; Harrison 167; Dyer 208–09; Stoler, *Allies*, 95.

6 Harrison 167–68, citing WBS to DDE, 1/6/44; EJK, "Notes About 'Quadrant' Which I've Forgotten to Write Down," n.d., KP (box 4) ("details"); DDE to JCS, 1/23/44, EP 3:1675; Leighton 2:311–12, 330; Matloff, *1943–45*, 415–16.

7 WSC to FDR, 12/26/43; FDR to WSC, 12/27/43; DDE to JCS, 1/23/44, EP 3:1675; Leighton 2:311–13; Matloff, *1943–45*, 415–16.

8 MP 4:313–14, citing BCS to Joint Staff Mission, 2/19/44; Matloff, *1943–45*, 415–16; Gilbert, *A Life*, 767 ("whale"); Roberts 461.

9 Baldwin, *Battles Lost and Won*, 228 ("All roads"); Harrison 168.

10 DDE to GCM, 2/19/44, EP 3:1735; Harrison 170–71.

11 JCS, minutes, 2/21/44, KP (box 9) ("tickled to death"); EP 3:1735.

12 DDE to GCM, 2/22/44, EP 3:1745 & n.1; Leahy to DDE, EP 3:1745–46 n.3; Harrison 172.

13 Thomas Handy, interview, 3/23/59, PL; Reginald Winn, memoirs, 2, GCML (Small MS Collections, box 3); Alex Danchev, "A Very Special Relationship," unpublished paper delivered at the Tenth Annual Meeting of the Society for Historians of American Public Relations, August 2–4, 1984; Croswell 255–57; Roberts 55.

14 EJK, interview, 11/27/50, KP (box 6); Cray 489 ("destroyed").

15 HLS, 9/1/42; 2/16/44; Harvey Bundy, interview, 10/7/59, PL ("Cambridge").

16 Harvey Bundy, interview, 10/7/59, PL; GCM to Charles Seymour, 2/13/44, MP 4:295–96; HLS, 2/16/44; GCM, address, 2/16/44, MP 4:306.

17 DDE to JCS, 3/9/44, EP 3:1763; DDE to GCM, 3/3/44, EP 3:1758, 3/20/44, EP 3:1775; 3/21/44, EP 3:1776–78; Matloff, *1943–1945*, 413–14, 422.

18 Brooke, 4/8/44; Pogue 3:341, quoting GCM to DDE, 3/16/44, 3/25/44, and Thomas Handy to DDE, 3/31/44.

19   Brooke, 4/1/44.

20   Pogue 3:341, quoting Joint Staff Mission to BCS, 4/1/44 ("It is"); Andrew Cunningham, 4/10/44; MP 4:404–05, citing WSC to GCM, 4/12/44; GCM to WSC, 4/13/44, MP 4:404–05.

21   MP 4:404–05 n.2, quoting WSC to GCM, 4/16/44.

22   WSC to FDR, 4/15/44; MP 4:423–24 n.1, quoting GCM to WSC, 4/18/44.

23   W. B. Smith, interview, 5/9/47, PL; Harrison, *Cross-Channel*, 218–20.

24   Butcher 509–10; Harrison, *Cross-Channel*, 218–20.

25   WSC to FDR, 5/7/44, Lowenheim 493; DDE to WSC, 4/5/44, EP 3:1809; Tedder 529–30; Harrison, *Cross-Channel*, 222–23.

26   DDE to WSC, 5/2/44, EP 3:1843–44 ("tied"); Tedder 522; Butcher 530; Cunningham, 4/3/44 ("sob").

27   WSC to FDR, 5/7/44, Lowenheim 494 ("The War Cabinet").

28   FDR to WSC, 5/11/44, Lowenheim 496 ("regrettable").

29   WSC, *Closing the Ring*, 468 ("decisive"); Walter Smith to GCM, 5/17/44, in MP 4:454–56 n.3; John E. Fagg, "Pre-Invasion Operations," in Craven and Cate 3:155–56; Pogue, *Supreme Command*, 132.

30   FDR to WSC, 3/21/44.

31   Leahy 199.

32   JCS, "Minutes of Meeting," 11/15/43, KP (box 9); William Leahy, interview, 7/15/47, PL; HLS, 6/14/44; Pogue 3:397, quoting FDR, in WBS for DDE to GCM, 5/14/44.

33   Beschloss 321, quoting DDE, recorded conversation with Sen. Walter George, 1/7/55, EL ("egomaniac").

34   DDE to GCM, 5/16/44, EP 3:1866–67 ("Vichy gang").

35   Pogue 3:398, quoting FDR to GCM, 6/2/44 ("awfully easy").

FORTY-ONE ★ SORROWS OF WAR

1   GCM to Miss Craig's Class, 3/15/44, MP 4:345–46.

2   Leonard T. Gerow, interview, 2/24/58, PL ("hell").

3   HLS, 11/25/43; DDE, interview, 6/28/62, PL ("alibi").

4   GCM to DDE, 8/25/43, MP 4:93–94, 9/1/43, MP 4:108–09; DDE to GCM, 8/27–28/43, EP 2:1357–58, 1364; Alexander, interview, 1/10/49–1/15/49, USAMHI (OCMH Collection, Sydney Matthews Papers, box 2).

5   John Eisenhower, *General Ike*, 92–94; GCM to DDE, 12/21/43, MP 4:210–11 n.1; DDE to GCM, 12/27/43, EP 3:1623; GCM to DDE, 12/28/43, MP 4:210–11; DDE to GCM, 12/29/43, EP 3:1630–31 ("in no event"); Pogue 3:428.

6   GCM to DDE, 1/19/44, MP 4:239.

7   Devers to GCM, 1/18/44, MP 4:238–39 n.1; GCM to DDE, 1/19/44, MP 4:239.

8   "Patton Foresees British-US Rule," NYT, 4/26/44.

9   "Congress Members Displeased," NYT, 4/26/44 ("balmy"); "Stimson Denies Tie to Patton Speech," NYT, 4/26/44 ("screwy"); GCM to DDE, 4/29/44, MP 4:442–44 ("nationalities").

10   "Stimson Denies Tie to Patton Speech," NYT, 4/26/44; GCM to DDE, 4/26/44, MP 4:437–38 ("killed").

11   GCM to DDE, 4/29/44, MP 4:442–44 ("If you").

12   DDE to GCM, 4/29/44, EP 3:1837 ("weary"), EP 3:1838 ("drastic").

13   DDE to GCM, 4/30/44, EP 3:1840 ("relieve").

14   GCM to DDE, 5/1/44 ("heavy burden"); DDE to GCM, 5/5/3/44, EP 3:1846 ("convictions").

15   Ross McIntire, "Clinical Record, William Franklin Knox," FDRL (McIntire Papers, box 6); Morison 597; Goodwin 462.

16   H. H. Carroll, "Clinical Notes," 4/25–28/44, FDRL (McIntire Papers, box 6); HLS, 4/27–28/44; Leahy 279.

17   HLS, 4/28/44 ("pathetic"), 5/1/44; "The Navy Loses a Leader," *Bureau of Naval Personnel Information Bulletin* (June 1944) 3.

18   "Knox Dies in Home," NYT, 4/29/44 ("Well done"); "The Navy Loses a Leader," *Bureau of Naval Personnel Information Bulletin* (June 1944) 2–4.

19   EJK, interview, 7/3/50, KP (box 7); EJK, interview, 8/26/50, KP (box 13) ("undercutting"); Buell 427–28.

20   EJK to Rawleigh Warner, 5/12/44, KP (box 5); Leahy 280; "Roosevelt Appoints Forrestal," NYT, 5/11/44; Buell 427.

21 EJK to Rawleigh Warner, 5/12/44, KP (box 5) ("After due").

22 EJK to Rawleigh Warner, 5/12/44, KP (box 5); EJK, "Notes Taken From Flight Log No. 3, Year 1944," n.d., KP (box 4) ("Although"); EJK, interview, 7/3/50, KP (box 7) ("dirty cracks").

23 "The General's Son," *Time*, 2/14/44; FDR-PL 2:1498; Elliott Roosevelt to Fritz Lanham, FDR-PL 2:1408–09; Reilly 170; Truman Smith, interview, 10/15/59, PL; "Harry Hopkins' Youngest Son is Killed in Battle," St. Petersburg (FL.) *Times*, 2/18/44.

24 Katherine Marshall 195 ("That horseshoe").

25 Katherine Marshall 195–96.

26 Pogue 3:347, quoting Cora Thomas, interview, 3/10/61.

27 Katherine Marshall 195; Cora Thomas, interview, 3/10/61, GCML.

FORTY-TWO ★ "DR. WIN-THE-WAR"

1 FDR, press conference, 12/28/43, FDR-PP, 1943, 570–71 ("Win-the-War").

2 ER, interview, 9/3/53, FDRL (Oral Histories, box 1).

3 "Our War Casualties Rise to 158,478," NYT, 2/9/44.

4 ER, "My Day," 6/26/44 ("There is").

5 FDR, message, 10/27/43, FDR-PP 1943, 449–53 ("I believe"); FDR, fireside chat, 12/24/43, FDR-PP 1943, 561 ("We hear").

6 R. B. Pitkin, "How the G.I. Bill Was Written," *American Legion* (January 1969).

7 "A Brief History of the G.I. Bill," *Time*, 5/29/08.

8 Leahy, interview, 5/24/48, FDRL (Oral History Collection, box 1).

9 FDR, fireside chat, 12/24/43, FDR-PP, 1943, 558 ("As long").

10 Hassett, 6/15/44; Goodwin 469–70.

11 ER, "My Day," 11/10/43 ("I watched"); Goodwin 470.

12 Smith, *FDR*, 608–09, citing David S. Wyman, *The Abandonment of the Jews*, 69–70 (New York: Pantheon Books, 1984) ("the most"); Beschloss 39.

13 Stenographer's Log, 12/8/42, FDRL; "11 Allies Condemn Nazi War on Jews," NYT, 12/18/45 ("bestial"); "Allies Describe Outrages on Jews," NYT, 12/20/44; Beschloss 61–63, quoting WSC to Eden, 7/11/44 ("horrible crime").

14 GCM to FDR, 12/29/43 and FDR to GCM, 1/10/44, MP 4:218–20; FDR, remarks, 6/5/42, 6/8/43, in FDR-PP, 1942, 258, FDR-PP, 1943, 242–43; Cornelius Bull, 2/18/44, KP (box 6); Glenn Perry to Edmond P. Bartnett, 2/19/44, KP (box 11); Glenn Perry to Keats Speed, 2/19/44, KP (box 11).

15 Usher's Log, 1/16/44, FDRL; "Report to the Secretary on the Acquiescence of this Government in the Murder of the Jews," 1/13/44, at www.jewishvirtuallibrary.org/jsource/Holocaust/treasrep.html ("We should"); Smith, *FDR*, 610–11; Beschloss 43–46.

16 Executive Order 9417, 1/22/44, FDR-PP, 1944, 48–50; Smith, *FDR*, 610–11.

17 FDR, address, 3/24/44, FDR-PP, 1944, 104–05 ("the wholesale"); Beschloss 59.

18 Beschloss 63–66.

19 Martin Gilbert, "The Contemporary Case for the Feasibility of Bombing Auschwitz," in Newfeld and Berenbaum, 66–67; Tami Davis Biddle, "Allied Air Power: Objectives and Capabilities," in Newfeld and Berenbaum, 49–51; Beschloss 66–67, citing McCloy, interview, 1986, Henry Morgenthau III Private Archive, Cambridge, MA (asserting McCoy did raise the issue with FDR); Goodwin 515–16.

20 Jones, *Manhattan*, 115–16, 254–56, 319.

21 EJK, "Notes from Air Log—August—December 1945," n.d., KP (box 4); GCM, interview, 3/29/54, GCML; *U.S. News & World Report*, 11/2/59, 50–56.

22 HLS, telephone transcript with Rep. May, 11/27/43; HLS, 11/27/43, 12/10/43.

23 HLS, 2/15/44; GCM, interview, 2/11/57, GCML; Pogue 4:12–13, citing Sam Rayburn, interview, 11/6/57.

24 HLS, 2/18/44; Pogue 4:12–13, citing Sam Rayburn, interview, 11/6/57; "Entrance," *Life*, 1/12/53 ("Any jackass").

25 HLS, 8/9–10/44.

26 HLS, 6/17/43, 3/12/44.

27 HLS to HST, 3/13/44, in HLS, 3/12–13/44 ("nuisance").

28 HLS, 12/23/43, in MP 4:208 ("bombshell").

29 Perkins 309–10; Smith, *FDR*, 394.

30 HLS, 12/23/43, 12/27–31/43; Leahy 256.

31 "Disorderly Draft," *Time*, 2/1/45 ("hanged").

32 HLS, 8/2/40, 12/23/43.

33   HLS, 12/23–29/43; FDR to Byrnes, 6/10/44, FDR-PL 2:1515.

34   HLS, 12/31/43; Byrnes 201 ("sleepless with worry").

35   *Time*, 1/3/44 ("like and trust"); Pogue 3:348–51, quoting *Army and Navy Journal*, 1/15/44 ("Marshall for president"); Hassett, 6/8/44.

36   Glenn Perry to Edmond P. Bartnett, 12/20/43, KP (box 11); "Costly Ignorance," *Time*, 1/10/44 ("He banged"); HLS, 12/31/43.

37   Pogue 3:350.

38   Pogue 3:350, citing *Army and Navy Journal*, 1/15/44, 574.

39   "3 Unions End Holdout on Rises," NYT, 1/15/44 ("over").

40   Rexford G. Tugwell, *The Brains Trust* (New York: Viking Press, 1968), 434 ("oracle's cave"); DAM, *Reminiscences*, 96; Cray 119–20.

41   "Dewey Leads Poll in '44 Election," NYT, 9/18/43; Pogue 3:323, 444–45, quoting William Sexton to GCM, 11/22/43 ("demoted").

42   "Drive Links Bricker, M'Arthur for '44," NYT, 10/24/43; "Many Offers of Help," NYT, 11/6/43; Pogue 3:445.

43   Miller to DAM, 9/18/43 and DAM to Miller, 10/2/43, in "Text of Letters on MacArthur for President," NYT, 4/14/44.

44   Miller to DAM, 1/27/44 and DAM to Miller, 2/11/44, in "Text of Letters on MacArthur for President," NYT, 4/14/44.

45   Vandenberg 84–86.

46   "MacArthur Will Not Accept Nomination," NYT, 4/30/44 ("I have"); "M'Arthur's Stand Not Unexpected," NYT, 5/1/44.

FORTY-THREE ★ HALCYON PLUS FIVE

1   Goodwin 505, quoting ER to FDR, 5/2/44, FDRL (Roosevelt Family Papers Donated by the Children, box 16).

2   Jordan 360–61.

3   HLS 6/5/44; Pogue 3:386, 388, citing DDE to GCM, 5/17/44, 6/4/44 ("confirmed").

4   Usher's Log, 6/2/44, FDRL; Goodwin 506–07, citing Harriman, *Special Envoy*, 311, and quoting Anna Halsted, interview, FDRL; Hassett, 6/8/44 ("physical").

5   Tully 265 ("The Boss").

6   ER, interview, 7/13/54, FDRL (Oral Histories, box 1) ("polio").

7   Usher's Log, 6/5/44, FDRL; FDR, speech, 6/5/44, FDR-PP, 1944, 147–49 ("poised").

8   DDE to GCM, 6/6/44 3:1914–15; Hassett 6/6/44.

9   HLS, 6/6/44.

10 Goodwin 509, quoting ER, interview, FDRL; George Elsey, list of EJK-FDR appointments, 5/11/48, KP (box 9); FDR, blood pressure record, 5/1/44–6/14/44, FDRL (Anna Halsted Papers, box 66).

11 FDR, speech, 6/6/44, FDR-PP, 1944, 152 ("My fellow Americans").

12 "Let Us Pray," NYT, 6/7/44 ("The president's).

13 HHA, "Trip to England," 6/8–21/44, LC (Arnold Papers, box 2); EJK, "Flight Log Book No. 3, Year 1944," KP (box 4); Whitehill 547–53; Pogue 3:390–91; McCarthy to Pogue, 11/8/57 and 7/9/71.

14 Cunningham 605 ("saturnine"); HHA, 6/9/44, LC (Arnold Papers, box 2); Roberts 491.

15 HHA, 6/10/44, LC (Arnold Papers, box 2); CCS, minutes, 6/10–11/44; GCM, press release, 6/15/44, MP 4:483–84; Pogue 3:392; Craven and Cate, 5:99–102; WSC to HLH, 8/2/44, FDRL (Hopkins Papers, box 334); Cray 459.

16 HHA, 6/11/44, LC (Arnold Papers, box 2); Betsy Matter, interview, 12/9/74, 12, KP (box 10).

17 Pogue 3:394, quoting Frank McCarthy to Pogue, 11/8/67.

18 HHA, 6/12/44, LC (Arnold Papers, box 2) ("mad house").

19 Photograph, 9/1/39, GCML (GCM Papers, box 16); EJK, "Notes from Flight Log No. 3, Year 1944," KP (box 4); Pogue 3:395.

20 EJK, "Notes from Flight Log No. 3, Year 1944," KP (box 4); Bradley, *Soldier's Story*, 289–91; Pogue 3:396.

21 HHA, 5/12/44, LC (Arnold Papers, box 2); EJK, "Notes from Flight Log No. 3, Year 1944," KP (box 4) ("We arrived"); Whitehill 550; Brooke, 6/12/44.

22 EJK, "Notes from Flight Log No. 3, Year 1944," KP (box 4); EJK, interview, 1/31/49, KP (box 8) ("I managed it"); HHA, 6/14/44, LC (Arnold Papers, box 2) ("bed").

23 WSC to FDR, 6/9–10/44, and FDR to WSC, 6/12/44, in Pogue, *Supreme Command*, 233; HLS, 6/11/44; HLS, 6/22/44; Smith, *FDR*, 614 n. (*"Allez"*).

24 GCM, interview, 10/29/56, GCML ("Brad").

25 HLS, 6/12/44 ("danger").

26 HLS, 6/14/44.

27 HLS, 6/14/44 ("little figure"), 6/15/44.

28 HLS, 6/14/44 ("This is").

29 HLS, 6/15/44.

30 HLS, 6/13/44, 6/15/44 ("fight"), 6/20/44 ("blank").

31 HLS, "Memorandum of Conference at the State Department," 6/27/44.

32 Brooke, 6/23/44 ("We had").

33 Butcher 608; WSC to FDR, 6/21/44; DDE to Henry Wilson, 6/16/44, EP 3:1930; DDE to GCM, 6/20/44, EP 3:1938–39; DDE to CCS, 6/23/44, EP 3:1943–45.

34 WSC to FDR, 6/28/44 ("Our first").

35 FDR to WSC, 6/28/44 ("On balance").

36 GCM, interview, 10/29/56, GCML ("sideboards"); HLS, 6/28–29/44; FDR to WSC, 6/29/44 ("setback"). FDR edited this slightly before sending.

37 Brooke, 6/30/44 ("wild schemes," "damned fools").

38 WSC to FDR, 7/1/44 ("grieved").

39 WSC to FDR, 7/1/44 ("diminished").

40 WSC to Ismay, 7/6/44, in WSC, *Triumph and Tragedy*, 691–92; Pogue 3:413 ("lying down"); Gilbert, *Road to Victory*, 843 ("stupidest").

41 DDE to WSC, 8/24/44, MP 4:2095 ("prosperous").

FORTY-FOUR ★ HATFIELDS AND McCOYS

1 Hoffman 46, 77, n.4, 281–82; Crowl 79, 85–86; Shaw 267–69.

2 CCS, minutes, 6/14/44; Shaw 231–33, 636.

3 Shaw 233.

4 Craven and Cate 5:17–18, 24–25.

5 Jisaburo Ozawa, interrogation, 10/16/45, *Interrogations of Japanese Officials*, 8–9; Richard S. Pattee to Walter Whitehill, 2/25/52, enclosing "Japanese Ships Lost in Battle of Philippine Sea," n.d., KP (box 13).

6   Richard S. Pattee to Walter Whitehill, 2/25/52, enclosing "Japanese Ships Lost in Battle of Philippine Sea," n.d., KP (box 13).

7   "The Generals Smith," *Time*, 9/12/44; Oman T. Pfeiffer, interview, 251–52, KP (box 10).

8   Crowl 191–94; Shaw 301–03; Pogue 3:448.

9   Goldberg 14–15.

10 "The Generals Smith," *Time*, 9/18/44; Oman T. Pfeiffer, interview, 258–60, KP (box 10); Crowl 191–94, citing Spruance to Nimitz, 6/29/44, and Smith, *Coral and Brass*, 171.

11 Crowl 191; Pogue 3:449.

12 "The Generals Smith," *Time*, 9/18/44; Crowl 194–96, citing San Francisco *Examiner*, 7/6/44; Pogue 3:448; EJK, interview, 7/3/50, KP (box 6); Frank Mason to Thomas Buell, 6/6/77, KP (box 7).

13 Thomas Handy, interview notes, 3/23/59, PL; EJK, "Notes Based on Flight Log No. 3, Year 1944," n.d., KP (box 4); GCM to Richardson, 8/5/43, MP 4:80–81 ("circumstances").

14 Crowl 195, citing Handy to GCM, 8/4/44; Thomas Handy, interview notes, 3/23/59, PL.

15 Pogue 3:450, quoting GCM to EJK, 11/22/44 ("healthy rivalry").

16 J. R. Topper to Thomas Buell, 9/27/74, KP (box 2).

FORTY-FIVE ★ MR. CATCH

1   Smith, *Triumph*, 7.

2   Pogue 3:441–43, quoting DAM to GCM, 6/18/44.

3   HLS, 6/22/44 ("butting"); Smith, *Triumph*, 4–5.

4   GCM to DAM, 6/24/44 ("A successful").

5   MP 4:492–95, quoting JCS to DAM and Nimitz, 6/12/44, and DAM to GCM, 6/18/44; Matloff, *1943–1944*, 481, 485, citing DAM to GCM, 7/8/44.

6   HLS, 3/6/44; EJK, interviews, 5/1/48, 7/4/50, KP (box 8); EJK, interview, 5/1/48, KP (box 13); EJK and Whitehill, "Mr. Roosevelt Versus the Philippines and Formosa," n.d., KP (box 8); Whitehill, extracts of JCS meetings, n.d., 6/14/44 entry, KP (box 4); Glenn Perry to Edmond. P. Bartnett, 4/16/44, KP (box 11); Matloff, *1943–1944*, 480–81.

7 Smith, *Approach*, 452; Buell 441; EJK and Whitehill, "Mr. Roosevelt Versus the Philippines and Formosa," n.d., KP (box 8).

8 EJK, "Notes Based on Flight Log No. 3, Year 1944," n.d., 7/17/44 entry, KP (box 4) ("All hands"); EJK and Whitehill, "Mr. Roosevelt Versus the Philippines and Formosa," n.d., KP (box 8).

9 EJK, interviews, 7/4/50, 11/29/50, KP (box 8) ("I was quite careful").

10 FDR to Robert E. Hannigan, 7/11/44, FDR-PP, 1944, 197 ("If the people"); Sherwood 809.

11 "Roosevelt Nominated for Fourth Term," NYT, 7/21/44; "Commander in Chief," NYT, 7/22/44.

12 HLS, telephone conversation with James Byrnes, 3/24/44; Tully 274 ("I became").

13 Howard Bruenn, examination notes, 3/27/44–5/26/44, FDRL (Bruenn Papers); Howard Bruenn, draft article, n.d., FDRL (Bruenn Papers); Lomazo and Fettman 102–03.

14 Hassett 239–41; Leahy 258; Lomazo and Fettman, 92–95, 97–104, 109–11; Smith, *FDR*, 603–05, 617–18; Reilly 197 ("ghouls"); Goodwin 516–17.

15 McCullough, *Truman*, 300–01.

16 FDR to Samuel D. Jackson, 7/14/44, FDR-PP, 1944, 199 ("personal friend"); Byrnes, *All in One Lifetime*, 222–25 ("You are").

17 Weintraub 43 ("Nice boy"); Usher's Log, 6/13/44, 7/7/44, 7/12–13/44; Smith, *FDR*, 617–19, citing Edward J. Flynn, *You're the Boss*, 194–96; FDR to Robert Hannegan, 7/19/44, FDR-PP, 1944, 200.

18 "Log of the President's Inspection Trip to the Pacific," 3/16/45, FDRL (Tully Papers, box 7); Leahy 293; Tully 130, 276; Leahy 291–93; Whitehill 566; Beschloss 5.

19 EJK, interviews, 7/4/50, 11/29/50, KP (box 8) ("Is that?"); EJK and Whitehill, "Mr. Roosevelt Versus the Philippines and Formosa," n.d., KP (box 8); Parrish 239.

20 James Roosevelt, *Affectionately, F.D.R.*, 351 ("Jimmy," "Help"); FDR to ER, 7/21/43, FDR, *Personal Letters*, 2:1525; "Log of the President's Inspection Trip to the Pacific," 3/16/45, FDRL (Tully Papers, box 7); Goodwin 528–29.

21 FDR, address, 7/20/44, FDR-PP, 1944, 204 ("I am now").

22 Rosenman, *Working*, 453; *Life*, 7/31/44; Weintraub, *Final Victory*, 152; Goodwin 530; Parrish 432–33.

23 "Log of the President's Inspection Trip to the Pacific," 3/16/45, FDRL (Tully Papers, box 7); Howard Bruenn, draft article, n.d., FDRL (Bruenn Papers); Reilly 60.

24 GCM to Robert Richardson, 7/20/44, MP 4:528–29; GCM to DAM, 7/6/44, 7/18/44, MP 4:528–29 n.1 ("Mr. Catch," "Leahy etc."); Reilly 191–92.

25 "Log of the President's Inspection Trip to the Pacific," 3/16/45, FDRL (Tully Papers, box 7); Rosenman 456 ("longest"); Leahy 293–94; Smith, *FDR*, 620.

26 "Pacific War Talks," NYT, 8/11/44; Rosenman 457 ("Hello, Doug.").

27 William Rigdon, "Log of the President's Trip to Quebec," FRUS, *Quebec*, 282; Pogue 3:451, citing Nimitz, interview ("car").

28 Pogue 3:451–52, citing Nimitz, interview, November 1960 ("Douglas"); Larrabee 347 ("I've been too busy").

29 Cannon 6, citing S. E. Morison to Cannon, 1/22/51 (claiming FDR told him) and Robert Richardson to GCM, 8/1/44.

30 Leahy, 7/27–28/44, LC (Leahy Papers, box 2); "Log of the President's Inspection Trip to the Pacific," 3/16/45, FDRL (Tully Papers, box 7); Leahy 294.

31 Wilson Brown, draft of *Four Presidents*, 174, FDRL (Brown Papers) ("never heard"); Leahy 294 ("After so much").

32 FDR to DAM, 9/15/44, FDR-PL 2:1541; Robert Richardson to GCM, 8/1/44, MP 4:528–29 n.4; "Pacific War Talks," NYT, 8/11/44; Hopkins, *Pacific War*, 242, quoting Weldon E. Rhoades, *Flying MacArthur to Victory* (College Station: Texas A&M University Press, 1987), 260.

33 EJK, interviews, 7/30/50, 8/27/50, KP (box 8) ("let me down," "people I dislike"); Charlotte Pihl, interview, 3/16/74, 19, KP (box 10) ("back stiff"); EJK and Whitehill, "Mr. Roosevelt Versus the Philippines and Formosa," n.d., KP (box 8) ("Nimitz was").

34 EJK, "Notes Based on Flight Log No. 3, Year 1944," n.d., 7/17/44 entry, KP (box 4) ("Of course").

35 GCM to Stanley Embick, 9/1/44, MP 4:567–69 ("half"); MP 4:567–69, citing George Lincoln to Thomas Handy, 8/31/44.

36 MP 4:567–69, citing JCS, minutes, 9/1/44; Cannon, *Return*, 7; Smith, *Triumph*, 9–10.

FORTY-SIX ★ TRAMPLING OUT THE VINTAGE

1   WSC to FDR, 7/16/44 ("When," "mosquitoes").

2   CCS, plenary session minutes, 9/13/44; HHA, 9/14–16/44, LC (Arnold Papers, box 3); EJK, interview, 7/7/47, PL.

3   Charles Portal, interview, 2/7/47, PL ("Admiral King"). CCS, minutes, 6/14/44; JCS, "CCS 452/27, 'British Participation in the War Against Japan,'" 9/13/44; EJK, interview, 7/7/47, PL; Perry, *Drew Bart*, 16 ("inefficient").

4   HHA, 9/14/44, LC (Arnold Papers, box 3) ("hell"); CCS, minutes, 9/14/44 ("practicability").

5   CCS, minutes, 9/14/44; Tim Wendel and Richard Langworth, *The Definitive Wit of Winston Churchill* (PublicAffairs, 2009) 202 ("bull" comment attributed to Churchill about John Foster Dulles).

6   CCS, minutes, 9/14/44; Buell 508.

7   Brooke, 9/14/44 ("We had"); Cunningham, *Cunningham Papers* 2:160 ("King made"); Cunningham, *Sailor's Odyssey*, 612 ("I don't think").

8   William R. Smedberg III, interview, 6/9/76, KP (box 11).

9   Paul D. Stroop, interview, 11/1/69, 205–06, KP (box 10); GCM, *Biennial Report, 1943–45*, 71; Cannon, *Return*, 8, quoting Halsey to Nimitz, 9/14/44 ("no shipping").

10   Garand and Strobridge 65.

11   A. S. McDill, interview, 11/30/48, KP (box 9); EJK, 12/1/48, KP (box 9); Hayes 621, quoting GCM to DAM, 9/13/44 ("highly to be").

12   GCM, *Biennial Report, 1943–45*, 71; Cannon, *Return*, 8–9; Buell 446–447.

13   CCS, minutes, 9/15/44 ("took note"); Buell 447.

14   Pogue 3:454; Buell 447, citing EJK-Nimitz, minutes, 9/29/44.

15   EJK, interview, 7/4/50, KP (box 8); Buell 447, citing EJK-Nimitz minutes, 9/29/44; EJK, interview, 7/4/50, KP (box 8).

16   Buell 448, citing EJK-Nimitz, minutes, 9/29/44.

17   Cannon 9, citing JCS to DAM, 10/3/44; Smith, *Triumph*, 16–17.

18   FDR-PL 2:1364; FDR to Harold Smith, 11/19/42, FDR-PL 2:1371; Beschloss 51, citing Morgenthau, 8/25/44 ("You and I"); Leahy 320.

19   Beschloss 71–75.

20 Blum 342 ("castrate"), 344 ("wards"); FDR, remarks, 8/23/44, FDR-PP, 1944, 233 ("prisoners").

21 Beschloss 95–96.

22 HLS, 8/21/44, 8/23/44.

23 HLS, "Memorandum of Problems of Germany," 8/25/44; HLS, 8/25–26/44, 10/28/43 ("eat"), 12/18/43; Harvey Bundy, "Memorandum of Conference with the President," 8/25/44; Bundy, *Active Service*, 567–68.

24 Blum, 351–52; Beschloss 100–01.

25 Blum, 352–55 ("flood and dynamite"); Bundy, *Active Service*, 570, quoting Hull, memorandum, 9/5/44; Beschloss 100–04.

26 HLS, 8/26/44, 9/4/44 ("Morgenthau is").

27 Blum 360 ("This Naziism"); HLS, 9/5/44; Beschloss 106

28 McCloy, 9/5/44 ("greatly distressed").

29 HLS, 8/25/44, 9/5/44 ("I feel"); HLS, memorandum, 9/5/44, in HLS, 9/5/44.

30 Usher's Log and Stenographer's Diary, 9/6/44, FDRL; Stenographer's Log, 9/6/44, FDRL; HLS, 9/6/44; Bundy, *Active Service*, 573; Beschloss 107–08.

31 HLS, 9/7–8/44 ("a Jew like Morgenthau"); Bundy, *Active Service*, 574.

32 Usher's Log and Stenographer's Diary, 9/9/44, FDRL; HLS, 9/9/44.

33 HLS, 9/9/44; Beschloss 118; FDR to Hull, 10/13/44, FDR-PL 2:1545–46.

34 HLS, 9/11/44 ("spirit"), 9/20/44; Beschloss 117–18.

35 H. Freeman Matthews, memorandum, 9/20/44, FRUS, *Malta-Yalta*, 135–36 ("chained," "angry"); John McCloy, meeting notes, 9/20/44, in HLS, 9/20/44; Morison 609, citing Hull, *Memoirs*, 2:1614; WSC, *Triumph and Tragedy*, 156–57; Beschloss 125 ("unnatural"); Manchester and Reid 874; Blum 369 ("He turned").

36 HHA, "Memorandum of Conversation," 9/14/44, in HHA, 9/14/44, LC (Arnold Papers, box 3); John McCloy, meeting notes, 9/20/44, in HLS, 9/20/44; H. Freeman Matthews, memorandum, 9/20/44, FRUS, *Malta-Yalta*, 135–36; WSC, *Triumph and Tragedy*, 156–57; Morison 609–10; Manchester and Reid, 874.

37 HLS, 9/16–17/44 ("I have yet"); John McCloy, meeting notes, 9/20/44, in HLS, 9/20/44; HLS to FDR, 9/15/44, in HLS, 9/16–17/44.

38 "Morgenthau Plan is Criticized," NYT, 9/28/44 ("hooligans"); "Morgenthau Plan Discussed," NYT, 10/3/44 ("Carthaginian"); "Dutch Look Beyond War," NYT, 10/3/44; Beschloss 139.

39 HLS, 9/23/44, 10/3/44; "President Denies Rift on Germany," NYT, 9/30/44; "F.D. Says Cabinet 'Split' Is Essentially Untrue," *Washington Daily News*, 9/30/44, FDRL (Hopkins Papers, box 332); "President Sees Increase in Nazi Influence," *Christian Science Monitor*, 9/29/44; "Administrative Puzzles," NYT, 10/3/44, FDRL (Hopkins Papers, box 332); Blum 365.

40 "F.E.A. Asked to Plan Curbs on Germany," *New York Herald-Tribune*, 9/30/44, FDRL (Hopkins Papers, box 332) ("demons"); Pogue, *Supreme Command*, 342; Beschloss 143, quoting Morgenthau, 9/28/44 ("loudspeakers").

41 "Text of Dewey's Speech," NYT, 10/19/44 ("overnight"); "Dewey Denounces 'Secret Diplomacy,'" NYT, 10/19/44; "Text of Address by Gov. Dewey," NYT, 11/5/44 ("ten fresh divisions").

42 Beschloss 146.

43 FDR, address, 10/21/44, FDR-PP, 1944–45, 352–53 ("enslaved").

44 "A Good Example of the Value of Publicity," NYT, 9/29/44 ("high adminis-tration sources"); "F.E.A. Asked to Plan Curbs on Germany," *New York Herald Tribune*, 9/30/44, FDRL (Hopkins Papers, box 332); Morison 610, citing John McCloy, notes, 9/20/44; "Morgenthau's Absurd Plan Out," *Philadelphia In-quirer*, 9/30/44, FDRL (Hopkins Papers, box 332); "F.D.R. Outlines Plan to Keep Reich Helpless," *New York Evening Post*, 9/29/44, FDRL (Hopkins Papers, box 332); "Morgenthau Plan Shelved," NYT, 9/28/44; McCloy, memorandum, 12/19/44; HLS, 12/19/44; Beschloss 169–70.

45 HLS, 10/3/44 ("He was frankly"); John McCloy, memorandum, 10/3/44 ("boner"); FDR and WSC, memorandum, 9/15/44, in HLS, 9/20/44 ("pastoral").

46 HLS, 10/27/44.

47 HLS, 11/4/44 ("punished"), 2/11/41 ("youngsters").

FORTY-SEVEN ★ OLD WOUNDS

1 Weintraub, *Final Victory*, 64–65.

2 HLS, 11/10/44.

3 Sherwood 819–20; Weintraub, *Final Victory*, 151 ("How can you").

4 GCM to Thomas Dewey, 9/27/44, MP 4:607–11; "Countercharge," *Time*, 10/2/44; Weintraub, *Final Victory*, 155.

5 "Remember Pearl Harbor," *Time*, 9/4/44; Pogue 3:470, citing Speech of Rep. Harness, *Cong. Record*, 9/11/44, 7648–51; Weintraub, *Final Victory*, 155, quoting Forrestal to FDR, 9/14/44 ("Information").

6   GCM, testimony, 12/7/45, PHH 3:1139; Pogue 3:470; GCM to HST, 9/22/45, MP 5:309–10.

7   GCM, testimony, 12/7/45, PHH 3:1139 ("pointed"); Pogue 3:471.

8   GCM to EJK, 9/25/44, MP 4:604 ("This letter," "dynamite").

9   Pogue 3:471.

10 "Dewey Declares Roosevelt Seeks to Sow Disunity," NYT, 9/27/44; Weintraub, *Final Victory*, 159.

11 GCM to Thomas Dewey, 9/25/44, MP 4:605 ("dear Governor"); GCM, testimony, 12/7/45, PHH 3:1133.

12 GCM to HST, 9/22/45, MP 5:309–10; MP 4:605 n.2, quoting Carter Clarke, "Statement," NARA ("Marshall"); Weintraub, *Final Victory*, 161 ("knew what was").

13 GCM to Dewey, 9/27/44, PHH 3:1132 ("my word"); GCM, testimony, 12/7/45, PHH 3:1135.

14 GCM, testimony, 12/7/45, PHH 3:1135–36.

15 GCM to Dewey, 9/27/44, PHH 3:1133 ("tragic").

16 MP 4:607–11 n.10, quoting Clarke, statement, n.d., NARA ("Well, colonel").

17 GCM to HST, 9/22/45, MP 5:309–10; GCM, testimony, 12/7/45, PHH 3:1135–36; "Editor Says Dewey Guarded War Data," NYT, 9/21/45; "Dewey Silent on Japanese Code," NYT, 9/22/45.

18 Buell 330 ("didn't like Stimson").

19 "Editor Says Dewey Guarded War Data," NYT, 9/21/45; "Dewey Silent on Japanese Code," NYT, 9/22/45.

20 "Dec. 7 to Nov. 7," *Time*, 10/30/44 ("damage"); Buell 329, 331 ("down the river").

21 "Dec. 7 to Nov. 7," *Time*, 10/30/44 ("election damage," "truth").

22 Leahy 323; Weintraub, *Final Victory*, 255.

23 Stenographer's Log, 11/4/44, FDRL; Leahy 323–25; "Roosevelt Strikes at Foes," NYT, 11/5/44 ("I can't talk").

24 Sherwood 820–21 ("damn"); Weintraub, *Final Victory*, 263; Goodwin 547–49.

25 "Torchlight Parade Honors President," NYT, 11/8/44; Hassett 292–92; Leahy 325–26.

26 CQ, *Presidential Elections*, 62, 114.

27 Hassett 294 ("I still think"); Goodwin 552–53.

FORTY-EIGHT ★ VOLTAIRE'S BATTALIONS

1 Hirshson 552, quoting Patton to James Doolittle, 10/19/44.

2 Matloff, *1943–1945*, 519.

3 HLS, 10/2/42, 10/12/42; Bundy, *Active Service*, 475; Matloff, *1943–1945*, 114; EJK to FDR, 2/12/42, FDRL (PSF, box 7); Pogue 3:354.

4 Crosswell 792–83.

5 HLS, 4/12/44; HLS to GCM, 5/10/44, in HLS 5/11/44; Morison 603; JCS to FDR, 11/18/44, FDRL (PSF, box 6); Greenfield, *AGF*, 169; Matloff, *1943–1945*, 114–15, citing Leahy to FDR, 9/30/42; Crosswell 787; Stoler, *Allies*, 98.

6 Leighton 2:297; Greenfield, *AGF*, 169; Pogue 3:354–55, 490; GCM, *Biennial Report, 1943–1945*, 103–05.

7 GCM to William Sexton, 11/22/43, MP 4:190–91 ("It is ridiculous").

8 HLS, "Notes After Cabinet Meeting," 2/18/44, HLS, 2/18/44; HLS 3/17/44; Matloff, *1943–1945*, 115–16; Lerwill 286; Crosswell 795; GCM, interview notes, 10/5/56, PL.

9 HLS, 4/13/44; GCM to FDR, 2/5/44, MP 4:266; Pogue 3:354; Matloff, *1943–1945*, 116–17, 388–89; Lerwill 286.

10 HLS, "Notes After Cabinet Meeting," 2/18/44, in HLS, 2/18/44; HLS, notes, "Marshall," n.d. (c. 12/44), SP; Palmer, Wiley and Keast 472–75.

11 HLS to GCM, 5/10/44, in HLS, 5/11/44; Crosswell 789–90.

FORTY-NINE ★ COUNTING STARS

1 21 Army Group, "General Situation," 12/16/44, USAMHI (Hansen Papers, box 2); Bradley, interview, 11/6/46, USAMHI (OCMH Collection, Pogue Interviews); Forrest Pogue, memorandum, "Interview of Mr. Pogue with General Bradley," 11/6/46, USAMHI (Hansen Papers, box 8); Edwin Sibert to Hanson W. Baldwin, 1/2/47, USAMHI (Hansen Papers, box 1); Kenneth Strong, interview, 12/12/46, USAMHI (OCMH Collection, "WWII—Supreme Command—1A2(b)—Forrest Pogue"); 12th Army Group, After-Action Report, 3:25, NARA (RG 331, entry 200A, box 266); Chet Hansen, 12/17/44, USAMHI (Hansen Papers, box 5); Everett Hughes, 12/16/44, LC (Hughes Papers, box 2); DDE to CCS, 1/20/45, EP 4:2447; McCloy, 12/18/44 ("A complete surprise").

2 GCM to DDE, 12/22/44, MP 4:707–08; HLS, 12/18–19/44.

3 HLS, 12/18–19/44.

4   HLS, 12/20/44 ("gamble"); McCloy, 12/20/44 ("disquieting").

5   Usher's Log, 12/20/44, FDRL; Leahy 332–33 ("For a few"); GCM to DDE, 12/22/44, MP 4:707–08; Suckley, 11/28/44–12/31/45; Hassett 300–10; GCM, interview, 11/14/56, GCML ("Roosevelt didn't"); HLS, 12/31/44 ("The anxiety").

6   MP 4:624–26; Brooke, 11/24/44; Montgomery, *Memoirs*, 255–72.

7   GCM, interview, 11/15/56, GCML ("I came"); Pogue, *Supreme Command*, 314–15; Pogue 3:481, citing WSC to FDR, 12/6/44, FDR to WSC, 12/9/44; GCM to DDE, 12/30/44, MP 4:720–21; Croswell 785; GCM to DDE, 12/30/44, MP 4:720–21 ("My feeling").

8   FDR to WSC, 12/6/44 ("For the time being"); WSC, *Triumph*, 231–33.

9   HLS, 12/31/44; FDR, address, 1/6/45, FDR-PP, 1944–45, 484 ("confidence").

10 EJK to Leahy and GCM, 11/17/42, KP (box 7) ("We should"); Leahy 260; Pogue 3:365, citing Miller G. White to William T. Sexton, 1/13/44.

11 Knox to EJK, 11/20/42, KP (box 7); EJK to GCM, 1/21/44, KP (box 7) ("Captain Admiral"); Knox to EJK, 1/44, KP (box 7) ("Personally I don't care").

12 HLS, 2/16–18/43, 2/1/44, 9/13/44; Leahy 260; Pogue 3:365, citing Miller G. White to William T. Sexton, 1/13/44; Buell 364, quoting EJK to GCM, 1/21/44.

13 HLS, telephone conversation with Rep. May, 1/13/44, 9/19/44; GCM, interview, 2/14/57 ("I didn't want"); HLS, 2/1/44; HLS to FDR, 2/16/43, HLS, 2/16/43; Pogue 3:365.

14 HLS, 2/1/44; HLS to Miller White, 2/1/44, HLS, 2/1/44.

15 Pogue 3:366–67, citing *Army and Navy Journal*, 3/4/44; New York *Herald-Tribune*, 4/4/44 ("Marshall Asks," "Friends of").

16 HLS, 9/14/44, 11/28/44 ("hellbent"), SP; Pogue 3:483.

17 HLS, 9/13/44.

18 HLS, 11/28/44, 12/12/44; "Senate Votes New Five-Star Rank," NYT, 12/16/44; Leahy 331; EJK, interview, 7/29–31/50, KP (box 7); Pogue 3:484; Buell 365.

19 HLS, 12/15/44.

FIFTY ★ THE TSARINA'S BEDROOM

1   WSC to FDR, 1/1/45, FRUS, *Malta-Yalta*, 26 ("Malta to Yalta"); FDR to Stalin, 7/17/44, 7/27/44, FRUS, *Malta-Yalta*, 3, 405; Stalin to FDR, 7/22/44, 10/18/44, FRUS, *Malta-Yalta*, 4, 9; Sherwood 843–45; Harriman to FDR, 9/24/44, FRUS, *Malta-Yalta*, 5; Leahy 243.

2   FDR to WSC, 11/2/44, FRUS, *Malta-Yalta*, 12–13; WSC to FDR, 11/5/44, FRUS, *Malta-Yalta*, 13–14; "Log of the President's Trip," FRUS, *Malta-Yalta*, 460 ("worse place").

3   FDR to WSC, 12/23/44, 1/3/45, FRUS, *Malta-Yalta*, 21, 26 ("descendants"); WSC to FDR, 12/31/44, FRUS, *Malta-Yalta*, 24 ("argonaut").

4   "Log of the President's Trip," FRUS, *Malta-Yalta*, 459; Leahy, 2/2/45, LC (Leahy Papers, reel 1); Anna Boettiger, 2/2/45, FDRL (Anna Halsted Papers, box 84); Leahy 342–43.

5   WSC, *Triumph and Tragedy*, 343 ("I watched").

6   FDR to WSC, 1/9/45, FRUS, *Malta-Yalta*, 32.

7   CCS, minutes, 1/30/45; EJK, "Notes Assisted from My 'Air-Log' of *1945*," n.d., 1/29/45 entry, KP (box 4); EJK, interview, 8/27/50, KP (box 9).

8   W. B. Smith, interview, 5/9/47, PL; Brooke, interview, 1/28/47, PL.

9   CCS, minutes, 1/30/45; EJK, "Notes Assisted from My 'Air-Log' of *1945*," n.d., 1/29/45 entry, KP (box 4).

10   EJK, "Notes Assisted from My 'Air-Log' of *1945*," n.d., 1/29/45 entry, KP (box 4); EJK, interview, 8/27/50, KP (box 4).

11   W. B. Smith, interview, 5/8/47, PL; W. B. Smith, interview, 7/29/58, PL; Crosswell 862, quoting WBS to Samuel E. Morison, 4/1/57, GCML ("Please leave"); GCM to DDE, 1/11/45, MP 5:27–28; DDE to GCM, 1/12/45, EP 4:2422; Leahy, 2/2/45, LC (Leahy Papers, reel 1); Brooke, 2/1/45; Danchev 653; Butcher 752; Cray 499–500; DDE, memorandum, 1/28/45, EP 4:2460.

12   Smith to Handy, 2/9/45, MP 5:41–46 ("a bitter"); CCS, minutes, 1/30/45; DDE to CCS, 1/20/45, EP 4:2450–54; DDE to WBS, 1/31/45, EP 4:2463; Brooke, 2/1/45; Butcher 753; W. B. Smith, interview, 5/8/47, PL; GCM, interview, 11/19/56, GCML.

13   EJK, "Air Log—1945—Continued. April–July," n.d., KP (box 4) ("Of course"); CCS, minutes, 2/1/45; GCM, interview, 11/19/56, GCML ("wrong foot").

14   Brooke, 2/1/45.

15   W. B. Smith, interview, 5/8/47, PL; EJK, interview, 7/7/47, PL; CCS 776/3, "Report to the President and Prime Minister," 2/2/45; Ismay 385 ("One can read"); Thomas Handy, interview notes, 8/21/56, PL ("If you will").

16   HLS, 2/17/45; GCM, interview, 11/19/56, GCML ("terrible"); Pogue 3:516; Brooke, diary notes, in Danchev 653 ("I did not").

17 Anna Boettiger, 2/2/45, FDRL (Anna Halsted Papers, box 84); Leahy, 2/2/45, LC (Leahy Papers, reel 2); Reilly 211; FRUS, *Malta-Yalta*, 548; EJK, "Notes Assisted from My 'Air-Log' of *1945*," n.d., 1/29/45 entry, KP (box 4).

18 EJK, "Notes Assisted from My 'Air-Log' of *1945*," n.d., 1/29/45 entry, KP (box 4); Leahy 345; Sherwood 849.

19 FRUS, *Malta-Yalta*, 549; EJK, "Notes Assisted from My 'Air-Log' of *1945*," n.d., 1/29/45 entry, KP (box 4); EJK, "Air Log—1945—Continued. April–July," n.d., 7/12/45 entry, KP (box 4); Crosswell 864.

20 GCM to Stilwell, 10/18/44, MP 4:631; Brooke 2/3/45.

21 "Log of the President's Trip," FRUS, *Malta-Yalta*, 549; Anna Boettiger, 2/3/45, FDRL (Anna Halsted Papers, box 84); Leahy, 2/3/45, LC (Leahy Papers, reel 1); Reilly 161, 209; Leahy 347; Paul D. Stroop, interview, 11/1/69, 195, KP (box 10).

22 Anna Boettiger, 2/3/45, FDRL (Halsted Papers, box 84); Brooke, 2/2/45; Danchev 653; Leahy 347; Pogue 3:518.

23 "Log of the President's Trip," 2/3/45, FRUS, *Malta-Yalta*, 549; Anna Boettiger, 2/3/45, FDRL (Halsted Papers, box 84); Leahy 347–48.

24 Anna Boettiger, 2/2/45, FDRL (Halsted Papers, box 84); "Log of the President's Trip," FRUS, *Malta-Yalta*, 550; EJK to Betsy Matter, 2/7/45, KP (box 7); Leahy 348; EJK, "Notes Assisted from My 'Air-Log' of *1945*," n.d., 1/29/45 entry, KP (box 4) ("Russian drivers"); "Log of the President's Trip," FRUS, *Malta-Yalta*, 550.

25 Leahy, 2/3/45, LC (Leahy Papers, reel 1); "Log of the President's Trip," FRUS, *Malta-Yalta*, 552; Leahy 348; "General Information Bulletin," n.d., FDRL (Halsted Papers, box 84).

26 "Log of the President's Trip," FRUS, *Malta-Yalta*, 550; Leahy, 2/3/45, LC (Leahy Papers, reel 2); Leahy 348–49; Pogue 3:521, quoting John E. Hull to Pogue, 12/22/69.

27 "General Information Bulletin," n.d., FDRL (Halsted Papers, box 84); Anna Boettiger, 2/2/45, FDRL (Halsted Papers, box 84); Paul D. Stroop, interview, 11/1/69, 195, KP (box 10); Kern, n.p.; Beria 103–04; Andrew and Mitrokhin 133; Costigliola 240, citing Richard Pack, Jr., "Suggestions," n.d., FDRL (Rigdon Papers).

28 Brooke, 2/4/45; Danchev 655; EJK, "Notes Assisted from My 'Air-Log' of *1945*," n.d., 2/3/45 entry, KP (box 4); EJK, interview, 8/14/49, KP (box 6); Anna Boettiger, 2/2/45, FDRL (Halsted Papers, box 84); Whitehill 588; Leahy 349 ("Salty"); James Powder, interview, 10/19/59, GCML ("What the hell"); "Marshall's Big Sergeant," *St. Petersburg Times* 10/18/59.

29   Frank McCarthy, interview, 10/17/57, PL; "General Information Bulletin," n.d., FDRL (Halsted Papers, box 84) ("Supplementary"); Paul D. Stroop, interview, 11/1/69, 196, KP (box 10); Costigliola 241 ("excepting only"); Buell 457.

30   "Log of the President's Trip," FRUS, *Malta-Yalta*, 552.

31   Bohlen, FDR-Stalin minutes, 2/4/45, FRUS, *Malta-Yalta*, 570.

32   Bohlen, FDR-Stalin minutes, 2/4/45, FRUS, *Malta-Yalta*, 570–71; Leahy 349–51; Pogue 3:521, quoting John E. Hull to Pogue, 12/22/69; Cray 508.

33   Bohlen, "Second Plenary Meeting," 2/5/45, FRUS, *Malta-Yalta*, 611–23; Leahy 321.

34   Bohlen, "Second Plenary Meeting," 2/5/45, FRUS, *Malta-Yalta*, 620–21; Molotov, "Soviet Proposal for Reparations from Germany," n.d., FRUS, *Malta-Yalta*, 707; Leahy 354–55.

35   Bohlen, "Second Plenary Meeting," 2/5/45, FRUS, *Malta-Yalta*, 611–12, 621 ("If you wished"); Matthews, minutes, 2/5/45, FRUS, *Malta-Yalta*, 627; "Protocol of the Proceedings," 2/11/45, FRUS, *Malta-Yalta*, 978.

36   Matthews, minutes, 2/5/45, FRUS, *Malta-Yalta*, 628; Bohlen, "Second Plenary Meeting," 2/5/45, FRUS, *Malta-Yalta*, 617 ("I can get"); Leahy, 2/3/45, LC (Leahy Papers, reel 1); Brooke, 2/5/45.

37   Cornelius Bull, 12/18/43, KP (box 6) ("The Russians"); Deane to GCM, 12/2/44; Pogue 3:530; Stoler, *Allies*, 213.

38   Gromyko 98 ("Why did nature").

39   Bohlen, "Fourth Plenary Meeting," 2/7/45, and "Fifth Plenary Meeting," 2/8/45, FRUS, *Malta-Yalta*, 772, 774–75; Tripartite Communique, 2/11/45, FRUS, *Malta-Yalta*, 971; FDR to Stalin, 2/10/45, FRUS, *Malta-Yalta*, 966; Bohlen, "Third Plenary Meeting," 2/6/45, "Fifth Plenary Meeting," 2/8/45, FRUS, *Malta-Yalta*, 666–67, 775; Leahy 357, 362, 375.

40   Costigliola 247.

41   Bohlen, "Third Plenary Meeting," 2/6/45, and Mathews minutes, 2/6/45, FRUS, *Malta-Yalta*, 668–69, 679.

42   FDR to Stalin, 2/6/45, FRUS, *Malta-Yalta*, 727–28; Stoler, *Allies*, 189.

43   Bohlen, "Fifth Plenary Meeting," 2/8/45, FRUS, *Malta-Yalta*, 778–79; Matthews, "Fifth Plenary Meeting," 2/8/45, FRUS, *Malta-Yalta*, 787–88; Leahy 249, 357–67.

44   CCS 777/2, "Reciprocal Agreement on Prisoners of War," 2/8/45; Andrew and Mitrokhin 134–35.

45 Tripartite Communique, 2/11/45, FRUS, *Malta-Yalta*, 972; Leahy 371 ("Mr. President").

46 Harriman, "Memorandum of Conversations," 2/10/45, FRUS, *Malta-Yalta*, 894–95; Bohlen, minutes, FDR-Stalin meeting, 2/8/45, FRUS, *Malta-Yalta*, 769–70 ("The French").

47 Stalin, FDR, WSC, Agreement Regarding Japan, 2/11/45, FRUS, *Malta-Yalta*, 984; Leahy, 2/3/45, LC (Leahy Papers, reel 1); Leahy 373.

48 GCM, "Suggested Topics for Overseas Press Club Dinner," 3/1/45, MP 5:66–71 ("always").

49 Cray 513, quoting Alger Hiss, "Two Malta Myths," *The Nation*, 1/23/82 ("Ed"); Cornelius Bull, 7/24/43, KP (box 6) ("too damn smart"); Buell 396; Stoler, *Allies*, 126–30.

50 Leahy, 2/3/45, LC (Leahy Papers, reel 1) ("One result").

51 Costigliola 233–34, quoting Charles Bohlen, *Witness to History*, 172, and Berezhkov to Arthur Schlessinger, Jr., n.d., FDRL (Misc. Collections) ("Everybody").

52 Howard Bruenn, "Clinical Notes," 2/8/45, FDRL (Bruenn Papers); Anna Boettiger, 2/3/45, FDRL (Halsted Papers, box 84); Costigliola 236–37.

53 Paul D. Stroop, interview, 11/1/69, 200–01, KP (box 10) ("rather embarrassing").

54 Paul D. Stroop, interview, 11/1/69, 200–01, KP (box 10) ("The patio").

FIFTY-ONE ★ "O CAPTAIN"

1 GCM to DDE, 3/6/45, MP 5:76–79 ("Making war").

2 DDE to GCM, 3/12/45, EP 4:2521 ("Misery loves").

3 DDE to GCM, 4/7/45, EP 4:2588–93; DDE to Bernard Montgomery, 3/28/45, 4/8/45, EP 4:2552, 2593–94; DDE to Deane for Stalin, 3/29/45, EP 4:2557–58.

4 Chester Hansen, 4/7/45, USAMHI (Hansen Papers, box 5); DDE to GCM, 4/7/45, EP 4:2588–93; Omar Bradley, "Ifs of History," n.d., USAMHI (Blair Collection, box 49); Kenneth Strong, interview, 5/14/63, Ohio University Library (Ryan Papers); Bradley, *Soldier's Story*, 535; EP 4:2117.

5 WSC to DDE, 3/31/45, EP 4:2563 n.2.

6 WSC to DDE, 3/30/45, WSC to FDR, 4/1/45, in DDE 4:2563 n. 2 ("If they also"); Brooke, 3/29/45.

7   DDE to GCM, 4/7/45, EP 4:2592–93 ("I am the first").

8   W. B. Smith, interview, 11/1/51, PL; Chester Hansen, 4/7/45, USAMHI (Hansen Papers, box 5); DDE to GCM, 3/30/45, EP 4:2560; Omar Bradley, "Ifs of History," n.d., USAMHI (Blair Collection, box 49); Kenneth Strong, interview, 5/14/63, Ohio University Library (Ryan Papers); Bradley, *Soldier's Story*, 535.

9   Truman Smith, interview, 10/15/59, PL; GCM, interview, 2/11/57, GCML; Pogue 3:556–57, 575; Manchester and Reid 922 ("iron curtain"); EP 4:2570 n.2.

10   GCM to Antonov, 2/20/45, MP 5:53–54; Antonov to GCM, 3/30/45, FDRL (Map Room Papers, box 9) ("to bluff").

11   Brooke, 3/10/45, 3/14/45.

12   HLS, 3/17/45 ("bodes evil").

13   FDR to Stalin, 3/31/45, FDRL (Map Room Papers, box 9) ("No negotiations"); Leahy 386–87; Pogue 3:564.

14   Stalin to FDR, 4/3/45, FDRL (Map Room Papers, box 9) ("You affirm").

15   FDR to Stalin, 4/4/45, FDRL (Map Room Papers, box 9) ("It would be"); HLS, 4/3/45; GCM, interview, 2/11/57, GCML.

16   Leahy, 4/4/45, LC (Leahy Papers, reel 1); Leahy 392–93; Pogue 3:565.

17   Stalin to FDR, 4/7/45, FDRL (Map Room Papers, box 9).

18   FDR to Harriman and Harriman to FDR, 4/12/45, FDRL (Map Room Papers, box 35) ("minor"); Leahy 393 ("minor"); Butler 316–18; Costigliola 310.

19   Leahy 244; Hassett, 3/30–31/45 ("He is slipping away"); Margaret Suckley, interview, 8/57, 3, FDRL; Goodwin 584, 596; Parrish 463.

20   Hassett, 4/12/45; "Last Words: 'I Have a Terrific Headache,'" NYT, 4/13/45; Suckley, 4/12/45; Photograph, 4/45, FDRL (Pare Lorentz Papers, box 57); "L.I. Artist Tells of Roosevelt's Last Hour," (Long Island) *Newsday*, 4/16/45.

21   Suckley, 4/12/45; Howard Bruenn, "Clinical Notes on the Illness and Death of President Franklin D. Roosevelt," FDRL (Bruenn Papers); Margaret Suckley, interview, 8/57, FDRL; "Franklin D. Roosevelt for History," *Sunday News*, 5/27/45 (copy of unfinished portrait).

22   "L.I. Artist Tells of Roosevelt's Last Hour," (Long Island) *Newsday*, 4/16/45; Margaret Suckley, interview, 8/57, FDRL; "Franklin D. Roosevelt for History," *Sunday News*, 5/27/45; Hassett 334.

23   Suckley, 4/12/45 ("fifteen minutes"); Goodwin 602.

24   "Franklin D. Roosevelt for History," *Sunday News*, 5/27/45.

25 Suckley, 4/12/45; Hassett, 4/12/45; Margaret Suckley, interview, 8/57, FDRL ("I have"); Reilly 230.

26 Suckley, 4/12/45; Hassett, 4/12/45; Margaret Suckley, interview, 8/57, 5, FDRL.

27 Suckley, 4/12/45; Margaret Suckley, interview, 8/57, 5–6, FDRL; "Franklin D. Roosevelt for History," *Sunday News*, 5/27/45

28 Suckley, 4/12/45; Howard Bruenn, "Clinical Notes on the Illness and Death of President Franklin D. Roosevelt," FDRL (Bruenn Papers); "Franklin D. Roosevelt for History," *Sunday News*, 5/27/45; Reilly 232.

FIFTY-TWO ★ TRUMAN

1 EJK, "Air Log—1945—Continued. April–July," n.d., KP (box 4); Frank McCarthy to GCM, 4/13/45, MP 5:141; Pogue 3:557; HLS, 4/15/45; King and Whitehill, 600; Hassett, 4/13–15/45.

2 Clifton Pritchett, "Plan for the Funeral Ceremonies of the Late President Franklin D. Roosevelt," 4/13/45, GCML (McCarthy Papers, box 26); HLS, 4/14/45.

3 Hassett 343–45; Leahy 403; Pogue 3:557; HLS, 4/15/45; King and Whitehill 600; Reilly 60–61.

4 ER to GCM, 4/14/45, MP 5:151.

5 HLS, 4/12/45 ("For all"); Leahy, 4/12/45, LC (Leahy Papers, reel 1) ("a personal").

6 Hassett, 4/13/45; McCullough 348.

7 HST, *Memoirs* 1:5 ("Is there anything").

8 McCullough 384 ("I am here"); William Hassett, interview, 10/18–19/63, HSTL, 3–4.

9 Hassett, 8/18/44; McCullough 327 ("tried to pour"); Omar T. Pfeiffer, interview (int. 4), 268, KP (box 10).

10 Morgenthau, 5/15/52, FDRL (Morgenthau Papers, box 516) ("Juggler"); Pittsburgh *Post-Gazette*, 2/23/44 ("To entrust"); Beschloss 224; HLS, 4/13/45 ("laboring under"), 4/18/45; Leahy, 4/12/45, LC (Leahy Papers, reel 1); Giangreco 50; McCullough 354.

11 HLS, 4/12/45; Leahy 407.

12 HLS, 4/13/45.

13   Leahy 129 ("[I]f the South"); HLS, 4/13/45; Leahy 407; Pogue 3:558; Giangreco
     50, citing *Yank*, 3/9/45; Roberts 553–54.

14   HLS, 3/19/41 ("I am really"), 4/13/45 ("We will not"); McCullough 434, quoting
     Moran, *Diaries*, 303 ("Watch"); Pogue 3:558.

15   HST, interview, 11/14/60, PL; Giangreco 98; McCullough 171, 255.

16   McCullough 402–04.

17   McCullough 399–400, quoting HST, 6/7/45 ("Mr. Prima Donna").

18   Bohlen, minutes, 4/23/45, FRUS, *Europe 1945*, 2:252–53; HLS, 4/23/45; Leahy
     412; Stoler, *Allies*, 237.

19   Bohlen, minutes, 4/23/45, FRUS, *Europe 1945*, 2:252–53 ("one way street");
     Leahy 409.

20   "Our Policy Stated," 6/24/41 ("If we see"); Stoler, *Allies*, 233.

21   HLS, 4/23/45; Bohlen, minutes, 4/23/45, FRUS, *Europe 1945*: 2:252–53; Forrestal,
     4/27/45, *Forrestal Diaries*, 49-50.

22   Bohlen, minutes, 4/23/45, FRUS, *Europe 1945*, 2:252–53 ("If the Russians"); For-
     restal, 4/27/45, *Forrestal Diaries*, 49-50.

23   HLS, 4/23/45 ("then").

24   Bohlen, minutes, 4/23/45, FRUS, *Europe 1945*, 2:254; HLS, 4/23/45 ("The
     Russians").

25   Pogue 3:580, quoting FRUS, *Europe 1945*, 5:254.

26   HLS, 4/25/45.

FIFTY-THREE ★ DOWNFALL

1   EJK, statement, 5/8/45, KP (box 6) ("Another hard job").

2   Craven and Cate 5:608–14.

3   U.S. Strategic Bombing Survey, "Incendiary Attacks on Japan" (April 1947) 2;
     Craven and Cate 5:615–17 ("sea of flames").

4   "Honshu Badly Hit," NYT, 9/3/45; Hopkins, *Pacific War*, 324; Craven and Cate
     5:621–23.

5   Thomas Handy, interview, 3/23/59, PL ("toughest").

6   HLS, 6/11/45; JCS 1388/4, "Details of the Campaign Against Japan," 7/11/45;
     Leahy 448–49; Pogue 4:16–18; EJK, "Air Log—1945—Continued. April–July,"

n.d., KP (box 4); Phelps Adams to Lloyd Graybar, n.d., KP (box 11); Thomas Handy, interview, 3/23/59, PL; Giangreco 95; D. M. Giangreco, "Casualty Projections for the U.S. Invasions of Japan, 1945–1946," *Journal of Military History* (July 1997): 21–82.

7 Leahy, 6/18/45, LC (Leahy Papers); Andrew McFarland, "Minutes of Meeting Held at the White House," 6/18/45, FRUS, *Potsdam*, 1:904–07; Robert Dennison, interview, 1/17/73, 84, KP (box 10); Giangreco 109–10; Hewlitt and Anderson 363–64.

8 Andrew McFarland, "Minutes of Meeting Held at the White House," 6/18/45, FRUS, *Potsdam*, 1:904–07; Leahy, 6/18/45, LC (Leahy Papers); John McCloy, interview, 3/31/59, PL; Cray 541.

9 JCS 1388/4, "Details of the Campaign Against Japan," 7/11/45.

10 Andrew McFarland, "Minutes of Meeting Held at the White House," 6/18/45, FRUS, *Potsdam*, 1:904; T. T. Handy to Leahy, 9/1/44, NARA (RG 218, "Chairman's File, Admiral Leahy," box 20); Truman, *Memoirs*, 1:417.

11 JCS 1388/4, "Details of the Campaign Against Japan," 7/11/45; EJK, interview, 11/29/50, KP (box 7); Hastings Ismay, interview, 12/20/46, PL.

12 Leahy 449; Leahy, 6/18/45, LC (Leahy Papers).

13 EJK to Hanson Baldwin, 6/12/50, KP (box 4); Leahy 449; Buell 463; EJK, interviews, 11/29/50, 8/28/50, KP (box 7) ("I have said"); EJK, "Notes from Air Log—August—December 1945," n.d., KP (box 4); Larrabee 198, quoting Albion and Connery, *Forrestal and the Navy*, 287–88.

14 GCM, speech, 6/11/45, MP 5:220–27 ("In a war").

15 GCM, interview, 2/11/57, GCML.

16 GCM, interview, 3/29/54, in *U.S. News & World Report*, 11/2/59, 50–56 ("We had to assume"); GCM, interview, 2/11/57, GCML.

17 Leahy 449; Leahy, 6/18/45, LC (Leahy Papers); Louis Morton, "Decision to Use the Atomic Bomb," 340 n.34.

18 McFarland, "Minutes of Meeting Held at the White House," 6/18/45, FRUS, *Potsdam*, 1:908 ("submerged class"); Hewlett and Anderson 363.

19 McFarland, "Minutes of Meeting Held at the White House," 6/18/45, FRUS, *Potsdam*, 1:907–09; Pogue 623 n.12; Forrestal, 5/8/47, in *Forrestal Diaries*, 70.

20 John McCloy, interview, 3/31/59, PL ("It was just"); McCloy 40–44; Hewlett and Anderson 364; Pogue 4:18, citing JCS, minutes, 6/18/45; Forrestal, 5/8/47, in *Forrestal Diaries*, 70.

FIFTY-FOUR ★ "COME AND SEE"

1 Morison 604, 618–19, 631, citing McCloy, interview, 10/15/59. Chapter title from *Revelation* 6:7.

2 HLS, 12/31/44, 1/3/45, 3/15/45; Pogue 4:17; Morison 628.

3 HLS to HST, 4/24/45, HST, *Memoirs*, 1:85 ("I think it"); HLS, 4/24/45.

4 HLS, 4/25/45; HLS to HST, 4/25/45, in HLS, 4/25/45 ("the most terrible").

5 HST, *Memoirs*, 1:87 ("I listened").

6 Bundy, *Active Service*, 613; Morison 619, citing John J. McCloy, interview, 9/14/55; HLS, 3/5/45; Leslie Groves, testimony, 4/15/54, *Oppenheimer Hearings*, 171 ("at the earliest"); Bundy, *Active Service*, 613.

7 HLS, 5/10/45 ("the locking"); Pogue 4:17, citing Groves to GCM, 12/30/44.

8 John McCloy, memorandum, 5/29/45, Amherst (McCloy Papers) ("We must").

9 William Smedberg, interview, 6/9/76, 24, KP (box 11) ("Smedberg").

10 William Smedberg, interview, 6/9/76, 24–25, KP (box 11) ("fault line").

11 HLS, 5/1–4/45, 5/31/45 ("I told them"); Bundy, *Active Service*, 616–17; Morison 624; Frank McCarthy, interview, 9/29/58, PL; Jones 530.

12 Jones 532, quoting "Notes on Interim Committee Meeting," 5/31/45 ("The visual"); Gordon Arneson to George Harrison, 6/6/45; Hewlett and Anderson, *New World*, 350–59.

13 Jones 532, quoting "Notes on Interim Committee Meeting," 5/31/45 ("make a"); Gordon Arneson to George Harrison, 6/6/45; Hewlett and Anderson, *New World*, 350–59.

14 John McCloy, interview, 3/31/59, PL; Bundy, *Active Service*, 617; Morison 625–29, citing Arthur Compton, *Atomic Quest*, 220, 238, and James B. Conant to McGeorge Bundy, 11/30/46; Leahy, 6/4/45, LC (Leahy Papers, reel 2); HLS, 3/15/45; Leahy 502–03 ("a professor's"); GCM, interview, 2/11/57, GCML; Smith, "Behind the Decision," 297.

15 HLS, 5/31/45, 6/1/45; Morton, "Decision," 338; HST, *Memoirs* 1:419–20; Bundy, *Active Service*, 617; Jones 533, citing Arthur Compton to HLS, 6/12/45 (attaching unsigned copy of Franck Report) and quoting "Notes on Interim Committee Meeting," 6/21/45.

16 Morton, "Decision," 338; McCloy, interview, 10/15/59, PL.

17 WSC, *Triumph and Tragedy*, 639 ("The decision"); HST, 7/25/45, HSTL.

18   GCM, interview, 2/11/57, GCML.

19   J. A. Derry to Leslie Groves, "Summary of Target Committee Meetings," 5/12/45; "Minutes of the Third Target Committee Meeting," 5/28/45; Jones 529; Hewlett and Anderson 365.

20   HLS, 6/6/45 ("I told him"); HLS to HST, 5/16/45, in HLS, 5/16/45.

21   Leslie Groves, interview, 5/7/70, PL; Bundy, *Active Service*, 625; John McCloy, interview, 3/31/59, PL; Jones 530.

22   George Harrison to HLS and HLS to Harrison, 7/21/45, FRUS, *Potsdam*, 1372 ("pet city"); HLS, 7/21/45.

23   Bundy, *Active Service*, 625; Pogue 4:18–19; Hewlett and Anderson 365; Morison 635.

24   Jones 530, quoting HLS to Harrison, 7/23/45 ("highest authority").

25   HLS, 7/24/45; Leslie Groves to HHA, 7/24/45; Jones 530, citing Thomas T. Handy to Carl A. Spaatz, 7/25/45; Hewlett and Anderson 365.

26   HLS, 7/2/45; Beschloss 244.

27   "Log of the President's Trip," 7/15/45, FRUS, *Potsdam*, 9; William Rigdon, interview, 7/16/70, HTSL, 21–22.

28   Leahy 452–60; EJK, "Air Log—1945—Continued. April–July," n.d., KP (box 4); George Harrison to HLS, 7/16/45, FRUS, *Potsdam*, 1360 ("Operated on"); HLS, 7/16/45.

29   HLS, 7/16–17/45; WSC, *Triumph and Tragedy*, 637–38.

30   Leslie Groves to HLS, 7/18/45, FRUS, *Potsdam*, 1361; HLS, 7/21/45; Frank McCarthy, interview, 9/29/58, PL.

31   HLS, 8/21/45 ("The president").

32   Frank Knox to Paul Mowrer, 6/16/42, LC (Knox Papers, box 3); HLS, 7/23/45; Leahy 474; Pogue 4:20.

33   Truman, *Memoirs*, 1:416 ("good use"); McCullough 442–43; Stoler, *Allies*, 133.

34   HLS, 7/23–24/45; Leahy 484; Stoler 256–57.

35   HLS, "Memorandum for the President," 7/2/45 ("mad fanatics"); Morison 633, citing Joseph Grew to HLS, 2/12/47; HST, *Memoirs*, 1:428; CCS, minutes, CCS-FDR-WSC meeting, 2/9/45, FRUS, *Malta-Yalta*, 826.

36   HLS, 7/2–3/45; HLS, "Memorandum for the President," 7/2/45.

37 HST, address, 6/7/45 (Universal International News); McCullough 436.

38 Togo to Sato, 7/12/45, FRUS, *Potsdam*, 873, 875–76; McCullough 427, quoting HST, 7/18/45 ("Believe Japs").

39 HLS, draft proclamation, n.d., FRUS, *Potsdam*, 894 ("may include").

40 "Proclamation by the Heads of Governments," 7/26/45, FRUS, *Potsdam*, 1474–76; Morison 633, quoting Joseph Grew to HLS, 2/12/47; Leahy 488–89; HLS, 7/24/45 ("I had felt").

41 HLS, 7/24/45.

42 HLS, 7/28/45.

43 John Stone to HHA, 7/24/45, NARA; Truman, *Memoirs*, 1:420–21; Morton, *Decision*, 350.

44 HLS, 8/3–4/45; 313th Bomb Wing to War Department, 8/4/45.

FIFTY-FIVE ★ "THIS IS A PEACE WARNING"

1 Leahy 501; Truman, *Memoirs*, 1:421–22 ("This is," "Please").

2 HLS, "Memorandum of Conference with the President, August 8, 1945," HLS, 8/9/45 ("When you punish").

3 Craven and Cate 5:721–23.

4 HLS, 8/10/45 ("There has been"); Truman, *Memoirs*, 1:427; Leahy 506.

5 EJK to Nimitz, 6/10/45, in Kort 267 ("This is"); HLS, 8/12/45–9/3/45; Leahy 508; Buell 470, quoting EJK to Nimitz, 8/14/45 ("Suspend all").

6 Robert Dennison, interview, 1/17/73, 91–92, KP (box 10) ("I did carry").

7 "Peace Be Now Restored," *Time*, 9/10/45.

EPILOGUE

1 HLS, 9/21/45 ("emotional"); Truman, *Memoirs*, 1:525–27.

2 Goodwin 624–25.

3 "Henry L. Stimson Dies at 83 in His Home on Long Island," NYT, 10/21/50; Morison 654.

4 Cray 555, 729.

5 Cray 731–32.

6   Cray 733–35.

7   William Smedberg, interview, 6/9/76, 32, KP (box 11); Alan McFarland to Thomas Buell, 9/3/74, KP (box 2).

8   Neil Dietrich, interview, 12/10/74, KP (box 11) ("King was").

9   Paul Pihl, interview, 3/16/74, 1, KP (box 10) ("armed truce"); Alexander S. McDill, interview, 11/30/48, KP (box 9); William Smedberg, interview, 6/9/76, 37, KP (box 11); R. D. Shepherd to Thomas Buell, 8/9/74, KP (box 2); EJK, "Notes from Air Log—August—December 1945," n.d., KP (box 4).

10   EJK, interviews, 7/3/50, 8/26/50, KP (box 7) ("I didn't like," "brains," "I hated"); Charlotte Pihl, interview, 3/9/74, 9–10, KP (box 10).

11   EJK, "Notes from Air Log—August—December 1945," n.d., KP (box 4) ("Mr. Forrestal"); Neil Dietrich, interview, 12/10/74, KP (box 11).

12   EJK, interview, 11/28/50, KP (box 13) ("I'll never forgive"); Buell 333.

13   Charlotte Pihl, interview, 3/9/74, 13, KP (box 10); Betsy Matter, interview, 12/9/76, 17, KP (box 10).

14   Charlotte Pihl, interview, 3/9/74, 16, KP (box 10); William Smedberg, interview, 6/9/76, 51, KP (box 11); Buell 480.

PHOTO INSERT CREDITS

Page One
    Background photograph of Admiral King (NHC)
    Inset photograph of General Marshall (GCML)
    Inset photograph of Churchill and FDR aboard *Augusta* (FDRL)

Page Two
    Photograph of FDR's cabinet (FDRL)
    Photograph of the Joint Chiefs of Staff (GCML)

Page Three
    Background photo of Marshall and Henry Stimson (GCML)
    Inset photograph of FDR and Churchill aboard *Prince of Wales* (GCML)
    Inset photograph of King taking oath (NARA)

Page Four
    Photograph of King and Knox (NARA)
    Photograph of FDR with globe (FDRL)
    Photograph of Admiral Leahy (NARA)

Page Five
    Photograph of King, Marshall, and Arnold (NARA)
    Photograph of Americans at Casablanca Conference (NARA)
    Inset photograph of paratroopers in England (NARA)

Page Six
    Photograph of Churchill's strategists aboard *Queen Mary* (IWM)
    Photograph of Generals Eisenhower and Marshall (NARA)
    Inset photograph of Marshall, Churchill, and General Montgomery (IWM)

Page Seven
    Background photograph of Willow Run plant (LC)
    Inset photograph of Marshall at West Point (Corbis)

Page Eight
    Photograph of King in Knox's office (NARA)
    Photograph of Chiang Kai-shek, FDR and Churchill (GCML)
    Photograph of Stalin, Roosevelt, and Churchill (FDRL)

Page Nine
    Background photograph of Kasserine Pass (NARA)
    Inset photograph of U.S. soldiers on troopship (NARA)
    Inset photograph of FDR (NARA)
    Inset photograph of Tuskegee Airmen (NARA)

Page Ten
    Background photograph of soldiers in Tunisia (NARA)
    Inset photograph of Marines on Tarawa (NARA)
    Inset photograph of Eisenhower and King on Omaha Beach (NARA)
    Inset photograph of King, Admiral Nimitz, and Major General Holland Smith (NARA)

Page Eleven
    Photograph of General MacArthur, FDR, Leahy, and Nimitz (NARA)
    Photograph of Frenchwoman and tank (NARA)
    Photograph of FDR accepting Democratic nomination (LIFE)

Page Twelve
    Photograph of the Philippine Sea (NARA)
    Photograph of Eleanor Roosevelt with troops (NARA)
    Photograph of Stimson and Marshall (NARA)

Page Thirteen
    Photograph of FDR with Henry Morgenthau (Corbis)
    Photograph of Marshall with Generals Walker and Patton (GCML)
    Photograph of King, Leahy, FDR, Marshall and General Kuter (NARA)

Page Fourteen
    Photograph of Stimson with Colonel Kyle (NARA)
    Photograph of FDR and Churchill with daughters (NARA)
    Photograph of King and Nimitz with James Forrestal (NARA)

Page Fifteen
    Background photograph of mushroom cloud over Nagasaki (NARA)
    Inset photograph of Patton, Stimson, and John J. McCloy (NARA)
    Inset photograph of "Fat Man" (NARA)

Page Sixteen
    Background photograph of Harry Truman, Stimson, and Marshall (Getty)
    Inset photograph of POWs at Aomori Prison (NARA)
    Inset photograph of MacArthur at surrender ceremony (NARA)

INDEX

Aachen, Germany, 405
ABDA theater, 127, 129, 134, 151
Acheson, Dean, 42, 83, 97, 207
Adams, Phelps, 212–13
Admiralty Islands, 170, 275, 319, 321, 373
Aegean Islands, 299, 305, 306, 308
Akagi (Japanese carrier), 169
Aleutian Islands, 162, 163
Alexander, Sir Harold, 233, 289, 293, 365, 419, 447
Allen, General Terry, 188
Alsop, Joe, 296
Alva (yacht), 116
ANAKIM, 229–30, 231, 262
Andaman Islands, 305, 312, 313
Anderson, Orvil, 233, 239
Antonov, General Aleksei, 432, 433
Antwerp, Belgium, 405
ANVIL, 284, 312, 327–28, 330–31, 335, 360, 366, 407
Anzio, Italy, 327
ARCADIA Conference, 121–32, 134, 328
Ardennes Forest, 410–11, 419
ARGONAUT, 416–19
Arnold, Henry H. "Hap," 23, 29, 89, 111n, 124, 125, 129, 148, 149, 154, 194, 254, 272, 282, 287, 298, 414, 443, 448, 452, 459
   antisubmarine warfare and, 244, 245
   Cairo Conference and, 304
   Casablanca Conference and, 224–27, 237
   death of, 471
   European strategy and, 158, 366
   health of, 257–58, 338, 416
   Joint Chiefs and, 130
   King and, 149, 204
   Marshall and, 23–24, 130, 258

North Africa and, 199, 203
OVERLORD and, 358, 360, 361
Pacific strategy and, 184, 189, 203–4, 205, 238
personality of, 24
Quebec Conference (1944) and, 387, 388
Roosevelt and, 24n, 52, 258
Stimson and, 257
strategic air campaign and, 24, 153, 237–39, 348, 368–69
Tehran Conference and, 309
Aruba, 99
Atlantic, Battle of the, 69, 152
Atlantic Charter, 90, 345
Atlantic Wall, 258
Atomic bomb, 164–66, 283–84, 329, 348–50, 453–60, 462, 463, 465–66, 470
Auchinleck, Claude, 180
Australia, 150–52, 161, 170, 344
AVALANCHE, 277

B-17 Flying Fortress, 51–53, 118n
B-24 Liberator, 53, 153, 205, 369
B-25 Mitchell, 156
B-29 Superfortress, 312, 316, 319, 320, 349, 368, 369, 389, 449, 458
Badoglio, Field Marshal Pietro, 291, 292
Baldwin, Hanson, 27, 202–3, 328
Balkans, 258–59, 264, 276–78, 280, 281, 294, 299, 305, 306, 308
Bard, Ralph, 191
Barkley, Alben, 378
Baruch, Bernard, 210, 377
Bastogne, Belgium, 411
Bataan Peninsula, 118, 120, 134, 135, 137, 179, 353, 383

Bavaria, 430, 432
Beaverbrook, Lord, 121, 127
Bechloss, Michael, 348*n*
Bell, Elliott, 401
Berezhkov, Valentin, 426
Bergman, Ingrid, 220
Beria, Lavrenti, 421
Berle, Adolf, 36
Berlin, Germany, 430, 432, 446–47
Berlin, Irving, 77, 78
Bernadotte, Count, 446
Biddle, Francis, 30, 169, 445
Bismarck (German battleship), 74*n*, 87
Bismarck Archipelago, 150–52, 170, 264, 275
Bismarck Sea, Battle of the, 243
Blackouts, 153*n*
Blitz, 51, 222, 331
Boettiger, Anna Roosevelt, 254, 357, 377, 419, 427
Boettiger, John, 357
Bohlen, "Chip," 426
Boland, Pat, 91
BOLERO, 158, 160, 175, 177, 178, 181, 182, 185, 198, 277
Bonus March of 1932, 101, 344
Bowman, Isaiah, 393
Bradley, Omar, 23, 414*n*, 445
  Eisenhower and, 233, 335, 418
  European campaign and, 289, 362, 405, 410, 416, 429–30, 444
  North Africa and, 233, 241, 242
  Sicily and, 275
Bratton, Rufus, 2, 107
Bricker, John, 353, 354
British forces
  Eighth Army, 233, 241, 275
  First Airborne Army, 405
  Twenty-first Army Group, 416, 417
Brooke, Field Marshal Sir Alan, 159, 181, 185, 187, 225, 227, 228, 230, 233, 259–63, 264, 281–83, 289, 299, 305, 307, 311, 365, 366, 387, 411–13, 416–17
Brown, Captain Charles, 202
Brown, Allen, 26, 340
Browning, Miles, 238
Bruenn, Howard, 358, 377, 419, 427, 435, 436
BUCCANEER, 305, 306, 312–13
Buckner, Simon Bolivar, Jr., 371
Budd, Ralph, 45
Bulge, Battle of the, 410–11, 419
Bull, Cornelius, 212, 221

Bullitt, Bill, 32
Bundy, Harvey, 329
Burma, 104, 180, 229, 231, 232, 237, 261, 265, 284, 304, 305, 312, 387, 426
Burma Road, 103, 105, 229, 251, 254, 262, 284, 304, 305
Bush, Vannevar, 349–50
Butcher, Harry, 303
Byrnes, James, 48, 195, 352, 378, 384, 455, 456, 460, 462, 466–67

CACTUS, 199
Cairo Conference (SEXTANT), 299, 302, 304, 310, 319, 320, 346
Camp David, 214*n*
Capra, Major Frank, 166–67
Caroline Islands, 135, 170, 231, 232, 263–65, 272, 284, 316
Casablanca (movie), 220
Casablanca Conference (SYMBOL), 222–32, 234–37, 242, 250, 257, 260, 271, 292
Celebes Islands, 373
Chamberlain, Neville, 14
Chennault, Claire, 253–55, 262, 296
Cherwell, Lord, 394
Chiang Kai-shek, 74, 94, 229, 251, 253–55, 296–97, 304, 312, 313, 316, 369*n*, 374, 419, 426, 462, 463
China, 94, 102, 103, 105, 259, 262, 265, 284, 304, 312–13, 320, 419, 446, 448, 460, 470
China-Burma-India theater, 252, 262
Christmas Island, 344
Churchill, Winston, 14, 29, 30, 38, 52, 67, 69, 73, 93, 95, 113, 134, 141, 158, 165, 177, 219*n*, 251, 288, 314, 347, 390, 412, 443
  ARCADIA Conference and, 121–23, 125, 127–30
  atomic bomb and, 283–84, 458
  Balkans and, 264, 276–78, 281, 294, 305, 306
  Berlin and, 430, 432
  Brooke and, 305–6
  Casablanca Conference and, 223, 228, 230, 232, 234, 235, 236
  "closing the ring" strategy and, 125
  Dill and, 264, 329, 330
  Eleanor Roosevelt and, 122–23
  European strategy and, 160, 179, 182, 183, 185–87, 221, 258–60, 264, 276, 277, 279, 281, 292–94, 299, 303, 305–6,

Churchill, Winston (*cont.*)
  European strategy and (*cont.*)
    308–10, 312, 327, 330, 331–32,
    360–62, 365–67, 419
  as First Lord of the Admiralty, 40
  health of, 128
  Italy and, 257, 258, 276, 277, 292–94,
    308, 327, 330, 419
  Katyn Forest massacre and, 268
  Leahy on, 264
  Malta Conference and, 416
  map room of, 121–22, 140
  Marshall and, 306, 331, 471
  Mediterranean peace feelers and, 433
  North Africa and, 125, 175, 178,
    180–82, 185
  OVERLORD and, 331–32, 360–62
  Placentia Bay Conference, 87–89
  Poland and, 424–25
  postwar Germany and, 393, 422–23
  press and, 169*n*
  Quebec Conference (1943) and, 283–84
  Quebec Conference (1944) and, 386–87,
    393
  religious beliefs of, 88
  Roosevelt and, 14, 29, 40–43, 62–63,
    86–90, 92, 121–23, 129–30, 141*n*,
    178–81, 264, 280, 283–84, 290, 292,
    365–67
  Russia and, 79–80
  ships-for-bases agreement and, 40–43
  Stalin and, 423
  Stimson on, 179
  "supreme commander" concept and,
    127–30
  Tehran Conference and, 299, 307, 308,
    310–11
  Tobruk, fall of and, 179180
  on U-boat peril, 152
  ultimatum to Japan and, 462
  unconditional surrender issue and, 236*n*
  United Nations and, 424
  visits to Washington, 121–23, 125,
    127–30, 178–80, 256, 258–60, 264–65
  Yalta Conference and, 415, 423, 427, 428
Civil War, 17, 92, 236, 303–4, 336*n*, 343,
  344, 444
Civilian Conservation Corps, 18, 47
Civilian Pilot Training Program, 57
Clapper, Raymond, 212
Clark, Bennett, 67
Clark, Mark, 149, 217, 232, 289, 292, 357

Clarke, Carter, 400, 401
Coffee, Robert, 302
Collins, J. Lawton "Lightning Joe," 335
Combined Chiefs of Staff, 129, 131–32, 134,
  187, 198, 230–32, 259, 260, 263–64,
  277, 282, 284, 299, 305, 311, 312, 314,
  319, 320, 326, 328, 360, 385, 387, 388,
  416–18, 430
Compton, Arthur, 456
Cooke, Admiral Charles, Jr. "Savvy," 116,
  226, 238, 239, 468
Coolidge, Calvin, 35, 39
Coral Sea, Battle of the, 161–62, 168, 179,
  202, 221, 401, 402
CORONET, 451
Corregidor, 118–20, 135, 137, 353, 383
Craig, Lillian, 334
Cudahy, John, 11, 12
Cunningham, Fleet Admiral Andrew B.,
  124*n*, 311, 331, 360, 361, 387, 413, 418
Curaçao, 99

D-Day landings, 325, 356–58
Darien, 425, 426
Darlan, Admiral Jean Louis Xavier François,
  207, 217, 234
Darlan, Alain, 217
Darlan Deal, 217, 218, 220
Davis, Colonel Benjamin, 57
Davis, Chester, 45
Davis, Elmer, 206, 268
De Gaulle, Charles André Joseph Marie,
  207, 234–36, 251, 332–33, 362–64,
  366, 367
Deane, John, 423
Declaration on Jewish Massacres, 347
Delano, Laura "Polly," 436
Delano, Sarah, 254
Delano, Warren, II, 254
Delphine (yacht), 116
Dennison, Robert, 468
Desert Victory (documentary), 242
Devers, General Jacob L., 289, 335, 336, 416
Dewey, Thomas E., 213, 295, 353–55,
  376–77, 381, 383, 395–405
DeWitt, John, 145, 146
Dietrich, Marlene, 253
Dill, Field Marshal Sir John, 124, 126,
  131, 132, 224, 260, 264, 299, 328–31,
  338, 471
Dill, Nancy, 126
Dodecanese Islands, 293, 305, 306

Dönitz, Grossadmiral Karl, 69, 447
Donnelly, Charles, 305
Donovan, William, 332–33, 364, 402
Doolittle, James, 156
Doolittle Raid, 156–57, 161, 168, 248
Douglas, Air Marshal Sholto, 305
Douglas, William, 378
DOWNFALL, 449–50
Draft, 46–50, 54, 58–59, 80, 84–86, 90–91, 219, 406, 407
DRAGOON, 360, 365–67
Dunlap, Abby, 247, 248
Dutch East Indies, 95, 97, 103, 105, 118, 134, 209, 272, 318, 319

Earle, Captain George, 268n
Early, Steve, 24, 56–57, 177, 288, 377
East China Sea, 448
Eccles, Marriner, 144
Eden, Anthony, 234, 268, 363, 364, 393, 425, 443
Edwards, Rear Admiral Richard, 116
Einstein, Albert, 164
Eisenhower, Dwight D., 16, 118–20, 123, 127, 214n, 230, 240, 246, 264, 280, 287–89, 291, 292, 313, 340, 391, 414, 445
    Berlin and, 430
    Bradley and, 233, 335, 418
    "broad front" strategy, 416–19
    as commander of OVERLORD, 314–15
    De Gaulle and, 333
    deserters and, 406
    Devers and, 335
    European strategy and, 157–58, 227, 325–28, 330–32, 335, 337, 338, 358, 361, 363, 365, 410–11
    final push in Europe and, 429–30, 432, 444
    on King, 149
    Marshall and, 25, 183, 188, 232, 233, 241, 315, 325, 330, 334–36, 412, 418, 429, 430, 432, 471
    Montgomery and, 411–12, 417, 418
    North Africa and, 188, 203, 206–7, 215, 217–18, 220, 232–33, 241–42
    OVERLORD and, 325–28, 330–32, 335, 337, 338, 358, 361, 363
    Pacific strategy and, 151–52, 182–83
    Patton and, 337–38, 418
    Philippines and, 135–36

promotion to general, 233
    Roosevelt and, 233, 303–4, 333, 412, 419
    selection of generals and, 335–36
    on Stark, 149
    Transportation Plan and, 331–32
    unconditional surrender and, 447
Eisenhower, John, 340
Eisenhower, Mamie, 325, 340
Elliott, Harriet, 45
Eniwetok, 320
Enola Gay (B-29), 463–64
Esperancilla, Irineo, 436
Espionage Act of 1917, 169
Essary, Fred, 48
Essex-class carriers, 169
EUREKA, 298, 299, 306–11, 328, 346, 421
European Jews, 347–48, 390
European Recovery Program (Marshall Plan), 470
Evatt, Herbert, 148

Fala (dog), 62, 112, 377, 379, 436
"Fat Man," 456, 463
Fermi, Enrico, 456
Fields, George, 12
Fiji Islands, 162
First French Army, 405
Fish, Hamilton, 50, 66, 295
Fletcher, Frank Jack, 161, 198, 199
Flynn, Edward, 378
Formosa, 170, 263, 319, 320, 322, 373–74, 376, 383–85, 388, 389
Forrestal, James, 191, 212, 243, 323, 338, 339, 399, 403, 414n, 443–46, 472, 473
Frankfurter, Felix, 34, 36–37, 218, 250, 393
Fredendall, Major General Lloyd, 188, 241–42
French Committee of National Liberation, 332, 333
French Indochina, 94, 95, 100, 102
Fuchs, Klaus, 284n

Galbraith, John Kenneth, 210
Garibaldi, Giuseppe, 271
German forces
    2nd SS Panzer Division, 410
    Fifth Panzer Army, 232
    537th Signal Regiment, 267
    Hermann Göring Division, 275
Gerow, Leonard T., 335
Ghormley, Robert, 152, 173–74, 198, 201, 203, 240

G.I. Bill of Rights, 345
Gilbert Islands, 27, 170, 263, 272, 284, 316, 320
Giraud, General Henri Honoré, 207, 217, 234–36, 250–51, 332
Goebbels, Joseph, 267
Goodenough Island, 317
GOOSEBERRY, 361
Göring, Hermann, 51
Graham, Frank H., 465
GRANITE, 316
Great Depression, 15–16, 44
Great White Fleet, 42
Grew, Joseph, 100
Gromyko, Andrei, 420, 423
Groves, Colonel Leslie J., 165, 455–60
Guadalcanal, 172, 174, 197–99, 201, 203, 204, 205, 211, 221, 227, 240, 272, 344
Guam, 118, 170, 194, 316, 368, 448
Guderian, Heinz, 29
Guillain-Barré syndrome, 14n
GYMNAST, 125, 175, 178, 181, 182, 185, 187–88

HABAKKUK, 282–83
Hachmeister, Louise, 38
Haile Selassie, Emperor, 351
HALCYON, 356, 357
Halifax, Lord, 121, 364
Halsey, William, 161, 238–40, 320–21, 374, 385, 388, 412, 413, 414n, 418n
Handy, General Thomas, 124, 305n, 371, 414, 463
Hannegan, Robert, 376, 378
Harding, Warren G., 250
Harmon, Ernest, 241
Harness, Forest, 399
Harriman, Averell, 433–35, 445, 446
Hart, Admiral Thomas, 119, 120, 238
Hassett, Bill, 122, 206, 223, 435
Hastie, William, 57, 58
Heller, Joseph, 413n
Hemingway, Ernest, 62, 340
Henderson, Leon, 45
Hill, T. Arnold, 54–55, 57
Hilldring, John, 208
Hillman, Sidney, 45, 58
Himmler, Heinrich, 446
Hirohito, Emperor, 94, 100, 106–7, 399, 462, 466–67
Hiroshima, 452n, 459, 463, 465–66

Hiryu (Japanese carrier), 169
Hitler, Adolf, 15, 20, 37, 61, 68, 79, 112, 175, 176, 227, 230, 256, 259, 308, 347, 348, 381, 395, 402, 409, 417, 424, 446
Hiyo (Japanese carrier), 369
HMS Bulolo, 226
HMS Duke of York, 121, 124
HMS Kelvin, 361, 362
HMS Kimberley, 367
HMS Orion, 416
HMS Prince of Wales, 87–88, 345
Hodges, Courtney Hicks, 335, 337, 405, 432
Hollandia, 320, 322
Homma, General Masaharu, 118, 134
Hong Kong, 118
Honshu, 369, 451, 452, 466, 467
Hoover, Herbert, 35, 37, 101, 111, 154, 208, 209, 250, 271, 344, 451
Hoover, J. Edgar, 98
Hopkins, Harry Lloyd, 3, 12, 22, 48, 62, 63, 73, 75, 76, 78, 80, 137, 140, 158, 177, 195, 207, 233, 258, 270, 283, 286–87, 379, 404, 415, 419
    ARCADIA Conference and, 122, 123, 127
    Casablanca Conference and, 222
    health of, 13
    Marshall and, 19, 27, 28, 289, 313–14
    materials stockpile campaign of, 14
    meetings in London and, 159, 160, 175, 186
    Pacific strategy and, 83, 134, 209
    physical appearance of, 13
    postwar Germany and, 392
    Roosevelt and, 13
    son of, 340
Hopkins, Stephen, 340
Horne, Vice Admiral Frederick, 323
Hull, Secretary of State Cordell, 2, 4–5, 12, 21, 48, 73, 75, 76, 83, 100, 102n, 103–6, 121, 217, 234, 251, 253, 338, 346, 348, 363–65, 379, 392, 394
Hürtgen Forest, 405
HUSKY, 275
Huston, Walter, 167

Ickes, Harold, 12, 30, 63, 76, 79, 82, 83, 95, 97, 207, 208, 270, 286–87, 353, 445
Indochina, 94, 95, 100, 102, 209
Ingersoll, Royal E., 413
Ingram, Jonas H., 418n
Inoue, Admiral Shigeyoshi, 161

Intelligence, 95, 102, 104–6, 160, 162, 168, 169, 202, 248, 256, 399–400, 402, 403

Ismay, General Hastings "Pug," 160, 180, 305n, 306, 367, 452n

Isolationism, 21, 30, 33, 38, 41, 46, 47, 50, 66, 74

Italy, 221, 227, 257–59, 261–62, 264, 271, 274, 276, 277, 280–82, 291–94, 299, 308, 327, 357, 365, 419, 433–34, 447

Iwo Jima, 389

Jackson, Robert, 12, 42

Jackson, Samuel, 378

Jacob, Ian, 228

Japanese Americans, 145–47, 407

Japanese forces
 Third Fleet, 134
 Fourteenth Army, 118, 134
 16th Division, 118, 388n

Java, 134

Java Sea, Battle of the, 194, 300n

Jodl, General Alfred, 447

Johnson, Edwin, 395

Johnson, Louis, 21, 31

Joint Army-Navy Board, 65, 97, 102

Joint Chiefs of Staff, 129–32, 152, 172, 175, 178, 182, 184, 194–97, 203, 204, 209, 226, 235, 238–40, 254, 257, 272, 273, 277, 279, 280, 298, 321, 322, 327, 328, 330–31, 361, 365, 373, 374, 385, 443–44, 451, 452, 461

Joint Strategic Survey Committee, 273

Jordan, Louis, 213

Jorgensen, Hans, 91

Joyce, General Kenyon, 25

Justice, Department of, 52–53

Kaga (Japanese carrier), 169

Kaiser, Henry J., 193–94

Kamikaze attacks, 449, 452

Kantaro, Suzuki, 463

Kasserine Pass, 241, 340

Katyn Forest massacre, 267–68, 347

Kelly, Ed, 378

Kennedy, Joseph, 66–67, 310, 340, 423

Kennedy, Joseph P., Jr., 340

Kenney, George, 238

Kesselring, Field Marshal Albrecht, 292, 432, 433

Kido Butai, 104, 106

Kimmel, Husband, 112, 398, 402–3, 473

King, Ernest J., 109, 124, 126, 137, 175, 176, 199, 266, 301, 302, 340, 369n, 464
 antisubmarine warfare and, 242–45
 Army and Navy promotions and, 412–14
 Arnold and, 204
 atomic bomb and, 456
 background of, 70
 bomber ownership issue and, 153–55
 Cairo Conference and, 304, 305
 Casablanca Conference and, 225, 226, 228–29, 231, 237
 Chiang and, 304–5
 as Chief of Naval Operations (CNO), 190–91, 323–24
 as Commander-in-Chief, U.S. Fleet, 112–15, 190, 323–24
 as commander of Atlantic Fleet, 73
 death of, 473
 Doolittle Raid and, 156
 Eisenhower on, 149
 European strategy and, 158, 177, 180, 186, 187, 189, 221–22, 228, 257, 299, 312, 326, 327, 328, 356, 358, 360–62, 366
 flagship of, 116–17
 Forrestal and, 323–24, 339, 472
 funeral of, 473
 Guadalcanal and, 198, 203
 Joint Chiefs of Staff and, 130–31
 Knox and, 73, 112–14, 191–92, 339, 472
 leadership style of, 71–72
 Leahy and, 195–96, 240, 287–88
 leisure activities of, 117, 247–48
 MacArthur and, 174, 238, 321–22, 374, 384
 Malta Conference and, 416, 417
 Manhattan Project and, 349
 Marshall and, 148–52, 155, 182, 183, 194, 198–99, 203–4, 224, 227, 240–41, 287
 Montgomery and, 418n
 naval policy and, 75–76, 92, 93, 193
 Nimitz and, 161–63, 384
 North Africa and, 187, 188, 228
 office routine of, 117
 OVERLORD and, 302–3, 326, 356, 358, 360–62
 Pacific strategy and, 150–52, 155, 160–62, 170, 172–74, 183, 184, 189, 198–99, 202–4, 206, 211, 221, 226, 228–31, 238–40, 262–63, 272–73, 281–82, 284, 300, 305, 312–13, 316–20, 361, 370–72, 374, 376, 383–85, 387–89, 444, 446, 448, 449, 452, 453

King, Admiral Ernest J (*cont.*)
  Pearl Harbor attack and, 6, 402–3, 473
  personality of, 5, 70, 116, 148–49, 152, 228
  Philippines and, 119–20
  physical appearance of, 5–6, 69
  postwar years and, 472–73
  Potsdam Conference and, 459
  press and, 113, 168–69, 202, 206, 212–13, 221, 317
  Quebec Conference (1943) and, 281–82
  Quebec Conference (1944) and, 387–89
  reputation of, 70–71, 116
  Roosevelt and, 112–14, 190, 192–94, 201–2, 211, 287, 323
  Savo Island and, 201, 202, 241
  ship building program and, 193–94
  staff of, 115–16, 149, 228
  Stark and, 113–15, 190
  Stimson and, 154, 178, 403, 473
  "supreme commander" concept and, 126–27, 129
  surrender of Japan and, 467–68
  Tehran Conference and, 309, 423
  Truman and, 444
  uniform design and, 72, 245–46
  women friends of, 247–48
  Yalta Conference and, 420, 421, 426
King, Mackenzie, 43
King, Mattie Egerton, 70–71, 117, 126, 247
Kinkaid, Thomas, 319, 320
Kipling, Rudyard, 343–44
Klein, Julius, 208–9
Knox, Annie, 338
Knox, Secretary of the Navy William Franklin, 3, 4, 8, 42, 54, 55, 73, 74, 76, 116, 190, 195, 212, 223, 270, 287, 317, 323–24, 413
  aid to Britain and, 64, 67
  antisubmarine warfare and, 243–44
  background of, 37
  becomes secretary of the navy, 38
  death of, 338–39
  European strategy and, 158, 178
  integration of armed forces and, 55–56, 58, 143–45
  Japanese war preparations and, 104–7
  King and, 73, 112–14, 191–92, 339, 472
  Pacific strategy and, 75, 189
  press and, 37, 169
  Senate confirmation of, 38–39
  Stimson and, 338–39
  war scenarios and, 65

Knudsen, William, 45, 139, 351
Konoye, Prince Fumimaro, 94, 100
Krock, Arthur, 396
Kurile Islands, 425, 426
Kuter, Laurence, 416, 421
Kwajalein, 330, 340
Kyoto, 458–59, 467
Kyushu, 368, 369, 451

La Guardia, Fiorello, 36
Landing craft issue, 311–12, 326–28, 330
Landon, Alf, 35, 36, 60, 101
Langer, William, 347
Lawrence, Ernest, 456
League of Nations, 345
Leahy, Admiral William D., 225*n*, 246, 259, 287, 288, 293, 294, 300, 302, 307, 322, 323, 379, 382–84, 404, 411, 426, 428, 433, 434, 443–46
  atomic bomb and, 457
  on Churchill, 264
  death of Roosevelt and, 442
  Japanese surrender and, 466–67
  Joint Chiefs meetings and, 196–97
  King and, 195–96, 240, 387–88
  Marshall and, 197
  Pacific strategy and, 203, 204, 240, 449, 451, 452
  physical appearance of, 195
  promotion to Admiral of the Navy, 412–14
  Quebec Conference (1943) and, 264, 265, 281, 282
  as senior commander of service chiefs, 195
  strategic bombing and, 239
  Yalta Conference and, 419, 425
Lee, Robert E., 17, 154, 236, 304, 445
LeHand, Marguerite "Missy," 60
LeMay, Curtis, 448, 449, 452, 458
Lend-Lease, 64, 66–67, 73, 80, 94, 177, 222, 390, 393, 394, 426
Lesinski, John, 267
Lewis, John L., 58, 111, 269
Leyte, 318, 374, 383, 385, 388, 389, 398, 448
Leyte Gulf, Battle of, 448
Lindbergh, Charles, 33, 46, 111
Lippmann, Walter, 146
"Little Boy," 455–56, 463
Livadia Palace, 420–21
Lodge, Henry Cabot, Jr., 56
Loesser, Frank, 13
Long, Huey P., 353

Loomis, Alfred, 164
Low, Captain "Frog," 243
Luce, Clare Booth, 294
Luzon, 273, 318, 319, 320, 322, 373–74,
    383–85, 389, 444, 449, 451

MacArthur, Arthur, 136, 137
MacArthur, Douglas, 123, 148, 152, 166,
    184, 194, 198, 204, 227, 231, 276, 280,
    287, 288, 330, 369n, 414, 416, 418n,
    429, 451
  command issues and, 238–40, 272–73
  Hollandia plan of, 320, 322
  King and, 174, 238, 321–22, 374, 384
  Manus Island and, 321–22
  Marshall and, 119, 120, 137, 231–322,
    238, 317–19, 373–74, 381
  New Guinea and, 203, 205
  Pearl Harbor meeting and, 379, 381–84
  personality of, 101, 238
  Philippines and, 101, 104, 118–20,
    134–38, 284, 318–19, 322, 373–74,
    383–84, 388, 389, 398, 422n, 448
  political ambitions of, 294–95, 353–55
  Rabaul and, 170, 172–74, 199, 203, 230,
    239, 240, 273
  Roosevelt and, 101, 136, 137, 287, 322,
    353–54, 373, 382, 383
  Stimson and, 137, 295, 321
  surrender of Japan and, 467–68
  Tulagi assault and, 172–74
MacArthur, Jean, 136, 137
MacLeish, Archibald, 218
MAGIC, 102, 106, 107, 141, 399n, 462
Makin, 316, 340, 370
Malaya, 97, 103, 104, 134
Malmédy, Belgium, 411
Malta Conference (ARGONAUT), 416–19
Manchuria, 36, 94, 95, 102, 266, 446, 452,
    460, 461
Manhattan Project, 165, 329, 348–50, 461n
Manly, Chesley, 98
Manus Island, 321
Mao Tse-tung, 94
Mariana Islands, 170, 263, 272, 273, 275,
    284, 316, 320, 322, 361, 368, 369,
    448, 449
MARKET-GARDEN, 405
Marshall, George Catlett, 1–2, 7, 29, 46, 61,
    70, 104, 109, 189, 232, 246, 254, 298,
    333, 447, 464
  aid to Britain and, 30–31, 51–53, 66, 181

  antisubmarine warfare and, 244, 245
  ARCADIA Conference and, 124,
    126–29
  Army and Navy promotions and, 412–14
  army background of, 18
  Arnold and, 23–24, 130, 258
  atomic bomb and, 165–66, 455–56, 458,
    460, 463
  becomes Army Chief of Staff, 19
  briefing system of, 22–23
  Brooke and, 159
  Cairo Conference and, 304–6
  Casablanca Conference and, 224–26,
    231, 237
  childhood of, 17
  Churchill and, 306, 331, 471
  "closing the ring" strategy and, 125
  daily routine of, 22, 117
  De Gaulle and, 363–65
  death of, 471
  Dewey and, 396, 399–403
  Dill and, 328–30, 471
  division of Pacific theater and, 151–52
  draft and, 47–48, 50, 84–86, 407
  education of, 17–18
  Eisenhower and, 25, 183, 188, 232, 233,
    241, 315, 325, 330, 334–36, 412, 418,
    429, 430, 432, 471
  European strategy and, 157–60, 175–78,
    180–82, 185–89, 203, 222–24, 226–27,
    230, 231, 256–57, 259–62, 271, 272,
    276–78, 281, 282, 293, 299, 306, 327,
    328, 330–31, 366, 408, 410, 411, 3635
  funeral of, 471
  Hopkins and, 19, 27, 28, 289, 313–14
  integration of armed forces and, 56, 57
  Japanese war preparations and, 2,
    103–5, 107
  King and, 148–52, 155, 182, 183, 194,
    198–99, 203–4, 224, 227, 240–41, 287
  Leahy and, 197
  leisure activities of, 26, 98–99, 228, 247
  MacArthur and, 119, 120, 137, 231–322,
    238, 317–19, 373–74, 381
  Malta Conference and, 416–19
  Manhattan Project and, 349–50
  Mediterranean peace feelers and,
    433–34
  military appropriations and, 20–22
  Montgomery and, 411–12, 418
  North Africa and, 125, 182, 186–88, 203,
    206, 218, 241–42

Marshall, General George Catlett (*cont.*)
  OVERLORD commander issue and,
    278–79, 286–89, 302–4, 313–14, 356,
    358, 360–62
  Pacific strategy and, 101–2, 150–52, 170,
    172–73, 183–85, 189, 198–99, 203–4,
    205, 230, 231, 238–40, 259, 262, 272,
    273, 371–72, 374, 384–85, 389, 432,
    449, 451–53, 461
  Patton and, 22, 336–38
  Pearl Harbor attack and, 2, 154,
    402–3, 473
  Pearl Harbor investigation and, 398
  personality of, 17, 19, 23, 26
  Philippines and, 101, 118–20, 123,
    134–37
  philosophy of war of, 133
  physical appearance of, 17
  popular media and, 166–67
  postwar years and, 470–71
  Potsdam Conference and, 459
  press and, 24–25, 242, 352
  Quebec Conference (1943) and, 281
  Quebec Conference (1944) and, 387,
    388
  Roosevelt and, 18, 19, 21, 26–28, 125–26,
    131, 136, 185, 192, 195, 197, 286–90,
    313–15, 442
  Roosevelt's death and, 442
  School of Military Government, 207–9
  selection of generals and, 334–36
  size of army and, 406–8
  staff of, 25–26, 115n
  Stalin and, 307
  Stilwell and, 297, 419n
  Stimson and, 39, 154, 288–89, 470
  strikes and, 351–53
  "supreme commander" concept and,
    126–29
  surrender of Japan and, 467
  Tehran Conference and, 307, 309
  Truman and, 444
  Truman Committee and, 68
  Vandenberg and, 294–95
  war production and, 67–68, 80, 140
  Washington Conference and, 259–62
  in World War I, 312
  Yalta Conference and, 421, 426
Marshall, Katherine Tupper Brown, 18,
  22, 39, 98, 126, 247, 290, 340, 341,
  358, 470
Marshall, Lily, 18

Marshall Islands, 135, 170, 231, 232, 263,
  264, 272, 275, 284, 316, 320, 340
Martin, Joseph, Jr., 90, 349
Matter, Betsy, 247–48
MATTERHORN, 369
Maximum ROUNDUP, 185
McCain, John, 244–45
McCarthy, Frank, 23n, 26, 224–25, 261,
  289, 315, 466
McCloy, John J., 39, 208, 329, 348, 391,
  393, 410, 453, 456–58, 467
McCormack, John, 85–86, 91, 349
McCormick, Robert, 37, 97–98, 267,
  295, 398
McCrea, John, 140, 180, 185, 191–93, 195,
  211, 300, 301
McIntire, Ross, 62, 413
McIntyre, Marvin, 130, 358n
McNair, General Leslie, 340
McNarney, Joseph T., 245, 261, 414
Mei-ling Soong (Madame Chiang), 253–54,
  296, 297, 304, 305
Meuse River, 37, 411
Midterm elections, 1942, 90, 189, 205, 208,
  213–14
Midway, Battle of, 168–70, 174, 179, 221,
  227, 266, 272, 401, 402
Midway Island, 82, 162–63
Miller, Albert, 354–55
Miller, Doris, 143, 144
Mindanao, 318, 322, 373, 374, 388, 449
Mitscher, Marc, 369
Molotov, Vyacheslav, 176–77, 179, 226, 306,
  420, 421, 425, 433, 445, 466
Montgomery, Bernard, 232–33, 241, 242,
  275, 327, 405, 411–12, 416–17, 430,
  444
Montgomery, Robert, 140
Moore, Captain Charles, 239
Moran, Lord, 128, 159, 298, 444
Morgan, Sir Frederick, 325
Morgenthau, Henry, Jr., 12, 73, 76, 80, 83,
  217, 218, 270, 347, 348, 362–63, 445
  aid to Britain and, 29, 30, 42, 51, 66,
    390, 394
  physical appearance of, 61
  postwar Germany and, 391–96
  Roosevelt and, 21–22, 390, 391
Mountbatten, Lord Louis, 160, 175, 178,
  282–84, 416
MULBERRY, 361
Murphy, Robert, 207, 217

Murrow, Edward R., 217
Mussolini, Benito, 16, 20, 32, 271, 291, 292, 351, 446n

Nagasaki, 452n, 459, 463, 466
Napoleon, 23, 80, 159
Nason, Mona, 23
National Defense Advisory Committee (NDAC), 45–46
National Guard, 46–48, 50
National Labor Relations Act, 45
National Recovery Act, 45, 89
Nelson, Donald, 140, 219
NEPTUNE, 325–27
Nesbitt, Mrs., 3, 12, 258, 349
Neutrality laws, 30, 73–74
New Britain, 240, 272
New Caledonia, 162, 205
New Deal, 13, 21, 37, 39n, 44, 45, 342–45, 398, 434
New Guinea, 134, 152, 170, 173, 203, 240, 243, 264, 275, 284, 317–20
New Hebrides, 150–52
Nimitz, Admiral Chester, 73, 112, 174, 184, 221, 330, 370–72, 379, 381–82, 416, 467, 468, 472
    Coral Sea and, 161–62
    Midway and, 163, 170
    Pacific strategy and, 152, 231, 238–40, 265, 272, 275, 316, 319–22, 373, 374, 376, 383–85, 388, 389
    promotion to Admiral of the Fleet, 413, 414
    Tulagi and, 172–73
    uniform issue and, 246
Nomura, Kichisaburo, 2, 102, 103, 106, 107, 399n
Norden bombsight, 145
North Africa, 175, 188, 189, 198, 199, 203, 206–7, 215–18, 220, 227, 228, 232–33, 241–42, 275, 303
Nye, Gerald, 33

OCTAGON, 386–89, 393
Office of Price Administration, 209–10, 213
Office of Production Management, 139
Okinawa, 389, 444, 451, 452
Olson, Culbert, 146
OLYMPIC, 451
Omaha Beach, 358, 361
O'Malley, Owen, 268
One World (book), 346

Oppenheimer, J. Robert, 456, 457
Osborn, Frederick, 318–19
Oshima, Baron Hiroshi, 399
Outer Mongolia, 426
OVERLORD, 264, 276–82, 284, 286–89, 294, 299, 302, 308–14, 325–28, 330–32, 335, 337, 338, 358–62, 407
Ozawa, Admiral Jisaburo, 369

P-40 Warhawk, 82
Palau Islands, 284, 316, 322, 388
Patterson, Cissy, 97, 98
Patterson, Robert, 39, 54–55, 287
Patton, George S., Jr., 36, 181, 188, 335, 340, 418n, 445, 463, 472
    Eisenhower and, 337–38, 418
    European campaign and, 405, 411, 432, 444
    Marshall and, 22, 336–38
    North Africa and, 22, 217, 222, 232
    personality of, 166
    press and, 336
    Roosevelt and, 336n
    Sicily and, 242, 275, 291
    slapping incident and, 336
    Stimson and, 337
    uniform of, 246
Payne, Marjorie, 340, 341
Pearl Harbor, 72, 75, 82, 381
    Japanese attack on, 1–8, 111, 112, 143, 144, 397–403, 473
Peleliu, 388
Pendergast, Tom, 68
Perkins, Frances, 12–14, 63, 103, 310, 445
Perry, Glen, 212–13
Pershing, John "Black Jack," 18, 19, 119, 289, 290, 294, 312, 313, 414, 445
Philippine Islands, 65, 97, 100–1, 103–5, 118–20, 134–38, 152, 170, 194, 229, 273, 275, 284, 316, 318–20, 322, 373–74, 383–84, 388, 389, 398, 422n, 448
Philippine Sea, Battle of the, 369–70, 448
Pihl, Captain Paul, 247
Pihl, Charlotte, 247, 384
Pink Star (freighter), 93
Pius XII, Pope, 291
Placentia Bay Conference, 86–89
Plan Dog, 135
Poland, 266–68, 424–25, 445–46, 460
Port Arthur, 426
Port Moresby, 160–62

Portal, Air Chief Marshal Sir Charles, 124, 126–27, 160, 225, 226, 228–31, 260, 261, 263, 311, 312, 418
Potsdam Conference, 452n, 459–61, 463, 466
Potsdam Declaration, 463
Pound, Admiral Sir Dudley, 124, 127, 162, 225, 227–29, 261, 263
Powder, Sergeant James, 26, 126, 421
Prelude to War (film), 167
Presidential elections
    1932, 35, 37
    1936, 35, 37, 45, 101
    1940, 49–50, 58–60, 67, 378
    1944, 295, 353–55, 376–81, 397–405, 405
Prettyman, Arthur, 436

QUADRANT, 280–85, 303, 386
Quebec Conference, 1944 (OCTAGON), 386–89, 393
Quebec Conference, 1943 (QUADRANT), 280–85, 303, 386
Queen Mary, 256, 258
Quezon, Manuel, 135, 136

Rabaul, 170, 172–74, 203, 230–32, 239, 263, 272, 273, 275, 284, 452
Randolph, A. Philip, 54–55, 57
Rankin, Jeannette, 111
Rasputin, 421
Rayburn, Sam, 19, 85, 91, 310, 349–50, 358, 378
Red Army, 80, 95, 158, 175, 222, 268, 308, 407, 420, 421, 424, 425, 430, 461
Reilly, Mike, 60, 122, 234–35, 300, 403
Remagen bridge, 429
RENO III, 318–19
Reparations, 422, 460
Revolutionary War, 344
Reynolds, Quentin, 95
Rhine River, 405, 408, 416, 418, 429
Richardson, James, 72
Richardson, Robert, Jr., 370–72, 382, 385
Rickenbacker, Eddie, 352
RINGBOLT, 199
Rogers, Will, 103
Rome, Italy, 291, 357
Rommel, Erwin, 29, 180, 232, 241, 326
Roosevelt, Eleanor, 30, 56, 57, 59, 77, 122–23, 143, 177, 253, 339, 346, 356, 404, 441

Madame Chiang and, 254
    marriage of, 32
    newspaper column of, 344
    physical appearance of, 32
    on Roosevelt's character, 34, 236, 290, 357
    Roosevelt's death and, 442
    servicemen and, 344
Roosevelt, Elliott, 87–89, 232, 302, 340
Roosevelt, Franklin Delano, 4, 9, 339, 469, 473
    addresses to Congress, 7, 15–16, 139
    aid to Britain and, 29–33, 40–43, 51–53, 61–64, 66–67, 73, 76, 85, 89, 90, 181
    aid to Giraud and, 250–51
    antisubmarine warfare and, 243
    ARCADIA Conference and, 121–24, 129–30
    Army and Navy promotions and, 412–14
    Arnold and, 24n, 52, 258
    as assistant secretary of the navy, 38, 40
    Atlantic Charter and, 90
    atomic bomb and, 164–65, 283–84, 454
    burial of, 441–42
    cabinet meetings of, 12–13
    Cairo Conference and, 299, 302, 304, 310, 346
    Caribbean cruise of, 62–63
    Casablanca Conference and, 222, 223, 225, 226, 230–32, 234–36, 271, 461
    cerebral hemorrhage of, 436
    Chiang Kai-shek and, 251, 253–55
    China and, 94, 102, 103, 265, 296, 304, 312, 419
    Churchill and, 14, 29, 40–43, 62–63, 86–90, 92, 121–23, 129–30, 141n, 178–81, 264, 280, 283–84, 290, 292, 365–67
    De Gaulle and, 234–36, 251, 332–33, 362–65
    death of, 436, 441
    direct line to military planners and, 26–27
    Doolittle Raid and, 156–57
    draft and, 46–50, 54, 58–59, 85
    duplicity of, 266
    Eisenhower and, 233, 303–4, 333, 412, 419
    Eleanor on character of, 34, 236, 290, 357
    European Jews and, 347–48
    European strategy and, 11–12, 158, 159, 176–82, 185, 186, 189, 256–57, 259, 264, 276, 278–79, 327, 328, 332, 356, 358, 360, 365–67, 419

fireside chats of, 64, 92, 112, 142, 210, 344, 357, 358, 360
"garden hose" speech by, 63
German declaration of war and, 112
on global strategy, 175–76
health of, 258, 356–58, 377, 379–80, 393, 398, 415, 419, 426–27, 435–36
Hopkins and, 13
informality of, 27
integration of armed forces and, 54–57, 143–45
Italy and, 292–93
Japanese Americans and, 147
Japanese war preparations and, 82–83, 95, 103, 105, 107
Joint Chiefs and Combined Chiefs of Staff and, 130–32
Katyn Forest massacre and, 267–68, 347
King and, 112–14, 190, 192–94, 196, 201–2, 211, 248–49, 287, 311, 323, 442
Knox and, 37–38
Leahy and, 196, 248–49, 311, 442
Lend-Lease and, 64, 66–67, 73, 80, 85, 94
MacArthur and, 101, 136, 137, 287, 322, 353–54, 373, 382, 383
Madame Chiang and, 254
Malta Conference and, 416, 419
map room of, 140–41
marriage to Eleanor, 32
Mediterranean peace feelers and, 433–34
midterm elections (1942) and, 205, 213–14
military appropriations asked for by, 15, 21–22
Morgenthau and, 21–22, 390, 391
morning routine of, 12
National Defense Advisory Committee and, 45–46
naval policy and, 38, 72–76, 89, 90, 92, 192–93
New Deal and, 44, 45, 342–45
North Africa and, 187, 206, 215, 217, 218, 278
OVERLORD and, 332, 356, 358, 360
Pacific strategy and, 134, 151, 175, 184–85, 189, 205–6, 211, 259, 376, 382–84, 425–26
paralysis of, 14–15, 27, 357
Patton and, 336n
Pearl Harbor attack and, 3, 5–8, 111, 398–99, 401–3

Pearl Harbor meeting and, 379, 381–84
personality of, 28
Philippines and, 136, 137
physical appearance of, 7, 377, 380–81, 393, 419, 427, 435–36
Placentia Bay Conference, 86–89
Poland and, 424–25
postwar Germany and, 390–96, 422–23
presidential elections and (see Presidential elections)
press and, 169n, 256, 266, 291, 342–43
Quebec Conference (1943) and, 283–84
Quebec Conference (1944) and, 386–87, 393
religious beliefs of, 89
Russia and, 80, 82, 175–77, 267–68
School of Military Government, 208–9
size of army and, 406, 408
sons of, 340
"stab in the back" address, 32–33, 38
Stalin and, 176, 267–68, 310, 423, 428
Stark and, 42–43, 47, 76
Stilwell on, 313
Stimson and, 36–39, 53–54, 82, 130, 179, 188–89, 270, 294, 396
strikes and, 269, 351–52
study of, 11
"supreme commander" concept and, 127, 129–30
Tehran Conference and, 298, 306–11, 346, 421
third term issue, 48–49
trip on USS Iowa, 300–2
Truman and, 443
unconditional surrender issue and, 235–36, 271, 292, 447, 461
United Nations and, 345–46, 423–25, 434, 435
unlimited national emergency declared by, 76–77, 85
veterans and, 344–45
war production and, 139–40, 209–10, 219
war scenarios and, 65
Washington Conference and, 258, 259, 264–65
women in life of, 247
Woodring and, 33–34
Yalta Conference and, 415, 420–28
Yamamoto's death and, 248–49
Roosevelt, Franklin D., Jr., 32, 234, 302, 340
Roosevelt, James, 340, 379–80

Roosevelt, John, 340
Roosevelt, Theodore, 35–37, 42
Roper, Elmo, 38
Rosenman, Sam, 47, 59, 76, 77, 382, 403
Rough Riders, 37
ROUNDHAMMER, 185, 262, 277, 279
ROUNDUP, 158, 160, 172, 175, 177,
    178, 181, 185–89, 199, 203,
    226, 227, 237, 257, 259, 261–62,
    312, 407
Rubber, 209–10
Rundstedt, Field Marshall Gerd von, 326
Russian Maritime Provinces, 103
Rutherford, Lucy Mercer, 435–36

S-1 section, 165
St. Vith, Belgium, 411
Saipan, 170, 316, 361, 368–71, 374, 376,
    381, 448, 449, 451, 452
Sakhalin Island, 425, 426
San Juan Hill, 37
Santa Cruz, 173
Sato, Naotake, 466
Savo Island, 199, 201–3, 241
Scharnhorst (German cruiser), 69
Scheer (German cruiser), 69
School of Military Government, 207–9
Schulman, Sammy, 235
Selective Service Act, 48, 58, 68, 84
Seligman, Morton, 169
Sequoia (yacht), 116
SEXTANT, 299, 302, 304, 310, 319,
    320, 346
Seymour, Charles, 329
Sherman, Forrest, 389
Sherman tank, 181
Sherwood, Robert, 13, 63–64, 74n, 76, 403
Shoho (Japanese carrier), 161
Shokaku (Japanese carrier), 161, 369
Short, Walter, 398, 402–3
Shoumatoff, Elizabeth, 435–36
Sicily, 223, 227, 232, 237, 242, 256, 259,
    271, 275, 291, 294
Siegfried Line, 405, 416
Sikorski, General Wladyslaw, 267–68
Singapore, 100, 105, 118, 123, 134
SLEDGEHAMMER, 158–60, 179, 180–82,
    185–86, 188, 189, 262, 284
Slovik, Eddie, 406
Smedberg, William, 456
Smith, Harold, 21
Smith, Holland M., 272, 368, 370–72, 376

Smith, Ralph, 370–71
Smith, Walter Bedell "Beetle," 23, 24n, 26,
    30, 188, 305n, 335, 336, 416–18
Solomon Islands, 150–52, 170, 172, 173,
    198, 199, 201, 203–5, 211, 212, 220,
    240, 264, 272
Somers, Andrew, 91
Somervell, Brehon, 68, 226, 286–88, 349,
    471
Somerville, Admiral Sir James, 282n
Soong, T. V., 255, 296, 297
Soryu (Japanese carrier), 169
South-East Asia Command (SEAC), 284
Spanish-American War, 37, 70, 93, 207
Spruance, Raymond, 238, 272, 320, 369,
    370, 371, 374, 414n
Stalin, Josef, 79, 80, 94, 95, 125, 141, 158,
    175–77, 179, 187, 219n, 227, 267–68,
    294, 328, 347, 356, 366, 386, 443
    Berlin and, 446–47
    health of, 415
    Mediterranean peace feelers and, 433–34
    Pacific strategy and, 425–26
    personality of, 423
    Poland and, 424–25
    postwar Germany and, 422
    Potsdam Conference and, 460–61
    Roosevelt and, 176, 267–68, 310, 423, 428
    Tehran Conference and, 298–99, 306–11
    Truman and, 445–46, 460–61
    United Nations and, 424
    Yalta Conference and, 415, 421–28
Stalingrad, 222, 227
Stark, Harold L. "Betty," 65, 69, 71, 73, 86,
    89, 101, 124, 125, 135, 192, 194, 195,
    398, 402–3, 443
    aid to Britain and, 41–42, 53, 66, 67
    draft and, 48, 50
    Eisenhower on, 149
    Japanese war preparations and, 2, 103–6
    King and, 113–15, 190
    Pearl Harbor and, 2, 7, 154, 473
    Roosevelt and, 42–43, 47, 76
    "supreme commander" concept and,
        126–27, 129
    war scenarios and, 65, 67
State, Department of, 66, 208, 233, 235,
    292, 347, 379, 390, 467
Stettinius, Edward, 45, 425, 426, 445, 446
Stilwell, Joseph "Vinegar Joe," 149, 241,
    253n, 254–55, 262, 296–97, 305, 313,
    419n

Stimson, Mabel, 36, 39, 40, 218, 247, 338, 339, 454, 458, 464, 470
Stimson, Secretary of War Henry Lewis, 4–5, 33, 61, 73, 76, 83, 97–98, 118, 125, 131, 195, 204, 217, 223, 264, 268, 294, 340, 348, 374, 412, 414, 433
   aid to Britain and, 51–53, 64, 66, 67
   aid to Giraud and, 250–51
   antisubmarine warfare and, 243–45
   Army-Navy relations and, 149, 154
   Arnold and, 257
   atomic bomb and, 165–66, 454–60, 463, 470
   background of, 35–36, 38, 164
   bomber ownership issue and, 154
   Churchill and, 179, 276–79
   De Gaulle and, 363–65
   death of, 470
   detroyers-for-bases agreement and, 42, 43
   Dill and, 328–30
   Doolittle Raid and, 157
   draft and, 48, 50, 58–59, 86, 90–91, 406
   European strategy and, 158, 160, 178, 181, 224, 271, 276–79, 293–94, 309, 408, 410, 411
   health of, 36–37
   integration of armed forces and, 56–58, 143
   Japanese Americans and, 145–47
   Japanese war preparations and, 103–5
   King and, 154, 178, 473
   Knox and, 338–39
   leisure activities of, 40, 247
   MacArthur and, 137, 295, 321
   Manhattan Project and, 349, 350
   Marshall and, 39, 154, 288–89, 470
   naval policy and, 74, 75
   North Africa and, 182, 218
   OVERLORD and, 356
   Pacific strategy and, 183–85, 449, 453, 461–62
   Patton and, 337
   Pearl Harbor investigation and, 398
   personality of, 158
   Philippines and, 123, 134–36
   physical appearance of, 35–36
   postwar Germany and, 391–95
   postwar years and, 469–70
   Potsdam Conference and, 459, 463
   Roosevelt and, 36–39, 53–54, 82, 130, 179, 188–89, 270, 294, 396
   Roosevelt's death and, 442
   School of Military Government, 207–9
   selection of generals and, 334
   Senate confirmation of, 38–39
   strikes and, 269, 351–52
   surrender of Japan and, 466–67
   Truman and, 350, 443–46, 455, 458, 459, 469
   Truman Committee and, 68
   ultimatum to Japan and, 462–63
   war production and, 53
   war scenarios and, 65
   working routine of, 39
Stimson Doctrine, 36
Stone, Harlan, 442
Strategic air campaign, 24, 153, 237–39, 284, 335, 348, 368–69
Stratemeyer, George, 239
Straus, Gladys Guggenheim, 21
Suckley, Daisy, 377, 436
Summersby, Kay, 303
Sumner, Jessie, 288
SUPER-GYMNAST, 125, 185
Supreme Headquarters, Allied Expeditionary Force (SHAEF), 326, 328, 331, 332, 410
Sutherland, Richard, 238, 319
SYMBOL, 223
Szilárd, Leó, 164

Taft, Robert, 76, 429
Taft, William Howard, 35, 38, 250
Taiho (Japanese carrier), 369
Tarawa, 316, 317, 323, 330, 340, 370
Tehran Conference (EUREKA), 298, 299, 306–11, 328, 346, 421
Telek (dog), 303
Thailand, 103–5
Theobold, Robert, 152
Thomas, Cora, 341
Tinian, 170, 316, 368, 369, 448, 463
Tobruk, 180
Togo, Shigenori, 102n
Tojo, General Hideki, 94, 100, 106, 157, 381
Toland, John, 102n
TORCH, 188, 189, 198, 199, 203, 206, 211, 242, 302, 312, 357
Towers, John, 319
Transportation Plan, 331–32
TRIDENT, 256–65, 272
Triumph and Tragedy (Churchill), 367n
Truk, 170, 229, 263, 316, 322, 368, 452

Truman, Harry S., 68, 165–66, 288,
  350, 384
  atomic bomb and, 453–57, 460, 462,
    463, 465–66
  background of, 444
  becomes president, 442
  briefing of, 443–44
  MacArthur and, 445
  1944 presidential election and, 378, 381
  Pacific strategy and, 449, 451–53
  personality of, 442
  Poland and, 445–46
  Potsdam Conference and, 459, 460
  Stalin and, 445–46, 460–61
  Stimson and, 350, 443–46, 455, 458,
    459, 469
  surrender of Japan and, 466–67
  ultimatum to Japan and, 462
Truman Committee, 68, 372, 444
Truscott, Lucian K., 188, 241, 275, 336
TUBE ALLOYS "S-1," 283
Tulagi, 172–74, 198, 199
Tully, Grace, 7, 140, 215, 357, 377
Tunisia, 206, 218, 227, 232–33, 241, 242,
  275, 303
Turkey, 258, 259, 293, 308, 309, 460
Turner, Richmond Kelly, 199, 201, 272, 371
Tydings-McDuffie Act of 1934, 135n

U-boats, 40, 69, 73–76, 89, 92, 93, 99,
  133, 152–53, 155, 193, 211, 220, 232,
  242–44, 327
Unconditional surrender, 235–36, 271,
  292, 447, 461–63, 467
United Nations, 345–46, 423–25, 428,
  434, 435, 445
United States armed forces
  First Airborne Army, 405
  1st Armored Division, 181, 340
  First Army, 335, 405, 411, 429, 430
  1st Infantry Division, 358
  1st Marine Division, 54, 58–59, 80,
    84–86, 90–91, 205, 272, 273
  2nd Armored Division, 181
  Second Army, 242
  II Corps, 241, 242, 275, 335
  2nd Marine Division, 272, 317, 370
  Third Army, 335, 405, 430
  Third Fleet, 320–21, 388
  3rd Infantry Division, 275
  4th Armored Division, 411
  4th Infantry Division, 358

4th Marine Division, 370
Fifth Army, 232, 292, 357
Fifth Fleet, 320
5th Marine Division, 379, 380
Sixth Army Group, 416
VI Corps, 292
Seventh Army, 275, 405
Seventh Fleet, 319, 320
Ninth Army, 411
9th Infantry Division, 232
Tenth Army, 444
Tenth Fleet, 244, 245, 369n
Twelfth Army Group, 416
14th Air Force, 46–50, 54, 58–59,
  84–86, 90–91, 219
Twentieth Air Force, 369, 448
27th Infantry Division, 273, 368,
  370, 371
82nd Airborne Division, 275, 411
101st Airborne Division, 411
442nd Infantry Regiment, 407
draft, 46–50, 54, 58–59, 80, 84–86,
  90–91, 219, 406, 407
integration issue, 54–58, 142–45
military appropriations, 15, 20–22
size of, 16, 20, 46, 47, 50, 54, 97, 140,
  219, 406–8
USS Arizona, 4–7, 112
USS Augusta, 5, 86–88, 113, 465
USS Baltimore, 379, 381, 382
USS California, 6
USS Dauntless, 116, 199, 298, 300
USS Enterprise, 161, 198
USS Greer, 92
USS Hornet, 156, 161, 206, 340
USS Houston, 300
USS Indianapolis, 463
USS Iowa, 300–1
USS Kearny, 93
USS Lexington, 70, 71, 161, 162, 273
USS Maryland, 5
USS McDougal, 88
USS Missouri, 467, 468
USS Nevada, 5, 6
USS Oglala, 6
USS Oklahoma, 3–4, 6, 112
USS Panay, 37
USS Pennsylvania, 5
USS Plymouth, 116
USS Potomac, 300
USS Quincy, 416, 419
USS Reuben James, 93

USS *Rocky Mount*, 368
USS *South Dakota*, 202
USS *Tennessee*, 4–6
USS *Texas*, 71–72, 245, 246
USS *Thompson*, 361, 362
USS *Tuscaloosa*, 62
USS *Utah*, 3, 6
USS *Wasp*, 198
USS *West Virginia*, 4, 6, 143
USS *William D. Porter*, 301
USS *Yorktown*, 161–63, 169

Van Buren, Martin, 36
Vandenberg, Arthur, 294–95, 355, 398
Vanderbilt, William, 116
Veterans, 344–45, 434
Victory Program, 97, 112
Vinson, Carl, 323–24, 339
Virginia Military Institute, 17–18
Vittorio Emanuele III, King, 291
Voroshilov, Marshal Kliment, 308

Wadsworth, Jim, 91
Wainwright, Jonathan, 471
Wake Island, 82, 118, 179, 194
Waldrop, Frank, 288
Wallace, Henry, 12, 13, 21, 295, 377, 378, 384
Wallenberg, Raoul, 348
Walsh, David, 33, 41, 46
War Manpower Commission, 269
War production, 53, 67–68, 80, 139–40, 209–10, 219
War Production Board, 140, 209, 210, 219, 312
War Refugee Board, 347–48
Ward, Colonel Orlando, 23
Warren, Earl, 146
Washington, George, 48, 59, 131, 195, 414*n*

Washington Conference (TRIDENT), 256–65, 272
WATCHTOWER, 198, 203
Watson, Edwin "Pa," 24, 62, 131, 192, 195, 357, 379–80, 404
Wavell, Archibald, 134, 261
Wedemeyer, Albert C., 23*n*, 98*n*, 226, 237, 238
Welles, Sumner, 30, 32, 42, 76, 83, 253
Wellington, Arthur Wellesley, Duke of, 91
Wheeler, Burton K., 38, 46, 66, 98*n*, 267
Whiskey Rebellion, 195
White, Harry Dexter, 395
White, Walter, 57.55
Whitehall, Commander Walter Muir, 473
Why We Fight series, 167
Wickersham, Cornelius, 207–8
Willkie, Wendell, 33, 49–50, 58–60, 67, 137, 194, 218, 247, 254, 295, 310, 346, 353–55, 402, 423
Willson, Russell, 116, 246
Wilson, Field Marshal Henry "Jumbo," 365–67, 416
Wilson, Woodrow, 38, 73, 90, 278, 345
Wise, Stephen, 346–47
Wolff, General Karl, 432
Woodring, Harry, 12, 21, 30, 31, 33–34
Works Progress Administration, 13, 286
World War I, 36, 39, 159–60, 235, 312, 314, 344, 391, 410, 444
Wright Brothers, 24

Yalta Conference, 420–28, 445
Yamamoto, Admiral Isoroku, 103, 133, 161–63, 169, 184, 240, 248–49, 403
Yap, 388

Zuikaku (Japanese carrier), 161

Photo by Keith W. Dunn

Jonathan W. Jordan is the author of the *New York Times* bestselling *Brothers Rivals Victors* and the award-winning *Lone Star Navy*.

CONNECT ONLINE

jonathanwjordan.com